# CLUSTERING ASPECTS OF NUCLEAR STRUCTURE

# CLUSTERING ASPECTS
# OF NUCLEAR STRUCTURE

Invited Papers presented at the 4th International Conference on
Clustering Aspects of Nuclear Structure and Nuclear Reactions,
Chester, United Kingdom,
23 - 27 July, 1984

EDITED BY

J. S. LILLEY

and

M. A. NAGARAJAN

*SERC Daresbury Laboratory,*
*Warrington, United Kingdom*

Sponsored by:
The British Council
The Institute of Physics
The International Union of Pure and Applied Physics

Organised by:
SERC Daresbury Laboratory

D. Reidel Publishing Company

A MEMBER OF THE KLUWER ACADEMIC PUBLISHERS GROUP

Dordrecht / Boston / Lancaster

**Library of Congress Cataloging in Publication Data**

CIP

International Conference on Clustering Aspects of
    Nuclear Structure and Nuclear Reactions (4th: 1984: Chester,
    Cheshire)
    Clustering aspects of nuclear structure.

    Includes index.
    1.   Cluster theory (Nuclear physics)–Congresses.   2.   Nuclear
structure–Congresses.   3.   Nuclear reactions–Congresses.   I.   Lilley, J. S.
(John S.)   II.   Nagarajan, M. A. (Mangalam A.)   III.   British Council.
IV.   Institute of Physics (Great Britain)   V.   International Union of Pure
and Applied Physics.   VI.   Title.
QC793.3.S8I56          539.7          85–2246
ISBN-13: 978-94-010-8868-8          e-ISBN-13: 978-94-009-5366-6
DOI: 10.1007/978-94-009-5366-6

Published by D. Reidel Publishing Company
P.O. Box 17, 3300 AA Dordrecht, Holland

Sold and distributed in the U.S.A. and Canada
by Kluwer Academic Publishers,
190 Old Derby Street, Hingham, MA 02043, U.S.A.

In all other countries, sold and distributed
by Kluwer Academic Publishers Group,
P.O. Box 322, 3300 AH Dordrecht, Holland

# CONTENTS

*Invited Paper not orally presented

FOREWORD

The Fourth International Conference on Clustering Aspects of Nuclear
Structure and Nuclear Reactions was held at Chester College, U.K.,
from 23-27 July 1984, and was organized by SERC Daresbury Laboratory.
It is one of a series of conferences on these topics the last being
held in Winnipeg, Manitoba in 1978.

The Conference consisted of 11 sessions which were covered mainly by
21 invited review talks. Over 130 contributed papers were received
and accepted by the Programme Committee, and a limited number of these
were selected for oral presentation during the sessions. In addition,
two evening sessions were arranged in which 60 poster displays were
presented. There were no parallel sessions. The contributed papers
together with short abstracts of the invited talks were collected
together in the form of a handbook which was given to all participants
on arrival.

The purpose of the conference was to review major experimental and
theoretical advances in the field. The programme dealt mainly with the
traditional clustering aspects of nuclear behaviour which have progres-
sed considerably in recent years including microscopic calculations of
nucleus-nucleus potentials, break-up and fragmentation, cluster transfer, quasimolecular resonance phenomena, and polarization studies with
cluster-type projectiles. Some emphasis was given to newer developments, and talks were presented on the use of algebraic techniques for
describing cluster configurations, the importance of channel coupling,
anomalons and quark aspects of nucleons and the nucleon-nucleon force.

The invited papers are presented in this volume. The table of
contents follows the format of the conference programme. This is
followed by the invited papers in the order in which they were
presented. A list of participants and subject index complete the
volume.

Speakers were asked to prepare manuscripts of their talks in camera-
ready format before the meeting and we are grateful to them not only
for their thoughtful, stimulating presentations, but also for the care
they took in providing their manuscripts in good time and according to
the publishers instructions.

The efforts of many groups and individuals contributed to the success
of the conference. In particular we wish to express our appreciation
to the Scientific Advisory Committee for their advice and wisdom in

the early stages, to the Programme Committee for their assistance in
formulating the programme and for their continued support throughout,
to the session chairmen and scientific secretaries, who ensured the
smooth running of the proceedings, and to all the delegates, who, by
their full participation and lively discussions have demonstrated the
continuing vigour of this field of physics.

We also gratefully acknowledge the sponsorship of the Institute of
Physics, the British Council and IUPAP, and thank the Royal Society
for additional financial support.  Our thanks also are due to Chester
College for their hospitality, to SERC Daresbury Laboratory for organi-
zational support and for providing the conference staff at Chester.
Finally, we owe a particular debt of gratitude to Mrs. Shirley Lowndes
and Mrs. Christine Lloyd (Jackson) for their continuous and efficient
work as mainstays of the conference organization.

J.S. Lilley and M.A. Nagarajan

International Scientific Advisory Committee

| | |
|---|---|
| A. Arima | M.V. Mihailovic |
| D.M. Brink | V.G. Neudatchin |
| D.A. Bromley | J.O. Newton |
| A. Budzanowski | D. Robson |
| B.G. Giraud | H. Tanaka |
| M. Harvey | Y.C. Tang |
| H.D. Holmgren | A. Van der Woude |
| D.F. Jackson | W.T.H. van Oers |
| B.K. Jain | A. Weiguny |
| King Sing Nan | K. Wildermuth |
| P. Kramer | |

Programme Committee

| | |
|---|---|
| D.M. Brink | G.C. Morrison |
| L.L. Green | M.A. Nagarajan |
| D.F. Jackson | W.R. Phillips |
| R.C. Johnson | P.J. Twin |
| J.S. Lilley | R.R. Whitehead |
| S.A. Lowndes | |

Organizing Committee

| | |
|---|---|
| J.S. Lilley | S.A. Lowndes |
| C.A. Lloyd | M.A. Nagarajan |

## Session Chairman

D.M. Brink

N.S. Chant

J.D. Garrett

B.K. Jain

G.C. Morrison

M.A. Nagarajan

W.R. Phillips

K.S. Rao

H. Rebel

D. Robson

N. Sarma

H. Tanaka

A. Weiguny

K. Wildermuth

## Scientific Secretaries

B.R. Fulton

D.A. Eastham

A.C. Merchant

H.G. Price

N. Rowley

I.J. Thompson

P.M. Walker

# CLUSTERING PHENOMENA IN THE NUCLEAR MANY-BODY SYSTEM

D. Allan Bromley
A.W. Wright Nuclear Stucture Laboratory
Yale University, New Haven, Connecticut

**ABSTRACT:**

Fragmentary evidence for nuclear clustering phenomena has been found throughout the periodic table, from helium to uranium and from quarks to transient supernuclei. But we still face more questions than answers; the field is a dynamic and promising one. We understand the origin of alpha particle clustering in light nuclei and there is growing evidence for a new dipole collectivity in nuclei closely related to such clustering near closed shells; but we do not understand helion or triton clustering or indeed whether it even exists. We do not know to what excitations or to what angular momenta clustering persists even in the most carefully studied cases.

Group theoretic techniques have proven to be powerful in correlating new data on both bound and unbound cluster states. Sharp cluster states have been found at excitations up to 70 MeV in medium mass nuclei and at angular momenta approaching the centrifugal disintegration threshold; these frontiers await new accelerators and techniques.

A new form of natural radioactivity has been observed in which carbon nuclei are emitted; this is almost certainly the first of many such new decay modes for cold heavy nuclei. It has also been recognized that in hotter nuclei, fragmentation involves emission of a wide range of cluster masses and may provide important insight on the equation of state of nuclear matter.

These and other aspects of clustering in the nuclear many-body system will be illustrated with examples drawn from the published literature and from the wide range of contributions submitted to this conference.

**CONTENTS:**

1

*J. S. Lilley and M. A. Nagarajan (eds.), Clustering Aspects of Nuclear Structure, 1–32.*

## INTRODUCTION:

As shown in Figure 1, the concept of clustering in nuclear systems is an old one dating back at least to 1937 with Wefelmeir[1], Wheeler[2] and von Weizsacker[3]. A postwar renaissance in these studies was initiated, from a theoretical viewpoint, by Wildermuth and Kanellopoulos[4] and from an experimental one by Phillips and Tombrello[5,6]. More recently the centroid of theoretical activity has moved to Japan where Arima[7-10], Ikeda[11], Horiuchi[8,10], Fujewara[12-13] and others have made very important contributions. The wealth of contributions to this Fourth International Conference on Clustering Aspects provides an excellent measure of current international interest in this field.

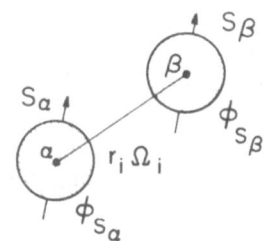

CLUSTERS IN NUCLEI

$$\psi_J = \overset{i_m}{\underset{i=0}{\Sigma}} \psi_{S,J}(\vec{r_i}) = \overset{i_m}{\underset{i=0}{\Sigma}} A_i U_i(r_i) \left\{ \left[ \phi_{S_\alpha}(\alpha)\, \phi_{S_\beta}(\beta) \right]^S Y_L(\Omega_i) \right\}^J$$

$U_i(r_i) = $ radial function for relative motion

$A_i = $ normalized antisymmetrization operator

W. Wefelmeier  Naturwiss. 25  525 (1937)
J.A. Wheeler  Phys. Rev. 32  1083 (1937)
C.F. von Weizsäcker  Naturwiss. 26  209 (1938)
D.M. Dennison  Phys. Rev. 378 (1954)
K. Wildermuth and Th. Kanellopoulos  Nuc. Phys. 7  150 (1958)
G.C. Phillips and T.A. Tombrello  Nuc. Phys. 19  555 (1960)
S. Saito  Prog. Theo. Phys. 41  705 (1969)
K. Ikeda, N.Takigawa and H. Horiuchi  Supp. Prog. Theo. Phys. Extra (1968)
Y. Fujiwara, H. Horiuchi and R.Tamagaki  Prog. Theo. Phys. 61  1629 (1979)

**Figure 1**

Obviously I neither can nor should attempt to cover even a fraction of all this new work in this opening talk; rather I shall attempt to provide an overview, selecting data and examples more or less at random to highlight open questions. Despite its relatively ancient history this field is just now coming of age with new experimental and theoretical approaches. There are an enormous number of open questions and many hold the promise of fundamental new understanding of the nuclear many-body system.

As illustrated schematically in Figure 1 a critical assumption is that the clusters we consider retain their identity for times long compared to characteristic nuclear motions. On that assumption we write the wave function as shown for a pure two cluster configuration. In general this is simply the leading term in a multi-cluster expansion.

If we choose harmonic oscillator internal wave functions $\varphi_{s,\alpha}$ and $\varphi_{s,\beta}$ for the two clusters having a common oscillator constant $\nu = M\omega/\hbar$, where M is the nucleon mass, and if the radial function $u_i(r_i)$ is also a harmonic oscillator function having $\nu = \mu\omega/\hbar$ where $\mu$ is the reduced mass of the clusters then these cluster model wave functions carry the $(\lambda, \mu)$ representation of Elliot's SU(3) model[14].

However, the mere fact that we can write down, or solve for, the wave function in such a cluster form does not necessarily imply that the state in question is of cluster character. The crucial difference between a shell and a cluster model depends upon the particular truncation of the shell model space in use. Specifically, typical shell model calculations truncate the higher oscillator quanta while typical cluster model ones truncate the lower spatial symmetries. For a state to properly be termed a cluster one the admixture of higher oscillator quanta in the relative motion wave function must be large and the root-mean-square separation between the clusters must be comparable to the sum of their radii. Experimentally this leads to two characteristic signatures for true cluster states; their moments of inertia are large and they exhibit reduced decay widths that are large fractions of the Wigner limit for the channel in which the clusters separate asymptotically. As we shall see below, both signatures do appear.

## ALPHA PARTICLE MODELS:

Reflecting the saturation and large binding energy of the alpha particle, essentially all early work focussed on it as the essential cluster. Indeed the experimental observation of natural alpha particle radioactivity provided strong support for this approach. Figure 2 shows the mechanical models[15] for a number of light 4n nuclei; using such models with minimal assumptions regarding bond constants, Dennison[16], and later Kameny[17], were surprisingly successful in reproducing much of the quantum spectrum of $^{16}O$. But there were obvious problems. The $^{16}O$ ground state behaved like a closed shell model state rather than a pyramidal alpha particle structure. In the closed 1s, 2p shell the four nucleon correlations are not such as to yield the spatial symmetry that distinguishes an alpha particle and the ground state correlations, reflecting the residual interactions that favor this special spatial symmetry, are simply not strong enough to result in alpha particle condensation in the ground state.

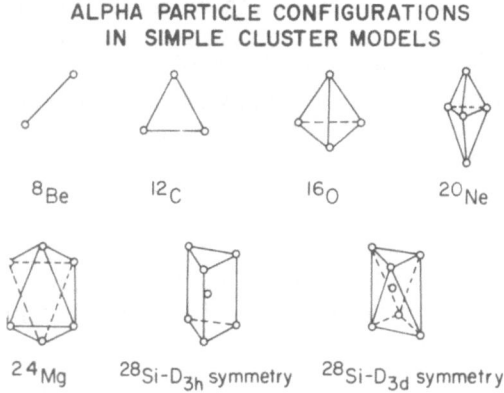

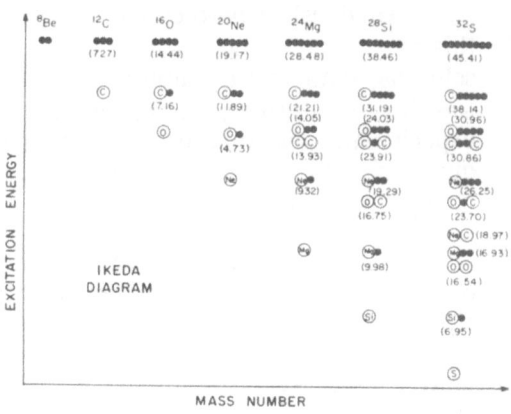

Figure 2                                              Figure 3

In 1956 Morinaga[18] suggested that true alpha particle condensation would occur only in excited configurations and he focussed upon the 7.66 MeV first excited $0^+$ state in $^{12}$C to which we shall return below. In 1968 Ikeda[11] and his coworkers systematized this idea into the Ikeda diagram shown in Figure 3. What are shown here are the threshold energies for separation of each of these 4n nuclei into particular, alpha particle containing, molecular or cluster states. It was Ikeda's suggestion that it would be reasonable to expect the cluster structures shown in the vicinity of the corresponding thresholds, e.g. the 7.66 MeV $0^+$ state of $^{12}$C—suggested by Morinaga[18] to have a linear alpha particle chain structure—near the 7.27 MeV threshold for this dissociation. Similarly in $^{24}$Mg, the $^{12}$C + $^{12}$C threshold occurs at 13.93 MeV and the lowest $^{12}$C + $^{12}$C resonances are indeed found in this energy region. Ikeda also noted a useful empirical rule: on average it appears to require ~ 2.4 MeV to separate an F body fermion system into an (F-4) ⊗ 4 one.

## STRUCTURE OF $^{20}$Ne:

By a substantial margin, $^{20}$Ne is the most thoroughly studied nucleus from a cluster viewpoint[19-21]. This is not surprising given its structure with two neutrons and two protons as valence particles outside the closed $^{16}$O core. In Figure 4 I show pertinent data and calculations. In the upper right of this figure I show 5 rotational bands labelled by the appropriate $K^\pi$ band head. These data come from a wide variety of experiments to some of which I shall return.

The ground state band, $(sd)^4$ has an SU(3) classification of (8,0) and, as in the case of $^{16}$O above, the bands members have reduced widths of less than 10% of the Wigner limit i.e. these are not well developed alpha particle cluster states. The $K^\pi = 0_4^+$ band, in contrast, shows a substantially greater moment of inertia and reduced widths greater than 30% of the Wigner limit; both are the signatures of a well developed cluster band. At the lower right I show the results of calculations of the reduced width amplitudes for the states in the $0_1^+$ (ground state) and $0_4^+$ (alpha particle cluster band), respectively, using an extended shell model basis. Clearly the alpha particle reduced width in the latter case is predicted to be enhanced and at larger radius.

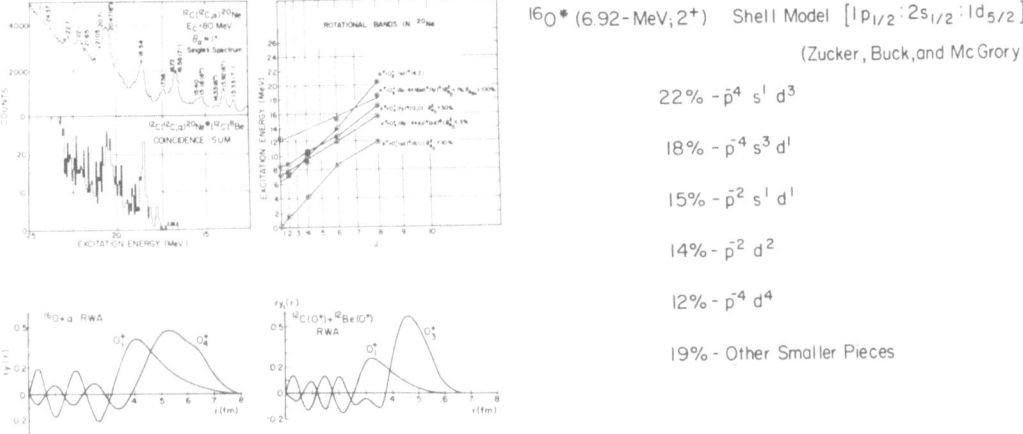

$^{16}O^*$ (6.92-MeV; $2^+$)    Shell Model $[1p_{1/2} \, 2s_{1/2} \, 1d_{5/2}]$

(Zucker, Buck, and McGrory)

22% - $\bar{p}^4 \, s^1 \, d^3$

18% - $\bar{p}^4 \, s^3 \, d^1$

15% - $\bar{p}^2 \, s^1 \, d^1$

14% - $\bar{p}^2 \, d^2$

12% - $\bar{p}^4 \, d^4$

19% - Other Smaller Pieces

**Figure 5**

**Figure 4**

Moving now to the $K^\pi = 0_3^+$ and $K^\pi = 0_6^+$ bands we again find $\theta_{\alpha,0_{1,2}}^2$ less than 3% and 1%, respectively, but for both large reduced widths for the $^{12}C + {}^8Be$ channel[21-22] are observed; in the $0_6^+$ case, in fact, this reduced width reaches values greater than 100% of the Wigner limit emphasizing the purity of these 8 particle 4-hole wave functions. The corresponding Tomoda-Arima[23] calculations for this reduced width amplitude are shown in the figure at lower left; again the $^8Be$ strength in the $0_3^+$ states is predicted to be much enhanced over that of the $0_1^+$ states and to be pushed to larger radius—both signatures of a $^8Be$ cluster configuration. At the upper left of this figure I show a typical set of spectra obtained by Hindi **et al.**[21] in their study of the $^8Be$ cluster states illustrating how the $8^+$ member of the $0_6^+$ band emerges when the spectrum of alpha particles feeding the state is examined in coincidence with the decay alpha particles from the ground state of $^8Be$. The $0_2^+$ band in $^{16}O$ is predominantly of 4 particle-4 hole character and is considered as very similar to the ground state band of $^{20}Ne$.

Over the years, substantial effort has been devoted to searches for a $10^+$ member of the $0_1^+$ ground state band but the results have been uniformly negative consistent with the shell model limit of $8^+$ for an $(sd)^4$ configuration. Given that the ground state is not a pronounced cluster structure this result is not surprising. In the case of the $0_4^+$ band, however, with a very strong $^{16}O + {}^4He$ structure, it is surprising that no $10^+$ member of the band, expected near 23 MeV, has ever been found. Here both $10^+$ and $12^+$ states would be allowed within a shell model framework and would be expected because of the band structure. A specific search for these states would be of considerable interest. Beyond that, the general question of how high in excitation it may be possible to trace rotational bands such as those shown in Figure 3, before the member states effectively dissolve into the underlying continuum, remains unanswered and provides a very real opportunity for the new generation of higher energy precision nuclear accelerators. These experiments will be difficult, will require use of highly selective reactions, but will be of substantial interest in tracing the evolution of familiar nuclear structure with increasing temperature of the many body system.

Figure 5 illustrates that shell model calculations such as those of Zucker, Buck and McGrory[24] shown here, for a low-lying state in $^{16}O$, lead to quite a complex total wave function. Figure 6, from calculations of Fujiwara[13] using an extended harmonic oscillator basis illustrates my earlier comment that good cluster states ($K^{\pi} = 0_4^+$) have wave functions that involve large numbers of oscillator shells as compared to states such as those of the $0_1^+$ ground state band in $^{20}Ne$ (see Figure 3) or the $0_2^+$ 4 particle-4 hole band in $^{16}O$ (see Figure 4). It should be noted, however, as emphasized by Weiguny[25-27], that while a very large basis is required for shell model calculation of cluster states, this basis can be truncated drastically if one moves from a single center to a two-center shell model which is intrinsically better matched to the problem.

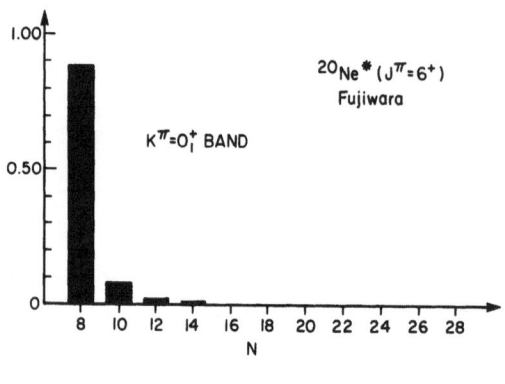

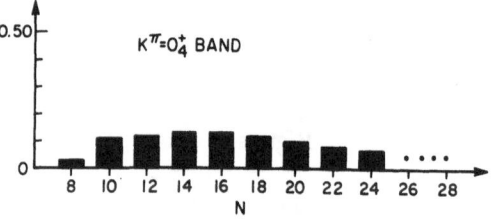

**Figure 6**

Finally, Figure 7 demonstrates that it is not only the positive parity bands in $^{20}Ne$ that display cluster structure. In the upper panel are displayed the electromagnetic deexcitation characteristics of the ground and lowest two negative parity bands. The width of the arrows is proportional to the reduced transition strength quoted in Weisskopf units. The lower panel again shows Tomoda-Arima[23] reduced width amplitudes for selected states; here the dotted curves are for pure SU(3) states while the solid curves are appropriate to a very large basis shell model. The enhancement of the alpha particle reduced width amplitude and its extension to much larger radius for states in the $0^-$, band is entirely consistent with the observation that the average alpha particle reduced width for these states exhausts 50% of the Wigner limit. These, too, are well developed cluster states. The much less striking difference between the SU(3) and the extended shell model for the ground state band is typical of non-cluster states.

In this special case of $^{20}Ne$, and to a somewhat lesser extent in neighbouring light nuclei, we have a reasonably good understanding of alpha particle cluster structure and its interrelationship with other nuclear degrees of freedom.

But this knowledge remains very much limited. Why do equivalently simple cluster models appear to have very much less validity in the upper end of the sd shell, for example? Is this a reflection of the changing spin orbit interaction? Where else in the periodic table does the alpha particle cluster have importance for nuclear structure?

## ALPHA PARTICLE STRUCTURE IN HEAVY NUCLEI:

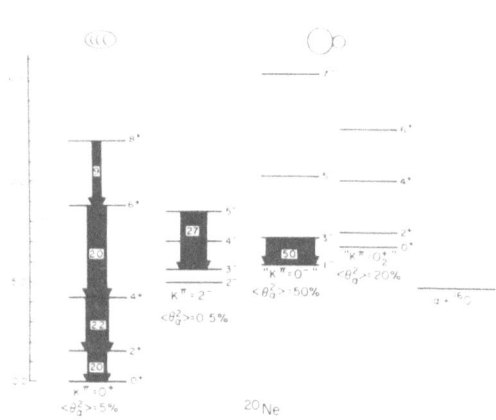

ARIMA-TOMODA: α TRANSFER FORM FACTORS FOR $^{20}$Ne

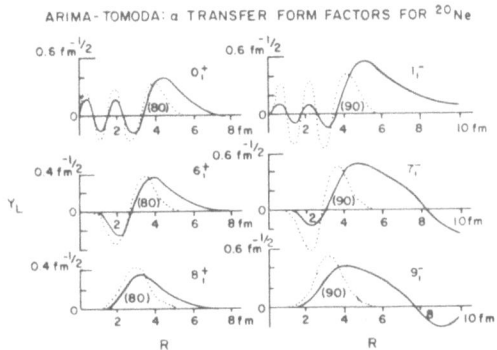

The fact that we observe spontaneous alpha particle emission from nuclei heavier than $^{208}$Pb and not equivalent helion or triton radio-activity implies again that the special alpha particle symmetry and stability play an important role. Figure 8 presents a compilation of experimental alpha particle reduced widths, normalized to that of $^{212}$Po, as prepared by Roeckl[28] and spanning a large section of the upper periodic table; of particular interest here is the large alpha particle reduced widths found just above magic numbers. I shall return to them below.

In Figure 9, however, we consider with what success it has been possible to reproduce the experimental data[29] for the reference nucleus of Figure 8, $^{212}$Po. The upper table compares calculated[30] channel radii, reduced widths and their ratio to the experimentally observed value. The pure shell model calculation predicts a reduced width too small by a factor of

**Figure 7**

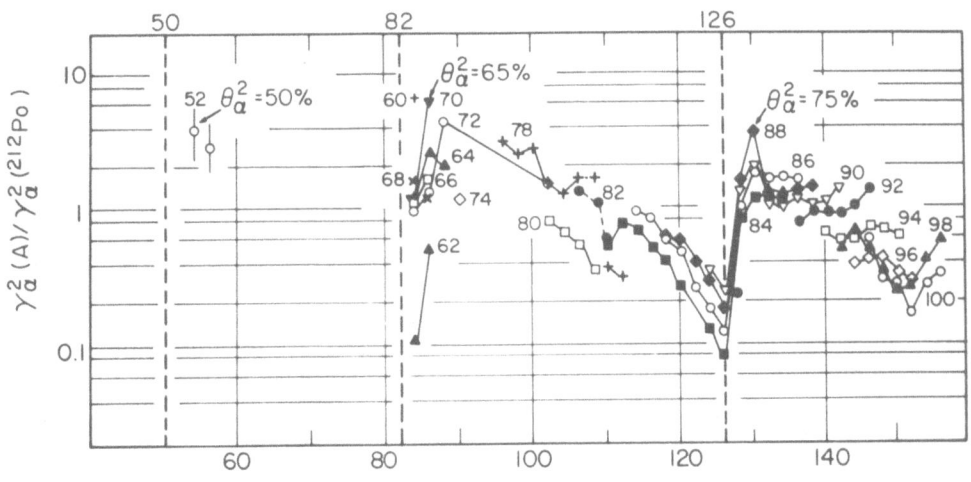

**Figure 8**

NEUTRON NUMBER OF ALPHA EMITTER

DATA COMPILED BY ROECKL *et al.*,GSI

Calculated Channel Radius $\gamma_c$, Reduced Width $\theta^2(\gamma_c)$,
and Its Ratio to the Observed Value for the $\alpha$ decay
$^{212}Po(0_1^+) \to ^{208}Pb(0_1^+) + \alpha$

| Configuration | $\gamma_c$ (fm) | $\theta^2(\gamma_c)_{cal}$ | $\dfrac{\theta^2(\gamma_c)_{cal}}{\theta^2(\gamma_c)_{obs}}$ |
|---|---|---|---|
| (a) $(\pi 0h_{9/2})^2(\nu 1g_{9/2})^2$ | 8.4 | $6.3 \times 10^{-6}$ | $3.4 \times 10^{-5}$ |
| (b) Glendenning and Harada | 8.5 | $4.4 \times 10^{-5}$ | $3.4 \times 10^{-4}$ |
| (c) $\psi_a^{(8)}\psi_b^{(8)}$ (up to $7\hbar\omega$) | 9.0 | $2.9 \times 10^{-4}$ | $1.3 \times 10^{-2}$ |
| (d) $\psi_a\psi_b$ (up to $13\hbar\omega$) | 9.6 | $1.3 \times 10^{-4}$ | $1/25$ |

Calculated Spectroscopic Amplitudes

for

$$^{212}Po(0_1^+) \longrightarrow {}^{208}Pb(0_1^+) + \alpha$$

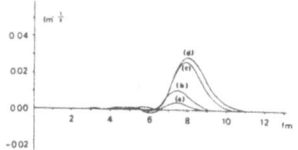

a.  Single shell model configuration $(\pi 0h_{9/2})^2 \otimes (\nu 1g_{9/2})^2$

b.  Glendenning and Harada wave function

c.  Configuration mixing up to $7\hbar\omega$ included

d.  Configuration mixing up to $13\hbar\omega$ included

**Figure 9**                    1. Tonozuka, Ph.D. Thesis
                                University of Tokyo, (1975)

3.4 x $10^{-5}$; use of a Glendenning-Harada[31] wave function improves this to 3.4 x $10^{-4}$, and of an extended shell model including up to $7\hbar\omega$, to 1.3 x $10^{-2}$. Even including $13\hbar\omega$, however, the calculations underestimate the alpha particle reduced width by more than a factor of 20. This is typical of all such calculations and there is no general agreement as to the reason for this continuing discrepancy. The lower panel of this Figure simply shows the corresponding calculated[30] spectroscopic amplitudes—at large radius as they should be—but still too small.

## DIPOLE COLLECTIVITY IN HEAVY NUCLEI:

Substantial new interest has been focussed on the question of alpha particle clustering in heavy nuclei as a consequence of the suggestion by Iachello and Jackson[32] that alpha clustering modes could coexist with the familiar quadrupole excitations of heavy nuclei. This is shown schematically for the case of $^{218}$Ra in Figure 10. Whereas the quadrupole deformation of the $^{218}$Ra ground state gives rise to the familiar K = 0 rotational band, the alpha particle cluster state, with an alpha particle external to an assumed $^{214}$Rn core, is characterized by the radius vector linking the cluster centroids and thus by a dipole degree of freedom. This asymmetric configuration leads to the rotational band shown with alternating even and odd parity states. Obviously the even states would be expected to mix with, and repel, their counterpart states in the quadrupole band whereas the negative parity states, with no counterpart, remain fixed in excitation. Moreover the model predicts that the dipole states should be connected by strongly enhanced E1 transitions. Ennis et al.[33] have studied $^{218}$Ra via the $^{208}$Pb($^{13}$C,3n)$^{218}$Ra and $^{13}$C($^{208}$Pb,3n)$^{218}$Ra reactions and have indeed found the low-lying negative parity sequence that had been missed in earlier studies and the $0^+$, $2^+$, $4^+$ sequence of states that had been pushed up in the spectrum as a result of mixing with the ground state band members. Moreover, from absolute lifetime measurements they have determined that these states are linked by E1 transitions which have reduced matrix elements ~ $10^{-2}$ Weisskopf units making them among the strongest electric dipole transitions ever observed from low-lying nuclear states and very strongly enhanced in terms of a new sum rule for electromagnetic transitions in such cluster states developed by Alhassid, Gai and Bertsch[34]. The Vibron model[35] then successfully reproduces all the new data on $^{218}$Ra and adjacent nuclei where similar phenomena have appeared.

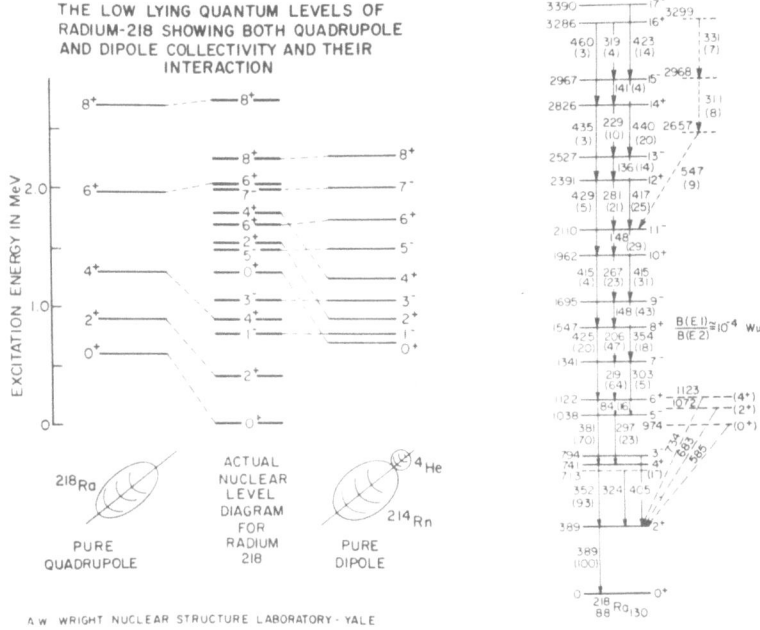

**Figure 10**

Detailed calculations on alpha particle clustering in heavy nuclei and on the mixing of dipole and quadrupole degrees of freedom, have been reported by Daley and Iachello[36]. And in a paper submitted to this conference Catara **et al** report on pioneering microscopic calculations in heavy nuclei that succeed in predicting two and four particle surface correlations that lead to a calculated surface alpha particle distribution function. Gambhir **et al**[38] have also reported indications for the existence of a superfluid condensate of alpha particles in a calculation starting from the interacting boson model, IBM-II, in which neutrons and protons are treated explicity. We thus appear to have one or more theoretical framework adequate for the qualitative calculation of alpha particle preformation in heavy nuclei. As yet, however, quantitative reproduction of experimental studies still eludes us. Mother Nature produces alpha particles more than twenty times more readily than do our models!

Obviously one of the very important open questions at present is the size of the region in the vicinity of closed shells where the new dipole phenomena are of importance. Several laboratories are actively involved in such studies in the actinide and lead regions. Also of importance is the question of whether three nucleon clusters, helions and tritons, can participate effectively in dipole modes in heavy nuclei—or, indeed, anywhere—as we shall discuss below.

## DIPOLE COLLECTIVITY IN LIGHT NUCLEI:

In parallel with the work noted above on $^{218}$Ra, Gai **et al.**[39] have reported equally detailed studies on the low-lying structure of $^{18}$O. This nucleus has long been recognized as featuring coexistence of simple shell model configurations with two valence neutrons in the sd shell and collective quadrupole states of substantially more complex microscopic structure. Gai et al.[39] selected $^{18}O$ for study as a possible candidate for dipole collectivity on the basis that to the extent that it could be considered as $^{14}C + {}^{4}He$, in analogy to $^{218}Ra = {}^{214}Rn + {}^{4}He$ (as shown in Figure 10) these participant nuclei are among the stiffest in the light nuclei and thus candidates for the O(4) limit of the U(4) group structure underlying the Vibron model.

Figure 11 summarizes the particular dynamic symmetries underlying this model. In the now familiar interacting boson model the fundamental entities are assumed to be $s(\ell = 0)$ and $d(\ell = 2)$ bosons (nucleon pairs) and the governing group in the spectrum generating algebra is thus U(6) with the three limiting chains shown. One of the great advantages of the model is that it permits calculations on all nuclei at, or between, these limits.

In contrast, and as outlined above, the Vibron model assumes $\sigma (\ell = 0)$ and $\pi$ $(\ell = 1)$ basic entities and the governing group is U(4). Here only two decomposition chains involve O(3) as is required for the model states to have good angular momentum. Physically the O(4) and U(3) limiting cases correspond to dinuclear molecular or cluster configurations having rigid and excited participant, respectively. Characteristic model spectra for the O(4)

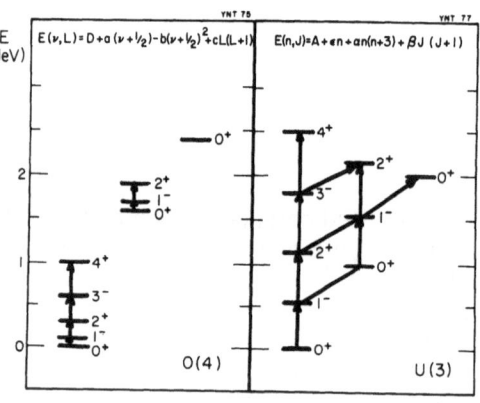

DYNAMIC SYMMETRIES

a) INTERACTING BOSON MODEL

$U(6) \begin{cases} U(5) \supset O(5) \supset O(3) \supset O(2) & \text{harmonic oscillator} \\ U(3) \supset O(3) \supset O(2) & \text{symmetric rotor} \\ O(6) \supset O(5) \supset O(3) \supset O(2) & \text{triaxial rotor} \end{cases}$

s bosons ($\ell = 0$) and d bosons ($\ell = 2$)

b) VIBRON MODEL – dinuclear molecular configurations

$U(4) \begin{cases} O(4) \supset O(3) \supset O(2) & \text{rigid participants} \\ U(3) \supset O(3) \supset O(2) & \text{excited participants} \end{cases}$

$\sigma$ bosons ($\ell = 0$) and $\pi$ bosons ($\ell = 1$)

$E(v,L) = D + a(v + \frac{1}{2}) - b(v + \frac{1}{2})^2 + cL(L+1)$

$E(n,J) = A + \epsilon n + an(n+3) + \beta J(J+1)$

**Figure 11**

and U(3) limits are shown at the bottom of this figure[35]; the O(4) limit leads to familiar vibration-rotation patterns whereas the U(3) limit leads to an anharmonic oscillator pattern. In both cases strongly enhanced E1 electromagnetic transitions are predicted; in the O(4) limit these are entirely intraband whereas in the U(3) limit both inter-and intraband transitions are included.

Figure 12 shows some of the results obtained by Ruscev et al.[39] via $^{14}C(\alpha,\gamma)^{18}O$ and $^{14}C(^{7}Li, t\gamma)^{18}O$ reaction studies. Here again E1 transitions exhausting large fractions of the molecular sum rule—and among the most enhanced ever found in nuclei—appear within a group of states having the $J^{\pi}$ sequence $0^+$, $1^-$, $2^+$, $3^-$, and $4^+$. Moreover, the latter two states have alpha particle reduced widths respectively 20% and 15% of the Wigner limit. E2 crossover transitions in this apparent rotational band are also enhanced as all molecular or cluster models would require.

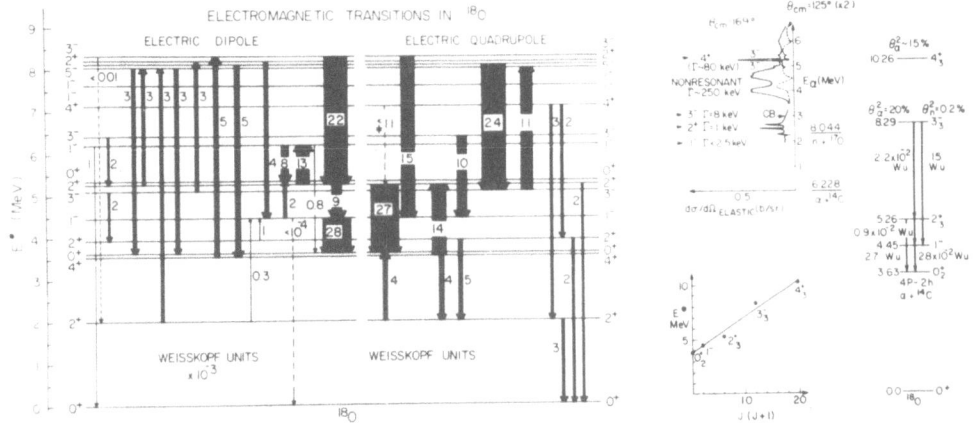

**Figure 12**

We thus appear to have yet a third coexisting degree of freedom among the low-lying states of $^{18}O$—dipole collectivity—very similar to that appearing in $^{218}Ra$.

In an attempt to establish the ubiquity of this new form of collectivity Gamadia **et al.**[40] have reported negative results of a search for dipole signatures in the vicinity of Z = 64, a semi-magic proton number and are currently working on the $^{52}Ti$ region. Determination of the extent to which clustering in the vicinity of closed shells occurs across the periodic table will be important not only for alpha particle clusters but also for both heavier and lighter, well-bound clusters such as h,t, $^{14}C$ and $^{16}O$ that have not been examined at all thus far.

Assenbaum, Langanke and Weiguny[27] have recently succeeded in reproducing, qualitatively, the observed enhanced E1 matrix elements in $^{18}O$ on the basis of detailed calculations within the framework of a $^{14}C + ^{4}He$ cluster model. In particular their model is a two-center one which permits more stringent truncation than does a single center one. Some of this work has been contributed to this conference.

Hayes **et al.**[41] are currently completing an ambitious microscopic one center shell model calculation for the states of $^{18}O$, but in a very large basis, and as soon as the results of these calculations become available it will be possible to examine the overlaps between the two-center cluster wave functions and those from the larger shell model basis to understand better the origin of the clustering and of the E1 enhancements.

It bears noting that Baye and Descouvemont[42], in a contribution to this conference, have also reported on a study of the $^{18}O$ structure using a generator coordinate approach; in contrast to Weiguny **et al.**[26-27], these authors find B(E2) values somewhat larger than those measured and B(E1) values larger than those measured by several orders of magnitude. Resolution of these differences will shed important light on the structure of even so well studied a nucleus as $^{18}O$ and provide new insight into the microscopic origin of clustering.

## ALPHA PARTICLE TRANSFER REACTIONS:

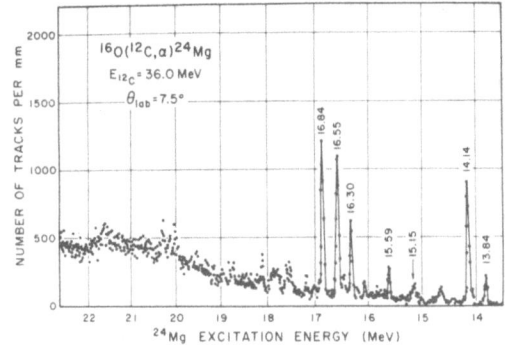

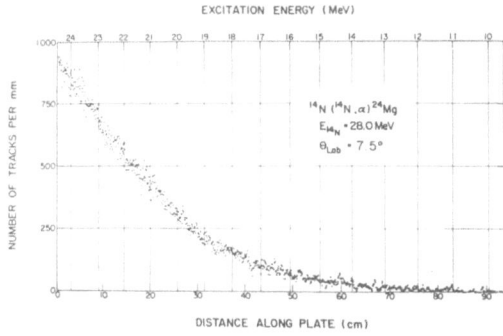

**Figure 13**

A very large fraction of all we know about alpha particle cluster states has been obtained in studies of alpha particle transfer reactions. Arima and Kubono[7] have provided a definitive review of this topic to which the reader is referred for details.

Perhaps the most striking feature of these reactions is their selectivity as illustrated, for example, in Figure 13 where I show alpha particle spectra from the reactions $^{14}N(^{14}N,\alpha)^{24}Mg$[43] and $^{16}O(^{12}C,\alpha)^{24}Mg$[44] respectively. In this latter case, indeed, it can be considered that it is a $^{12}C$ that is being transferred rather than an alpha particle but even in the general case this sort of selectivity is typical. Figure 14 shows typical[45] angular distributions measured with ($^{6}Li,d$) and ($^{16}O,^{12}C$) reactions to the low states of $^{44}Ti$ and illustrates how markedly the gross characteristics of these angular distributions depend upon the reaction involved. In the latter case, bell-shaped angular distributions, centering on the grazing angle and typical of general heavy ion induced reactions, are found whereas, in the former, distributions much more characteristic of a simple direct transfer are found. Here the solid lines are those obtained in standard DWBA calculations (the dashed lines are those calculated for reactions in which the emergent $^{12}C$ is left in its $2_1^+$ excited state). As is clear here, DWBA codes succeed reasonably well in reproducing the shapes of the angular distributions even when they differ as much as here illustrated. Unfortunately, however, even after decades of dedicated study[46], the DWBA analyses still are unable to reproduce absolute cross sections, frequently to within an order of magnitude!

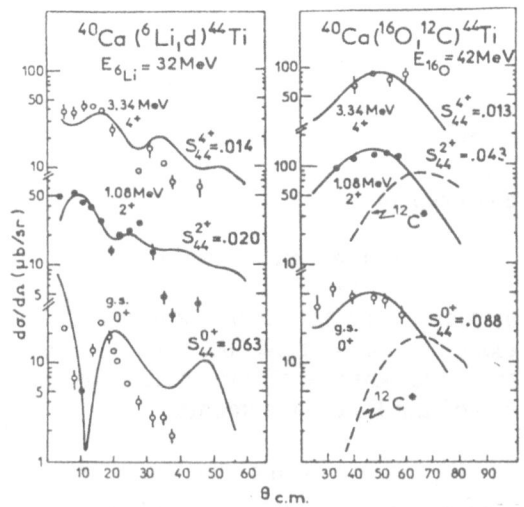

**Figure 14**

In part this very probably reflects the fact that the alpha transfer mechanism is a very large radius one so that the analyses become hypersensitive to the large radius behaviour of the optical model potentials used. Results of Chua and Bromley[47] suggest that traditional optical potential shapes are much too simple leading to correspondingly simplified large radius wave function tails. This however remains largely an open question and there are a number of contributions to this conference that address detailed optical model selection questions.

## DO ALPHA PARTICLES PREEXIST IN NUCLEI?:

A longstanding question has been that of whether alpha particle clusters have a continuing existence in nuclei[48]—describable by cluster wave functions—from which these clusters could be knocked loose in some appropriate reaction. Much effort has been devoted to data collection and analyses of such reactions as $(\alpha, 2\alpha)$[49] and $(p, \alpha)$[50-51]. Here again DWBA calculations are still of little help in reproducing absolute reaction cross sections. It has been suggested[52] that the reproduction of continuum alpha particle spectra from $(p, \alpha)$ reactions at medium energies, using a preequilibrium exciton model indicated the presence of preformed alpha particles in the targets and that the neglect of direct knockout amplitudes in the DWBA was responsible for its failure to give absolute cross sections. In a contribution to this conference Bonetti and Colli-Milazzo[53] report on studies on the $(p\alpha)$ reaction at 23.1 MeV on $^{150}$Sm. Full-finite range DWBA calculations failed for both assumed pickup and knockout processes underestimating the absolute cross sections by factors of 20 and 50 respectively; beyond that the predicted shapes of the angular distributions did not reproduce those of the data. This behavior is typical of all optical models. Using the Iachello-Jackson model[32], on the other hand, where the system is assumed to be excited into a collective oscillation of an alpha particle-plus-core system from which the former is then emitted, these authors have succeeded in fitting not only the shape but also the absolute magnitude of the cross section. This suggests that preformed alpha particles are emitted from lanthanide nuclei via collective mechanisms; this would further imply that the alpha cluster states in these nuclei lie at relatively high excitations consistent with the negative results of Gamadia et al.[40] in their search for dipole collectivity in the region of Z = 64. In a contribution to this conference Gladioli et al.[54] also show that, using standard DWBA techniques, it is impossible to use comparison between pickup and knockout reactions to establish preexistence of alpha particles in target nuclei inasmuch as, when properly executed, the calculations for the two processes make identical predictions.

## HEAVY ION-ALPHA PARTICLE CORRELATION STUDIES:

With the advent of sophisticated data acquisition and analysis systems a great many laboratories have begun to take advantage of multiple particle coincidence techniques to obtain nuclear structural information. Rae[55] will report later in this conference on the extensive Oxford University program in this area, and Figure 15 shows only a sample of the data obtained[56]. In the upper panel,

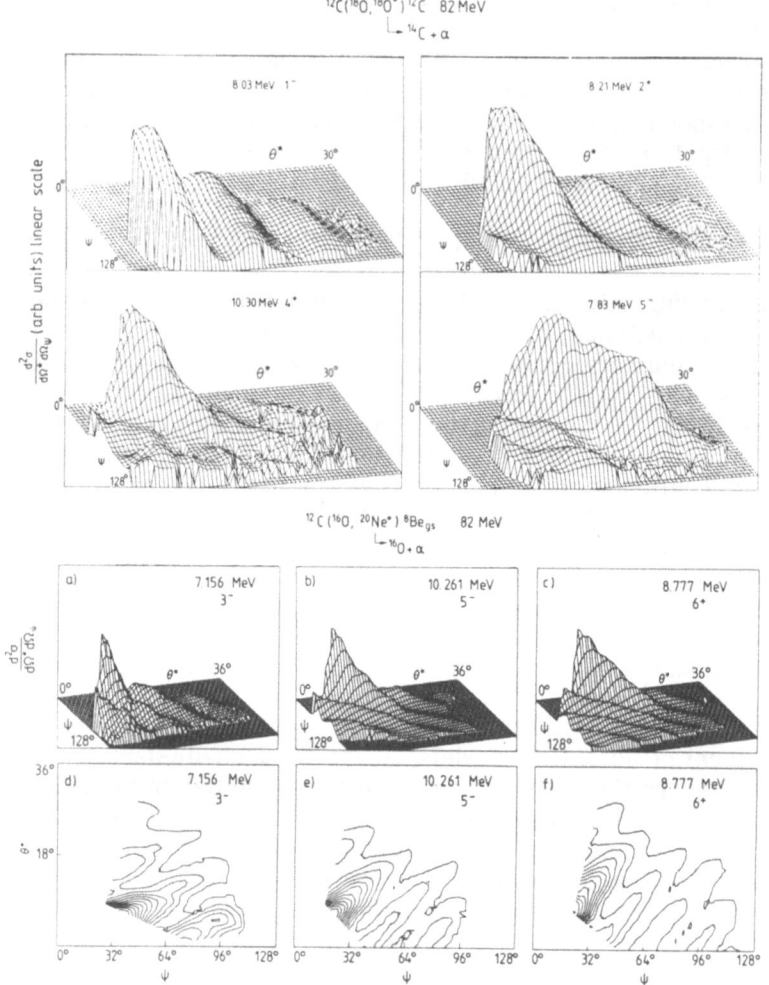

$^{12}C(^{18}O,^{18}O^*)^{12}C$  82 MeV
    └─ $^{14}C + \alpha$

$^{12}C(^{16}O, ^{20}Ne^*)^{8}Be_{gs}$  82 MeV
    └─ $^{16}O + \alpha$

**Figure 15**

inelastic excitation of $^{18}O$ in interactions with $^{12}C$ leads to subsequent decay of the excited $^{18}O$ states into the $^{14}C + ^{4}He$ channel. From data such as these in which both the $^{14}C$ and $^{4}He$ are detected, it is possible to determine not only the spin of the $^{18}O$ state involved but also much information concerning the reaction mechanism. The lower panel presents similar data for the $(^{16}O, ^{20}Ne)$ alpha particle transfer reaction leaving $^{8}Be$ in its (unobserved) ground state together with contour plots of the double differential cross section surfaces. This constitutes a powerful and efficient spectroscopic approach.

## HELION AND TRITON TRANSFER REACTIONS:

Thus far I have focussed exclusively on phenomena involving the alpha particle. But what of the helion and triton--both also very tightly bound nuclei?

Figure 16, from Nagatani **et al.**[54], shows that the selectivity in reactions involving transfer of three nucleons is just as great as was the case for alpha particle transfer; here transfers to mirror final - states yield essentially identical spectra. What may be even more remarkable is the fact that this very selective population of certain states is essentially invariant under change of reaction or energy. Specifically, all of the following helion transfer studies report essentially identical spectra: $^{12}C(^6Li,t)^{15}O$ [Yale; 40 MeV][59]; $^{12}C(^6Li,t)^{15}O$ [Indiana; 99 MeV][60]; $^{12}C(^{10}B,^7Li)^{15}O$ [Texas A&M; 100 MeV][57]; $^{12}C(^{11}B,^8Li)^{15}O$ [Oxford; 114 MeV][58]; $^{12}C(^{12}C,^9Be)^{15}O$ [Oxford; 114 MeV][58]. Figure 17, from Rae **et al.**[58], shows corresponding spectra corresponding to helion and triton transfer to $^{16}O$ and $^{15}N$ targets. Again the reactions are very highly selective and the yield appears to depend strongly upon the angular momentum of the residual state populated.

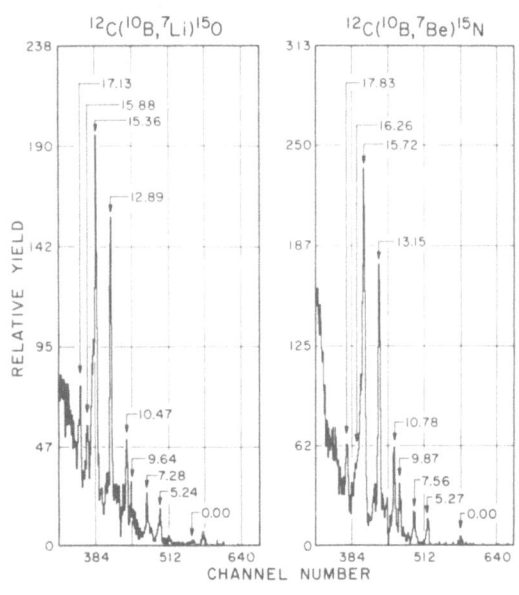

**Figure 16**

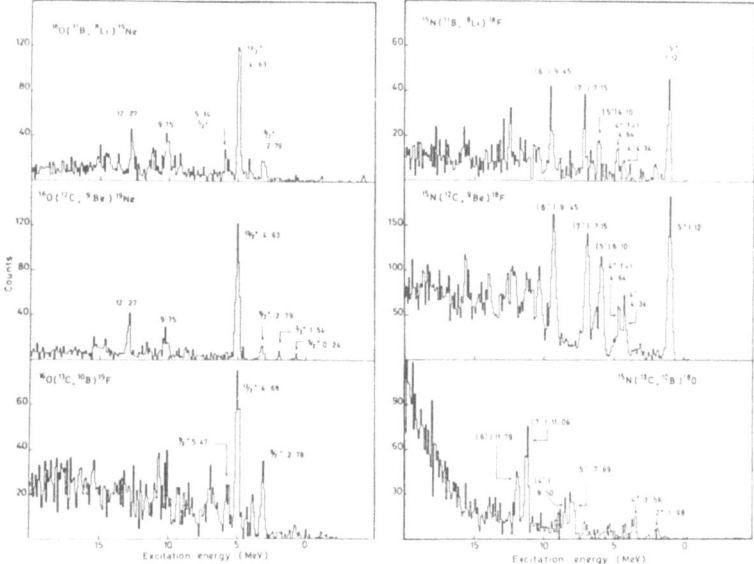

**Figure 17**

Finally, Figure 18 shows data of Martz **et al.**[59]. The left panel compares three and four nucleon transfer spectra leading to A = 19 final nuclei; the right panel compares the predictions of a folded potential model directly with the triton

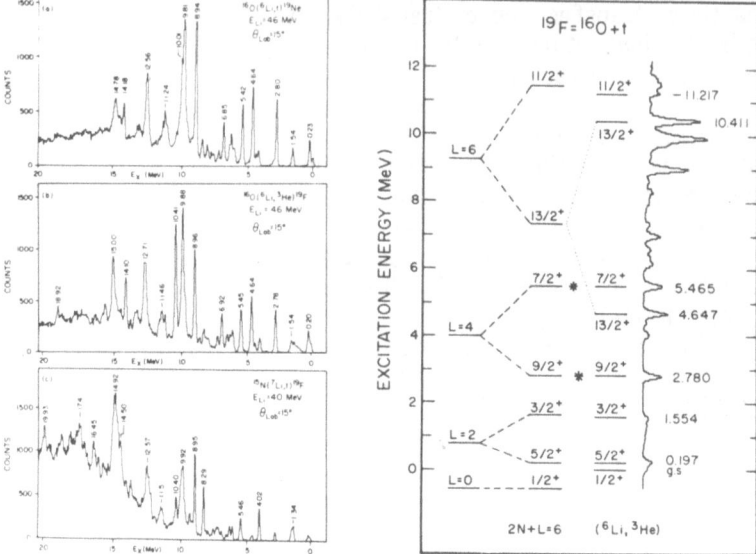

**Figure 18**

transfer data in the center panel and shows that there is at least a qualitative agreement between the two. More detailed examination, however, shows that we still have no solid evidence suggesting that these transfer reactions populate triton or helion cluster states in the final nuclei involved and, indeed, the observed selectivity can largely be explained on angular momentum ground: that these reactions selectively populate yrast states, the highest spin states available at any given excitation.

This poses major open questions. Are there well developed triton and helion cluster states? Systematic reduced width measurements are badly needed. And if there are such states will they display the kind of dipole collectivity recently found in alpha particle cluster states? It may be that the lesser symmetry and stability of the three nucleon systems mitigates against development of this degree of freedom. There is much work to be done to understand even these very simple light systems. And what of three body cluster states in heavy nuclei? Would such states, if they exist, exhibit dipole collectivity? This too remains an open question.

## A NEW NATURAL RADIOACTIVITY:

The question of possible cluster preformation in nuclei immediately brings us to that of natural radioactivity. As one of the oldest of nuclear phenomena, alpha radioactivity has been thoroughly examined and understood. More recently isolated examples of direct proton radioactivity—as for example[61-63] 53mCo—have been reported. Earlier this year Rose and Jones[64], at Oxford reported on

A NEW NATURAL RADIOACTIVITY

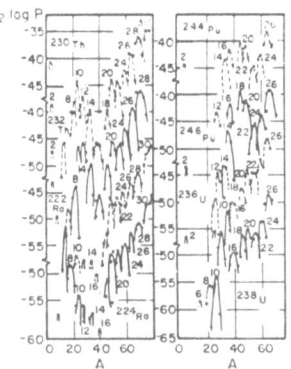

1980

A. Sándulesco, D.N. Poenaro
and W. Greiner

Sov. Jour. Part. Nucl. 11(6), 528, (1980)

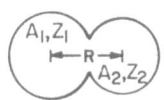

$$\eta = (A_1 - A_2)/(A_1 + A_2)$$
$$\eta_z = (Z_1 - Z_2)/(Z_1 + Z_2)$$

$V(R, \eta)$ minimized re $\eta_z$

1984

H.J. Rose and G.A. Jones

Nature 307 245 (1984)

$$^{223}Ra \rightarrow ^{209}Pb + ^{14}C \quad Q = 32 \text{ MeV}$$

Branching Ratio $^{14}C/^4He = (8.5 \pm 2.5) \times 10^{-10}$

$$\gamma^2_{^{14}C}/\gamma^2_{^4He} \sim 10^{-5} - 10^{-6}$$

A. A. Ogloblin et. al. Private communication (1984)

**Figure 19**

the observation of spontaneous emission of $^{14}C$ from $^{223}Ra$ with the characteristics shown in Figure 19. These results have subsequently been confirmed both by Ogloblin in Moscow[65] and by Vergnes et al.[66] at Orsay; this latter group, using a totally different detection technique, in addition to 13 $^{14}C$ events in a 5 day run, report a single event which they believe to be spontaneous emission of $^{16}O$ from $^{223}Ra$.

Also shown in Figure 19 is a section of the predictions of Sandulescu, Poenaro and Greiner[67-68], published in 1980, of such phenomena. This work stemmed from Greiner's quantization[69] of the mass and charge asymmetry parameters, $\eta = (A_1 - A_2)/(A_1 + A_2)$ and $\eta_z = (Z_1 - Z_2)/(Z_1 + Z_2)$ respectively and his development of the fission potential surface $V(\eta, R)$ which was subsequently minimized with respect to $\eta_z$ and the remaining parameters of his model before being used to solve for the mass yield in spontaneous fission of heavy nuclei. As shown here, at the lower right of the figure, predictions were made across the periodic table; here I show only those in the range $2 \leq Z \leq 26$. It bears noting that alpha particle emission from $^{236}U$ was predicted to have roughly the same emission probability as silicon or magnesium while carbon emission was predicted to be some 13 to 14 orders of magnitude less probable as compared to the observed branching ratio of 5 to 6 orders of magnitude. This discrepancy is hardly surprising in view of the relative crudeness of the model and with an experimental value for this reduced width ratio, so far in the wings of the fission mass distribution, it will be possible to bring the model into agreement with data without difficulty.

With this in hand, great interest now focusses on the search for spontaneous emission of other species of heavy fragments from actinide nuclei. Even a few additional points on this figure would be of great importance to fine tuning our quantitative understanding of cluster formation[68] in heavy nuclei.

## PROJECTILE BREAKUP:

It must be emphasized that in spontaneous fission emission of fragments, the emission is from a cold parent nucleus. In this sense it is quite different from the emission of fragments from nuclei that have been heated in a nuclear collision.

Perhaps the most gentle such collision is that involved in elastic scattering and over the years substantial effort has been devoted to

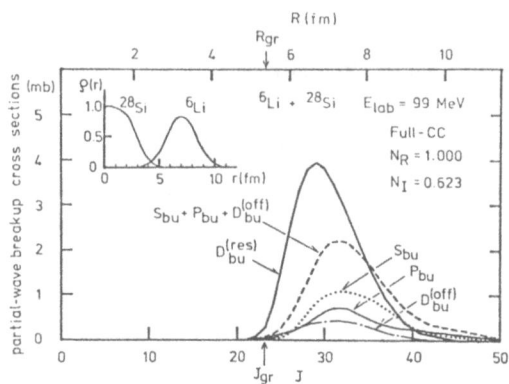

**Figure 20**

the study of $^6$Li breakup under such conditions. Figure 20 shows calculations of Sakuragi, Yahiro and Kamimura[70-71] on the breakup of $^6$Li in its scattering on $^{28}$Si. Several points emerge: a) the breakup interaction is a very peripheral one (the inset shows the matter distributions of $^6$Li and $^{28}$Si at a 7 Fm separation where the partial breakup cross sections peak); b) the breakup involves very high partial waves; c) breakup into the D wave resonance dominates the process; d) there is strong interference between one and two step processes in the breakup. In a contribution to this conference Pfaneta **et al.**[72] report on a study of the $^6$Li breakup at 156 MeV and conclude that the inelastic mode exhausts a much smaller part of the total fragmentation cross section than would be predicted by standard DWBA analyses. Shimoda **et al**[73] in yet another contribution find, at 176 MeV, strong coupling of the inelastic and breakup channels on $^{12}$C targets as well as large sequential breakup amplitudes in agreement with Sakuragi **et al**[71].

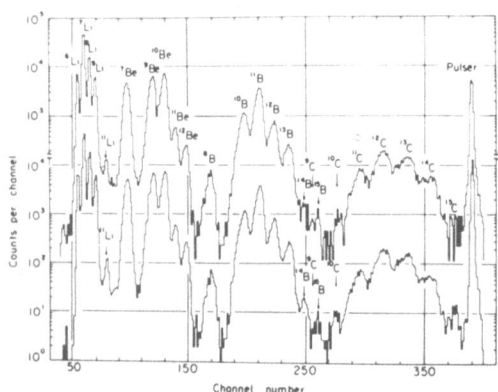

ISOTOPIC YIELD FROM $^{238}$U+p AT 5.3 GeV
Poskanzer et al.

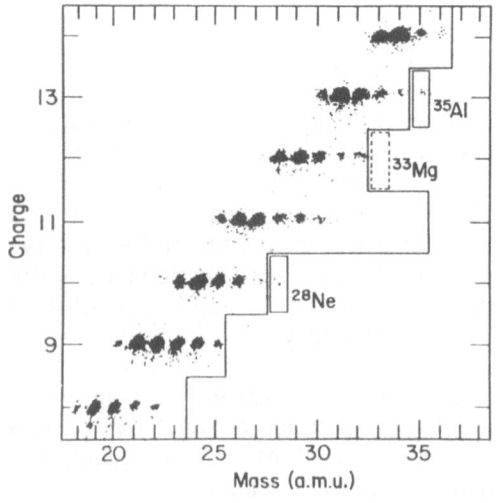

**Figure 21**

## FRAGMENTATION

Turning to more violent collisions, Figure 21 shows a section of the mass yield from breakup of $^{238}$U by 5.3 GeV protons as reported by Poskanzer et al.[74] and from ~ 2 GeV/A argon incident on a carbon target as reported by Scott et al.[75] It is clear that copious production of a great many light products is involved here, however, these data provide no evidence for clustering prior to the collision and are generally consistent with statistical considerations during the decay of a hot nucleus. Friedman and Lynch[76] have concluded from their statistical calculations that emission of intermediate mass fragments should be a general property of such hot nuclei, as the temperature of nuclei is raised to 5-10 MeV (50-100 billion degrees!) The yield of fragments in the carbon-nickel region is predicted to be relatively large. Figure 22 shows recent results of Chitwood et

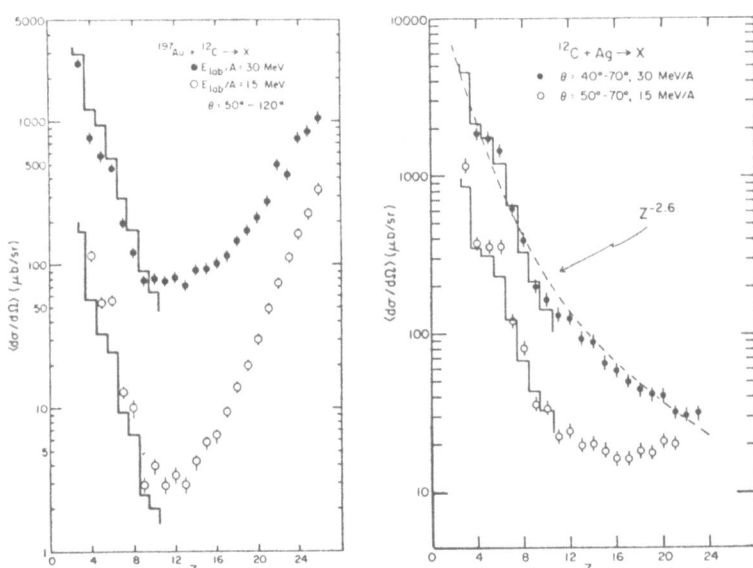

**Figure 22**

al.[77] at Michigan State from bombardment of Au and Ag targets with 15 and 30 MeV/A carbon beams (the histograms are the results of simple statistical calculations); temperatures between 3.5 and 4.5 MeV would be expected here. As this figure demonstrates, there is substantial yield of intermediate mass fragments; the rise in yield beyond Z~13 in the Au case reflects the tail of the more familiar fission mass yield curve. It is important to note that the statistical calculations are able to reproduce the increase by an order of magnitude in fragment yields in going from 15 to 30 MeV/A. Furthermore, these studies show that the decay by particle emission begins well before the excitation energy can be distributed throughout the hot nucleus so that temperatures substantially greater than the above mentioned equilibrium values are involved. It is already clear that such studies can provide an important window on the behavior of high temperature nuclear matter.

## NUCLEAR MOLECULAR PHENOMENA IN LIGHT SYSTEMS:

Thus far we have considered the decomposition of cluster states in nuclei and I now turn to the opposite extreme. What can we learn by putting together cluster states? The study of nuclear molecular resonances has a long history but only quite recently have we begun to understand the phenomena involved in any detail.

Figure 23, in the top panel, shows the presently known low energy excitation function for the $^{12}C$ + $^{12}C$ system[78], the most thoroughly studied of all such systems. Despite much theoretical effort, as more and more experiments added more and more resonances to this excitation function, no detailed understanding emerged. In 1981, however, and as sketched in Figure 11, in analogy with the U(6) dynamic symmetry underlying the interacting boson model for bound nuclear states, Iachello[79,35] suggested that the U(4) symmetry, in addition to its role in reproducing dipole collectivity in bound nuclear states, could provide a framework within which to understand the scattering resonances as well; this represented the first attempt to apply spectrum generating algebraic techniques to a scattering problem, i.e. to unbound nuclear states. Since it was well demonstrated that in these resonances the $^{12}C$ nuclei retained their identity, Erb and Bromley[80] chose to use the O(4) limit of the U(4) model and achieved the results shown in the lower panel of this figure. All the 39 known resonances at low energies fit neatly into a vibration-rotation pattern with none left over.

The fact that the dissociation energy found in this model was ~ 7 MeV suggested to Erb and Bromley that the 7.66 MeV $0^+$ state of $^{12}C$

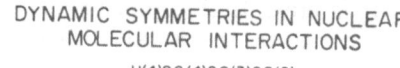

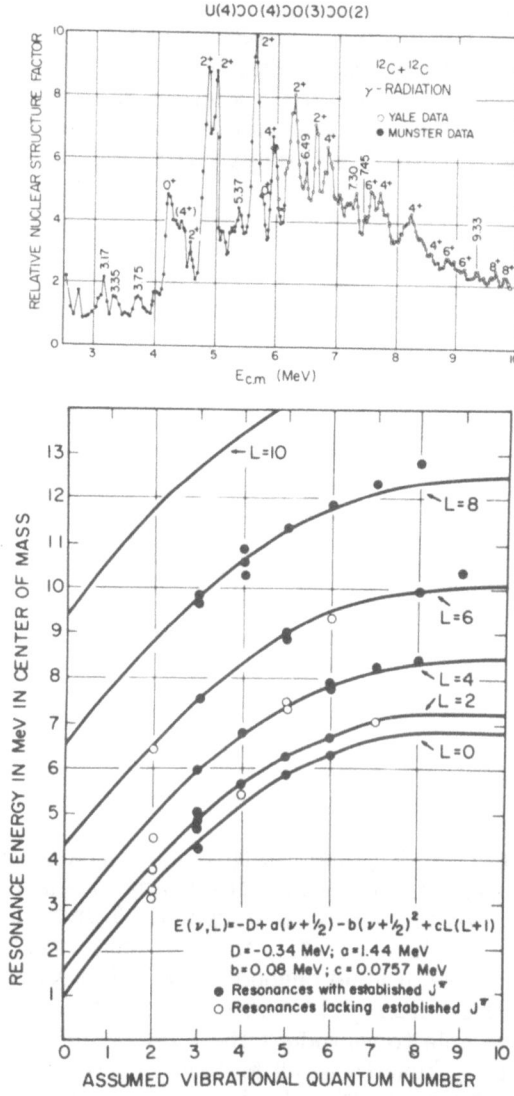

**Figure 23**

was playing a role in the molecular phenomena rather than the $2_1^+$ state at 4.43 MeV as had been uniformly assumed in earlier work.

Building on this hypothesis Strayer et al.[81] used a constrained time dependent Hartree-Fock approach to examine this system in greater detail. As shown in the upper left, of Figure 24 the $^{12}$C ground state is a soft, weakly oblate spheroid

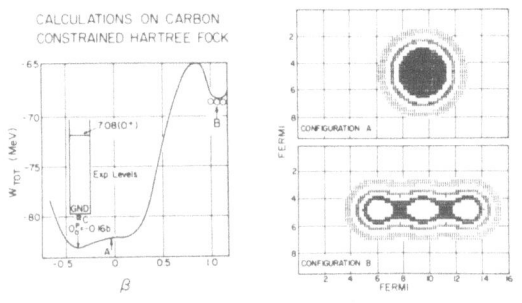

while the $0_2^+$, 7.66 MeV state is a linear alpha particle array; the upper right panel shows the corresponding density contours emerging from the TDHF. In the lower panel we show what happens at certain energies in the TDHF calculation if we assume that one of the $^{12}$C nuclei is excited to its $0_2^+$ state in an early phase of the collision and subsequently interacts with the remaining unexcited $^{12}$C collision partner. As shown here, the $^{12}$C nuclei oscillate through one another in a fashion reminiscent of the familiar ammonia molecule inversion. The period of this oscillation in the TDHF calculations is 2.3 x $10^{-21}$ seconds and the corresponding phonon energy, 1.79 MeV.

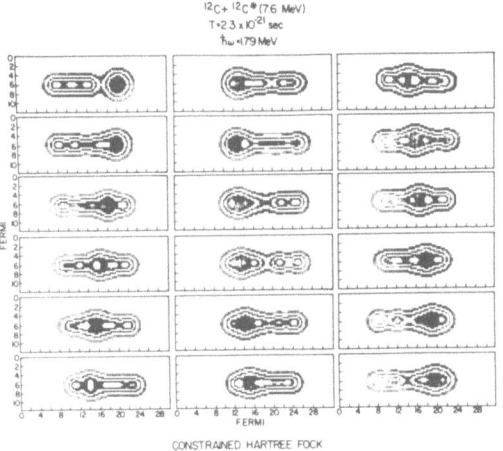

It must be emphasized that this is an extremely crude calculation but it is intriguing enough to suggest further work. It is also true that if such a mechanism is actually present the O(4) model limit used by Erb and Bromley is invalid. Much remains to be done here.

**Figure 24**

In Figure 25 I show these $^{12}$C + $^{12}$C data on a different plot of excitation energy vs J(J+1). At low energies we see the rotational bands of Figure 23 again; the moment of inertia corresponding to the slope of these loci is that of two touching $^{12}$C nuclei with a center separation of 6.7 Fm. At energies above the Coulomb barrier the moment of inertia corresponds to that of overlapping $^{12}$C nuclei and a center separation of 5.7 Fm.

One of the problems with the $^{12}$C + $^{12}$C system is that its symmetry suppresses all negative parity states. For that reason Ruscev et al.[39] undertook a detailed study of the asymmetric $^{16}$O + $^{12}$C system, focussing particularly on the lower energy regions that had not previously been examined in detail. Their results are shown in the lower panel of Figure 25. All resonances below the Coulomb barrier were found to have an average reduced width with respect to the entrance channel of 63% of the Wigner limit. As shown, they fall into two rotational bands of alternating parity and appear to be well reproduced by the O(4) limit of the U(4) model. The moment of inertia suggests a very small nuclear overlap with a center separation of 9.6 Fm.

Above the Coulomb barrier, the apparent moment of inertia is quite different with an average entrance channel reduced width less than 5% and a center separation of 5.3 Fm. These states are clearly not describable with the O(4) limit nor indeed is the U(3) limit any more successful. Ruscev **et al.** have suggested, in analogy with the IBA, that it will be interesting to see whether it is possible to reproduce these data with numerical calculation intermediate between the simple model limits and such calculations are in progress.

In a contribution to this conference Cseh[82)] reports on an initial attempt to make such calculations for the $^{12}C + ^{12}C$ system and finds that he obtains the best reproduction of the available data closer to the U(3) than to the O(4) limit used by Erb and Bromley. Although Cseh's calculations are still crude, in that the quadrupole degree of freedom is entirely ignored, they point the way to much future work.

The upper energy resonances also show alpha particle reduced widths less than 1% of the Wigner limit so the alpha particle degree of freedom, frequently suggested in connection with these data, cannot be present in any simple manifestation such as $^{24}Mg + ^4He$ or $^{12}C + ^{12}C + ^4He$. This is entirely consistent with the measurements of Gersch, Schobbert and Wohlforth[83)] on the 11.1 MeV $5^-$ resonance in $^{16}O + ^{12}C$; despite intensive search no evidence was found for the $^{12}C + ^{12}C + ^4He$ three-nuclear configuration.

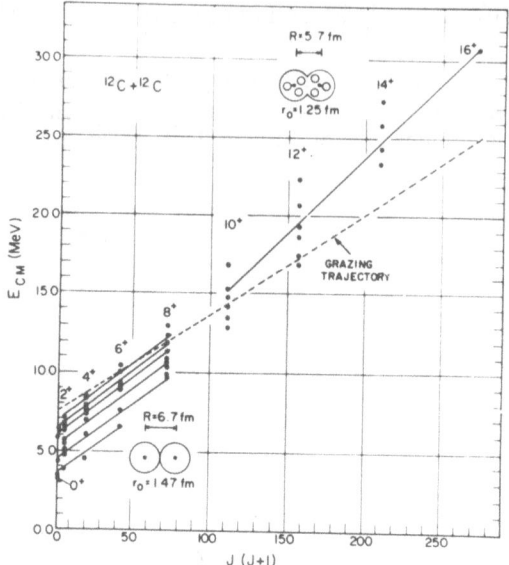

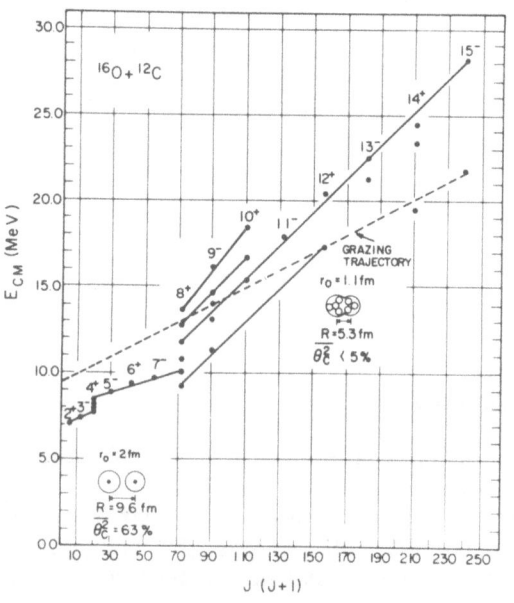

**Figure 25**

A number of groups have searched, without success, for the electromagnetic decay of the high energy resonances in both these systems[84-85] and have suggested that the absence of the expected photons vitiates the idea of a molecular configuration. But this is clearly wrong. Where the strongly enhanced transitions would be expected is where an O(4) description is valid, namely, below the Coulomb barrier and here the gamma ray measurements have not yet been attempted. Such measurements are now underway on the $^{16}O + {}^{12}C$ system at Yale. Kato et al.[86] have also reported calculations on E2 enhancements, using a band crossing model for the $^{16}O + {}^{12}C$ system, and find essentially no anomaly in the band crossing region.

It bears emphasis, however, that while the E1 transitions that the O(4) model predicts as particularly enhanced are completely forbidden[87] in a symmetric system such as $^{12}C + {}^{12}C$, they are also forbidden in first order in any self-conjugate system such as $^{16}O + {}^{12}C$. For that reason Sterbenz et al.[88] are studying the $^{18}O + {}^{14}C$ system and if suitable resonances are found, will focus on their allowed E1 electromagnetic decay.

## NUCLEAR MOLECULAR PHENOMENA IN INTERMEDIATE SYSTEMS:

In recent years improved accelerating facilities and experimental techniques have led to the observation of resonant phenomena in a great many nuclear systems and, indeed, in a number of systems that had previously been reported as resonance free. Figure 26 summarizes data of Betts and his collaborators[89-90]

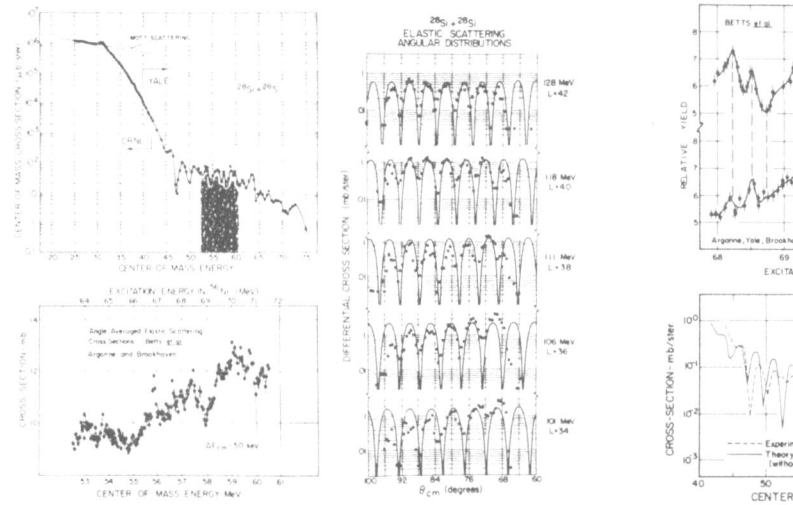
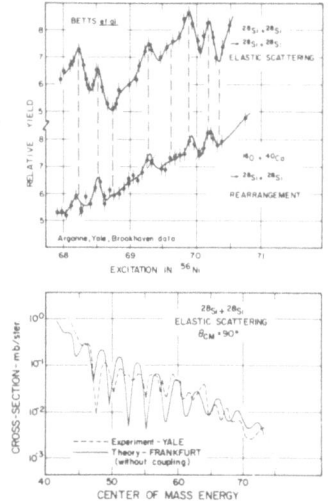

**Figure 26**

at Yale and Argonne, as well as calculations of the Frankfurt group[91], on the $^{28}$Si + $^{28}$Si system. The upper left panel shows the 90° elastic scattering excitation function as measured at Chalk River and at Yale. The lower left panel shows a much higher resolution section of this function measured by Betts **et al.**[90] showing sharp structure superimposed upon the potential scattering gross structure. Angular distributions measured by DiCenzo **et al.**[92] in the center panel establish the angular momenta of the resonances unambiguously. These are currently the highest angular momenta so characterized in nuclear physics; they are of particular interest in that crude liquid drop models suggest that the compound system here, $^{56}$Ni, should become unstable with respect to centrifugal disruption at about J = 50. Studies in this region await the large tandem accelerators operating in the 25 MV terminal range.

The upper right panel shows data of Dichter **et al.**[93] on the $^{16}$O + $^{40}$Ca channel together with the $^{28}$Si + $^{28}$Si data. The fact that there appears to be a one-to-one correspondence between the resonances in the two systems provides convincing evidence that we are seeing cluster states at some 70 MeV of excitation in the 56-body system, $^{56}$Ni, having high spin and substantial reduced widths connecting then to **both** the $^{28}$Si + $^{28}$Si and $^{16}$O + $^{40}$Ca channels. Saini and Betts[94] have shown, on the basis of detailed statistical calculations, that the fine structure in these excitation functions is not of statistical origin and indeed does correspond to long-lived states in $^{56}$Ni. That sharp cluster states should persist to such high excitation is contrary to much conventional wisdom.

The lower right panel shows coupled channel calculation results of Thiel, Greiner and Scheid[91] which reproduce the general features of the excitation function.

Very similar effects have been reported in the $^{24}$Mg + $^{24}$Mg system[95] although these are not as striking as in $^{28}$Si + $^{28}$Si. Figure 27 shows, on the left,

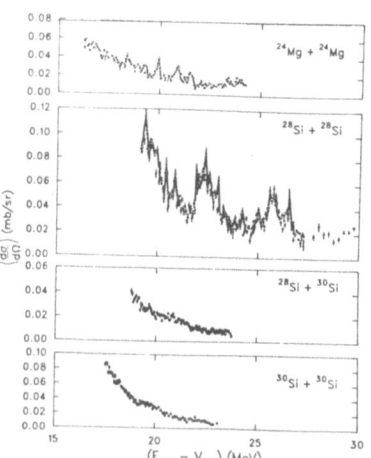
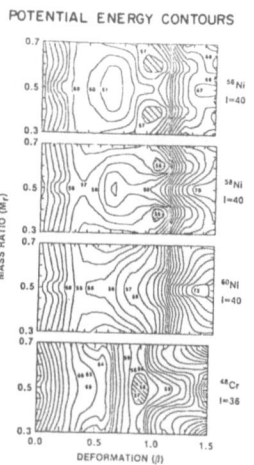

**Figure 27**

angle-averaged, back-angle elastic scattering cross sections[95-96] for four different systems $^{24}$Mg + $^{24}$Mg, $^{28}$Si + $^{28}$Si, $^{28}$Si + $^{30}$Si and $^{30}$Si + $^{30}$Si corresponding, respectively, to the compound systems $^{48}$Cr, $^{56}$Ni, $^{58}$Ni and $^{60}$Ni.

The right panel of this Figure shows potential energy contours calculated by Leander and Larsson[97-99] for these compound nuclei and the appropriate range of angular momenta. It is interesting to note that in $^{48}$Cr and $^{56}$Ni, those compound systems where sharp resonance phenomena are observed, pronounced minima are found in the potential surfaces at very large deformations, $\beta\sim0.9$ while in both $^{58}$Ni and $^{60}$Ni these super-deformed wells are much shallower or absent entirely. It would be extremely attractive to be able to relate the observed fine structure to these shell phenomena in the compound system but as yet this must remain as an interesting speculation requiring much more investigation, both experimentally and theoretically. How would states in the super-deformed minima be expected to reveal their presence? To what extend does this situation parallel that of the outer minimum in the fission barrier where the level density in the inner deformation well modulates that in the outer one in (n,f) excitation functions, for example?

With higher energy heavy ion beams of heavier species having adequate energy resolution it will be very important to establish at what excitation energy the sharp states disappear from experimental excitation functions in different mass regions. Thus far in heavy ion physics, whenever sharp resonances have been searched for with adequate resolution they have been found. New accelerators now under construction or becoming operational greatly extend the regime open to such studies.

**NUCLEAR MOLECULAR EFFECTS IN SUPERHEAVY NUCLEI:**

One of the most interesting results to emerge from the new field of high energy atomic physics is that of Schweppe **et al.**[100] and of Berdermann **et al.**[101] involving emission of positrons in supercritical Coulomb fields occurring during collisons of heavy nuclei where $Z_T = Z_1 + Z_2 \gtrsim 170$. The results of Schweppe **et al.** are shown in Figure 28; of particular interest is the sharp positron line at $\sim 300$ keV in the upper left panel.

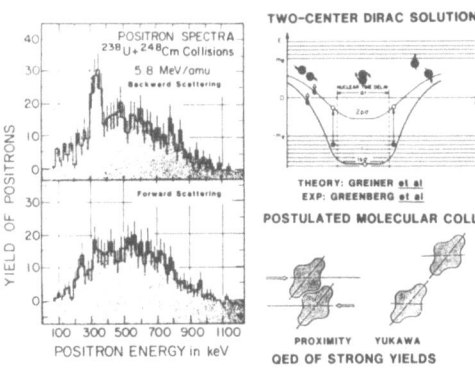 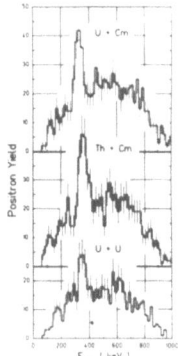

**Figure 28**

Emmision of unaccompanied positrons corresponding to ionization of the vacuum has been studied extensively and was originally suggested by Greiner and his collaborators[102-105] at Frankfurt. What was unexpected was the sharp line indicating the presence of a long-lived ($\sim 10^{-19}$ sec) configuration in the collision as illustrated schematically in the upper right panel. Greiner and Greenberg[106] have suggested that the positron line reflects formation, in a small fraction of the collisions, of a nuclear molecular complex and calculations of the collision potentials suggest that it is the nose-to-nose configuration shown at the lower right that most probably binds. Recent results, shown on the extreme right of this figure, suggest that the positron line energy may actually be the same for U + U, U + Th and U + Cm collisions; if this is confirmed then it would suggest that there is a particularly stable molecular configuration that is reached in all three systems—after rapid emission of an appropriate number of neutrons and alpha particles. Experiments to search for such particles in coincidence with the positron line are just now being mounted.

Whatever the result, the suggestion is that one may be dealing with an entirely new kind of extreme cluster phenomenon in these very heavy nuclear collisions. Although substantial progress has been made in understanding the atomic aspects of these collisions, the nuclear aspects remain virgin territory.

## CLUSTERING IN THE QCD DOMAIN:

It is becoming increasingly clear that quantum-chromodynamics has at least the potential of reproducing phenomena in the strong interaction domain with something of the success of quantum electrodynamics in the electromagnetic one. New accelerators in the interface between nuclear and particle physics hold high promise of illuminating this centrally important question.

One of the earliest quantitative attempts to apply QCD concepts to the nuclear problem was that of Robson[107] who developed a colored quark exchange model and from it a so-called tetrahedral cluster model of many light nuclei. Among these models, one of the most explicit predictions[108] was that of a new, sharp $3^-$ level at about 9.9 MeV in $^{16}O$. Basically Robson's model for $^{16}O$ comprises four clusters in a tetrahedral equilibrium configuration which can be excited into normal vibrational modes. This is very much reminiscent of the old alpha particle models of Dennison[16] and Kameny[17] and as shown schematically in Figure 2. But Robson's model requires[109] a $J^{\pi} = 0^+$, $3^-$, $4^+$, $6^+$, $7^-$, $8^+$ ground state band as shown in the left panel of Figure 29. This predicted $3^-$ state has been searched for in inelastic alpha particle scattering[110], in inelastic electron scattering, and in a wide variety of heavy ion induced reactions[112-116]; all these searches yielded negative results and very convincing evidence against this rather explicit model prediction.

Despite this initial set back, many investigators have been attempting—both experimentally and theoretically—to mount attacks on this fundamentally important role of QCD in nuclear physics. The right panel of Figure 29 shows the experimental data of Hennino **et al.**[117] on the $^6Li(p, \Delta^{++})^6He$ reaction at 1.04 GeV together with the quark model calculations of Jain and Gupta[118]. These authors find that the reaction is dominated by the interaction of two 3-quark clusters

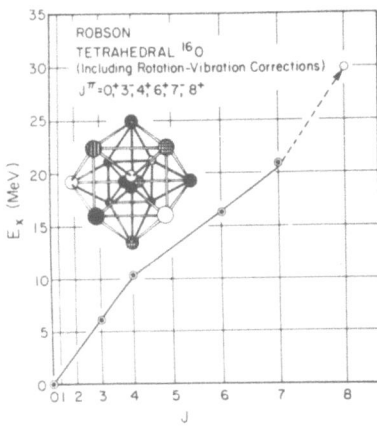
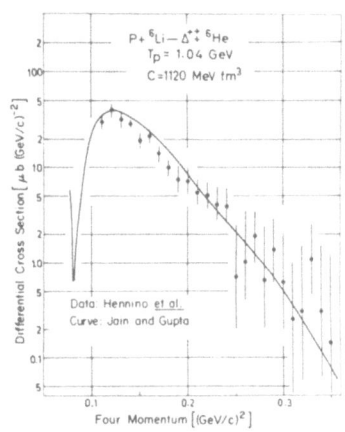

**Figure 29**

which interact to flip both spin and isospin of one quark in each cluster without any change of radial or orbital angular momentum. Although a very simple interaction, it remains extremely unclear why it would be selected by QCD considerations.

Clearly what we require is more high quality experimental data in this interface between nuclear and particle physics. The large number of theoretical contributions to this conference on quark topics in nuclear physics—some of which take over such familiar cluster approaches as the resonating group method directly[119]—show that theorists are ready and waiting.

## CONCLUSIONS:

Cluster phenomena are a ubiquitous feature of nuclear structure and dynamics. But even as old and familiar a cluster as the alpha particle has new and unexpected facets of behaviour as witness the recently discovered dipole collectivity in nuclei. And we still do not know whether or not true helion and triton cluster states exist or whether the states selectively populated in reactions which transfer such clusters are simply yrast states. And after years of study we are still unable to reproduce, to within orders of magnitude, the absolute cross sections of cluster transfer reactions.

We have no idea how high in excitation we can expect to see sharp cluster states nor do we know whether they exist across the periodic table or only in particular regions. We still do not have a clear understanding of the mechanism whereby nuclear molecular resonances are populated and we are still searching for the electromagnetic branches of their deexcitation.

I believe that we are at the threshold of a major new development in the nuclear-particle physics interface where we bring our techniques, concepts and theories from nuclear physics to bear on the underlying quark-gluon constituents of nuclei. This is one of our most active frontiers and one where cluster techniques and ideas can confidently be expected to be important.

Although we are now involved in the Fourth International Conference on Cluster Phenomena, in a very real sense this is a field which is only now coming into its adolescence. We often forget just how limited and fragmentary has been our exploration of the rich domain that the nucleus presents. With new tools and new techniques we can look forward to data of ever increasing precision and completeness and to data in entirely new energy regimes but I expect that we will continue to be surprised even more frequently than in the past. Indeed I shall be very much disappointed if we are not.

## ACKNOWLEDGEMENTS:

I am much indebted to my colleagues Moshe Gai and Peter Parker for discussions concerning material in this paper and to my students, over the years, and most recently John Ennis, Anna Hayes, Mario Ruscev, Edward Schloemer and Stephen Sterbenz for their results, many of them unpublished, that I have included herein. I am also indebted to Akito Arima, George Bertsch, Russell Betts, Karl Erb, Walter Greiner, Franco Iachello, William Rae and Michael Strayer for providing data and suggestions. And I am grateful to the Conference organization for giving me access to the contributions submitted to it; I have found them fascinating and my apologies go to all those whose results I should have liked to include in this review but for which space and time simply were not available.

Finally, my special thanks go to Mrs. Rita Bonito and to Mrs. Sandra Sicignano for their help in preparing camera ready copy.

## REFERENCES:

1. W. Wefelmeir, Naturwiss **25** 525 (1937).
2. J.A. Wheeler, Phys. Rev. **32** 1083 (1937).
3. C.F. von Weizsacker, Naturwiss **26** 209,225 (1938).
4. K. Wildermuth and Th. Kanellopoulos, Nucl. Phys. **7** 150 (1958).
5. G.C. Phillips and T.A. Tombrello, Nucl. Phys. **19** 555 (1960).
6. T.A. Tombrello and G.C. Phillips, Nucl. Phys. **20** 648 (1960).
7. A. Arima and J. Kubono, **Treatise on Heavy Ion Physics Vol. I** ed. by D.A. Bromley, Plenum Press, New York 1984.
8. A. Arima, H. Horiuchi and T. Seke, Phys. Lett. **25B** 129 (1967).
9. A. Arima, V. Gillet and J. Ginnochio, Phys. Rev. Lett **25** 1043 (1970).
10. A. Arima, H. Horiuchi, K. Kubodera and N. Takigawa, **Advances in Nuclear Physics Vol. 5** ed. by M. Baranger and E. Vogt, Plenum Press, New York (1972).
11. K. Ikeda, N. Takigawa and H. Horiuchi, Suppl. Prog. Theo. Phys. (Japan) Extra Number (1968).
12. Y. Fujiwara, H. Horiuchi and R. Tamagaki, Prog. Theo. Phys. (Japan) **61** 1629 (1979).
13. Y. Fujiwara, Prog. Theo. Phys. (Japan) **62** 122,138 (1979).
14. J.P. Elliott, Proc. Roy. Soc. (London) **A245** 128 (1958).
15. D.M. Brink, H. Friedrick, A. Weiguny and C.W. Wong, Phys. Lett. **33B** 143 (1970).
16. D.M. Dennison, Phys. Rev. **57** 454 (1940). Phys. Rev. **96** 378 (1954).
17. S.L. Kameny, Phys. Rev. **103** 358 (1956).

18. H. Morinaga, Phys. Rev. **101** 254 (1956).
19. Y. Fujiwara, H. Horiuchi, K. Ikeda, M. Kamimura, K. Sato, Y. Suzuki and E. Negaki, Prog. Theo. Phys. Suppl. **68** 109 (1980).
20. S.J. Sanders, L.M. Martz and P.D. Parker, Phys. Rev. **C20** 1743 (1979).
21. M.M. Hindi, J.H. Thomas, D.C. Radford and P.D. Parker, Phys. Rev. **C27** 2902 (1983).
22. M.M. Hindi, J.H. Thomas, D.C. Radford and P.D. Parker, Phys. Lett. **99B** 33 (1981).
23. T. Tomoda and A. Arima, Nuc. Phys. **A303** 217 (1978).
24. A.P. Zucker, B. Buck and J.B. McGrory, Phys. Rev. Lett. **21** 39 (1968).
25. A. Weiguny, private communication (1982).
26. A. Weiguny, K. Langanke and H.J. Assenbaum, Contribution No. 95; **4th Int. Conf. on Clustering Aspects,** Chester 1984.
27. H.J. Assenbaum, K. Langanke and A. Weiguny, Preprint University of Munster (1984).
28. E. Roeckl, Nucl. Phys. **A400** 131c (1983).
29. R.M. DeVries, D. Shapira, W.G. Davies, G.C. Ball, J.S. Foster and W. Mclatchie, Phys. Rev. Lett. **35** 835 (1975).
30. I. Tonozuka and A. Arima, Nucl. Phys. **A323** 45 (1979).
31. N.K. Glendenning and K. Harada, Nucl. Phys. **72** 481 (1965).
32. F. Iachello and A.D. Jackson, Phys. Lett. **108B** 151 (1982).
33. J.F. Ennis, M. Gai, M. Ruscev, E.C. Schloemer, B. Shivakumar, S.M. Sterbenz, N. Tsoupas and D.A. Bromley, Contribution No. 9; **4th Int. Conf. on Clustering Aspects,** Chester (1984). See also M. Gai, J.F. Ennis, M. Ruscev, E.C. Schloemer, B. Shivakumar, S.M. Sterbenz, N. Tsoupas and D.A. Bromley, Phys. Rev. Lett. **51** 646 (1983).
34. Y. Alhassid, M. Gai and G.F. Bertsch, Phys. Rev. Lett. **49** 1482 (1982).
35. F. Iachello, Nucl. Phys. **A396** 233c (1983).
36. H. Daley and F. Iachello, Phys. Lett. **131B** 281 (1983).
37. F. Catara, A. Insolia, E. Maglione and A. Vitturi, Contribution No. 17; **4th Int. Conf. on Clustering Aspects,** Chester (1984).
38. Y.K. Gambhir, P. Ring and P. Schruck, Phys. Rev. Lett. **51** 1235 (1983).
39. M. Ruscev, M. Gai, E.C. Schloemer, J.F. Ennis, K.N. Gamadia, J.F. Shriner Jr., S.M. Sterbenz, N. Tsoupas and D.A. Bromley, Contribution No. 12; **4th Int. Conf. on Clustering Aspects,** Chester (1984). M. Gai, M. Ruscev, A.C. Hayes, J.F. Ennis, R. Keddy, E.C. Schloemer, S.M. Sterbenz and D.A. Bromley, Phys. Rev. Lett. **50** 239 (1983).
40. K.N. Gamadia, M. Gai, J.F. Ennis, S.M. Sterbenz, N. Tsoupas, D.A. Bromley, P.M. Stwertka, T.M. Cormier, M. Herman and N. Nicholis, Bull. Am. Phys. Soc. **20** 718 (1984).
41. A.C. Hayes, J. Millener and D.A. Bromley, Phys. Rev. (to be published) 1984.
42. D. Baye and P. Descouvemont, Contribution No. 72; **4th Int. Conf. on Clustering Aspects,** Chester (1984).
43. R. Middleton, J.D. Garrette and H.T. Fortune, Phys. Rev. **C4** 1987 (1971).
44. R. Middleton, J.D. Garrette and H.T. Fortune, Phys. Rev. **24** 1436 (1970).
45. R.M. DeVries, Phys. Rev. Lett. **30** 666 (1973).
46. G.R. Satchler, **Direct Nuclear Reactions** Oxford, London and New York, (1983).
47. L.T. Chua and D.A. Bromley, Il. Nuovo Cimento **47A** 443 (1978).

48.     B.L. Cohen, **Proc. Rutherford Jubilee Int. Conf.**, Manchester (1962).
49.     H.B. Pugh, J.W. Watson, D.A. Goldberg, P.G. Roos, D.L. Bonbright and R.A.J. Riddle, Phys. Rev. Lett. **22** 408 (1969).
50.     C.B. Fulmer and J.B. Ball, Phys. Rev. **B40** 330 (1965).
51.     P.D. Kunz, T. Kamamuri and H. Shimaoka, Nucl. Phys. **A376** 401 (1982).
52.     L. Colli-Milazzo, G.M. Braga-Marcazzan, M. Milazzo and C. Signorini, Nucl. Phys. **A218** 274 (1974).
53.     R. Bonetti and L. Colli-Milazzo, Contribution No. 92; **4th Int. Conf. on Cluster Aspects**, Chester (1984).
54.     E. Gladioli, E. Gladioli-Erba, P. Guazzoni, P.E. Hodgson and L. Zetta, Contribution No. 54; **4th Int. Conf. on Cluster Aspects**, Chester, (1984).
55.     W.D.M. Rae, Invited paper, **4th Int. Conf. on Cluster Aspects**, Chester (1984).
56.     R.K. Bhowmik, W.D.M. Rae and B.R. Fulton, Phys. Lett **136B** 149 (1984).
57.     K. Nagatani, D.H. Youngblood, R. Kenefick and J. Bronson, Phys. Rev. Lett. **31** 250 (1973).
58.     W.D.M. Rae, N.S. Godwin, D. Sinclair, H.S. Bradlow, P.S. Fisher, J.D. King, A.A. Pilt and G. Proudfoot, Nucl. Phys. **A319** 239 (1979). See also N. Anyas-Weiss, J.C. Cornell, P.S. Cornell, P.S. Fisher, P.N. Hudson, A. Menchaca-Rocha, D.J. Millener, A.D. Panagiotou, D.K. Scott, D. Strottman, D.M. Brink, B. Buck, P.J. Ellis and T. Engeland, Phys. Reports **12C** No. 3 (1974).
59.     L.M. Martz, S.J. Sanders, P.D. Parker and C.B. Dover, Phys. Rev. **C20** 1340 (1979).
60.     H. Nann, Indiana University Progress Report (1983) and private communication to P.D. Parker (1983).
61.     K.P. Jackson, C.U. Cardinal, H.C. Evans, N.A. Jelley and J. Cerny, Phys. Lett. **33B** 281 (1970).
62.     J. Cerny, J.E. Estere, R.A. Gough and R.G. Sextro, Phys. Lett. **33B** 284 (1970).
63.     J. Cerny, R.A. Gough, R.G. Sextro and J.E. Esterl, Nucl. Phys. **A188** 666 (1972)—See also Hofmann **et al. Proc. Int. Conf. Nucl. far from Stability**, Helsingor (1981).
64.     H.J. Rose and G.A. Jones, Nature **307** 245 (1984).
65.     A.A. Ogloblin, private communication to D.A. Bromley 1984.
66.     M. Vergnes **et al.** University of Paris, Orsay preprint (1984).
67.     A. Sandulescu, D.N. Poenaru and Walter Greiner, Sov. J. Part. Nucl. **11** (6), 528, (1980).
68.     D.N. Poenaru, M. Ivascu, A. Sandulescu and W. Greiner (in press).
69.     A. Sandulescu, R.K. Gupta, W. Scheid and W. Greiner, Phys. Lett. **B60** 225 (1976).
70.     Y. Sakuragi, M. Yahiro and M. Kamimura, Prog. Theo. Phys. (Japan) **68** 322 (1982).
71.     Y. Sakuragi, M. Yahiro and M. Kamimura, Prog. Theo. Phys. (Japan) **70** 1047 (1983).
72.     R. Pfaneta, B. Neumann, H. Rebel, M. Albinska, J. Albinski, J. Buschmann, H.J. Gils, K. Gratowski and H. Klewe-Nebenius, Contribution No. 59; **4th Int. Conf. on Cluster Aspects**, Chester 1984.

73. T. Shimoda, N. Ikeda, K. Katori, T. Fukuda, J. Shimoura, T. Mori, T. Konno and H. Ogata, Contribution No 87; **4th Int. Conf. on Cluster Aspects,** Chester (1984).

74. A.M. Poskanzer, G.W. Butler and E.K. Hyde, Phys. Rev. **C3** 882 (1971).

75. D.K. Scott, private communication to D.A. Bromley (1981).

76. W.A. Friedman and W.G. Lynch, Phys. Rev. **C20** 16 (1983).

77. C.B. Chitwood, D.J. Fields, C.K. Gelbke, W.G. Lynch, A.D. Panagiotou, M.B. Tsang, H. Utsunomiya and W.A. Friedman, Phys. Lett. **131B** 289 (1983).

78. D.A. Bromley, **Nuclear Molecular Phenomena,** ed. by N. Cindro North Holland Publ. Co., Amsterdam, (1978).

79. F. Iachello, Phys. Rev. **C23** 2778 (1981).

80. K.A. Erb and D.A. Bromley, Phys. Rev. **C23** 2781 (1981).

81. M.R. Strayer, R.Y. Cusson, A.S. Umar, P.G. Reinhard, D.A. Bromley and W. Greiner, Phys. Lett. **135B** 261 (1984).

82. J. Cseh, Contribution No 63; **4th Int. Conf. on Cluster Aspects,** Chester (1984).

83. H.U. Gersch, H. Schobbert and D. Wohlforth, J. Phys. G: Nucl. Phys. **9** 939 (1983).

84. V. Metag, A. Lazzarini, K. Lesko and R. Vandenbosch, Phys. Rev. **C25** 1486 (1982).

85. R.L. McGrath, D. Abriola, J. Karp, T. Renner and S.Y. Zhu, **Lecture Notes in Physics 156,** ed. by K.A. Eberhard Springer, 290 (1982).

86. K. Kato, S. Okabe and Y. Abe, Contribution No. 98; **4th Int. Conf. on Cluster Aspects,** Chester (1984).

87. M. Gellmann and V.W. Telegdi, Phys. Rev. **91** 169 (1953).

88. S.M. Sterbenz, M. Gai, K.N. Gamadia, P. Cottle and D.A. Bromley (to be published) (1985).

89. R.R. Betts, S.B. DiCenzo and J.F. Petersen, Phys. Lett **100b** 117 (1981).

90. R.R. Betts, B.B. Back and B.G. Glagola, Phys. Rev. Lett. **47** 23 (1981).

91. A. Thiel, W. Greiner and W. Scheid, Phys. Rev. **C29** 864 (1984).

92. S.B. DiCenzo, Ph.D. Dissertation, Yale University (1981).

93. B. Dichter, R.R. Betts, S. Saini, S.J. Sanders, O. Hansen and R.W. Zurmuhle (to be published) (1984).

94. S. Saini and R.R. Betts, Phys. Rev. **C29** 1769 (1984).

95. R.W. Zurmuhle, P. Kutt, R.R. Betts, S. Saini, F. Haas and O. Hansen, Phys. Lett. **129B** 384 (1983).

96. R.R. Betts, S. Saini, S.J. Sanders, B. Dichter, O. Hansen, R.W. Zurmuhle and P. Kutt (to be published–1984) and private communication to D.A. Bromley–1984.

97. G. Leander and S.E. Larsson, Nucl. Phys. **A239** 93 (1975).

98. S.E. Larsson, G. Leander, I. Ragnarsson and N.G. Alenius, Nucl. Phys. **A261** 77 (1976).

99. T. Bengtsson, M. Faber, M. Ploszajczak, I. Ragnarsson and S. Aberg, Lund Institute of Technology Report MPh–84/01 (1984).

100. J. Schweppe, A. Gruppe, K. Bethge, H. Bokemeyer, T. Cowan, H. Folger, J.S. Greenberg, H. Grein, S. Ito, r. Schule, D. Schwalm, K.E. Stiebing, N. Trautmann, P. Vincent and M. Waldschmidt, Phys. Rev. Lett. **51** 2261 (1983).

101.    M. Clemente, E., Berdermann, P. Kienle, H. Tsertos, W. Wagner, C. Kozhuharov, F. Bosch and W. Koenig, Phys. Lett. **137B** 41 (1984).
102.    J. Reinhardt, B. Mueller and W. Greiner, Phys. Rev. **A24** 103 (1981).
103.    B. Mueller, G. Soff, W. Greiner and V. Ceaucescu, Zeets f. Phys. **A285** 27 (1978).
104.    J. Reinhardt and W. Greiner, Rep. Prog. Phys. **40** 219 (1977).
105.    J. Rafelski, L.P. Fulcher and A. Klein, Phys. Reports **38C(5)** 227 (1978).
106.    J.S. Greenberg and W. Greiner, Physics Today **35** 24 (1982).
107.    D. Robson, Nucl. Phys. **A308** 381 (1978).
108.    D. Robson, Phys. Rev. **C25** 1108 (1982)
109.    D. Robson, Phys. Rev. Lett. **42** 876 (1979).
110.    A.D. Frawley, J.D. Fox, K.W. Kemper and L.C. Dennis, Phys. Rev. **C25** 2935 (1982).
111.    W. Bertozzi, private communication (1984).
112.    K.D. Hilderbrand, H.H. Gutbrod, W. von Oertzen and R. Bock, Nucl. Phys. **A157** 297 (1970).
113.    A.N. Bice, A.C. Shotter, D.P. Stahel and J. Cerny, Phys. Lett. **101B** 27 (1981).
114.    F.D. Becchetti, J. Janecke and C.E. Thorn, Nucl. Phys. **A305** 313 (1978).
115.    D. Overway, J. Janecke, F.D. Becchetti, C.E. Thorn and G. Kekelis, Nucl. Phys. **A366** 299 (1981).
116.    D. Overway, F.D. Becchetti and J. Janecke, Contribution No. 84; **4th Int. Conf. on Cluster Aspects,** Chester (1984).
117.    T. Hennino, D. Bachelier, O.M. Bilaniuk, J.L. Boyard, J.C. Jourdain, M. Roy-Stephan, P. Radvanyi, M. Bedjidian, E. Descroix, P. Foessel, S. Gardien, J.Y. Grossiord, A. Guichard, M. Gusakow, R. Haroutounian, M. Jacquin, J.R. Pizzi and A. Garin, Phys. Rev. Lett. **48** 997 (1982).
118.    B.K. Jain and S.K. Gupta, Contribution No. 60; **4th Int. Conf. on Cluster Aspects,** Chester (1984).
119.    J. Burger and H.M. Hofmann, Contribution No. 105; **4th Int. Conf. on Cluster Aspects,** Chester (1984).

# MICROSCOPIC POTENTIALS

# MICROSCOPIC STUDY OF LOCAL NUCLEUS–NUCLEUS POTENTIALS

H. Horiuchi
Department of Physics
Kyoto University
Kyoto, 606
Japan

ABSTRACT. By using the WKB approximation as a mediator, we convert with high accuracy the non-local potentials of the resonating group method into their equivalent local potentials. On the basis of the analyses of the localized potentials we discuss the basic properties of the inter-nucleus interaction in the local potential picture. At the same time, by reviewing various microscopic theories to calculate the inter-nucleus potential, we compare our theory with other microscopic theories and discuss the reasons why there exist large difference for some potential properties between different theories. The problems we treat include (1) potential depth in relation to the Pauli-forbidden region and to the validity of the adiabatic treatment, (2) energy-dependence and mass-number-dependence in relation to the various exchange contributions, (3) comparison of potentials with light and heavy projectiles.

## 1. Introduction

It is now widely accepted that the optical potential gives a successful way of the description of the interaction process between composite nuclei. At the same time, however, it is well known that in most cases the optical potential can not be determined unambiguously. Therefore it is highly desirable that we have any microscopic theory which can derive at least the basic properties of the inter-nucleus potential such as depth, range, energy-dependence etc. and which enables us to extract the physical information from the phenomenologically determined potentials.

In the theoretical side, there have been proposed many kinds of methods of calculating the inter-nucleus potential. However, even when we take into consideration that the inter-nucleus potential can be defined more than one way, there exist contradictions among the results about the basic potential properties obtained by different theoretical methods. For example, the folding model[1] gives us the deep inter-nucleus potential and the exchange potential by this model is attractive by the same reason as the Fock part of the nucleon-nucleus

35

*J. S. Lilley and M. A. Nagarajan (eds.), Clustering Aspects of Nuclear Structure, 35–52.*

potential is attractive. Since the exchange effect becomes small as the incident energy gets higher, the potential by the folding model become shallower as the energy gets higher. On the other hand, many other microscopic theories, like the two-center shell model and the method of moving wave packet[2),3)] which take the Pauli principle into account exactly give us the shallow inter-nucleus potential. This means that the exchange potential is repulsive. The repulsive nature of the exchange potential is explained as to express the effect of the Pauli exclusion principle which forbids two nuclei to occupy the same spatial position. In this case, the weakening of the exchange effect due to the increase of the incident energy leads to the energy-dependence that the potential becomes deeper as the incident energy gets higher.

The above example shows typically that the microscopic theory of the inter-nucleus potential is now in a quite confused situation in spite of its fundamental importance. In view of this status of the theory, the aim of this paper is two-fold. The first and main aim is to discuss the basic properties of the local inter-nucleus potentials which are obtained by the "RGM + WKB" theory.[4)] This theory converts with high accuracy the non-local potentials of the resonating group method (RGM)[5)~8)] into their equivalent local potentials by the aid of the WKB technique.[9),10)] The second aim is to compare (the results of) the "RGM + WKB" theory with (those of) the other microscopic theories. The reasons why we get different results for some potential properties between different theories are discussed by taking accout of the difference of the basic view point of the individual theories.

The "RGM + WKB" theory has, of course, its own restricted range of applicability, yet it gives us fruitful and reliable knowledge about the basic properties of the inter-nucleus interaction expressed in the local potential picture. It is largely due to the RGM framework itself which is quite satisfactory for the microscopic study of the inter-nucleus interaction by the following reasons: 1) The inter-nucleus relative motion is treated in a fully quantum-mechanical framework which is unambiguous in kinematics. For example, the relative kinetic energy operator is unambiguously introduced, the specification of the channel is clear and the total center-of-mass coordinate is exactly eliminated. 2) The framework is fully microscopic; namely the total antisymmetrization is exactly executed and the dynamics is described by the use of the nucleon-nucleon effective force. 3) The framework is a unified one in the sense that we treat both scattering and bound states by the same equation, which enables us to study the inter-nucleus interaction from negative energy up to high scattering energy consistently.

In this paper, we first ( §2) discuss various microscopic theories to calculate the local inter-nucleus potential. It is for the sake of a review of this field and to make clear the basic stand-point of the "RGM + WKB" theory. In § 3, the problem of the potential depth is treated in relation to the Pauli-forbidden region in the phase space and to the validity of the adiabatic approach. In § 4, the contributions from various types of exchange process are analysed, by which we can study the mechanism of the energy-dependence and the mass-number-dependence of the potential. In § 5 we compare the potentials by the

"RGM + WKB" method between the system of light-ion projectile and that of heavy-ion projectile. In § 6 we give concluding remarks.

## 2. Various microscopic methods to calculate the local inter-nucleus potential

### 2.1. Method of energy curve

In this method, the inter-nucleus potential is given by energy of the total system as a function of a parameter of distance between colliding nuclei. The energy-density method and the two-center shell model are typical examples.[11]

In the energy-density method, if we are concerned with the case of the quite high scattering energy, the spatial density of the total system can be given simply by the sum of densites of two scattering nuclei. But when the scattering energy per nucleon is near or less than the Fermi kinetic energy, the total density is no more a simple sum of densities of scattering nuclei. This is due to the change of the internal wave functions of scattering nuclei due to the Pauli exclusion principle. Therefore the total density need to be constructed by the aid of some many-body wave function which takes account of the Pauli principle correctly.

The microscopic theories which treat the interaction process of two nuclei (or clusters) may be divided into two types, adiabatic and diabatic approaches, according to the types of the many-body wave functions adopted. This classification of the theories need not be restrictedly used within the method of energy curve but is applicable to other methods we discuss later. The two-center shell model and the constrained Hartree-Fock approach are the typical examples of the adiabatic approach. On the other hand, the so-called Brink wave function[12] which is used in the generator coordinate method (GCM) is an example of the diabatic wave function. This is because, when the Brink wave function is rewritten into a Slater determinant of orthonormalized single particle wave functions, these single particle orbits do not cause any dynamical mixing among themselves. Since the RGM is equivalent to the GCM, the RGM can be said to belong to a diabatic approach. Recently Nörenberg[13] argues that the diabatic process is more probable than the adiabatic one in usual nuclear collisions.

Whichever type of many-body wave functions, adiabatic or diabatic, we may adopt, there exists a fundamental problem in calculating the inter-nucleus potential by the method of energy curve. Namely, we need a justification why the total energy of the system can be regarded to be an inter-nucleus potential. The usual answer for this problem is based on the general theory to construct the collective Hamiltonian under the assumption of the slow velocity of the collective coordinate. This assumption is called the adiabaticity assumption. The cranking method and the adiabatic time-dependent Hartree-Fock theory belong to this category.[14] According to this kind of theory, the collective Hamiltonian is written as $T(D,\dot{D}) + V(D)$ where D stands for the inter-nucleus distance parameter and $\dot{D}$ the time derivative of D. V(D) is just the energy curve (namely the expectation value of the total Hamiltonian by the many-body model wave function with distance parameter D) and $T(D,\dot{D})$ is the kinetic energy containing $\dot{D}$ up to quadratic order according to the adiabaticity assumption. We consider, however, that this justification of the method of energy curve has doubtful and ambiguous points as are explained below. First the adiabaticity assumption (namely slow velocity assumption) itself is quite doubtful. In fact, as is discussed in the next subsection (§2.2), a recent theory[15] in the framework of the time-dependent variational principle shows clearly that the adiabaticity assumption is not justified in the case of the inter-nucleus relative motion. And furthermore, as is discussed in §3, one of the important conclusions of the "RGM + WKB" theory is that the inter-nucleus velocity is very large in the interaction region of the relative distance even in the case of zero incident energy. Secondly when we express the relative kinetic energy by $(\mu/2)\dot{D}^2$ as in the ordinary scattering theory with $\mu$ being the reduced mass, the term $\Delta T \equiv T(D,\dot{D}) - (\mu/2)\dot{D}^2$ should be treated as the potential energy to be added to $V(D)$. Thus the inter-nucleus potential is not simply given by $V(D)$ and in order to evaluate the additional potential $\Delta T$ we need to determine the local velocity $\dot{D}$ at each spatial point self-consistently.

## 2.2. Method of moving wave packet

This method constructs the collective Hamiltonian of the inter-nucleus relative motion without the slow-velocity (adiabaticity) assumption. The many-body model wave function contains in it the relative coordinate parameter D and the relative momentum parameter P. The conservation condition of the energy and the angular momentum determines the relation between D and P by which one constructs the inter-nucleus potential. Since the relative-motion part of this kind of many-body wave function has the form of moving wave packet with the mean separation distance D and with the mean relative momentum P, we call this kind of theory the method of moving wave packet.

Fliessbach[2] was the first who proposed this kind of self-consistent theory which is free from the doubtful adiabaticity assumption used in the method of energy curve. The many-body wave function he adopts has the form

$$\Phi(\vec{D},\vec{K}) \equiv \left(\frac{A_P + A_T}{A_P}\right)^{-1/2} \mathcal{A}\left[e^{i\vec{K}(\vec{X}_P - \vec{D}_P)}\psi_P(\vec{D}_P)e^{-i\vec{K}(\vec{X}_T - \vec{D}_T)}\psi_T(\vec{D}_T)\right]$$

$$= \left(\frac{A_P + A_T}{A_P}\right)^{-1/2}\mathcal{A}\left[\varphi(\vec{r}-\vec{D},\vec{K})\phi_P\phi_T\right]\omega_0(\vec{X}_G),$$

$$\varphi(\vec{r}-\vec{D},\vec{K}) \equiv \left(\frac{2\gamma}{\pi}\right)^{3/4}\exp\left[-\gamma(\vec{r}-\vec{D})^2 + i\vec{K}(\vec{r}-\vec{D})\right],$$

$$\omega_0(\vec{X}_G) \equiv \left(\frac{3(A_P + A_T)\nu}{\pi}\right)^{3/4}\exp\left[-(A_P + A_T)\nu\vec{X}_G^2\right],$$

$$(2.1)$$

where

$$\vec{D} = \vec{D}_P - \vec{D}_T \; , \; A_P\vec{D}_P + A_T\vec{D}_T = 0 \; , \; \gamma \equiv \mu_o \nu \; , \; \mu_o \equiv \frac{A_P A_T}{A_P + A_T} \; ,$$

$\vec{X}_G \equiv$ total center-of-mass coordinate,

$\vec{X}_i \equiv$ center-of-mass coordinate of i=P or T,

$\psi_i(\vec{D}_i)$ = harmonic oscillator shell model wave function centered at $\vec{D}_i$ for i=P or T

$\phi_i$ = internal wave function for i=P or T which is obtained from $\psi_i$ by dropping the center-of-mass wave function,

$A_P, A_T$ = mass numbers of the projectile (P) and the target (T) nuclei, respectively.

When we express the energy expectation value $E(\vec{D}, \vec{K})$ by this $\Phi(\vec{D}, K)$ as        (2.2)

$$E(\vec{D}, \vec{K}) = \langle \Phi(\vec{D}, \vec{K}) | H | \Phi(\vec{D}, \vec{K}) \rangle / \langle \Phi(\vec{D}, \vec{K}) | \Phi(\vec{D}, \vec{K}) \rangle \; , \qquad (2.3)$$

the local momentum $\hbar\vec{K}(\vec{D})$ is defined as the solution of the following equation of energy conservation

$$E(\vec{D}, \vec{K}(\vec{D})) = E(\infty, \vec{k}) \; ,$$
$$\frac{\hbar^2}{2\mu} k^2 = E = \text{incident energy.} \qquad (2.4)$$

Once $\vec{K}(\vec{D})$ is calculated, the local potential $V^{eff}(D)$ is given by

$$V^{eff}(D) = E - \frac{\hbar^2}{2\mu} K^2(\vec{D}) \; . \qquad (2.5)$$

It is easy to verify the following relation[16)]

$$V^{eff}(D) = E(\vec{D}, \vec{K}(\vec{D})) - E(\infty, \vec{K}(\vec{D})) \; . \qquad (2.6)$$

This method initiated in Ref.2 was further pursued by other authors for example in Ref.3. Some authors[17),18)] used a simplified version of this method by substituting the asymptotic momentum $\hbar k$ in place of the local momentum $\hbar\vec{K}(\vec{D})$ in Eq.(2.6);

$$\tilde{V}^{eff}(D) = E(\vec{D}, \vec{k}) - E(\infty, \vec{k}) \; . \qquad (2.7)$$

Needless to say, the potential $(V^{MEC}(D))$ by the method of energy curve is obtained just by setting $\vec{k}=0$ in Eq.(2.7);

$$V^{MEC}(D) = E(\vec{D}, \vec{k}=0) - E(\infty, \vec{k}=0) \; . \qquad (2.8)$$

As is mentioned in §1, the numerical calculations show that $V^{eff}(D)$ is quite shallow compared with the double folding potential (namely the direct potential without any exchange effect). This gross feature is common to $\tilde{V}^{eff}(D)$ and $V^{MEC}(D)$. When the incident energy gets higher, the effect of the Pauli principle becomes weak and therefore both $V^{eff}(D)$ and $\tilde{V}^{eff}(D)$ change toward the limit of the direct potential without exchange effect. Namely, both $V^{eff}(D)$ and $\tilde{V}^{eff}(D)$, when viewed grossly, become deeper as the energy gets higher.

The method of moving wave packet by Fliessbach is closely related to the "RGM + WBK" method,[4] but the potentials calculated by these two methods are entirely of different character except the region of quite large separation distance. We will discuss the relation between two methods in §2.4.

Recently an important progress has been made in regorously formulating the method of moving wave packet. This progress is due to the work by Saraceno et al.[15] which is based on the time-dependent variational principle. According to this work, the potential $V^{eff}(D)$ discussed above can be by no means regarded to be the correct inter-nucleus potential. This is because the parameters $\vec{D}$ and $\hbar\vec{K}$ in the Fliessbach's method can not be regarded to be canonically conjugate each other. When $\vec{D}$ and $\hbar\vec{K}$ are not the canonically conjugate coordinate and momentum, the definition of the potential by Eq.(2.5) is meaningless. Here, however, it is to be noticed that Eq.(2.4) is still valid because it is not relevant to whether $\vec{D}$ and $\hbar\vec{K}$ are canonically conjugate or not.

The canonically conjugate coordinate $\vec{\delta}$ and momentum $\hbar\vec{\kappa}$ are given by

$$\vec{\omega} = \sqrt{u}\ \vec{z}\ ,$$

$$\vec{\omega} \equiv \sqrt{\gamma}\left(\vec{\delta} + \frac{i}{2\gamma}\vec{\kappa}\right)\ ,\quad \vec{z} \equiv \sqrt{\gamma}\left(\vec{D} + \frac{i}{2\gamma}\vec{K}\right)\ ,$$

$$u \equiv \partial\log\hat{N}/\partial(\vec{z}^{*}\cdot\vec{z})\ ,$$

$$\hat{N} \equiv \exp(\vec{z}^{*}\cdot\vec{z})\langle\Phi(\vec{D},\vec{K})|\Phi(\vec{D},\vec{K})\rangle\ . \tag{2.9}$$

In the asymptotic region of $\vec{z}^{*}\cdot\vec{z}\rightarrow$ large, we have $\vec{\omega}\rightarrow\vec{z}$. If the anti-symmetrization were absent, $\hat{N}$ would become $\exp(\vec{z}^{*}\cdot\vec{z})$ and we would have $\vec{\omega}=\vec{z}$. The equation to determine the local momentum $\hbar\vec{\kappa}(\vec{\delta})$ is just Eq.(2.4) as mentioned above. Namely if we rewrite the solution $\vec{K}=\vec{K}(\vec{D})$ of Eq.(2.4) in terms of $\vec{\delta}$ and $\vec{\kappa}$ with the aid of Eq.(2.9) we get the desired trajectory $\vec{\kappa}=\vec{\kappa}(\vec{\delta})$.

The most important conclusion of this new theory of Saraceno et al. is the existence of the region in the phase space of $\vec{\delta}$ and $\hbar\vec{\kappa}$ into which any trajectory $\vec{\kappa}=\vec{\kappa}(\vec{\delta})$ can not enter. This is because, by the definition of Eq.(2.9), $\vec{\omega}$ must obey the following restriction

$$\vec{\omega}^{*}\cdot\vec{\omega} \geq N_{F}+1\ . \tag{2.10}$$

$N_{F}$ is the non-negative interger which is known in the RGM theory as the maximum Pauli-forbidden number of harmonic oscillator quanta for the inter-nucleus relative motion.[5]~[8] The same equation as Eq.(2.10) is also obtained in the "RGM + WKB" theory and we discuss the implication of Eq.(2.10) in detail in §3. Here we only indicate that Eq.(2.10) means the denial of the adiabaticity assumption because in the interaction region where $\delta$ is small the momentum $\hbar\vec{\kappa}$ can not be small.

## 2.3. Double folding model

The inter-nucleus potential by this model consists of two parts, the direct potential (the double folding potential in a narrow sense) and the exchange potential coming from the knock-on exchange process.[1] Usually the knock-on exchange potential is effectively replaced by the direct potential[1] calculated by using the zero-range effective two-nucleon force.[19] As is known, this model with the M3Y interaction has had considerable success in interpreting the elastic data[1],[20] some of which are sensitive to the potential at small distance.[1],[20]

The folding model may look quite crude since it assumes that among many exchange terms only the knock-on exchange term is effective. Especially for the small distant part of the potential, one usually considers that the many-nucleon exchange terms are overwhelming and the folding model has no meaning at all. But, as we discuss in § 5, the "RGM + WKB" theory teaches us that even in the small distance region the one-nucleon exchange contribution is largest and in the case of light projectile even dominant among many exchange terms. The microscopic problem to be investigated for the folding model is therefore in some sence of quantitative nature: What is the percentage of the knock-on exchange term in the total one-nucleon exchange contribution? Furthermore, what is the percentage of the one-nucleon exchange term in the total exchange contribution? These problems are discussed in § 5.

## 2.4 "RGM + WKB" method

This method[4] constructs the local inter-nucleus potential by converting equivalently the non-local potential of the RGM (resonating group method). The conversion is done by using the WKB approximation as a mediator, which is discussed in detail in Refs.9) and 10). The definitions of the equivalent local potential[21] and the incidental Perey factor[21] are the same as those by Austern[22] and Fiedeldey[23]. The idea to use the WKB approximation in treating the non-local potential was early proposed by Austern.[22] In all the systems we have treated, we have found that the local potential thus constructed reproduce very accurately the scattering phase shifts by the original RGM non-local potential.[24]

The many-body wave function of the RGM for the single-channel system has the form $\mathcal{A}\{\chi(\vec{r})\phi_P\phi_T\}$ and the relative wave function $\chi(\vec{r})$ obeys the equation

$$\int \{H(\vec{\rho},\vec{\rho}') - EN(\vec{\rho},\vec{\rho}')\}\chi(\vec{\rho}')d\vec{\rho}' = 0 ,$$

$$\left\{\begin{matrix}H(\vec{\rho},\vec{\rho}')\\N(\vec{\rho},\vec{\rho}')\end{matrix}\right\} = \left\langle\delta(\vec{r}-\vec{\rho})\phi_P\phi_T\Big|\left\{\begin{matrix}H-E_P-E_T\\1\end{matrix}\right\}\Big|\mathcal{A}\{\delta(\vec{r}-\vec{\rho}')\phi_P\phi_T\}\right\rangle, \quad (2.11)$$

where $E_P$ and $E_T$ are binding energies of P (projectile) and T(target). The application of the WKB approximation to Eq.(2.11) gives us the following Hamilton-Jacobi equation which determines the local momentum $\vec{p}(\vec{\rho})$;[9],[10]

$$H^W(\vec{P}, \vec{p}(\vec{P})) - E\, N^W(\vec{P}, \vec{p}(\vec{P})) = 0 .$$

(2.12)

Here we use the notation $O^W(\vec{P}, p)$ which expresses the Wigner transform of the non-local operator $O(\vec{P}, \vec{P}')$;

$$O^W(\vec{P}, \vec{p}) = \int d\vec{s}\; e^{\frac{i}{\hbar}\vec{s}\cdot\vec{P}}\, O(\vec{P} - \frac{\vec{s}}{2}, \vec{p} + \frac{\vec{s}}{2}) .$$

(2.13)

Once the local momentum $\vec{p}(\vec{P})$ is obtained, the equivalent local potential $V^{eq}(\rho)$ is calculated by

$$V^{eq}(\rho) = E - \frac{1}{2\mu}\, p^2(\vec{p}) .$$

(2.14)

Discussions of the basic properties of $V^{eq}(\rho)$ thus calculated are given in §3 ~ §6.

Here we only discuss the relation of this "RGM + WKB" method with the method of moving wave packet by Fliessbach[16]. This is done by using the following important relation

$$\langle \Phi(\vec{D}, \vec{K}) | O | \Phi(\vec{D}, \vec{K}) \rangle = \int d\vec{p}\, d\vec{k}\; A(\vec{p} - \vec{D}, \vec{k} - \vec{K})\, O^W(\vec{p}, \hbar\vec{k})$$

$$\equiv \langle O^W(\vec{p}, \hbar\vec{k}) \rangle_{\vec{D}, \vec{K}} ,$$

$$A(\vec{p}, \vec{k}) \equiv (\frac{2\gamma}{\pi})^{3/2} e^{-2\gamma P^2} (\frac{1}{2\pi\gamma})^{3/2} e^{-k^2/2\gamma} ,$$

(2.15)

where $O$ is an arbitrary operator and $O^W(\vec{P}, p)$ is the Wigner transform of the RGM kernel $O(\vec{P}, \vec{P}') = \langle \delta(\vec{r} - \vec{P}) \phi_P \phi_T | O | A\{\delta(\vec{r} - \vec{P}') \phi_P \phi_T\} \rangle$. Eq.(2.15) shows that $\langle \Phi(\vec{D}, \vec{K}) | O | \Phi(\vec{D}, \vec{K}) \rangle$ is just the phase-space average of $O^W(\vec{P}, p)$. The widths of the averaging kernel $A(\vec{p}, \vec{k})$ in spatial and momentum degrees of freedom are just the uncertainty widths of the wave packet $\mathcal{Y}(\vec{r} - \vec{D}, \vec{K})$ in spatial and momentum degrees of freedom, respectively. By using Eq.(2.15), we can easily show that Eq.(2.4) can be rewritten as

$$\langle H^W(\vec{p}, \hbar\vec{k}) - (E + \frac{\omega}{4}\hbar\omega) N^W(\vec{p}, \hbar\vec{k}) \rangle_{\vec{D}, \vec{K}(\vec{D})} = 0 .$$

(2.16)

This shows clearly that the equation of the method of moving wave packet to determine the local momentum is obtained just by taking the phase-space average of the Hamilton-Jacobi equation of the "RGM + WKB" method. Therefore the method of moving wave packet is an approximation of the "RGM + WKB" theory. In Ref.16 it is clarified that this approximation is not good since the widths of the averaging operation by $A(\vec{p}, \vec{k})$ are not sharp enough compared with the spatial and momentum ranges of such RGM non-local potentials which come from the many-nucleon exchange processes.

## 3.    Depth of the potential

In the RGM theory, it is well known that there exist the so-called Pauli-forbidden states $\chi_F(\vec{r})$ which the inter-nucleus relative motion can not occupy.[5)~7),25),26)] $\chi_F(\vec{r})$ are defined as the functions which satisfy $\mathcal{A}\{\chi_F(\vec{r})\phi_P\phi_T\} = 0$, and therefore they are redundant solutions of the RGM equation of motion given in Eq.(2.11) at any energy E. $\chi_F(\vec{r})$ are expressed by the harmonic oscillator wave functions $R_{n\ell}(r, \delta)Y_{\ell m}(\hat{r})$ with $N = 2n+\ell \leq N_F$ where $\delta$ is given in Eq.(2.2). $N_F$ is the largest number of the harmonic oscillator quanta of the Pauli-forbidden states; for example $N_F=11$ for $\alpha + {}^{40}Ca$ system and $N_F=22$ for ${}^{16}O + {}^{16}O$ system.

When the RGM equation of motion is treated by the WKB method, the harmonic oscillator states are transformed into the harmonic oscillator trajectory $p^2/2\mu + \rho^2\mu\omega^2/2 = \hbar\omega(2n+\ell +3/2)$. Therefore the region of the phase space of $\vec{p}$ and $\vec{\rho}$ defined by $p^2/2\mu + \rho^2\mu\omega^2/2 \leq \hbar\omega(N_F + 3/2)$ is the Pauli-forbidden region into which the physical trajectory can not enter. The trajectory inside the Pauli-forbidden region is the redundant solution of the Hamilton–Jacobi equation of Eq.(2.12) at any energy E. In summary the Pauli-forbidden region in the phase space is the semi-classical image of the functional space spanned by the Pauli-forbidden states.

The physical solution of the Hamilton-Jacobi equation, Eq.(2.12), should be found outside of the Pauli-forbidden region. Therefore the local momentum $\vec{p}(\vec{\rho})$ should satisfy the inequality

$$p^2(\vec{\rho})/2\mu + \rho^2\mu\omega^2/2 > \hbar\omega\left(N_F + \frac{3}{2}\right),  \tag{3.1}$$

which, when expressed in terms of $V^{eq}(\rho)$ by using Eq.(2.14), takes the form

$$V^{eq}(\rho) < E - \hbar\omega\left(N_F+\frac{3}{2}\right) + \rho^2\frac{\mu\omega^2}{2}.  \tag{3.2}$$

It should be noted that Eq.(3.1) is the same as Eq.(2.10) derived by Saraceno et al. Eq.(3.1) shows that in the interaction region where $\rho$ is small the local momentum $p(\vec{\rho})$ can not be small. This fact teaches us that the adiabaticity assumption on the inter-nucleus velocity has no a priori justification even at zero incident energy.

Eq.(3.2) demands that $V^{eq}(\rho)$ is necessarily deep except at very high energy where E is comparable with $\hbar\omega(N_F+ 3/2)$. This is the kinematical requirement due to the avoidance of the Pauli-forbidden region and is independent of the choice of the effective two-nucleon force.

However at high energy comparable with or larger than $\hbar\omega(N_F+ 3/2)$, the Pauli-forbidden region has no more sizable effect to the depth of $V^{eq}$ and the behaviour of $V^{eq}$ depends now largely on the choice of the effective nuclear force. The high energy limit of $V^{eq}(\rho)$ is the direct potential $V_D(\rho)$ and it is well-known that $V_D(\rho)$ depends strongly on the effective force; $V_D(\rho)$ for the Volkov force[27)] is deeply attractive while $V_D(\rho)$ for the Brink-Boeker force[28)] is entirely repulsive. The fact that the Pauli-forbidden region loses its effect on $V^{eq}$ at high energy is quite natural since in quantum theory the orthogonality to the Pauli-forbidden states is automatically satisfied by the high energy relative wave function.

When the number of the physical bound states of the system is $n_B$

for given $\ell$ (angular momentum), we can prove that $V^{eq}(\rho)$ has such depth that it supports just $(n_B + n_F)$ bound states, where $n_F$ is the number of Pauli-forbidden states for $\ell$, namely $n_F = (N_F - \ell)/2$. The proof is done by using the Bohr-Sommerfeld-Ishihara quantization rule for bound states and the condition of Eq.(3.1). This result is in good accordance with the local potential model proposed previously by many authors[29)~33)] on the basis of the RGM theory. The argument of these authors is as follows: If the local potential is so deep that all the Pauli-forbidden states are reproduced in good accuracy as the bound states of this potential which are located below the physical (true) bound states of this potential, the orthogonality condition of the relative wave function to the Pauli-forbidden states can be automatically satisfied. As Neudatchin et al. stress,[30)] this deep potential is satisfactory also in view of the generalized Levinson theorem because this potential gives us $\delta_\ell(0) - \delta_\ell(\infty) = (n_B + n_F)\pi$.

## 4. Dependence on energy and mass-number

In Fig.1 we show the equivalent local potential $V^{eq}(r)$ of the $\alpha + {}^{40}Ca$ systems by decomposing it into the direct potential $V_D(r)$ and the exchange potential $\Delta V_\ell(r) \equiv V^{eq}(r) - V_D(r)$. Here and also later in this paper we only treat the case of zero angular momentum ($\ell = 0$) because the angular-momentum-dependence of $V^{eq}$ is very small[24)] in all the systems we have treated except the systems with strong parity-dependence like d+$\alpha$ and t+$\alpha$.[34)] In this paper we do not discuss the problem of the parity- and angular-momentum-dependence.[9),10),34)]

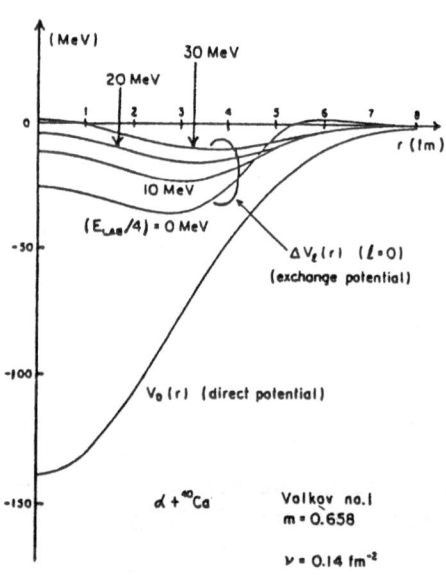

In the short distance region, the exchange potential is attractive and as the incident energy gets higher it becomes less attractive. On the other hand in the tail region, the exchange potential is even repulsive at very low incident energy and as the energy gets higher it becomes more attractive. But over some 20 MeV/u the tail part of the exchange potential changes to become less attractive as the energy gets higher. This feature of the energy-dependence of $V^{eq}(r)$ is quite general[35)] and is valid to all the systems we have treated except the systems with strong parity-dependence.[34)]

Fig.2a shows the energy-dependence of the volume integral per nucleon-pair of $V^{eq}(r)$ ($j_V \equiv 4\pi \times \int V^{eq}(r)r^2 dr/A_P A_T$) for three systems

Fig.1 $V^{eq}(r)$ of $\alpha + {}^{40}Ca$ system decomposed into the direct($V_D$) and exchange( $V$ ) potentials.

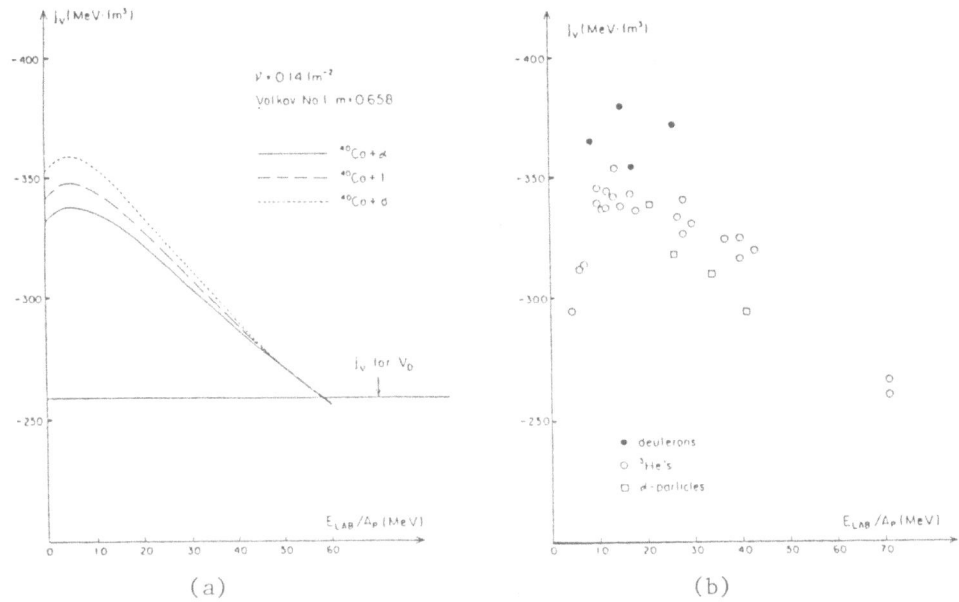

Fig.2    (a) Energy-dependence of the volume integrals per nucleon pair of the equivalent local potentials for scattering of d, t and $\alpha$ on $^{40}$Ca. (b) Energy-dependence of the volume integrals per nucleon pair of the real parts of the optical potentials for the scattering of d, t and $\alpha$ on nuclei in Ca~Ni region.

of d $+^{40}$Ca, t $+^{40}$Ca and $\alpha +^{40}$Ca. We notice three characteristic features: (i) In the energy region, $E \lesssim 10$. MeV/u, the absolute values of the three $j_V$ increase as E gets higher. (ii) In the energy region, $E \gtrsim 10$ MeV/u, the absolute values of the three $j_V$ decrease as E gets higher. (iii) At the same incident energy per nucleon, the absolute value of $j_V$ decreases as the mass number of the light ion projectile increases, but above some 50 MeV/u the $j_V$ become almost of the same value. These characteristic features of $j_V$ are also seen in the Z $+^{16}$O systems with Z denoting d, t and $\alpha$.[35] The first feature is due to the large energy-dependence of the tail part of $V^{eq}$ in the low energy region mentioned before for Fig.1. The second feature is due to weakening of the effect of tail part of $V^{eq}$.[36)~38)]
     According to the recent compilations of the systmatics of the light-ion optical potentials, the volume integral $j_V$ of the real part of the optical potential shows the quite similar dependence on energy and on the light-ion-projectile mass number as our theoretical potential $V^{eq}$. In Fig.2b we show an example of such compilation which is taken from Ref.36. Although our present aim is not a detailed data fitting, we can see that the correspondence of $j_V$ between theory and experiments is good semi-quantitatively. If we include the Coulomb force, the local momentum become smaller and the surface exchange potential would become less attractive than here, which would generate more steep energy-dependence of $j_V$ in the low energy region.
     The origin of the projectile-mass-number-dependence of $j_V$ which is

quoted as the third characteristic feature of $j_V$ in Fig.2 is as follows. When the projectile approaches the target nucleus, the projectile nucleons feel the Pauli exclusion exerted by the target nucleons and their mutual interaction energy changes from the value in the free state. The amount of this defect of the interaction energy of projectile nucleons is proportional to the number of the interaction bonds, $A_p(A_p-1)/2$. Thus this effect appears in $V^{eq}$ as the projectile-mass-number-dependence proportional to $A_p(A_p-1)$. Since, in calculating $j_V$, we divide the integral $\int_o V^{eq}(r)r^2 dr$ by $A_p$, there results the $A_p$-dependence of $j_V$ proportional to $(A_p-1)$. That this argument is correct is demonstrated in Fig.3 whose explanation is as follows. The exchange potential $\Delta V_\ell(r)$ can be decomposed into the kinetic exchange potential $t^{ex}(r)$ and the interaction exchange potential $v^{ex}(r)$ $(\Delta V_\ell(r) = t^{ex}(r) + v^{ex}(r).)$ $v^{ex}(r)$ can be further decomposed into $v_{HF}^{ex}(r)$ and $v_{INT}^{ex}(r)$ $(v^{ex}(r) = v_{HF}^{ex}(r) + v_{INT}^{ex}(r))$ where $v_{HF}^{ex}(r)$ comes from the Hartree-Fock potential energy of projectile nucleons which occupy the valence orbits around the target nucleus and $v_{INT}^{ex}(r)$ comes from the interaction among projectile nucleons in such valence orbits. $v_{INT}^{ex}(r)$ is just the term which expresses the defect of

the projectile interaction energy due to the Pauli exclusion discussed above. What are shown in Fig.3 are the volume integrals per nucleon pair of these three exchange potentials $t^{ex}(r)$, $v_{HF}^{ex}(r)$ and $v_{INT}^{ex}(r)$. We see clearly that only $j_V$ by $v_{INT}^{ex}(r)$ has the $A_p$-dependence.

Fig.3 teaches us, moreover, the detailed contents of the energy-dependence of $j_V$ shown in Fig.2a: For exmaple, the increase of the absolute magnitude of $j_V$ in the low energy region is due to the decrease of the repulsive contributions from $v_{INT}^{ex}(r)$ and $t^{ex}(r)$. Especially the rapid decrease of $t^{ex}(r)$ is of large effect. We have mentioned before that the low energy behaviour of $j_V$ is governed by the large energy-dependence of the tail part of $\Delta V_\ell(r)$. In fact we have ascertained that the large energy-dependence of the tail part of $\Delta V_\ell(r)$ is governed by the large energy-dependence of the tail part of $t^{ex}(r)$.[35],[4]

When the projectile is much lighter than the target, to represent the projectile nucleons as

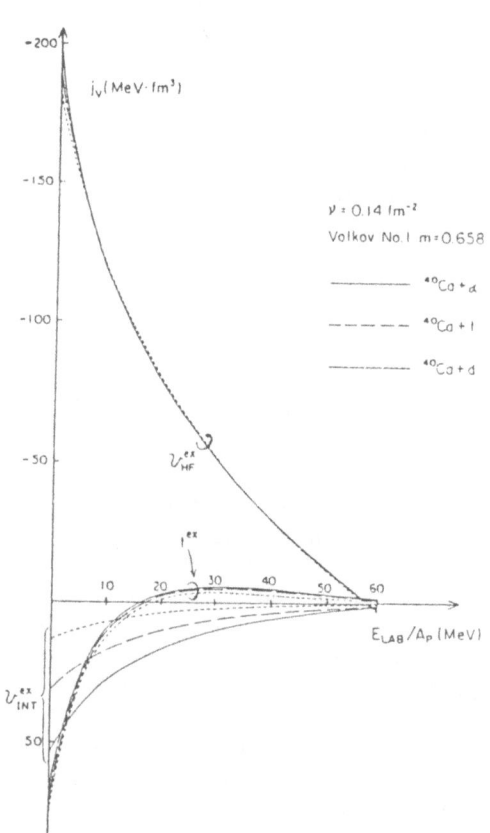

Fig.3  Volume integrals per nucleon pair of the exchange potentials $t^{ex}$, $v_{HF}^{ex}$ and $v_{INT}^{ex}$ of $Z+^{40}Ca$ systems ($Z=d$, $t$ and $\alpha$).

the valence nucleons around the target nucleus is quite natural. [39],[40] Therefore this representation has been used also by other authors. Their results of the studies of the energy-dependence and $A_p$-dependence of the light-ion potential are similar to ours qualitatively but the recognition of the important role of the kinetic exchange is absent.

The dependence of the inter-nucleus potential on the target mass-number in a wide mass-number range is an important problem. A new important ingredient in the study of this problem is the proper treatment of the density-dependence of the effective two-nucleon force. But we omit here the discussion of this problem since our study of this problem is now in a preliminary stage.

## 5. Comparison of potentials with light and heavy projectiles

As a typical example of the potential between two (light) heavy ions, we have studied the properties of the equivalent local potential of the $^{16}O+^{16}O$ system.[46] When we compare this $^{16}O+^{16}O$ potential with the light-ion potentials calculated for the target nuclei, $^{16}O$ and $^{40}Ca$, we find that some potential properties common to the light-ion potentials do not necessarily hold for the $^{16}O+^{16}O$ potential. Below we discuss two of such properties of basic importance.

### 5.1 Dominance of one-nucleon exchange ?

In order to understand the basic properties of the non-locality of the RGM kernels (non-local potentials), the classification of the RGM kernels by the number of exchanged nucleons is essential. This is because the spatial and energy ranges of the RGM kernels[41]~[44],[10] depend most strongly on the number of exchanged nucleons. The first systematic analyses of the RGM non-local potentials on the basis of the properties of their spatial and energy range parameters were done by Tang and his collaborators.[44] The spatial and energy range parameters of the "RGM + WKB" method are essentially the same as those of Tang's method[44] and many importnat results obtained by Tang's method are inherited in the studies by the "RGM + WKB" method.

In the tail region of the potential, it is clear that only the exchange potentials coming from a-few-nucleon exchange processes are effective. (Here we do not treat the systems with $A_p \approx A_T$ but with $A_p \neq A_T$. In these systems the core exchange and similar many-nucleon exchange processes are also effective in the tial region.[42],[44],[10],[34]) In the inner region of the potential, many kinds of many-nucleon exchange processes are possible and it seems that these processes are overwhelming. We must note, however, that the many-nucleon exchange potentials have large non-locality range. The large non-locality range means directly that the energy range is small. Now we recall that, as we discussed in §3, the potential is deep and so in the inner spatial region the kinetic energy of the relative motion is quite high. This fact leads to the conclusion that the many-nucleon exchange potentials can not be so effective even in the inner spatial region due to their small energy range.

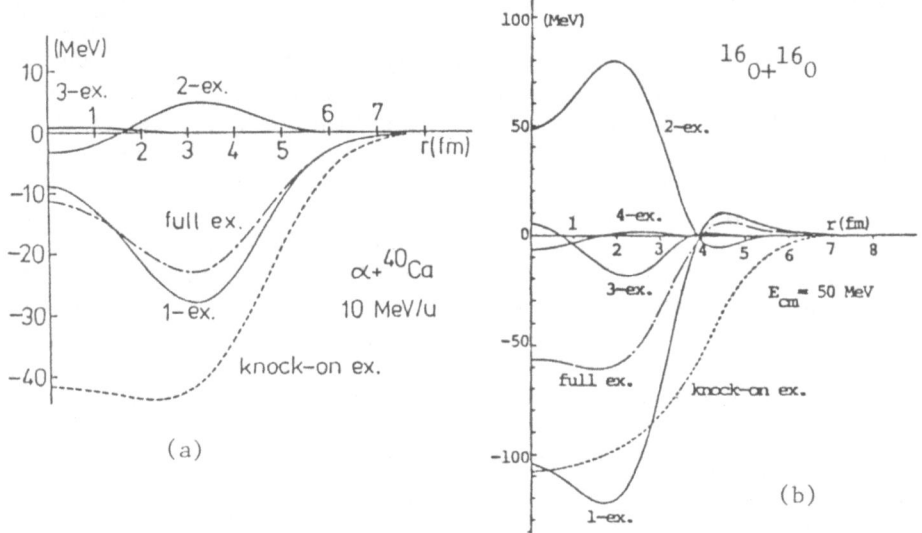

Fig.4    The exchange potential and its decomposition according
to the number of nucleons exchanged. Here is also shown the
knock-on exchange potential.   (a) is for $\alpha + {}^{40}Ca$ and (b) for
${}^{16}O + {}^{16}O$ systems.

The numerical calculations of the light-ion potentials for the
targets ${}^{16}O$ and ${}^{40}Ca$ show us the exchange potentials are almost exhausted by the one-nucleon exchange potential in the whole spatial
region.[10),35)]    An example is shown in Fig.4a which displays the decomposition of the exchange potential of the $\alpha + {}^{40}Ca$ system at 10 MeV/u
according to the number of nucleons exchanged. However, in the ${}^{16}O + {}^{16}O$
system, as is shown in Fig.4b, in addition to the one-nucleon exchange
potential which is of course the largest component, the two-nucleon
exchange potential has non-negligible magnitude.[46)]   Here the oscillator
parameter $\nu$ is 0.195 fm$^{-2}$ and the Volkov No.2 force with m=0.65 is
used following Ref.45). Although this result is not pleasing, we need
to keep this fact in mind in treating the heavy-ion potentials.

In Fig.4a and 4b, we also show the knock-on exchange potentials.
We see that in both systems the knock-on exchange potentials resemble
neither the one-nucleon exchange potential nor the full exchange potential.   When the energy gets higher, since the knock-on exchange potential has the largest energy range among the exchange potentials,[45),10)]
the difference between the knock-on exchange and the full exchange
potentials becomes smaller.[10),46)]

## 5.2  Perey factor

In the RGM study we often use $\sqrt{N}\chi$ as the relative wave function
instead of $\chi$, where N is the norm kernel given in Eq.(2.11).   In the
asymptotic region N=1 and $\sqrt{N}\chi = \chi$.   To discuss the relative motion, $\sqrt{N}\chi$
is more adequate than $\chi$ because the orthonomalization relation of

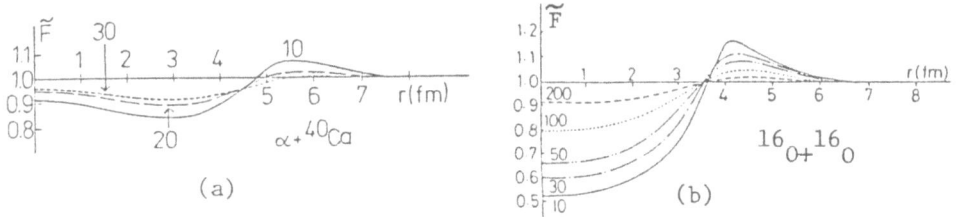

Fig.5  Perey factor $\widetilde{F}(r)$ for $\sqrt{N}\chi$. Figures indicate the energy in center-of-mass frame. (a) is for $\alpha + {}^{40}$Ca and (b) is for ${}^{16}$O+${}^{16}$O.

$\sqrt{N}\chi_E$ ($\chi_E$ standing for $\chi$ at the energy E) can be set to be ordinary, $\langle \sqrt{N}\chi_E | \sqrt{N}\chi_{E'} \rangle = \delta_{EE'}$ while for $\chi_E$ we have $\langle \chi_E | N | \chi_{E'} \rangle = \delta_{EE'}$. 6),32),47),48) The equation of motion for $\sqrt{N}\chi$ is $\{(1/\sqrt{N})H(1/\sqrt{N}) -E\}\sqrt{N}\chi = 0$. (In this notation, Eq.(2.11) is written as $(H-EN)\chi=0$.) When we apply the WKB prescription to this equation, the equation to determine the local momentum is again given by Eq.(2.12).[49] This means that the equivalent local potential to the non-local potential for $\sqrt{N}\chi$ is the same as that to the non-local potential for $\chi$.

When we denote the wave function generated by the equivalent local potential as $\chi^L(\vec{\rho})$, the Perey factor $\widetilde{F}(\rho)$ in the relation $(\sqrt{N}\chi)(\vec{\rho}) = \widetilde{F}(\rho)\chi^L(\vec{\rho})$ is given by

$$\widetilde{F}(\rho) = \sqrt{1 - \partial V^{eq}(\rho)/\partial E} , \qquad (5.4)$$

while the Perey factor $F(\rho)$ in the relation $\chi(\vec{\rho}) = F(\rho)\chi^L(\vec{\rho})$ is given by $F(\rho) = \widetilde{F}(\rho)/\sqrt{N^W(\vec{\rho},\vec{\rho}(\vec{\rho}))}$. Numerical calculations show us that $\widetilde{F}(\rho)$ for the light-ion potentials for the targets ${}^{16}$O and ${}^{40}$Ca are quite near unity.[50] An example is shown in Fig.5a for the $\alpha +$ ${}^{40}$Ca system. This means that the non-local potential for $\sqrt{N}\chi$ has effectively only weak non-locality so far as the Pauli-forbidden region is duely avoided. This conclusion gives us a quantitative foundation[50] for the orthogonality condition model[25),26)] in the case of the light-ion projectile.

On the other hand, as shown in Fig.5b,[46] in the ${}^{16}$O+${}^{16}$O system, $\widetilde{F}(\rho)$ is fairly smaller than unity in the inner spatial region except at high energy. This small Perey factor is one of the important factors which cause the strong damping of the inner amplitude of $\sqrt{N}\chi$ of the ${}^{16}$O+${}^{16}$O system found in Ref.45). The large deviation of $\widetilde{F}(\rho)$ from unity means that the orthogonality condition model with a purely local effective potential is difficult to justify for this heavy-ion system.

6.   Concluding remarks

In this paper we have first reviewed various microscopic methods to calculate the local inter-nucleus potential putting the stress on the discussion of the fundamental standpoint (or the assumption) of the methods. The adiabaticity (namely slow-relative-velocity) assumption on which the methods of energy curve is based is discussed in detail and is concluded to be quite doubtful. This is because both in the newly-formulated "method of moving wave packet"[15] and in our "RGM + WKB" method, the relative momentum in the interaction region can not be small even at zero incident energy, which is due to the existence of the Pauli-forbidden region in the phase space of relative coordinate and momentum. The mutual relation of the newly-formulated "method of moving wave packet" and the "RGM + WKB" method is not clear and its clarification is left in the future.

The "RGM + WKB" theory teaches us that the inter-nucleus potential is composed of the direct potential and a-few-nucleon (mainly one-nucleon) exchange potentials in the whole spatial region because the many-nucleon exchange potentials are suppressed to be inactive due to the large local momentum. The double folding model can be consdered therefore to be the model which assumes that the contribution of a-few-nucleon exchange processes is dominated by the contribution of the knock-on exchange process. Since numerical analyses show that the exchange potential is not simply exhausted by the knock-on exchange potential we consider that it is appropriate to regard the exchange potential used in the analyses by the double folding model as a simulation potential of the full one-nucleon exchange potential plus additional corrections including the polarization potential.

The low-energy behaviour and the projectile-mass-number-dependence of the volume integrals per nucleon pair of the light-ion potentials are quite interesting since, according to the "RGM + WKB" theory, they are goverened by the exchange processes other than the knock-on exchange process. Good semi-quantitative correspondence between theory and experiments displays the reliability and the usefulness of the "RGM + WKB" theory.

The comparison of the potentials with light and heavy projectiles given in this paper suggests the existence of non-small difference between the two. But to get more definite knowledge we need to extend our study to various kinds of heavy-ion-projectile systems.

We are now further extending the "RGM + WKB" theory in order to study microscopically the inter-nucleus coupling potential connecting different channels in the coupled channel formalism. For details of the formulation the reader is referred to Refs.51) and 52). In the contribution to this conference[52] we have reported the preliminary results of the study of the $\alpha + {}^6$Li system where ${}^6$Li can be in two states $0^+$ and $2^+$ neglecting the spin-orbit coupling. This preliminary study shows already a promising scope of the microscopic study in this new wide field.

Acknowledgment

The author thanks Dr. K. Aoki, Mr. T. Wada and Mr. K. Yabana on whose collaborations this paper is based. He also thanks Miss Y. Takeshita for her typing the manuscript.

References

1) G.R. Satchler and W.G. Love, Phys. Reports 55 No.3 (1979), 183.
2) T. Fliessbach, Z. Phys. 247 (1971), 117.
3) G.H. Göritz and U. Mosel, Z. Phys. A277 (1976), 243;
   P.G. Zint, Z. Phys. A281 (1977), 373.
4) H. Horiuchi, K. Aoki and T. Wada, Proc. 1983 RCNP Int. Symp. on
   Light Ion Reaction Mechanism (Osaka, Japan, 1983), p.806.
5) K. Wildermuth and Y.C. Tang, *A Unified Theory of the Nucleus*
   (Vieweg, Braunschweig, Germany, 1977).
6) Prog. Theor. Phys. Suppl. No. 62 (1977).
7) Y.C. Tang, M. LeMere and D.R. Thompson, Phys. Reports 47 No.3
   (1978), 167.
8) Proc. Int. Symp. on Nuclear Collisions and Their Microscopic
   Description (Bled, Yugoslavia, 1977), Fizika 9, Suppl. 3 and 4.
9) H. Horiuchi, Prog. Theor. Phys. 64 (1980), 184.
10) K. Aoki and H. Horiuchi, Prog. Theor. Phys. 68 (1982), 1658.
11) For example, A. Richter and C. Toepffer, *Heavy Ion Collisions*,
    ed. R. Bock (North Holland, 1979), Vol.1, Chapt. 1.
12) D.M. Brink, Proc. Int. School of Phsics "Enrico Fermi", Course
    36 (1966), 247.
13) W. Nörenberg, Proc. Int. Conf. on Heavy Ion Physics (Catania,
    Italy, 1983), Nucl. Phys. A 409 (1983), p.191c.
14) For example, P. Ring and P. Schuck, *The Nuclear Many-Body Problem*
    (Springer Verlag, Texts and Monographs in Physics, New York,
    Heidelberg, Berlin, 1980)
15) M. Saraceno, P. Kramer and F. Fernandez, Nucl. Phys. A 405 (1983),
    88.
16) K. Aoki and H. Horiuchi, Prog. Theor. Phys. 69 (1983), 1154.
17) D.M. Brink and F. Stancu, Nucl. Phys. A 243 (1975), 175.
18) T. Izumoto, S. Krewald and A. Faessler, Nucl. Phys. A 341 (1980),
    319.
19) G. Bertsch, J. Borysowicz, H. McManaus and W.G. Love, Nucl. Phys.
    A 284 (1977), 399.
20) G. Satchler, Proc. Int. Conf. on Heavy Ion Physics (Catania,
    Italy, 1983), Nucl. Phys. A 409 (1983), p.   .
21) F.G. Perey, *Direct Interactions and Nuclear Reaction Mechanism* ,
    eds. E. Clementel and C. Villi (Gordon and Breach, Science
    Publishers, Inc., New York, 1963), p.125.
22) N. Austern, *Direct Nuclear Reaction Theories* (Wiley-Interscience,
    John Wiley & Sons, Inc., New York, 1970), see 5.2; Phys. Rev. 137
    (1965), B752.
23) H. Fiedeldey, Nucl. Phys. A96 (1967), 463.
24) K. Aoki and H. Horiuchi, Prog. Theor. Phys. 66 (1981), 1508; 1903;
    67 (1982), 1236.
25) S. Saito, Prog. Theor. Phys. 40 (1968), 893; 41 (1969), 705.
26) S. Saito, S. Okai, R. Tamagaki and M. Yasuno, Prog. Theor. Phys.
    50 (1973), 463.
27) A.B. Volkov, Nucl. Phys. 74 (1965), 33.
28) D.M. Brink and E. Boeker, Nucl. Phys. A 91 (1967), 1.

29)  S. Okai, S. Saito and R. Tamagaki, Prog. Theor. Phys. $\underline{47}$ (1972), 484.

30)  V.I. Kukulin, V.G. Neudatchin and Yu.F. Smirnov, Nucl. Phys. $\underline{A245}$ (1975), 429.

31)  B. Buck, C.B. Dover and J.P. Vary, Phys. Rev. $\underline{C11}$ (1975), 1803.

32)  B. Buck, H. Friedrich and C. Wheatley, Nucl. Phys. $\underline{A275}$ (1977), 246.

33)  D.M. Brink, *Nuclear Physics with Heavy Ions and Mesons*, Vol.I, Course 1 (Les Houches XXX, 1977, North Holland).

34)  K. Aoki and H. Horiuchi, Prog. Theor. Phys. $\underline{69}$ (1983), 857.

35)  K. Aoki and H. Horiuchi, Prog. Theor. Phys. $\underline{68}$ (1982), 2028.

36)  M. Matoba, M. Hyakutake and I. Kumabe, Proc. 1983 RCNP Int. Symp. on Light Ion Reaction Mechanism (Osaka, Japan, 1983), p.47.

37)  S.K. Gupta and K.H.N. Murthy, Z. Phys. $\underline{A307}$ (1982), 187.

38)  D.K. Srivastava, N.K. Ganguly and D.N. Basu, Phys. Lett. $\underline{125B}$ (1983), 260.

39)  D.G. Perkin, A.M. Kobos and J.R. Rook, Nucl. Phys. $\underline{A245}$ (1975), 343.

40)  T. Fliessbach, Nucl. Phys. $\underline{A315}$ (1979), 109.

41)  H. Friedrich, Nucl. Phys. $\underline{A224}$ (1974), 537.

42)  D. Baye, J. Deenen and Y. Salmon, Nucl. Phys. $\underline{A289}$ (1977), 511.

43)  H. Horiuchi, Prog. Theor. Phys. Supple, No.62 (1977), 90.

44)  M. LeMere and Y.C. Tang, Phys. Rev. $\underline{C19}$ (1979), 391; M. LeMere, D.J. Stubeda, H. Horiuchi and Y.C. Tang, Nucl. Phys. $\underline{A320}$ (1979), 449.

45)  T. Ando, K. Ikeda and A. Tohsaki-Suzuki, Prog. Theor. Phys. $\underline{61}$ (1979), 101.

46)  T. Wada and H. Horiuchi, contributions to this conference.

47)  T. Fliessbach, Z. Phys. $\underline{A272}$ (1975), 39; T. Fliessbach and H.J. Mang, Nucl. Phys. $\underline{A263}$ (1976), 75.

48)  A. Arima, *Heavy Ion Collisions*, ed. R. Bock (North Holland, 1979), Vol.1. Chapt. 3.

49)  H. Horiuchi, Prog. Theor. Phys. $\underline{69}$ (1983), 516.

50)  H. Horiuchi, Prog. Theor. Phys. $\underline{71}$ (1984), 535.

51)  K. Yabana and H. Horiuchi, contributions to this conference.

52)  K. Yabana and H. Horiuchi, Prog. Theor. Phys. $\underline{71}$ (1984), 1275; preprints of Kyoto University, KUNS 742 and KUNS 744.

MICROSCOPIC THEORY OF THE IMAGINARY PART OF THE NUCLEUS-NUCLEUS
POTENTIAL

G. Pollarolo
Istituto di Fisica Teorica dell'Università di Torino, Italy
and INFN Sezione di Torino, 10125 Torino, Italy

ABSTRACT. A model to microscopically construct the absorptive part of
the optical potential between heavy ions is reviewed. The main contribution to this potential comes from the single nucleon transfer
channels and from the nuclear excitation of surface modes. The first
mechanism gives rise to a long range component with diffusivity
$a \simeq 0.6$ fm. whilst the second gives rise to a short range component
with $a \simeq 0.3$ fm. The application of this model to the study of the
elastic scattering angular distribution of the $^{16}O + ^{28}Si$ will be discussed and it will also be shown how this model can account for the
evolution of the backward rise in the elastic angular distributions of
the different Si isotopes. The origin of the imaginary coupling potential for inelastic scattering will also be outlined.

1.  INTRODUCTION

Among the large variety of reactions between heavy ions, grazing collisions contitute one of the main sources to obtain information on the
structure of the two colliding partners. The analysis of these reactions relies on the knowledge of the optical potential. This is used
to construct the wave functions of relative motion and it is usually
inferred by fitting elastic scattering data.
     Since the reaction can proceed through many channels this poten-
tial has to be complex, its imaginary part describing the depopulation
of the entrance channel due to the existence of reaction channels.
     From the empirical analysis one has learnt that the elastic
scattering data can be described with local optical potentials whose
real parts are energy and angular momentum independent and display a
smooth variation from system to system. The imaginary parts must be
energy dependent and present large variations from system to system.
     The folding models provide a good microscopic description of the
real part. The tail of it can, in fact, be very well reproduced in
terms of an effective nucleon-nucleon interaction and the ground state
densities of the two ions. The imaginary part lacks, unfortunately, a

53

*J. S. Lilley and M. A. Nagarajan (eds.), Clustering Aspects of Nuclear Structure, 53–68.*
© *1985 by D. Reidel Publishing Company.*

microscopic description and this introduces ambiguities in the
extracted nuclear structure quantities.

In section 2 a model[1] will be reviewed which, starting from a
semi-classical description of the reaction, allows for the construction
of the absorptive part of the optical potential in terms of the single
nucleon transfer channels and the nuclear excitation of the surface
modes. The resulting potential has two components: a long range part
($a \simeq 0.6$ fm) arising from the particle transfer channels and a short
range part ($a \simeq 0.3$ fm) arising from the excitation of surface modes.

In section 3, as an application, the elastic scattering angular
distributions of $^{16}O + ^{28}Si$ will be discussed stressing the importance
of the transparency of the potential and the coupling to the rotational
states of $^{28}Si$. The evolution of the backward rise in the scattering
of the Si isotopes with $^{16,17}O$ will also be discussed.

The origin of the imaginary coupling potential for inelastic
scattering will be discussed in section 4.

## 2.   THE MODEL

In the limit where the wavelength of relative motion is very small it
is appropriate to use a semiclassical description of the reaction. The
interaction $V(\bar{r})$ between the two impinging particles is thus a function
of time as the two ions move on the classical trajectory $\bar{r}(t)$.

The probability for the population of channel $\beta$ in first order
perturbation theory can thus be written as:

$$P_\beta = \left| c_\beta(+\infty) \right|^2 \tag{1}$$

with the amplitude $c_\beta(+\infty)$ given by:

$$c_\beta(+\infty) = \frac{1}{i\hbar} \int_{-\infty}^{+\infty} \langle\beta|V(\bar{r})|0\rangle \, e^{iQ_{\beta 0}t/\hbar} \, dt \tag{2}$$

where the time integral has to be performed along the classical
trajectory (see fig.1).

In the above expression $|0\rangle$ an $|\beta\rangle$ are the channel wave functions
for the entrance channel and for the $\beta$ channel respectively. These are
defined by:

$$|\beta\rangle = \psi_b(\xi_b)\psi_B(\xi_B)e^{i\delta_\beta} \tag{3}$$

$\psi_i(\xi_i)$ being the intrinsic wave function of nucleus i and $\delta_i$ the
"semiclassical phase". $Q_{\beta 0}$ indicates the Q-value of the transition.

Since we are interested in the imaginary potential the probability
of remaining in the elastic channel has to be evaluated. This, in the
limit of <u>many weak independent transitions</u> can be written as:

$$P_0 = \pi_\beta(1 - \left| c_\beta(+\infty) \right|^2)$$

$$\simeq \exp\left\{-\Sigma \left| c_\beta(+\infty) \right|^2\right\} \tag{4}$$

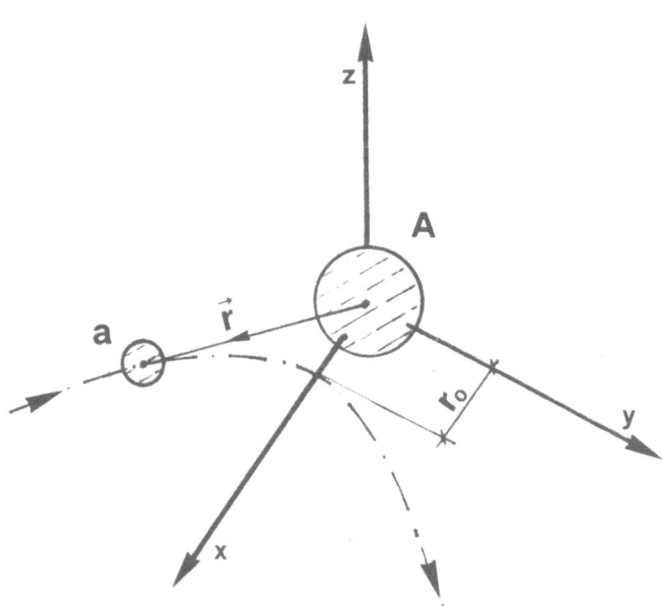

Figure 1.  Schematic representation of the reference frame where the
time integral is performed.  With a and A are indicated the projectile
and target respectively.

In the optical model the decrease of the probability of remaining
elastic, due to the coupling with reaction channels, is described
through an imaginary potential.  In terms of it the probability of
remaining elastic can be written as:

$$P = \exp\left\{ \frac{2}{\hbar} \int_{-\infty}^{+\infty} W(r(t))dt \right\} \tag{5}$$

where, as above, the time-integral has to be performed along the
classical trajectory.

From the comparison of relation (4) and (5) it follows:

$$\int_{-\infty}^{+\infty} W(r(t))\, dt = -\frac{\hbar}{2} \sum_{\beta} \left| c_\beta(+\infty) \right|^2 \tag{6}$$

It is from this relation that one is able to extract an expression for
$W(r)$.  Clearly, as it stands, this formula is inadequate to specify
$W(r)$, so to be useful it has to be supplemented with an additional
hypothesis.  An attractive one is suggested by the empirical observation that most of the elastic scattering data can be described by an
ℓ-independent local potential.  Before showing how this condition is
imposed one has to recall some useful properties of the matrix elements
(form factors) entering in the calculation of the transition amplitudes
$c_\beta(+\infty)$.

## 2.1. Particle Transfer Form Factor

In the case of particle transfer the matrix element in (2) is related to the evaluation of the following integral[2]:

$$\langle \beta | V | 0 \rangle = f_{\beta 0}^{(tr)}(\bar{r}) = \Sigma \int d^3 r_1 \; \phi_\beta^*(\bar{r}_{1A}) \; V_{1A}(\bar{r}_{1A}) \; \phi_0(\bar{r}_{1b}) \qquad (7)$$

performed over the co-ordinates of the transferred particle (labelled with the index 1) at a given distance $\bar{r}$ between the centres of mass of the two ions. In (7) this matrix element is written in the prior representation. The meaning of the different co-ordinates is clear from fig.2. $\phi_0$ and $\phi_\beta$ indicate the single-particle wave functions in the entrance (a+A) and exit (b+B) channels respectively.

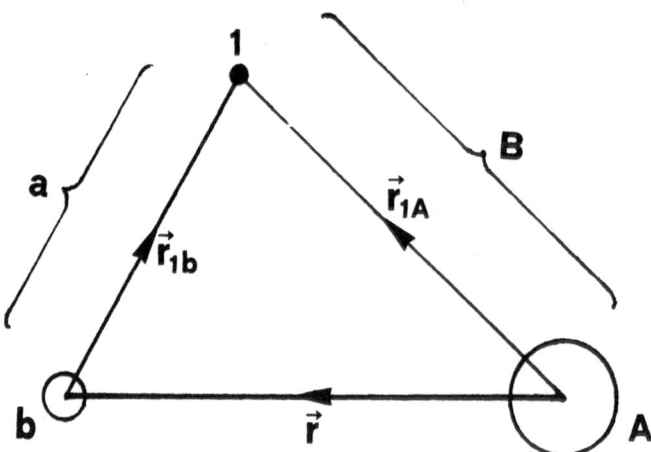

Figure 2. Co-ordinates used in defining the form factors for one particle transfer. The label 1 indicates the transferred particle while b and B indicate the residual nuclei in the exit channel.

With some manipulations it is easy to show that:

$$f_{\beta 0}^{(tr)}(\bar{r}) = \sum_{\lambda \mu} f_\lambda^\beta(r) \; Y_{\lambda \mu}(r) \simeq \sum_{\lambda \mu} k_\beta(\lambda) \; e^{-\frac{r-R_0}{a_\beta}} \; Y_{\lambda \mu}(r) \qquad (8)$$

where $a_\beta$, defining the range of the form factor, is related to the binding energy of the transferred particle. A good estimate of it is:

$$a_\beta \simeq 1.2 \text{ fm} \qquad (9)$$

## 2.2. Inelastic Form Factor

In the case of inelastic excitation, for the matrix element in (2) we can use the macroscopic model for which:

$$\langle \beta | v | 0 \rangle = f_{\beta 0}^{(In)}(\bar{r}) \simeq \sum_{\lambda \mu} \frac{\partial u(r)}{\partial r} Y_{\lambda \mu}(r)$$

$$\simeq \sum_{\lambda \mu} k_{\beta}^{In}(\lambda) \, e^{-\frac{r - Ro}{a}} Y_{\lambda \mu}(r)$$

(10)

where $U(r)$ represents the ion-ion potential with diffusivity:

$$a \simeq 0.6 \text{ fm}$$

(11)

## 2.3. Derivation of $W(r)$

From the above considerations on the form factors, it is clear that the main contribution to the matrix element (2) comes from a small region around the turning point $r_0$ of the classical trajectory. Making use of the following parametrization of the trajectory:

$$r(t) = r_0 + \frac{1}{2} \ddot{r}_0 \, t^2$$

(12)

$$\phi(t) = \dot{\phi}_0 t$$

the relation (6) can be rewritten as follows:

$$\sqrt{\frac{2 r_0}{\ddot{r}_0}} \int_1^{+\infty} \frac{W(r_0, x)}{\sqrt{x - 1}} \, dx = \left(\frac{\hbar}{2}\right) \sum_{\beta} \left| c_{\beta}(+ \infty) \right|^2$$

(13)

where $\ddot{r}_0$ represents the radial acceleration at the turning point and the corresponding angular velocity (cfr. also Fig.1).

From the observation that the matrix element $c_{\beta}(+ \infty)$ can be written as:

$$c_{\beta}(+ \infty) \simeq \bar{k}_{\beta}(\Omega_{\beta 0}, \lambda_{\beta}) \, e^{-r_0/a_{\beta}}$$

(14)

and from the requirement of the ℓ-independence we can solve approximately the resulting integral equation and obtain:

$$W(r) = \sum_{\beta} \bar{k}_{\beta 0}^2(\Omega_{\beta 0}, \lambda_{\beta}) \left| f_{\beta}(r) \right|^2$$

(15)

The constant $\bar{k}_{\beta}$ depends on the Q-value of the reaction, on the angular momentum transfer and on the bombarding energy.

It is useful to split the sum in (15) as a sum over the inelastic channels and one over the mass transfer channels. Doing this one can write:

$$W(r) = W_{in}(r) + W_{tr}(r)$$

(16)

where the component coming from the inelastic excitation is short range with a diffusivity $a \simeq 0.3$ fm while the one coming from particle transfer is long range with $a \simeq 0.6$ fm.

In terms of the quantum numbers of the single particle levels defining the form factors for one-particle transfer channels one has for $W_{tr}(r)$ the expression:

$$W_{tr}(r) = \sum_{a_1 a_1' \lambda} \sqrt{\frac{a_{tr}(a_1 a'_1)}{16\pi |\overset{..}{r}_o| \hbar^2}} \times$$

$$\times \left\{ (2j_1'+1) U^2(a_1) V^2(a_1') \left| f_\lambda^{(a_1 a_1')}(r) \right|^2 g_\lambda(Q) \right\} \tag{17}$$

The quantities $V^2$ are the occupation probabilities that the orbitals $a_1'$ are occupied in a. The quantity $U^2 = 1 - V^2$ is the corresponding probability that the state is empty. The function $g_\lambda(Q)$ determines the adiabatic cut-off. It weights the contribution of the different channels to $W(r)$.

The component of the imaginary potential arising from the nuclear inelastic scattering to vibrational states is

$$W_{in}(r) = \sqrt{\frac{a\pi}{\hbar^2 |\overset{..}{r}_o|^2}} \, \sigma^2 \left( \frac{\partial U(r)}{\partial r} \right)^2 \tag{18}$$

where U is the nuclear part of the ion-ion potential while

$$\sigma^2 = \sum_{\lambda, i} \frac{2\lambda+1}{4\pi} \left[ \frac{\hbar\omega_\lambda(i)}{2c_\lambda(i)} (R_i)^2 \right] g_\lambda(\hbar\omega_\lambda) \tag{19}$$

is the square of the zero-point fluctuation of the surfaces of the two nuclei, weighted by the adiabatic cut-off.

Although the expression of $W_{tr}(r)$ involves only form factors for one-nucleon transfer the effect of the multinucleon transfer channels and fusion on $W(r)$ is also included insofar as these processes proceed through a successive mechanism. The transfer of clusters is not included in this treatment. Their effect on the far tail of W should be negligible since the range of these form factors is much shorter than the one of one-particle transfer.

The actual value of the imaginary potential (16) will depend, for each target and projectile combination, on the specific distribution of single-particle states around the Fermi surface and on the detailed properties of the collective surface vibrations of the two ions. It seems thus unlikely that $W(r)$ can be calculated as a folding of the density of one of the two ions and the nucleon-nucleus imaginary potentials with the other ions.

## 2.4. Coulomb Interaction

In the previous derivation the contribution coming from the Coulomb interaction has been left out in the matrix elements in (2). Due to its long range nature ($\sim 1/r^{\lambda+1}$) the approximation used above to derive $W(r)$ cannot be applied.

At low bombarding energy the excitation due to Coulomb interaction can, however, be neglected for many projectile and target combinations. Clearly this is not the case for rotational states. They lead to a much larger probability to remain in the ground state than equation

(4)[3]. The transition amplitude $c_\beta(+\infty)$ displays an oscillatory behaviour as a function of the impact parameters and does not lead to a simple parametrization of $W(r)$. The rotational degrees of freedom should therefore be treated explicitly by coupled channel calculations.

## 3. APPLICATIONS

Before discussing the applications of the outlined model for the imaginary part of the optical potential to the evaluation in actual cases of elastic angular distributions, some words have to be said about the real part of the nuclear potential.

The double folding potential obtained by O. Akyüz and A. Winther[4] in its Woods-Saxon parametrisation[3] has been chosen:

$$U(r) = \frac{-V_o}{1 + \exp\left(\frac{r-Ro}{a}\right)} \tag{20}$$

with

$$V_o = 16\pi\gamma a \; R_a R_A/(R_a + R_A)$$

$$R_o = R_a + R_A$$

$$R_i = 1.20 \; A_i^{1/3} - 0.09 \; \text{fm.} \tag{21}$$

$$a = 0.885 \; [1 + 0.53 \; (A_a^{1/3} + A_A^{1/3})]^{-1} \; \text{fm.}$$

where $\gamma \simeq 1$ MeV/fm$^2$ is the surface tension. The parameters of this potential have been fixed so that it has the same tail as the double folding potential and the maximum force given by:

$$\left.\frac{\partial U(r)}{\partial r}\right|_{max} = 16 \; \pi\gamma R_a R_A/(R_a + R_A) \tag{22}$$

in accordance with the proximity approximation[5].

In a recent publication[6], the above model has been successfully applied to the study of elastic angular distributions of $^{16}$O on several targets and at several bombarding energies. In all cases, except $^{28}$Si, the dominant contribution to $W(r)$ comes from the particle transfer channels. The contribution due to the nuclear excitation of vibrational states can be neglected in most cases. In the framework of the macroscopic model the above calculations have also been successful in describing inelastic angular distributions. These have been obtained with the deformation parameters used in the calculation of $W_{in}(r)$. However these applications will not be reviewed here.

## 3.1 Elastic Angular Distributions of $^{16}$O + $^{28}$Si

The elastic angular distributions for the reaction $^{16}$O + $^{28}$Si have been obtained over the full angular range for a large variety of energies ranging from slightly to well above the Coulomb barrier[7]. For bombard-

ing energies $E_{cm}$ = 35 MeV these angular distributions exhibit in the
backward hemisphere an oscillatory behaviour and a rise up to a few per
cent of the Rutherford cross section (backward rise phenomenon).

Because $^{28}$Si is strongly deformed, $\beta_2 \simeq 0.35$, the Coulomb interaction should play an important role in the inelastic excitation of the
rotational states of $^{28}$Si. Thus these states have to be treated in a
coupled channel formalism.

In such a calculation for the imaginary potential one has to use
only the component arising from particle transfer. The other is in
fact taken into account explicitly.

In fig.3 the experimental energies of the single-particle orbitals
of $^{16}$O and $^{28}$Si entering in the calculation of $W_{tr}(r)$ are shown. To
obtain single-particle wave functions needed to construct the form
factors, a spherical Woods-Saxon potential of parameters a = 0.65 fm
and $r_0$ = 1.25 fm was used. The depth was adjusted to fit the
experimental binding energies. Unit spectroscopic factors were used.

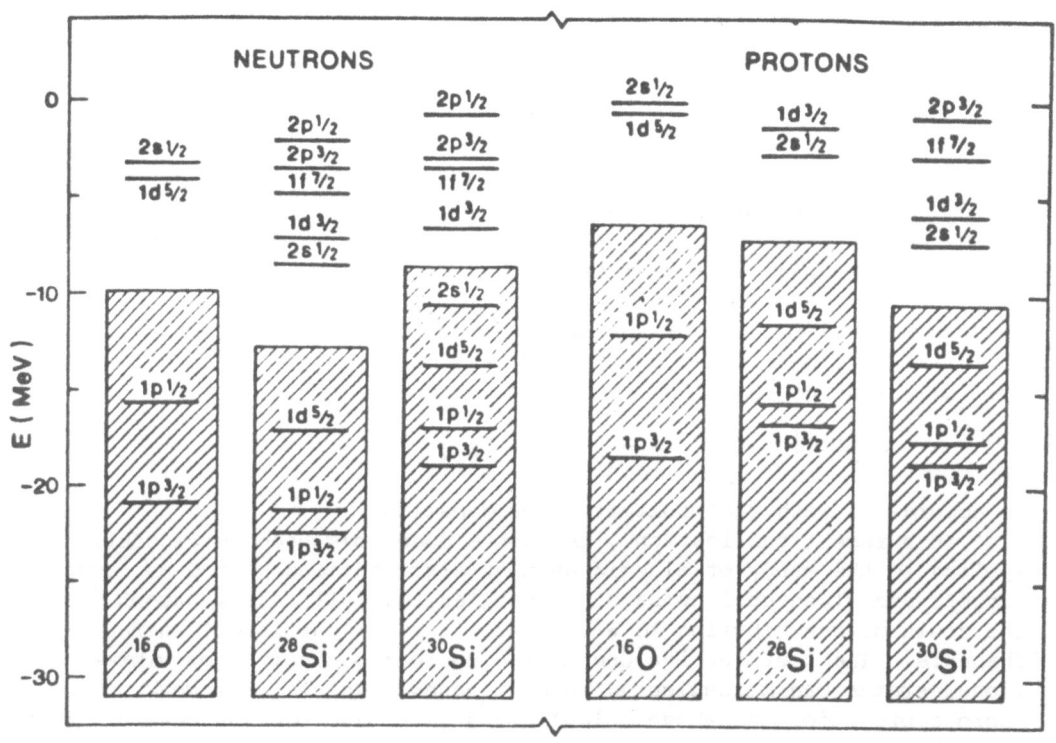

Figure 3. The experimental spectrum associated with the single-
particle levels of neutron and protons in $^{16}$O and $^{28,30}$Si.

Parametrising the resulting potentials as

$$W_{tr}(r) = \frac{W(E)}{1 + \exp\left(\dfrac{r - r_o(A_a^{1/3} + A_A^{2/3})}{a}\right)} \qquad (23)$$

the obtained parameters are shown in Table I for the different bombarding energies.

Table I.  The absorptive potential due to the transfer of particles for the scattering of $^{16}O + ^{28}Si$.

| $E_L$(MeV) | $W(E)$ | $r_o$ | $a$ |
|---|---|---|---|
| 33 | -0.50 | 1.08 | 0.55 |
| 36 | -1.60 | 1.08 | 0.525 |
| 55 | -5.0 | 1.17 | 0.5 |
| 81 | -6.0 | 1.17 | 0.53 |
| 141 | -7.5 | 1.17 | 0.53 |

This table shows that, at low bombarding energies, the absorption coming from particle transfer is very weak leading to a situation of transparency. To stress this, fig.4 compares at $E_{cm}$ = 21.1 MeV the absorption $W_{tr}(r)$ with the one coming from the excitation of the $2^+$ and $3^-$ in $^{28}Si$ treated as vibrational states.

Such a situation thus seems to be ideal in order to learn something about the "interior" of the ion-ion potential. Figs.5 and 6 show the results of coupled-channel calculations for the two lowest bombarding energies ($E_{cm}$ = 21.1 MeV and $E_{cm}$ = 22.7 MeV).

The calculation (8) was carried out using the real part of the nuclear potential of ref.4 in its Woods-Saxon parametrisation and for the imaginary part the one of Table I (remember only $W_{tr}$ is needed in these cases). For the Coulomb component, the standard prescription with a small Coulomb radius ($r_{oc}$ = 0.8 fm) has been used. Both real and imaginary potentials were deformed according to the shape of $^{28}Si$ ($\beta_2$ = 0.38). The $2^+$, $4^+$ and $6^+$ members of the rotational band in $^{28}Si$ were included in the calculation with the corresponding reorientation terms. The excitation of $^{16}O$ has no effect on the calculated angular distribution.

The good fit of figs. 5 and 6 could be obtained only by adding to the surface imaginary potentials of Table I a very short range volume term with W = - 2.5 MeV, a = 0.2 fm and $R_0$ = 5 fm. This is the only ingredient of the calculation whose parameters have been adjusted to the data. From the short range nature of this potential one may argue that its origin may be traced to massive transfer, for example, $\alpha$-transfer.

From the observation of the transmission coeficients (see ref.8) there is no reflection of the low partial waves indicating that the angular distribution is sensitive to the real potential up to $\simeq$ 5 fm (the volume term of W behaves in fact like an absorber).

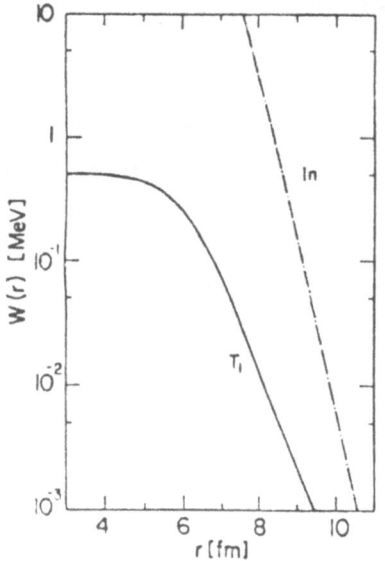

Figure 4.  The absorptive potential
arising from the particle transfer
processes (T) and from the
inelastic excitation of $^{28}$Si (In)
are compared for the scattering
$^{16}$O+$^{28}$Si at $E_{cm}$ = 21.1 MeV.  The
inelastic component is evaluated
considering the states in $^{28}$Si as
vibrational.

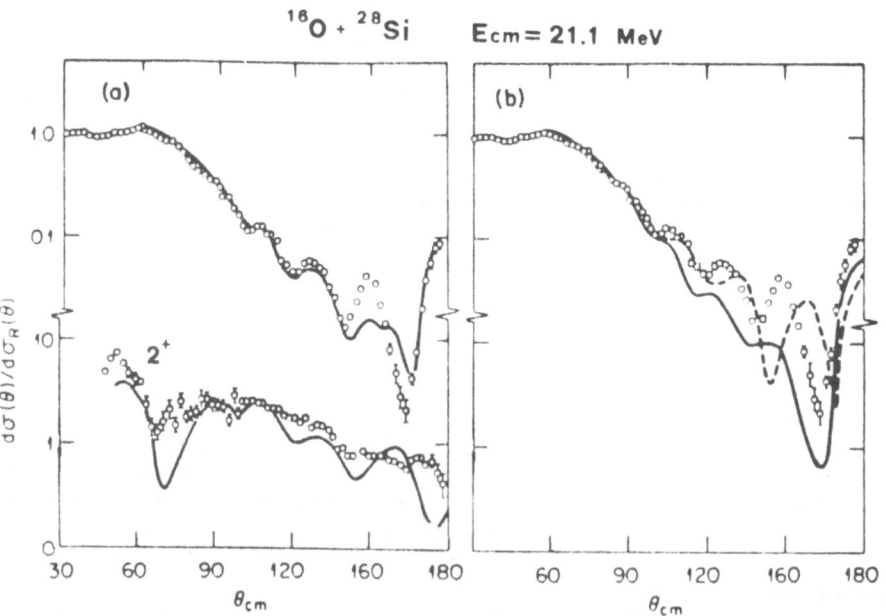

Figure 5.  The ratio of elastic to Rutherford angular distribution for
the $^{16}$O+$^{28}$Si scattering at $E_{cm}$ = 21.2 MeV.  The data are from ref.(7).
In (a) the full drawn curves show the result of the coupled-channel
calculation where the $2^+$, $4^+$ and $6^+$ members of the rotational band of
$^{28}$Si are taken into account.  Also shown is the calculated angular
distribution for the excitation of the $2^+$ in $^{28}$Si.  In (b) with the
full drawn line is shown the result of a calculation when all the
couplings have been neglected.  With the dashed line is indicated the
result when the coupling to the $6^+$ state in $^{28}$Si has been ignored.

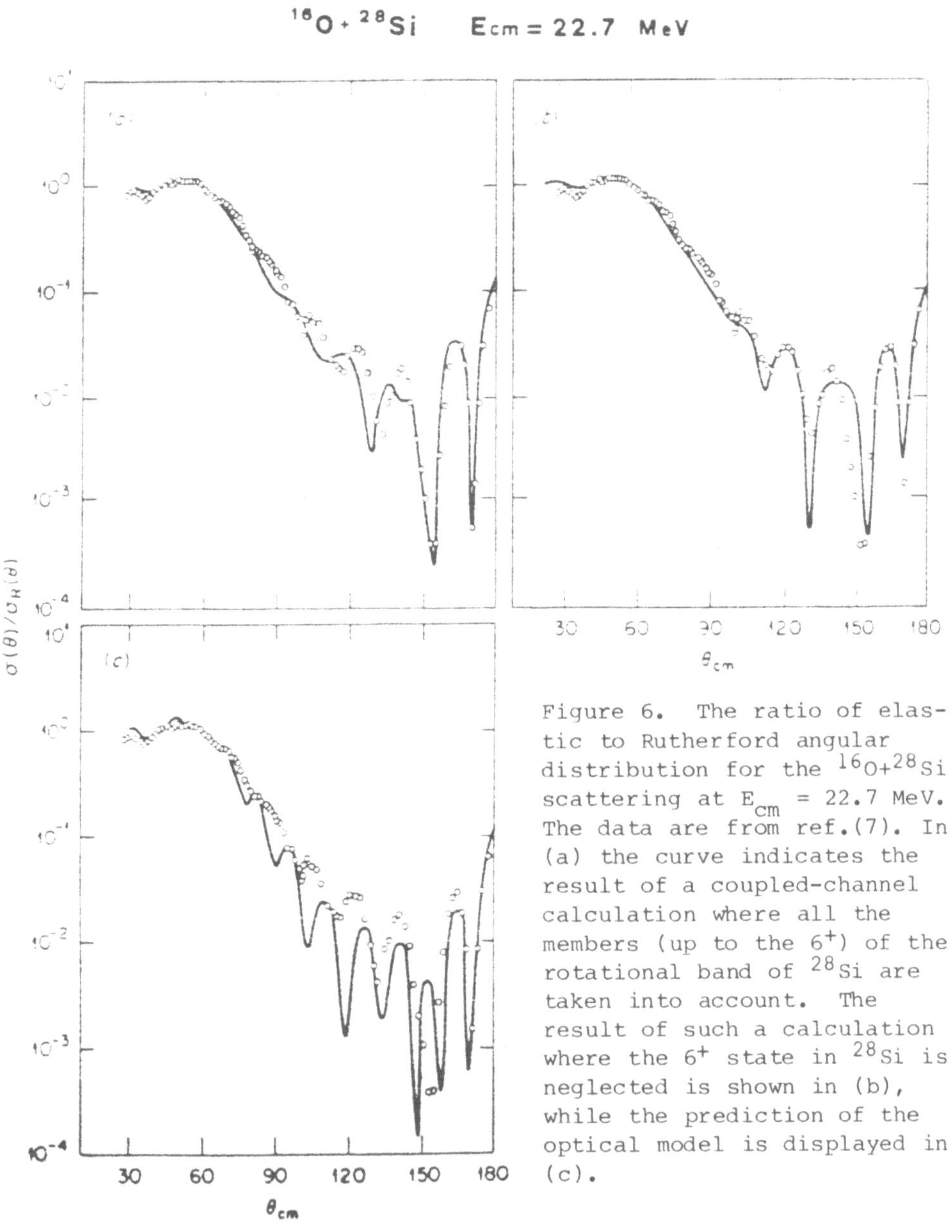

$^{16}O + ^{28}Si$     $E_{cm} = 22.7$  MeV

Figure 6.  The ratio of elastic to Rutherford angular distribution for the $^{16}O + ^{28}Si$ scattering at $E_{cm}$ = 22.7 MeV. The data are from ref.(7). In (a) the curve indicates the result of a coupled-channel calculation where all the members (up to the $6^+$) of the rotational band of $^{28}Si$ are taken into account. The result of such a calculation where the $6^+$ state in $^{28}Si$ is neglected is shown in (b), while the prediction of the optical model is displayed in (c).

Making use of a somewhat different real potential, obtained from
the analysis of 1 particle transfer channel in the neighbouring nuclei
it was possible to obtain[9] a comparable fit to the experimental data.
Also in this case it has been necessary to add a volume absorptive term
to the imaginary potential of Table I with parameters $W_0$ = -26.5 MeV,
a = 0.2 fm and $R_0$ = 5.56 fm. In this last calculation the backward
rise is a dramatic effect of the coupling among the elastic channel and
the rotational states of $^{28}$Si.

For both calculations it is essential to have a very weak absorption in the surface region and both show the sensitivity to the real
potential inside the Coulomb barrier. This has been also demonstrated
in recent calculations by A.M. Kobos and G.R. Satchler[10] where they
analysed the "full" bombarding energy range of the scattering
$^{16}$O + $^{28}$Si in the framework of the optical model. Their imaginary
potentials have an energy dependence that is in good agreement with the
one in Table I.

### 3.2. Excitation Functions of $^{16}$O on Si Isotopes

The excitation functions at $\theta$ = 180° have been measured for the
scattering of $^{16}$O on the Si isotopes[7].

It is seen that the rise of the cross section decreases with the
increasing number of neutrons. At $E_{cm}$ = 25 MeV the cross section of
$^{28}$Si is 8% of Rutherford while it is only 2% in the case of $^{29}$Si and
only 0.8% for the $^{30}$Si. This suggests that the potentials become less
transparent once neutrons are added to the $^{28}$Si core.

From the previous application one has learnt that the degree of
transparency of a potential is governed by the value of $W_{tr}(r)$. It is
thus very interesting to see what the outlined model predicts in these
cases.

The calculation of $W_{tr}$ for these cases has been done[11] using the
experimental single particle energies displayed in fig.3. Both neutron
and proton stripping and pick-up reactions have been considered. In
all cases occupancies equal to 1 and 0 have been assumed except in the
case of $^{29}$Si where the last occupied orbit was given a fractional
occupancy.

The resulting imaginary potentials $W_{tr}(r)$ are displayed in fig.7a.
As is seen a substantial change of $W_{tr}$ is obtained by adding neutrons
to the $^{28}$Si as was expected. It is interesting to point out that in
the case of $^{30}$Si the main contribution to $W_{tr}(r)$ arises from the proton
pick-up reactions.

In fig.7b is shown the dramatic change in $W_{tr}(r)$ when one neutron
is added to the $^{16}$O in the case of scattering with $^{28}$Si.

### 4.    IMAGINARY FORM FACTOR FOR INELASTIC EXCITATIONS

In the macroscopic model the form factor for inelastic excitation of a
nuclear state $\lambda\mu$ is related to the optical potential through the simple
relation:

$$f_{\lambda\mu}(\bar{r}) = \beta_\lambda \left[ \frac{\partial U(r)}{\partial r} Y^*_{\lambda\mu}(r) + i \frac{\partial W(r)}{\partial r} Y^*_{\lambda\mu}(r) \right] \tag{24}$$

where $\beta_\lambda$ is the deformation parameter of the state in question, U and W are the real and the imaginary parts of the optical potential.

It has been shown[12] that the real part of this form factor can be constructed microscopically as the matrix element:

$$f_R(\bar{r}) = \langle\lambda\mu | \hat{F} | \tilde{0}\rangle \tag{25}$$

of the operator $\hat{F}$ defined by:

$$\hat{F} = \sum_{ph} f_{ph}(\bar{r}) \beta^+_p \beta_h \tag{26}$$

$$f_{ph}(\bar{r}) = \sum \int d^3 r_1 \ \phi^*_p(\bar{r}_1) V_{1a}(\bar{r}_{1a}) \phi_h(\bar{r}_1) \tag{27}$$

and the $\beta$'s are the annihilation and creation operators.

In the above expressions, $|\tilde{0}\rangle$ and $|\lambda\mu\rangle$ are the wave functions of the ground and excited states respectively. With $\phi_i(\bar{r}_i)$ we have indicated the single-particle wave function in the nucleus A (for target excitation). The matrix element $f_{ph}(r)$ thus, describes the "scattering" of the nucleon 1 from the state h to the state p due to the field $V_{1a}(r_{1a})$ provided by the projectile (see fig.8a).

From what we have learnt above, the W(r) is in most cases determined by the particle transfer channels. It thus seems natural to try to ascribe the imaginary part of the form factor $f_I(r)$ to the successive transfer back and forth of a nucleon[13] (see fig.8b). This seems also to be confirmed by the experimental observation that the excitation of the first excited state $(1/2^+)$ in $^{17}O$ in the reaction with $^{208}Pb$ is essentially governed by this process[14].

To show how this two step process can give rise to an imaginary form factor one follows the derivation of W(r) and for simplicity one considers the excitation of a pure particle-hole (ph) in the target.

The transition amplitude $c_{ph}(+\infty)$ for the creation of a pure particle hole state (ph) can be written as:

$$c_{ph}(+\infty) = \frac{1}{i\hbar} \int_{-\infty}^{+\infty} dt \ f_{ph}(\bar{r}) \ e^{iQ_{ph}t/\hbar} +$$

$$+ \left(\frac{1}{i\hbar}\right)^2 \sum_{j'} \int_{-\infty}^{+\infty} dt \ f^{(s)}_{pj'}(\bar{r}) \ e^{iQ_{pj'}t/\hbar} \int_{-\infty}^{t} dt' \ f_{j'h}(\bar{r}) \ e^{iQ_{j'h}t'/\hbar} \tag{28}$$

where the first term represents the direct excitation of the (ph) state while the second represents the successive transfer. j' indicates the quantum numbers of the intermediate single-particle states in nucleus a. The form factors appearing in the last term are the usual one-particle form factors with range $a \simeq 1.2$ fm while the one appearing in the first term has the range of the shell-model potential $V_{1a}(r)$ ($a \simeq 0.6$ fm).

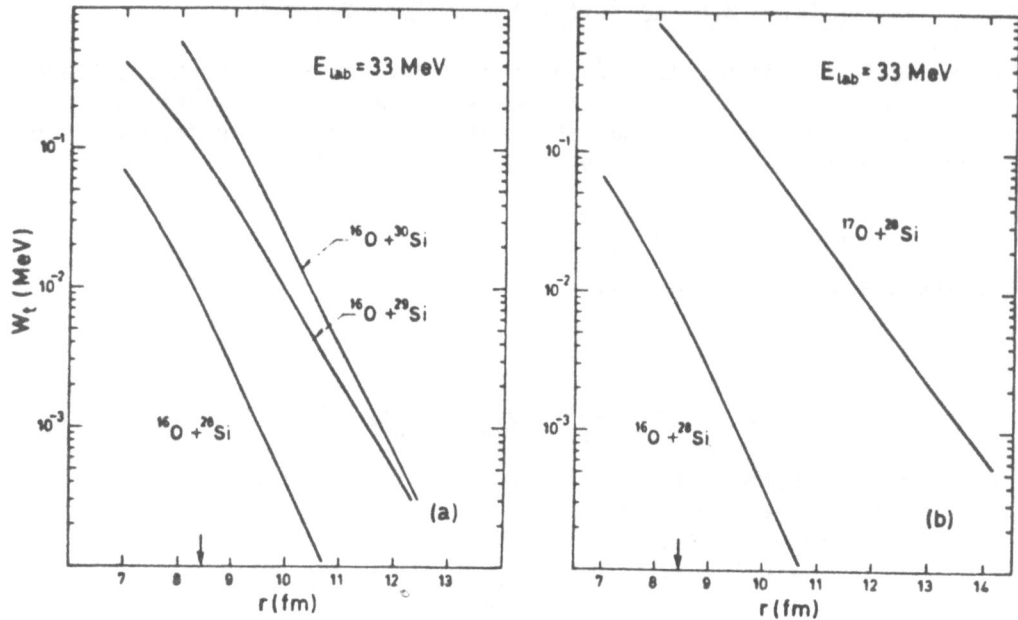

Figure 7.  Absorptive potential arising from particle transfer
processes associated to the scattering (a) $^{16}O + ^{28,29,30}Si$ and (b)
$^{16,17}O + ^{28}Si$ at a bombarding energy of 33 MeV.

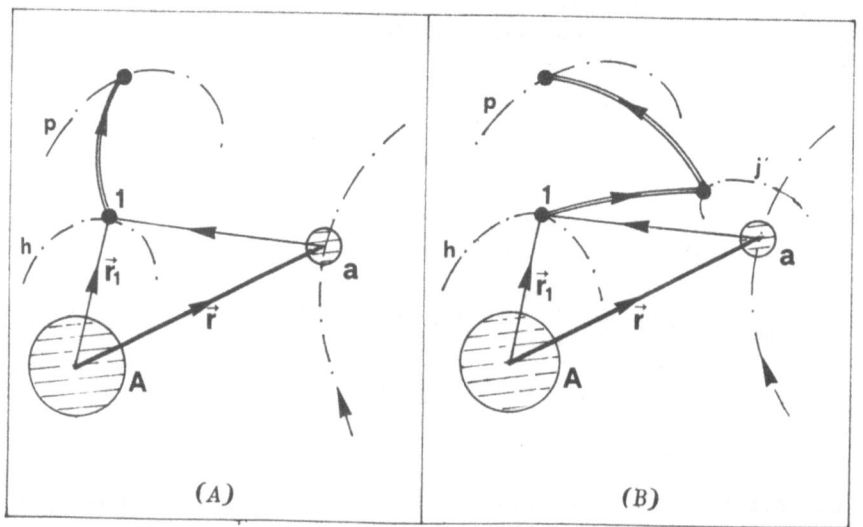

Figure 8.  Schematic representation of a target (A) particle hole (ph)
state created through a direct process (a) and a two-step transfer
process (b).  The intermediate single-particle states of the projecile
are labelled by j'.

Using the relations:

$$\theta(t-t') = \frac{1}{2} \left[1 + \varepsilon(t-t')\right]$$

$$\varepsilon(t-t') = \frac{i}{\pi} \mathscr{P} \int_{-\infty}^{+\infty} \frac{dq}{q} e^{iq(t-t')}$$

(29)

one can write:

$$c_{ph}(+\infty) = \frac{1}{i\hbar} \int_{-\infty}^{+\infty} dt\, f_{ph}(\bar{r})\, e^{iQ_{ph}t/\hbar}$$

$$+ \frac{1}{2} \left(\frac{1}{i\hbar}\right)^2 \sum_{j'} \int_{-\infty}^{+\infty} dt\, f^{(s)}_{pj'}(\bar{r})\, e^{iQ_{pj'}t/\hbar} \int_{-\infty}^{+\infty} dt\, f_{j'h}(\bar{r})\, e^{iQ_{j'h}t/\hbar}$$

$$+ \frac{i}{2\pi\hbar^2} \int_{-\infty}^{+\infty} \frac{dq}{q} \sum_{j'} (Q_{pj'} + q\,;\, Q_{j'h} - q)$$

(30)

It is the second term in this expression that gives rise to $f_I(r)$ while the last term gives a correction to $f_R(r)$, i.e. to $f_{ph}(r)$. After some manipulation borrowed from the derivation of $W(r)$, it is possible to write approximately:

$$c_{ph}(+\infty) = \frac{1}{i\hbar} \int_{-\infty}^{+\infty} dt\, \left[ f^R_{ph}(\bar{r}) + i\, f^I_{ph}(\bar{r}) \right]\, e^{iQ_{ph}t/\hbar}$$

(31)

where

$$f^I_{ph}(\bar{r}) \simeq \sum_{j'} f^{(s)}_{pj'}(\bar{r})\, f^{(P)}_{j'h}(\bar{r})\, G(Q_{pj'}, Q_{j'h})$$

(32)

and $f_{ph}$ represents the real part of the form factor, coming from the "direct" process, plus the contribution from the two step processes (last term in (30)). The function $G(Q_{pj'}, Q_{j'h})$ is a generalization of the cut-off function introduced in the expression of $W(r)$.

It is interesting to note the similarity of this expression with the one of the imaginary potential and that the range of this imaginary coupling is of the order of 0.6 fm.

## 5.  CONCLUSIONS

Although some open questions remain, the above model based solely on nuclear structure information gives us confidence that a detailed understanding of the grazing collisions can be obtained.

The calculation of $W(r)$ should be improved to incorporate the experimental single-particle spectroscopic factors as obtained from the measured single-particle transfer cross sections. Since these spectro-scopic factors enter not only as a multiplicative factor but also through the derived absorptive potential, this improvement can only be done self-consistently.

REFERENCES

1.  Broglia R A, Pollarolo G and Winther A, Nucl. Phys. **A361**
    (1981) 307.

2.  Broglia R A and Winther A, Phys. Rep. **4c** (1972) 153.

3.  Broglia R A and Winther A, Heavy Ion Reactions (Benjamin/Cummings,
    Reading, Mass., 1981).

4.   Akyuz O and Winther A, in Proc. Enrico Fermi Int. School of
    Physics, 1979, Course on Nuclear Structure and Heavy Ion
    Reactions, ed. R A Broglia, C H Dasso and R Ricci (North-Holland,
    Amsterdam, 1981).

5.  Blocki J, Randrup J, Swiatecki W J and Tang F, Ann. Phys. **105**
    (1977) 427.

6.  Pollarolo G, Broglia R A and Winther A, Nucl. Phys. **A406**
    (1983) 369.

7.  Braun-Munzinger P and Barrette J, Phys. Rep. **87C** (1982) 209
    and references therein.

8.  Pollarolo G and Broglia R A, Il Nuovo Cimento. **81A** (1984)
    278.

9.  Kahana S, Pollarolo G, Barette J, Broglia R A and Winther A, Phys.
    Lett. **113B** (1983) 283.

10. Kobos A M and Satchler G R, Oak Ridge preprint 1984.

11. Quesada J M, Broglia R A, Bragin V and Pollarolo G, NBI-84-20
    preprint.

12. Broglia R A, Dasso C H, Pollarolo G and Winther A, Phys. Rep.
    **48C** (1978) 351.

13. Dasso C H, Landowne S and Pollarolo G, in preparation.

14. Lilley J S, Fulton B R, Banes D, Cormier T M, Thompson I J,
    Landowne S and Wolter H M, Phys. Lett. **128B** (1983) 153.

CLUSTERING ASPECTS OF LIGHT, MEDIUM AND HEAVY NUCLEI

# CLUSTER STRUCTURE OF LIGHT NUCLEI

Brian Buck
Department of Theoretical Physics
1 Keble Road
Oxford OX1 3NP
England

ABSTRACT. A simple treatment of numerous cluster structures in light nuclei is outlined. The model is applied to explain the properties of rotational bands in several nuclei and the effects of the spins and internal excitations of the subsystems are discussed. The talk concludes with a detailed account of how the data on states of $^7$Li and $^7$Be can be understood.

It has become very clear in the last ten years that many states in light nuclei have a particularly simple structure. The basic hypothesis is that two real nuclei, with essentially unperturbed free space properties, can orbit each other to form loosely bound states or sharp resonances. My aim in this talk is to persuade you that the observable characteristics of such states can be accurately described by means of a very simple model indeed. Most often one of the constituent nuclei is an alpha particle, a helion, or a triton, while the core system is some other light nucleus which may not necessarily be in its ground state. Harald Friedrich has suggested[1] that it may well be more realistic to postulate undistorted alpha particles inside a nucleus than undistorted nucleons.

The cluster states in question frequently group naturally into bands with an underlying rotational spectrum, are strongly populated in transfer reactions, have enhanced in-band electromagnetic transitions and, when unbound, have large cluster decay widths. All such properties are qualitatively and in most cases quantitatively explained by a model with two sub-nuclei interacting via a deep local potential, which for practical purposes can be taken in the modified Gaussian form

$$V(R) = -V(1 + \gamma R^2)\exp(-\alpha R^2),$$

where R is the separation of the centres of mass. The potential parameters are largely independent of energy and angular momentum for levels in a given rotational band, but they may depend on parity. The procedure is then just to solve the relative motion wave equation

J. S. Lilley and M. A. Nagarajan (eds.), Clustering Aspects of Nuclear Structure, 71–84.
© 1985 by D. Reidel Publishing Company.

$$[- \frac{\hbar^2}{2\mu} \nabla^2 + V(R)] \bar{\Psi}_{-NL} = E \bar{\Psi}_{-NL}$$

for positive or negative E with suitably chosen values of the principal quantum number N and angular momentum L. For states in a single band, the allowed N and L are related to the single particle quantum numbers $n_j$ and $\ell_j$ of the cluster nucleons by having a constant value of

$$2N + L = \sum_{j=1}^{n_c} (2n_j + \ell_j),$$

where $n_c$ is the number of cluster nucleons and I have assumed the commonest case in which the cluster has zero internal excitation quanta. The cluster nucleons are regarded as occupying low lying orbitals above the Fermi surface of the core nucleus and the above relation is suggested by the ordinary shell model

These working assumptions have recently received strong support from the studies of Horiuchi[2] based on the microscopic Resonating Group Method (RGM). That technique is of course designed to incorporate all Pauli principle effects for nucleons, but it leads to complicated integral equations which are difficult to interpret physically. However, Professor Horiuchi has shown by WKB methods that the nonlocal kernels of the RGM may be transformed, to good accuracy, into equivalent local potentials between the two subsystems. The main effects of antisymmetry are taken care of by showing that there must be a suitable number of excitation quanta in the relative motion of the clusters and that the effective local potential must be deep. Thus the cluster and core have large relative velocities when interacting and this ensures that the cluster nucleons occupy orbitals above the Fermi surface of the core nucleus. Space exchange effects also appear in the equivalent potential which thereby becomes effectively parity dependent. These conclusions were derived from a detailed analysis of RGM calculations which use rather simplified assumptions about the internal wave functions of the clusters; but it is not a very great conceptual leap to suppose that a similar state of affairs prevails when the subsystems are real nuclei.

As a first example of how well this simple model works, I take the prototype system of $^8$Be treated as two interacting alpha particles. There are no bound states, but the phase shifts show clear evidence of resonances in s, d and g waves (symmetry allows only even relative angular momenta). The data for a long time seemed to demand complicated interactions with repulsive cores, but a satisfyingly straightforward explanation of the observations finally emerged. The 0s, 1s and 0d states of relative motion, which would correspond to bound states, are forbidden by the Pauli principle for nucleons and the first allowed states (2N + L = 4) form a band of resonances with L = 0, 2 and 4. The data are well described by a pure Gaussian interaction independent of energy and angular momentum, together with a Coulomb term. This potential is shown in Figure 1. The phase shifts are reproduced to fairly high energies as illustrated in Figure 2 and the $0^+$ 'ground state'

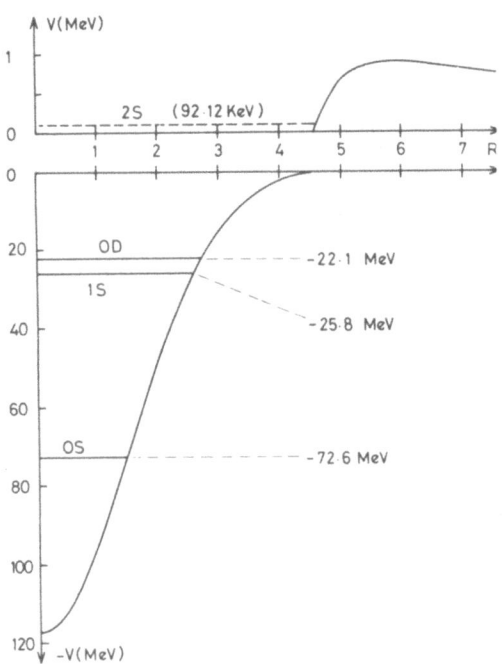

Fig. 1.    The α-α potential with positions of $0^+$ resonance and forbidden
          bound states.

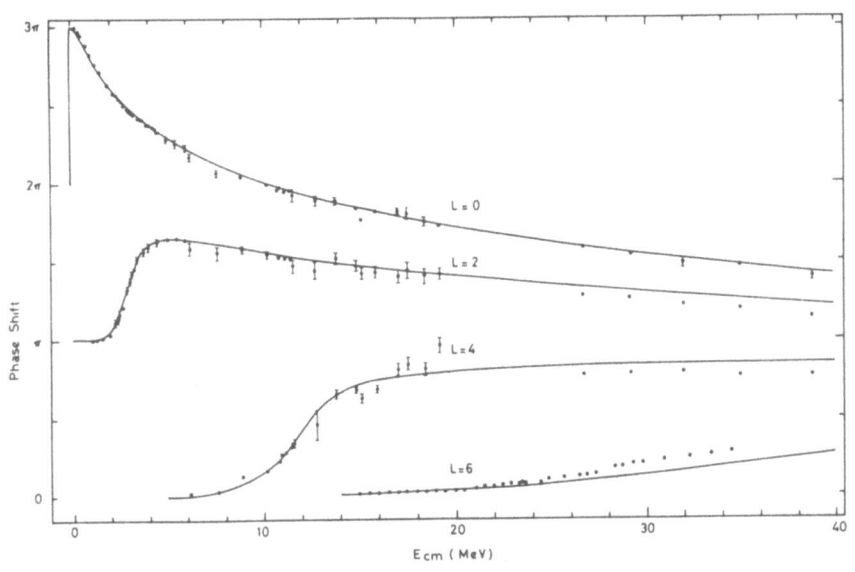

Fig. 2.    The α-α phase shifts calculated with the interaction of Fig. 1.

at 92 keV is predicted as a simple potential resonance having a width
of only 6 eV, in good agreement with experiment.

As another instructive example of the interaction of spin 0 sub-
nuclei I consider a revised version[3] of an earlier calculation[4] of the
well known $0^+$ and $0^-$ rotational bands in $^{16}O$, with bandheads at
$E_x$ = 6.05 and 9.63 MeV respectively. These are thought to be orbital
states of an alpha particle relative to the ground state of $^{12}C$. In
shell model approximation the four cluster nucleons are either all in
the 1s0d shell or three of them are and one goes into the 1p0f shell,
giving rise to 2N + L = 8 or 9 for the $\alpha$ - $^{12}C$ relative motion quantum
numbers. Thus the $0^+$ band has states with L = 0,2,4,6 and 8, while the
$0^-$ band sequence is L = 1,3,5,7 and 9. The experimental bound state and
resonance energies are reasonably well fitted by a modified Gaussian
potential, as appears in Figure 3, though with different parameters for
the two bands. This is an example of parity dependence. The calculated
widths of the unbound states, in the $\alpha_0$ channel, show a really
remarkable agreement with experiment.

Table 1.  Alpha Widths in $^{16}O$

| $J^\pi$ | $E_x$(MeV) | $\Gamma_\alpha^{cal}$(keV) | $\Gamma_\alpha^{exp}$(keV) |
|------|--------|-----------|-----------|
| $4^+$ | 10.35 | 23 | 27 ± 4 |
| $6^+$ | 16.27 | 375 | 392 ± 20 |
| $1^-$ | 9.63 | 451 | 400 ± 10 |
| $3^-$ | 11.60 | 824 | 800 ± 100 |
| $5^-$ | 14.66 | 577 | 632 ± 20 |
| $7^-$ | 20.86 | 645 | 540 ± 100 |

Results of comparable quantitative accuracy are obtained also for the
corresponding states in $^{20}Ne$ = $\alpha$ + $^{16}O$(GS).

Thus we see that in two quite different cases our deep, diffuse interactions reproduce the essential features of the data, namely energies
and widths. This kind of shape must therefore be preferred over more
conventional forms, such as the Wood-Saxon potentials, which fail
miserably on both counts. In our original work the potential shapes
were generated by means of a double folding calculation - they are, of
course, very similar in form to the Gaussian-type shapes used in the
results I have shown you. It seems to me that we have here a puzzling
problem which is worth some attention. Is it possible to understand
exactly why these bell-shaped potentials imply such accurately
rotational sequences? I have mentioned this problem on several previous
occasions, but it is still unsolved and it still continues to intrigue
me. As a numerical experiment, I calculated[5] the energies of all
single particle states in a pure Gaussian interaction. Every state is a
member of a beautiful truncated rotation sequence according to the
formula

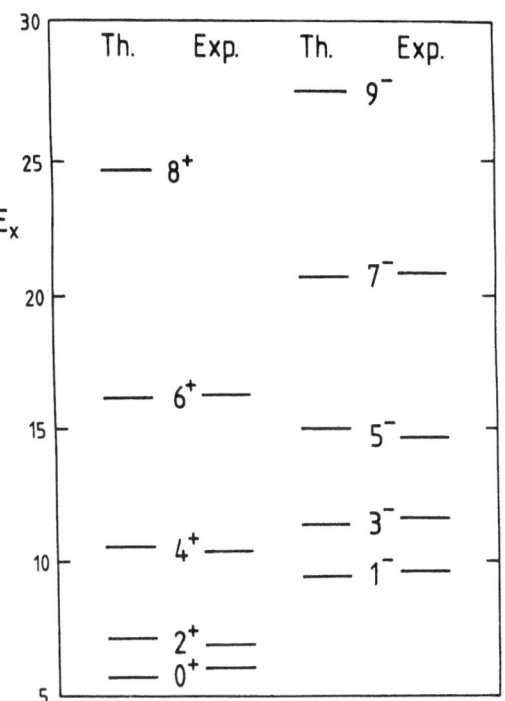

Fig. 3.    Theoretical and experimental energies in two rotational bands
           of $^{16}O = \alpha + {}^{12}C(GS)$. (From reference 3)

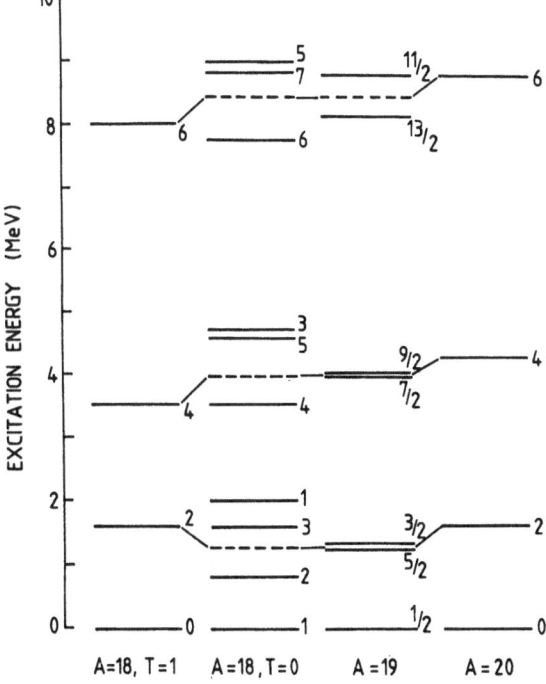

Fig. 4.    Alpha particle cluster bands with 2N+L = 8 in the nuclei
           $^{18}O$, $^{18}F$, $^{19}F$ and $^{20}Ne$.

$$E(N,L) = F(2N + L) + \beta L(L + 1)$$

where the function F controls the spacings between the bands (as labelled by the values of 2N + L) and the parameter $\beta$ is the same for all bands. Other potentials of similar shape but different analytical form give comparable though not quite such spectacular results. A proper mathematical explanation of this phenomenon is still lacking though John Spiers and I, and independently Neil Rowley, have found some suggestive results.

Having checked that the model gives a highly satisfactory account of dinuclear states for spin 0 clusters, we can turn to consideration of what happens when the subsystems have spin. For the moment I will restrict myself to alpha particle states in sd-shell nuclei and in Figure 4 I show some examples of spectra in which the core nucleus has spin $\frac{1}{2}$ or 1. For spin $\frac{1}{2}$ each rotational state with L > 0 splits into a doublet with angular momenta $J = L \pm \frac{1}{2}$, and such doublets can easily be picked out in the experimental spectrum of, for example, $^{19}$F. These states are well described[6] by the structure $\alpha + ^{15}N(\frac{1}{2}^-)$ and the lowest state (L = 0) of the resulting band is the first excited level of $^{19}$F, which has $J^\pi = \frac{1}{2}^-$. It is noticeable that the splitting of the other states is generally quite small and irregular, implying a rather small effective spin-orbit interaction, which may even change sign within a band. Indeed the observations are consistent with essentially zero spin-orbit force, the actual splittings being determined by weak mixing effects with non-cluster configurations.

When the core nucleus has spin $1^+$, as in the states of $^{18}$F with the structure $\alpha + ^{14}N(GS)$, the lowest level, with L = 0, has $J^\pi = 1^+$. This is the well known 'intruder' state at 1.7 MeV excitation in $^{18}$F. Members of the relative motion cluster band with orbital L > 0 group into characteristic triplets. In each case the state with J = L comes lowest and is well separated from the remaining close doublet with angular momenta $J = L \pm 1$. Clearly this is not the pattern produced by a spin-orbit force and indeed the spectrum can be explained without invoking such a force, which is consistent with the conclusion reached in the spin $\frac{1}{2}$ example of $^{19}$F. However, when the core nucleus has spin $S \geqslant 1$ interactions more complicated than spin-orbit are possible. For the particular system of an alpha particle orbiting the $1^+$ ground state of $^{14}$N, it is found that a tensor force proportional to $(\underset{\sim}{S}.\underset{\sim}{R})^2$, where $\underset{\sim}{R}$ is the relative coordinate, reproduces exactly the observed splitting pattern. In our original treatment[7] of $^{18}$F, we replaced the diagonal radial matrix elements of the tensor force by a single constant and neglected the coupling between relative motion states with different L-values. This worked extremely well, but eventually we felt it was necessary to check our rather sweeping assumptions by a more microscopic approach. Pioneering calculations by Neil Rowley and Alan Merchant[8] of the interaction of an alpha particle with the two holes in the p-shell of $^{14}$N have justified the phenomenological treatment in its main details. They showed conclusively that spin-orbit forces are small and that there exists an induced tensor force of the right sign and magnitude to explain the spectra, but that the off-diagonal radial matrix elements of the tensor force can not be neglected. However, by coupling in some

excited states of the $^{14}N$ core nucleus they were able to produce an even
better account[9] of the alpha cluster states of $^{18}F$.

The idea of considering several excited states of the real core
nucleus simultaneously with the ground state is a very powerful one and
a large extension of the number of nuclear levels described by the
cluster model is thereby achieved. I illustrate this here only for the
example of my favourite nucleus $^{16}O$. I showed you earlier a recent
calculation of the $0^+$ and $0^-$ bands of $^{16}O$ treated as $\alpha + {}^{12}C(GS)$ with
the alpha particle in relative motion states given by $2N + L = 8$ and $9$.
But the deformed nucleus $^{12}C$ has a $J^\pi = 2^+$ state at $E_x = 4.5$ MeV which
is well separated from other levels and which clearly may couple to the
relative orbital motion to produce many more alpha states in $^{16}O$. Such
states, with appropriate values of $J^\pi$, may well mix significantly with
the previous bands based on the ground state of $^{12}C$, so the problem
becomes one of coupled channels. The latest calculation along these
lines is by Richard Baldock and Alberto Rubio[10] and you will see in
Figure 5 that there is a satisfying measure of correspondence between
theory and experiment. In fact the calculation produces candidates for
a large fraction of the known $T = 0$ states in $^{16}O$ with appreciable
$\alpha$-width and the computed widths in the $\alpha_0$ and $\alpha_1$ channels come out in
excellent agreement with observation. Now it is quite hard to see any
simple patterns of levels in this diagram, so in Figure 6 I show the
positive parity states on a plot of excitation energy $E_x$ against $J(J+1)$.
Here it is easy to see the $0^+$ band based on the $^{12}C$ ground state
together with three bands arising from the $2^+$ excitation of $^{12}C$. The
picture clearly brings out the strong coupling effects inherent in our
model and the results are very reminiscent of typical spectra in heavy
rotational nuclei. There is a low $0^+$ 'ground state' band, a parallel
$2^+$ $\gamma$-band, a $0^+$ $\beta$-band and a staggered $K=1$ band. Similar bands appear
also in the negative parity spectrum. Inclusion· of the $0^+$ second
excited state of $^{12}C$ leads to the appearance of two further bands in $^{16}O$
with $K = 0^+$ and $0^-$.

I will now consider in some detail the properties of states in $^7Li$
and $^7Be$. These nuclei are of particular interest since the ground and
first excited states have strongly developed cluster structure. There
is a wealth of experimental data for comparison with models and also a
large amount of recent theoretical and experimental activity. Both
Akito Arima[11] and Tang[12], and their co-workers, have presented
detailed RGM calculations for mass 7 states in the last year or so, and
have achieved reasonable overall agreement with observed properties.
Here I shall show that the data can be explained rather convincingly on
the simplest possible hypothesis, i.e., that we have a real alpha
particle and a real triton or helion, with undisturbed free space
properties, interacting through a deep local potential.

I begin by specifying the basic structure of the first four states
of $^7Li$ (or $^7Be$). In lowest approximation we have three 0p-shell nucleons
outside the closed 0s-shell of $^4He$ and assume that, with small
admixtures of higher configurations, they correlate to form a real $^3H$
(or $^3He$) nucleus with essentially zero internal excitation quanta.
This leaves $2N + L = 3$ quanta for the relative motion of the centres of
mass of alpha and triton (or helion) so that the orbital angular

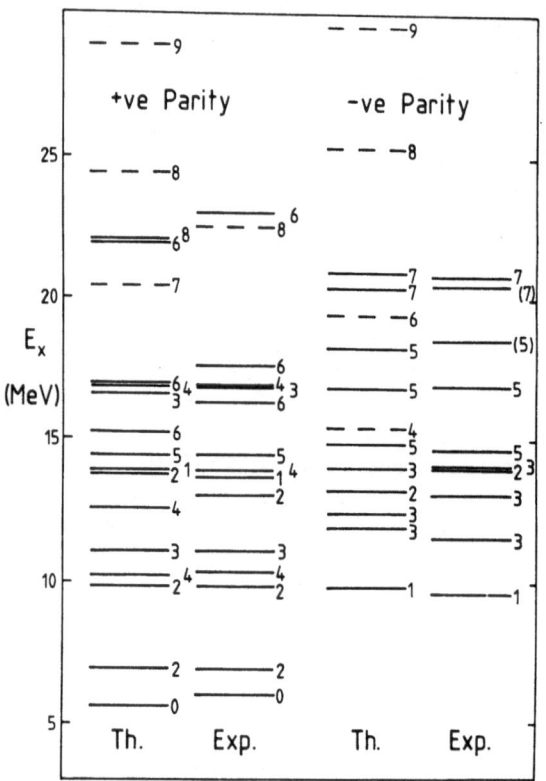

Fig. 5.    Calculated and observed T = 0 levels in $^{16}O$.

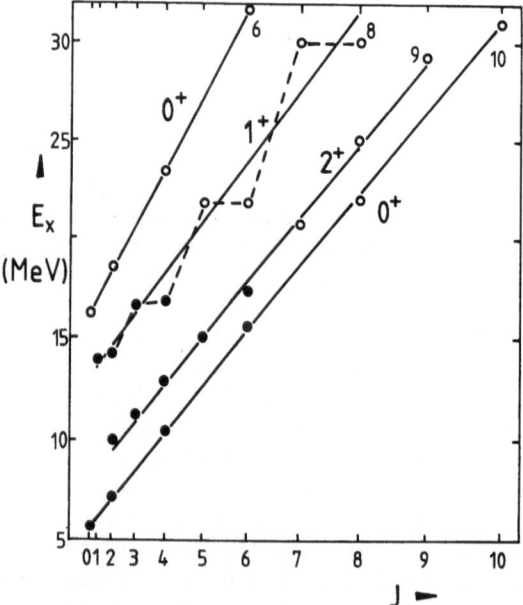

Fig. 6.    Positive parity rotational bands in $^{16}O$.

momentum can be L = 1 or 3.  Spin-orbit coupling for the A=3 subnucleus with $J^\pi = \frac{1}{2}^+$ then gives rise to the ground state doublet of A=7 with $J^\pi = 3/2^-, 1/2^-$ and an excited doublet with $J^\pi = 7/2^-, 5/2^-$, in that order. A single spin-orbit force is insufficient to give correctly both doublet splittings since the L = 3 states probably mix slightly with higher $7/2^-$ and $5/2^-$ states of different structure.  The energies of the cluster states relative to the break-up thresholds are shown in Figure 7. These energies and many other observables can be fitted  nicely by using as interaction a deep Gaussian central potential, a simple uniform sphere Coulomb term and a spin-orbit force of the usual derivative form with different depth for L = 1 and L = 3.

At this stage it is useful to note that several observables for A=7 can be calculated using only the basic structural hypotheses together with the highly plausible assumption that the radial wave functions of the L = 1 spin-orbit doublet states have overlaps very close to unity.  This last statement is well verified by later calcula- tions.  The $3/2^-$ ground state of $^7$Be, for instance, decays by electron capture to the $J^\pi = 3/2^-$ ground state and $J^\pi = 1/2^-$ first excited state of $^7$Li.  The (ft)-values of these decays are given in our model by the very simple expressions:

$$(ft) = \frac{B}{1 + \frac{5}{9}|G|^2} \quad ; \quad (ft)^* = \frac{B}{\frac{4}{9}|G|^2} \quad ,$$

where $|G|$ = 2.088 is the measured axial vector matrix element for the β-decay of the triton and the experimental constant B = 6170s is in essence the conserved vector current coupling strength as determined from superallowed $0^+ \to 0^+$ transitions in several nuclei.  Inserting the numbers we easily find the following results:

Table 2.  Electron Capture: $^7$Be$(\varepsilon)^7$Li, $^7$Li$^*$

|            | Theory | Experiment |
|------------|--------|------------|
| log(ft)    | 3.256  | 3.220      |
| log(ft)$^*$| 3.503  | 3.454      |

The empirical values of log(ft) quoted here are from Walliser et. al.[12] and are somewhat uncertain since they involve estimation of the atomic electron density around $^7$Be.  The branching ratio for electron capture to the first excited state of $^7$Li is very much better determined[13] and the observed value is 10.39 ± 0.6%.  Our simple equations given above lead to a calculated branching ratio of 10.1%[14].  The agreement with experiment is quite remarkable considering that this entire calculation can be done on the back of an envelope.

The ground state magnetic moment of $^7$Li and the B(M1↑) values for $3/2^- \to 1/2^-$ transitions may be calculated with equal ease.  The dipole

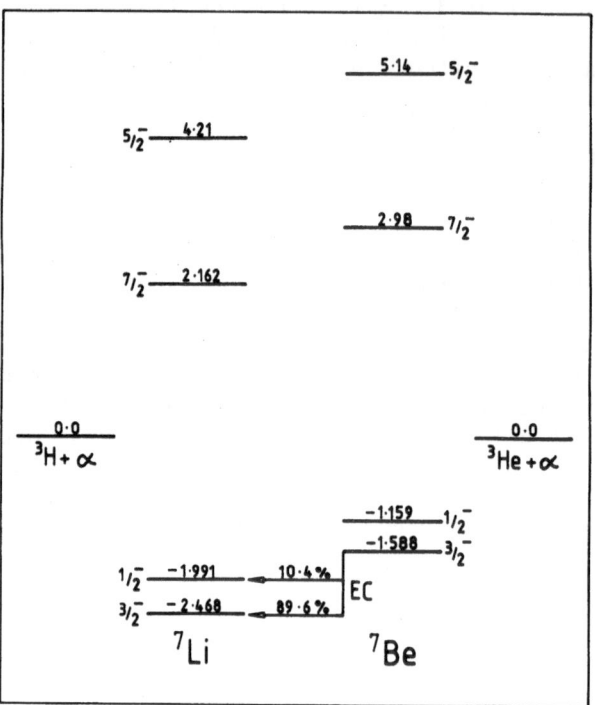

Fig. 7.    Energies of lowest states in $^7$Li and $^7$Be.

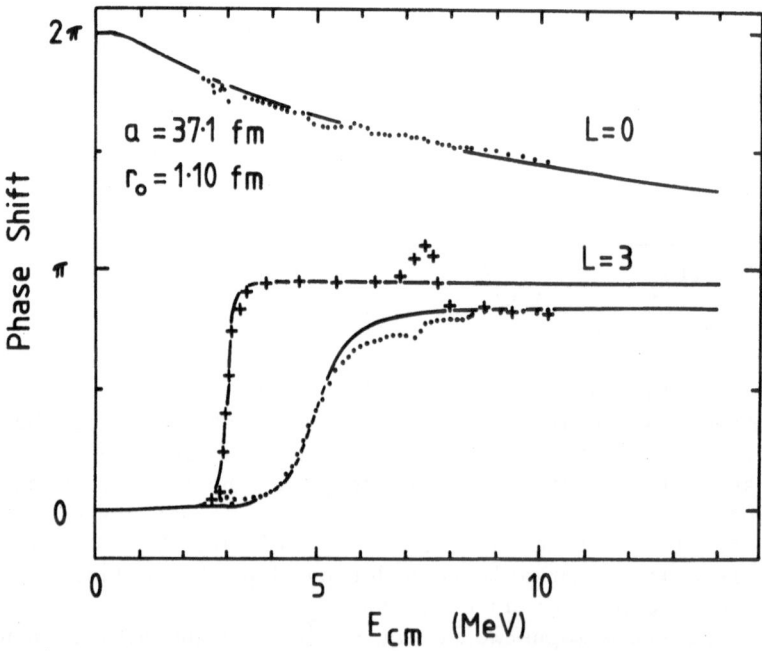

Fig. 8.    Theoretical and experimental phase shifts for s-wave and f-wave
scattering of $^3$He by $^4$He.

operator suitable for these states is given in units of the nuclear
magneton $\mu_N$ by

$$\underline{\mu}_7 = \underline{\mu}_3 + \alpha \underline{L},$$

where $\underline{\mu}_3$ is the magnetic moment operator of the A=3 nucleus and

$$\alpha = \frac{A_2}{A_1 + A_2} \cdot \frac{Z_1}{A_1} + \frac{A_1}{A_1 + A_2} \cdot \frac{Z_2}{A_2}$$

is an effective orbital g-factor straightforwardly derived from the
masses and charges of the two clusters. In the $J^\pi = 3/2^-$ ground state
of $^7$Li the observable magnetic moment is then just the measured
intrinsic moment of the triton ($\mu_3 = 2.979$) added to the appropriate
orbital g-factor ($\alpha = 0.405$), yielding $\mu_7 = \mu_3 + \alpha = 3.384$ nm. This
predicted value is reasonably close to the observed moment of 3.256 nm,
the deviation of 0.13 nm being of the order expected from mesonic
effects. The results for the magnetic dipole transitions come out
similarly well. Again assuming unit radial overlaps we easily find that

$$B(M1:3/2^- \to 1/2^-) = (2\mu_3 - \alpha)^2/4\pi ,$$

in units of $\mu_N^2$. Comparison with direct M1 excitation results or with
lifetime observations gives an almost incredibly good match between
theory and experiment as shown below.

Table 3. $B(M1:3/2^- \to 1/2^-)$ for A=7

| Nucleus | Theory | Experiment |
|---------|--------|------------|
| $^7$Li  | 2.45   | 2.48 ± .12 |
| $^7$Be  | 1.87   | 1.87 ± .20 |

Calculation of the charge radius, the magnetisation radius, the
electric quadrupole moment and the magnetic octupole moment of $^7$Li
requires more information about the radial wave function for relative
motion of the clusters. But all that is needed, apart from the known
properties of the subsystems, is the value of the mean square of the
separation radius between their centres of mass. Assuming as before
that the L = 1 radial functions for J = L ± ½ are nearly identical the
same quantity, $\langle R^2 \rangle$, suffices also for the evaluation of the B(E2↑)
transition strength in the ground state doublet. All the necessary
expressions are given below except that for the magnetisation radius
which is quite complicated. They are derived for $^7$Li in terms of the
masses and charges of the triton and alpha particle, their mean square
charge radii $\langle r^2 \rangle_3$ and $\langle r^2 \rangle_4$, the magnetic moment of the triton $\mu_3$ and
the mean square cluster separation radius $\langle R^2 \rangle$ :–

Charge Radius    $<r^2>_7 = \frac{1}{3}<r^2>_3 + \frac{2}{3}<r^2>_4 + \frac{34}{147}<R^2>$

Quadrupole Moment    $Q = -\frac{68}{245}<R^2>$

Octupole Moment    $\Omega = -\frac{48}{245} \mu_3 <R^2>$

$B(E2:3/2^- \rightarrow 1/2^-) = \frac{1}{\pi} (\frac{17}{49})^2 <R^2>^2$

A weighted average of the experimental results on the charge radius, quadrupole moment and B(E2↑) yields a best fit $<R^2> = 13.5 \pm .3$ fm$^2$ and insertion of this into the above expressions gives the theoretical results shown in Table 4. Clearly we have a reasonably consistent picture here and the derived value of $<R^2>$ then leads to predictions of the mean square magnetisation radius $<r_m^2>_7$ and octupole moment $\Omega$ in fair accord with experiment.

Table 4.  Radii, Moments and B(E2↑) in $^7$Li

| Quantity | Calculated | Observed | (Ref) |
|---|---|---|---|
| $\sqrt{<r^2>_7}$ fm | 2.42 ± .04 | 2.41 ± .1 | (15) |
| $Q$    fm$^2$ | -3.74 ± .08 | -3.66 ± .03 | (16) |
| B(E2)e$^2$fm$^4$ | 7.0  ± .3 | 6.7  ± .2 | (17) |
|  |  | 7.4  ± .1 | (18) |
|  |  | 8.3  ± .6 | (19) |
| $\sqrt{<r_m^2>_7}$ fm | 2.78 ± .05 | 2.70 ± .15 | (20) |
|  |  | 2.98 ± .05 | (17) |
| $\Omega/\mu_7$    fm$^2$ | -2.42 ± .05 | \|2.3\| ± .5 | (20) |
|  |  | \|2.9\| ± .1 | (17) |

Knowledge of $<R^2>$ for $^7$Li and the binding energies of the ground state doublets of $^7$Li and $^7$Be now constitutes enough data to determine without ambiguity the four parameters V, $V_{so}$, $\alpha$ and $R_c$ of our assumed potential model:

$V(R) = -(V + 4\alpha V_{so}\underline{L}\cdot\underline{\sigma})\exp(-\alpha R^2) + V_c(R)$

$V_c(R) = Z_1 Z_2 e^2/R, \ Z_1 Z_2 e^2/2 R_c [3-(R/R_c)^2]; \ R \gtrless R_c$

Taking $<R^2> = 13.5$ fm$^2$ and the energies shown in Figure 7 an excellent fit is obtained with the values V = 86.08 MeV, $V_{so} = 0.957$ MeV, $\alpha = 0.163$ fm$^{-2}$ and $R_c = 3.248$ fm.

As mentioned earlier, the spin-orbit depth has to be increased (to
$V_{SO}$ = 2.14 MeV) to fit the positions of the L = 3 doublet states. With
this last adjustment the phase shifts of the $J^\pi$ = 7/2⁻ and 5/2⁻ f-wave
scattering states of ³He + ⁴He are reproduced accurately as demonstrated
in Figure 8. The RGM calculations for this system predict a strong
parity dependence in the effective potential between ³He and ⁴He and
this is confirmed in our model by considering the $J^\pi$ = $\frac{1}{2}^+$ s-wave phase
shifts. These can only be fitted by reducing the central depth to
V = 67.67 MeV and we find that our calculation then predicts an s-wave
scattering length of a = 37.1 fm and an effective range of $r_o$ = 1.1 fm
in good agreement with a direct effective range plot of the data.

I have now assembled all the necessary ingredients for a calculation of the radiative capture of ³He by ⁴He to the L = 1 states of ⁷Be.
At low energies this is basically an electric dipole process from
incident s-waves with rather small contributions from d-waves and other
multipoles. Our calculation, in common with several others, gives the
correct ratio of capture cross sections to the ground and first excited
states of ⁷Be and reproduces the shapes as a function of incident energy.
The greatest deviations among the various calculations and experiments
are in the absolute magnitudes. Rather than discuss this in detail
here, I will concentrate on the value at low energy of the astrophysical
S-factor

$$S_E = E \exp(2\pi\eta)\sigma_{cap}(E) = S_E(3/2^-) + S_E(1/2^-)$$

where $\eta$ is the usual Sommerfeld parameter. At E = 0 it is possible to
do simple exact calculations of $S_o(3/2)$ and $S_o(1/2)$. Using the s-wave
potential given above, a single integration of the zero energy wave
equation

$$[-\frac{\hbar^2}{2\mu}\frac{d^2}{dR^2} + V(R)]U_o(R) = 0,$$

followed by a matching of the wave function to a combination of limiting
forms of the Coulomb functions,

$$U_o(R) \xrightarrow{R \to \infty} R[\frac{I_1(2\sqrt{2\varepsilon R})}{\sqrt{2\varepsilon R}} - 4\varepsilon a \frac{K_1(2\sqrt{2\varepsilon R})}{\sqrt{2\varepsilon R}}],$$

yields directly the scattering length a. In this expression $I_1$ and $K_1$
are modified Bessel functions and $\varepsilon = \mu Z_1 Z_2 e^2/\hbar^2 = 0.237$ fm⁻¹. The
$S_o(j)$ are then found by computing the dipole integrals in the formula

$$S_o(j) = Const. \times (2j+1)(B_j)^3 |<U_j|R|U_o>|^2$$

where $B_j$ and $U_j$ are the binding energies and radial wave functions for
the first two states of ⁷Be. Taking into account the uncertainty in the
value of $<R^2>$ (= 13.5 ± .3 fm² for ⁷Li), which was used to determine the
potential paramters, we arrive at the following results:

Table 5.  Astrophysical S-Factors for $^3$He + $^4$He

| Quantity | | Theory | Experiment[21] |
|---|---|---|---|
| Sum | $S_0$(keV.b) | .47 $\pm$ .02 | .53 $\pm$ .03 |
| Ratio | $S_0(1/2)/S_0(3/2)$ | .433 $\pm$ .001 | .41 $\pm$ .03 |

In conclusion, I would say that so much diverse data can be usefully correlated by this elementary model, and in such a physically appealing way, that I almost begin to believe in it myself.  Other applications are in the pipeline and I think that some quite surprising candidates for the cluster treatment may emerge soon.

I am grateful to my many collaborators and especially to Richard Baldock for his unstinting help in the preparation of this paper.

## Acknowledgement

Some figures are reproduced with kindly permission from North-Holland Physics Publishing Amsterdam.

## References

1) H. Friedrich, Conf. Cont. No.102.
2) K. Aoki and H. Horiuchi, Prog. Theor. Phys. 69, 1154 (1983).
3) B. Buck and J.A. Rubio, J. Phys. G. (in press).
4) B. Buck, C.B. Dover and J.P. Vary, Phys. Rev. C11, 1803 (1975).
5) B. Buck, Lecture Notes in Physics No.33, Springer-Verlag (1975).
6) B. Buck and A.A. Pilt, Nucl. Phys. A280, 133 (1976).
7) B. Buck, H. Friedrich and A.A. Pilt, Nucl. Phys. A290, 205 (1977).
8) B. Buck, A.C. Merchant and N. Rowley, Nucl. Phys. A327, 29 (1979).
9) N. Rowley and A.C. Merchant, Phys. Lett. 97B, 341 (1980).
10) R.A. Baldock, B. Buck and J.A. Rubio, Nucl. Phys. (in press).
11) T. Kajino, T. Matsuse and A. Arima, Nucl. Phys. A413, 323 (1984); T. Kajino and A. Arima, Phys. Rev. Lett. 52, 739 (1984).
12) H. Walliser, Q.K.K. Liu, H. Kanada and Y.C. Tang, Phys. Rev. C28, 57 (1983); H. Walliser, H. Kanada and Y.C. Tang, Nucl. Phys. A419, 133 (1984).
13) F. Ajzenberg-Selove, Nucl. Phys. A413, 1 (1984).
14) B. Buck and S.M. Perez, Phys. Rev. Lett. 51, 1496 (1983).
15) C.W. de Jager, H. de Vries and C. de Vries, At. Nucl. Data 14, 479 (1974).
16) Sheldon Green, Phys. Rev. A4, 251 (1971).
17) G.J.C. van Niftrik et. al., Nucl. Phys. A174, 173 (1971).
18) A. Bamberger et. al., Nucl. Phys. A194, 193 (1972).
19) O. Häusser et. al., Nucl. Phys. A212, 613 (1973).
20) R.E. Rand, R. Frosch and M.R. Yearian, Phys. Rev. 144, 859 (1966).
21) J.L. Osborne et. al. Nucl. Phys. A419, 115 (1984).

# ON MICROSCOPICAL CALCULATIONS WITH LIGHT NUCLEI

M.V. Mihailović
J. Stefan Institute, E. Kardelj University of Ljubljana
61000 Ljubljana, Jamova 39, Yugoslavia

ABSTRACT.   Two subjects are discussed:
(1) Three cluster structures of light nuclei and the effective interaction in the cluster model. — The three cluster model for light nuclei is developed and applied to the study of low—lying states of nuclei $^7$Li and $^6$Li. It is shown that $^7$Li can be successfully described by a superposition of too—cluster configurations ($^4$He $-$ $^3$H) and three—cluster configurations ($^4$He $-$ $^2$H) ($I_d = 0,1$) $-$ n). The ground state properties of $^6$Li, and the thresholds $^4$He + $^2$H and $^4$He + p + n, are described by superposition of configurations ($^4$He $-$ p $-$ n). Properties of the effective nucleon $-$ nucleon interaction are presented.
(2) Variational calculation of reaction parameters. — It is shown first that  existing variational methods might predict false resonances; to avoid them is sometimes tedious and time consuming. A new variational method eliminates false predictions when the analyticity of the Jost function and the conservation of the norm is guaranteed with additional constraints.

## 1.   INTRODUCTION

This talk will be concentrated on two subjects:
        (1) Three cluster structures of light nuclei and the effective interaction.— The discussion will be based on results of calculations of properties of nuclei $^7$Li and $^6$Li. For the $^7$Li there are three different calculations [1,2,3], the description of $^6$Li is attempted fully microscopically in the three— cluster model [4] for the first time.
        (2) Variational calculation of reaction parameters. — The use of the variational methods, which made larger calculations feasible, is at an unsatisfactory stage: (i) it is shown that all existing formulations [5—12] may predict false resonances, (ii) it is found that violation of the principle of causality causes false predictions. A variational formulation with constraints is proposed which does not provide false resonances in elastic and inelastic scattering (with no rearrangement) of nuclei.

*J. S. Lilley and M. A. Nagarajan (eds.), Clustering Aspects of Nuclear Structure, 85–100.*

## 2. STABILITY OF CLUSTER CONFIGURATION AND EFFECTIVE INTER-ACTION IN CLUSTER MODEL

Up to relatively recently light nuclei were described usually as single cluster configurations (mostly two—cluster). Parameters of the interaction and the oscillator parameter of the well have been adjusted to obtain some properties of the ground and low—lying states, as well as some scattering parameters. Simplicity of the corresponding wave function was the main justification for such model.

Two questions were open in such an approach: (i) How stable are results of these calculations in light of the fact that the main characteristic of light nuclei is the presence of low—lying thresholds? (ii) Could one improve the description by including low—lying channels and at the same time employ a unique set of parameters defining the nucleon — nucleon interaction?

I would like to show some results related to these questions which are obtained in three calculations on $^7$Li and the results of a calculation on $^6$Li:

(1) Description of $^7$Li in the generator coordinate (GC) basis of two—cluster configurations ($^4$He $-$ $^3$H) and ($^6$Li $-$ n) (Mihailović and Poljšak, [1]),

(2) Description in the GC basis of two—cluster ($^4$He $-$ $^3$H) and three—cluster configurations ($^4$He $-$ $^2$H $-$ n) (Beck, Krivec, and Mihailović, [2]),

(3) Description in the resonating group (RG) basis of two—cluster configurations ($^4$He $-$ $^3$H) and ($^6$Li $-$ n) (Fujiwara and Tang, [3]).

### 2.1. The interplay of cluster configurations in $^7$Li

#### 2.1.1. The nucleus $^7$Li described by interplay of configurations ($^4$He $-$ $^3$H) and ($^6$Li $-$ n)

The lowest thresholds in $^7$Li are ($^4$He $+$ $^3$H) and ($^6$Li $+$ n) which suggests that both configurations ($^4$He $-$ $^3$H) and ($^6$Li $-$ n) will influence properties of the ground and low—lying states of the $^7$Li.

The basis is formed of two—centre wave functions in the generator coordinate representation. One starts from Slater determinants of rank 7

$$\phi_{c\kappa}(x,S_c) = \mathcal{A}\{\phi_{He}(x_1,...,x_4,S_{He})\,\phi_H(x_5,...,x_7,S_H)\}, \quad c = (^4He - ^3H), S_c = S_{He} - S_H, (1)$$

where $\kappa = 1,2$ means that only two wave functions corresponding to the configuration $(1s)^3$ of $^3$H and the configurations $(1s)^4$ of $^4$He are included in the basis, and

$$\phi_{c\kappa}(x,S_c) = \mathcal{A}\{\phi_{Li,\kappa_1}(x_1,...,x_6,S_{Li})\,\phi_{n\kappa_2}(x_7,S_n)\} \tag{2}$$

$$c = (^6Li - n), \kappa = (\kappa_1,\kappa_2) = (n_p,\sigma_p,n_n\sigma_n,\kappa_2) = 1,...,72, \quad S_c = S_{Li} - S_n .$$

The cluster $^6$Li is described in the space of 36 states of the shell—model configuration $(1s)^4 (1p)^2$.

The basis of GC functions is derived from the functions (1) and (2) by projecting them into the eigenspace of the angular momentum and parity

$$\phi_{c\kappa}^{JM\pi}(x,S) = P^\pi P_{MK}^J \phi_{c\kappa}(x,S) \tag{3}$$

$$= (2J+1)/(8\pi^2)\int d\Omega D_{MK}^J(\Omega)^* R(\Omega)[\phi_{c\kappa}(x,S) - (-)^{\pi'}\phi_{c\kappa}(x,-S)],$$

$\pi' = \pi + A_1 + A_2$, where $A_i$ $(i = 1,2)$ is the number of nucleons in the p–shell of the i–th cluster. $S = S_A - S_B$.

The model Hamiltonian used in the calculation is

$$H = (\frac{\hbar}{2m} \sum_i v_i^2 - T_{c.m.}) + \sum_{i<j} V_{ij}^C + \sum_{i<j} V_{ij}^N \tag{4}$$

The first term is the kinetic energy operator, $V_{ij}^C$ is the Coulomb and $V_{ij}^N$ is the effective nucleon–nucleon potential. For the latter the central potential with the radial parts being two Gaussian functions same as those of Volkov [13] has been using.

Results:

(1) *Single two–cluster structure:*

(i) The structure $(^4He - ^3H)$. — The energy of the ground state $J = 1/2^-, 3/2^-$ is $-35.9$ MeV for the potential V2 and $\beta_{min} = 0.50$ fm$^{-2}$. The minimum lies at $S_c$ $R(^4He) - R(^3H)$. The position of the minima $(-29.3$ MeV) of the excited state $J = 5/2^-, 7/2^-$ is at a separation $S_c$ which is smaller than that for the ground state, which suggests a smaller deformation of the excited states in comparison with the ground state. The same deficiency is obtained in PHP calculation [14].

When the Hill–Wheeler equation is solved for the ground state the binding was $-37.54$ MeV, or $1.64$ MeV below the minimum of the single configuration.

(ii) The structure $(^6Li - n)$. — The most symmetric configurations $(n_p, \sigma_p, n_n \sigma_n, S_n) = (z+z+-)$ and $(z+z-+)$ are most bounded, the minima are shallow and correspond to smaller values ($\leqslant 2.5$ fm) of the separation $S_c$, and to the oscillator parameter $\sim 0.40$ fm$^{-2}$ $< (\beta_{min} (^4He - ^3H))$. By solving the eigenvalue problem for the energy for each $S_c$ separately in the basis

$$\left\{ \phi_{c\kappa}^{JM\pi} (x, S_c) |\kappa = (x+x+-), (x+x-+), (z+z+-), (z+z-+) \right\} \tag{5}$$

one gets for the best binding the value which is at least about 6 MeV above the most bound configuration $(^4He - ^3H)$.

(2) *Superposition of different two–cluster structures:*

The energy surfaces as functions of the separations $S_{(^4He-^3H)}$ and $S_{(^6Li-n)}$ calculated by solving the eigenvalue problem for the energy in the basis

$$\left\{ \phi_{(^4He-^3H)}^{JM} (x, S_{(^4He-^3H)}), \phi_{(^6Li-n)}^{JM} (x, S_{(^6Li-n)}) \right\} \tag{6}$$

is depicted in Fig. 1. For $\kappa = (z+z+-), (z+z-+)$ $E_{gs}$ is 1.12 MeV lower than the minimum energy in the case of the single structure $(^4He-^3H)$. More interesting is the fact that coupling of the configurations $(^4He - ^3H)$ and $\sum (^6Li-n)\kappa$ makes equal the positions of the minima in the ground state $(1/2^-, {}^\kappa 3/2^-)$ and the excited states $(5/2^-, 7/2^-)$ in contrast to the result of the single two– cluster configuration.

The diagonalisation of the Hamiltonian in the GC two–parametric basis:

$$\left\{ \phi_{(^4He-^3H)}^{JM\pi} (x, S_{(^4He-^3H)}), \phi_{(^6Li-n,)}^{JM} (x, S_{(^6Li-n)}), \kappa = (z+z+-), (z+z-+), \right.$$

$$S_{(^4He-^3H)} = 1.44, 2.2, 3.4, 4.5, 5.5, 6.5 \text{ fm}, \tag{7}$$

$$\left. S_{(^6Li-n)} = 0.2, 1.5, 2.5, 3.5, 4.5 \text{ fm} \right\}$$

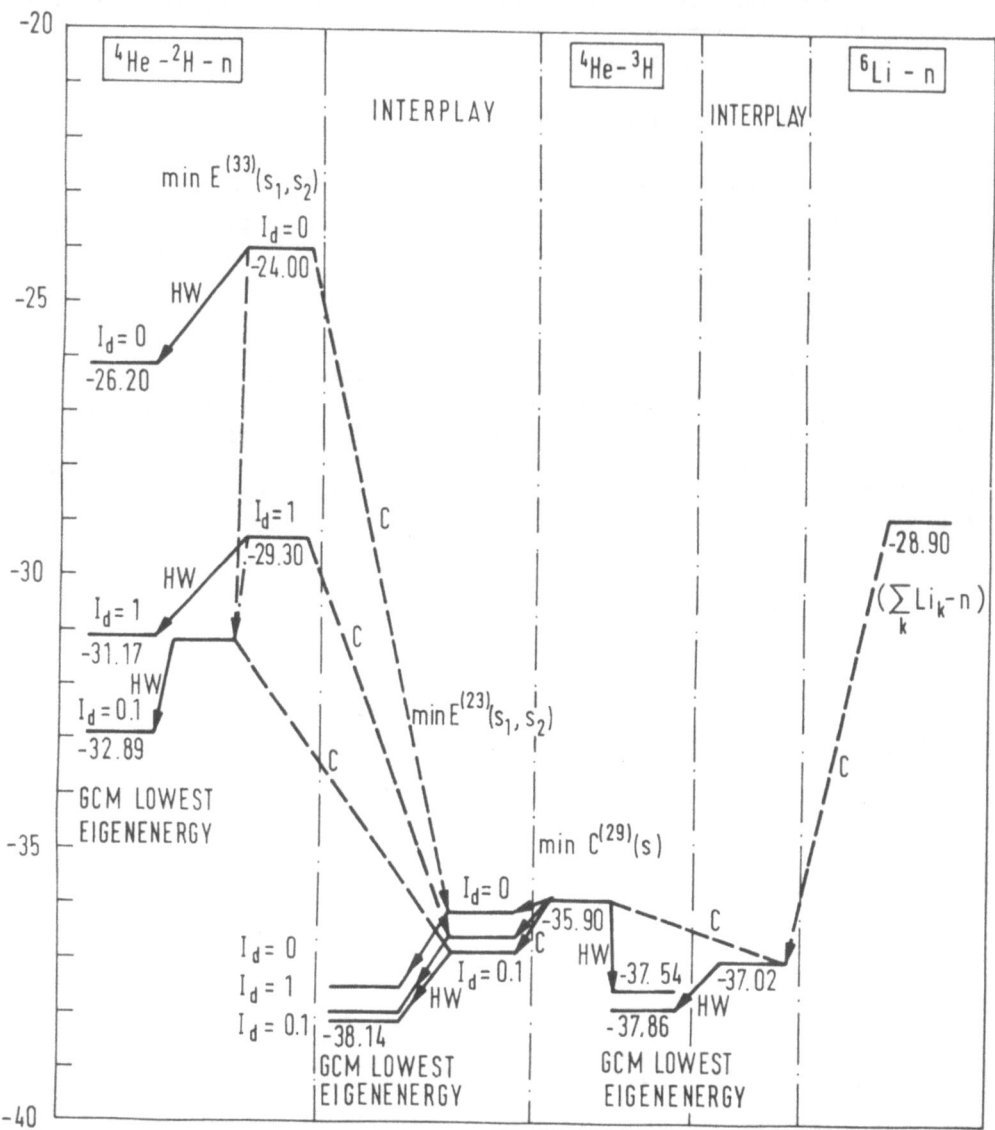

Figure 1. A diagram presenting the influence of the two mechanisms, the molecule-like vibrational degree of freedom and the interplay of different cluster structures, on the ground state of $^7$Li is shown. The symbol HW means that the solution is obtained by solving the corresponding Hill–Wheeler equation, and the coupling of different cluster configurations and/or structures in indicated by the symbol C. The symbol $I_d = 0$, I means that the state corresponds to the mixture of both three-cluster configurations ($^4$He $- ^2$H($I_d$) $-$ n).

The energy of the ground state $E_{gs}$ $(J = 1/2^{-}, 3/2^{-}) = -37.86$ MeV, when V2 with $M^C = 0.6$ is used. This energy is only $\sim 2.0$ MeV less, than the experimental value, the difference can be rather easily repaired by introducing the LS potential and by adjusting the spin–isospin parameter $M^C$. However, it is questionable what is the effect of including the tensor interaction.

Finally, let us note that the superposition of different cluster structure improves significantly the low–lying spectrum of the $^7$Li.

### 2.1.2. The nucleus $^7$Li described by interplay of two– and three–cluster configurations [2]

The more general model for $^7$Li is that which includes the coupling of the two–cluster configuration $(^4\text{He} - ^3\text{H})$ and three– cluster configurations $(^4\text{He} - ^2\text{H}_{I_H} - n_{I_n})$, $I_H = 0,1$ and $I_n = \frac{1}{2}$ .

The wave function of the generator coordinate type for two–cluster configuration is given earlier, eqs. (1) and (3). The three– cluster wave function (clusters A,B and C) is constructed from the linear combination of determinants

$$\phi^{(3)}_{I_i; I\nu}(x, s_1, s_2) = \left\{ \mathcal{A} [\phi_{A\alpha}(x_A, s_A)\,\phi_{B\beta}(x_B, s_B)\,\phi_{C\gamma}(x_C, s_C)] \right\}_{I\nu} \tag{8}$$

where $\quad s_1 = s_A - s_B, \quad s_2 = s_C - (A\,s_A + B\,s_B)/(A + B)$

$$A\,s_A + B\,s_B + C\,s_C = 0, \quad x = (x_A, x_B, x_C), \quad x_A = (x_1, ..., x_A) ,$$

$$x_B = (x_{A+1}, ..., x_{A+B}), \quad x_C = (x_{A+B+1}, ..., x_{A+B+C}) ;$$

the symbols $\alpha, \beta, \gamma$ denote the spin configurations of nucleons in the cluster A, B and C, respectively, the symbols $I_i$ stands for all intermediate spin quantum numbers from coupling spins of particles A, B and C to the total spin $I\nu$.

The wave function of the GC type is constructed in three steps by superposing first the functions (8)

$$\psi^{(3)}(x) = \sum_{I_i, I\nu} \int ds_1 \int ds_2 \, f^{(3)}_{I_i; I\nu}(s_1, s_2)\, \phi^{(3)}_{I_i; I\nu}(x, s_1, s_2) ;$$

then expressing $f^{(3)}$ and $\phi^{(3)}$ in terms of bipolar harmonics

$$B_{L_1 L_2 LM}(s_1, s_2) = \sum_{M_1 M_2} (L_1 M_1, L_2 M_2 \mid LM)\, Y_{L_1 M_1}(\hat{s}_1)\, Y_{L_2 M_2}(\hat{s}_2):$$

$$f^{(3)}(s_1, s_2) = \sum_{L_1 L_2 LM} \frac{1}{s_1 s_2} f^{(3)}(L_1 L_2 LM, s_1 s_2)\, B_{L_1 L_2 LM}(\hat{s}_1, \hat{s}_2) ,$$

$$\Phi^{(3)}(x, s_1, s_2) = \sum_{L_1 L_2 LM} \frac{1}{s_1 s_2} \Phi^{(3)}(x, L_1 L_2 LM, s_1 s_2)\, B^*_{L_1 L_2 LM}(\hat{s}_1, \hat{s}_2) ,$$

$$\psi^{(3)}(x) = \sum_{I_i I\nu L_1 L_2 LM} \int_0^\infty ds_1 \int_0^\infty ds_2 \, f^{(3)}_{I_i I}(L_1 L_2 LM, s_1 s_2)\, \phi^{(3)}_{I_i I\nu}(x, L_1 L_2 LM, s_1 s_2) \tag{9}$$

Finally, the basis of the GC functions is obtained by projecting (9) into the eigenspase of the angular momentum and parity operators:

$$\sum_{\kappa \kappa'} P^\pi P^J_{\kappa'' \kappa} \psi^{(3)}(x) = \sum_{I_i I\nu L_1 L_2 LMK'} (I\nu, LM \mid JK') \int_0^\infty ds_1 \int_0^\infty ds_2\, f^{(3)J\pi}_{I_i I}(L_1 L_2 L, s_1 s_2)$$

$$\times \Phi^{(3)}_{I_i I\nu}(x, L_1 L_2 LM, s_1 s_2) . \tag{10}$$

Finally, the basis of the GC functions is obtained by projecting (9) into the eigen-space of the angular momentum and parity operators:

$$\sum_{KK'} P^{\pi} P^{J}_{K'K} \psi^{(3)}(x) = \sum_{l_i l \nu L_1 L_2 LM K'} (l\nu, LM | JK') \int_0^{\infty} ds_1 \int_0^{\infty} ds_2 \, f^{(3)J\pi}_{l_i l}(L_1 L_2 L, s_1 s_2)$$

$$\times \Phi^{(3)}_{l_i l \nu}(x, L_1 L_2 LM, s_1 s_2), \tag{10}$$

with

$$f^{(3)J\pi}_{l_i l}(L_1 L_2 L, s_1 s_2) = \sum_{\nu M K} f^{(3)}_{l_i l \nu}(L_1 L_2 LM, s_1 s_2)(l\nu, LM | JK) \frac{1}{2}[1 + (-)^{(1-\pi)/2 - L_1 - L_2}] \tag{11}$$

The trial function of the form

$$\psi^{JM\pi} = P^{JM\pi}[\psi^{(2)}(x) + \sum \psi^{(3)}(x)]$$

leads to coupled Hill–Wheeler equations

$$\sum_{l_i i} \int_0^{\infty} ds' f^{(2)J\pi}_{l_i l}(L, s') < \Phi^{(2)}_{l_i l}(L, s) | \bar{H} | \Phi^{(2)}_{l_i l}(L, s') >$$

$$+ \sum_{l_i L'_1 L'_2} \int_0^{\infty} ds_1 \int_0^{\infty} ds_2 \, f^{(3)J\pi}_{l_i l}(L'_1 L'_2 L, s'_1 s'_2) < \Phi^{(2)}_{l_i l}(L, s) | \bar{H} | \Phi^{(3)}_{l_i l} L'_1 L'_2 L, s'_1 s'_2) > = 0,$$

$$\sum_{l_i i} \int_0^{\infty} ds' f^{(2)J\pi}_{l_i l}(L, s') < \Phi^{(3)}_{l_i l}(L_1 L_2 L, s_1 s_2) | \bar{H} | \Phi^{(2)}_{l_i l}(L, s') > \tag{12}$$

$$+ \sum_{l_i L'_1 L'_2} \int_0^{\infty} ds'_1 \int_0^{\infty} ds'_2 \, f^{(3)J\pi}_{l_i l}(L'_1 L'_2 L, s'_1 s'_2) < \Phi^{(3)}_{l_i l}(L_1 L_2 L, s_1 s_2) | \bar{H} | \Phi^{(3)}_{l_i l}(L'_1 L'_2 L, s'_1 s_2) > = 0,$$

which has to be satisfied for all values of $l_i, L_1, L_2, s, s_1, s_2$. Here $\bar{H} = H - EN$, where H is the hamiltonian operator and N is the identity operator.

The equation (12) is solved approximately: (i) the integrations are replaced by finite sums, (ii) only a few pairs $(L_1, L_2)$ are chosen for a given set $(J, \pi, l)$. So the secular equation is of the form

$$\begin{bmatrix} \bar{H}^{(22)} & \bar{H}^{(23)}_0 & \bar{H}^{(23)}_1 \\ \bar{H}^{(32)} & \bar{H}^{(33)}_{00} & \bar{H}^{(33)}_{01} \\ \bar{H}^{(32)}_1 & \bar{H}^{(33)}_{10} & \bar{H}^{(33)}_{11} \end{bmatrix} \begin{bmatrix} f^{(2)} \\ f^{(3)}(l'_d = 0) \\ f^{(3)}(l'_d = 1) \end{bmatrix} = 0, \tag{13}$$

where the following notation is used; the amplitudes f are vectors

$$f^{(2)} = \begin{bmatrix} f^{(2)}(s_1) \\ \vdots \\ f^{(2)}(s_{N(2)}) \end{bmatrix}, \quad f^{(3)}(l_d) = \begin{bmatrix} f^{(3)}_{l_d}(s_{11}, s_{21}) \\ \vdots \\ f^{(3)}_{l_d}(s_{1N(3)}, s_{2N(3)}) \end{bmatrix}, \quad l_d = 1, 0,$$

and the quantities $\bar{H}^{(22)}, \bar{H}^{(23)}_{l_d}$ and $\bar{H}^{(33)}_{l_d l'_d}$ are matrices with matrix elements

$$< \Phi^{(2)}_i(L, s_i) | \bar{H} | \Phi^{(2)}_{i'}(L, s_{i'}) >, \quad (i, i' = 1, ..., N^{(2)}),$$

$$< \Phi^{(2)}_i(L, s_i) | \bar{H} | \Phi^{(3)}_{l_d i'}(01 L, s_{1i'_1}, s_{2i'_2}) >, \quad (i = 1, ... N^{(2)}, i'_1, i'_2 = 1, ... N^{(3)}), \; l'_d = 1, 0,$$

$$< \Phi^{(3)}_{l_d i}(01 L, s_{1i_1}, s_{2i_2}) | \bar{H} | \Phi^{(3)}_{l'_d i'}(01 L, s_{1i'_1}, s_{2i'_2}) >, \quad (i_1, i_2, i'_1, i'_2 = 1, ... N^{(3)}), l_d, l'_d = 1, 0,$$

respectively.

Results:

Single three– cluster configuration $(^4\mathrm{He} - {}^2\mathrm{H}\,(I_d = 1) - n)_{3/2^-,\,1/2^-}$    has the minimum at $-29.3$ MeV which is below both energies that corresponds to any single cluster configuration $(^6\mathrm{Li} - n)(-28.0$ MeV) as well as to that corresponding to the cluster configuration $(^6\mathrm{Li} - n)$ in which the fragment $^6\mathrm{Li}$ is described by a shell model function composed of $(1s)^4 (1p)^2$ determinants. The solution of the Hill–Wheeler equation $H^{(33)}_{11} f^{(3)}_1 = 0$ in a basis of 25 functions provides $-31.17$ MeV for the ground state and $-18.08$ MeV for the first excited state.

Three–cluster configuration $(^4\mathrm{He} - {}^2\mathrm{H}(I_d = 0) - n)_{3/2^-,\,1/2^-}$    provides less binding $-26.23$ MeV and $-14.75$ MeV for the lowest and the first excited states, respectively.

Coupling two three–cluster configurations $(^4\mathrm{He} - {}^2\mathrm{H} - n)$,

$$
\begin{bmatrix}
\overline{H}^{(33)}_{00} & \overline{H}^{(33)}_{01} \\
\overline{H}^{(33)}_{10} & \overline{H}^{(33)}_{11}
\end{bmatrix}
\begin{bmatrix}
f^{(3)}_0 \\
f^{(3)}_1
\end{bmatrix}
= 0
$$

in a basis of 25 vectors, gives $-32.89$ MeV and $-22.70$ MeV for the ground and the first excited state.

Coupling all three configurations by solving eq. (13) in the GC basis

$$
\left\{ \Phi^{JM\pi}_{(4\mathrm{He}-3\mathrm{H})}(x,s),\ \Phi^{JM\pi}_{(4\mathrm{He}-2\mathrm{H})(I_d = 1)-n}(x,s_1,s_2),\ \Phi^{JM\pi}_{(4\mathrm{He}-2\mathrm{H}(I_d=0)-n)}(x,s_1,s_2) \right.
$$

$$
s = 0.5,\ 1.5,\ 2.5,\ 3.5,\ 4.5,\ 5.5,\ 6.5 \text{ fm},
$$
$$
s_1 = 1.0,\ 2.0,\ 3.0,\ 4.0,\ 5.0 \text{ fm}, \quad s_2 = 0.5,\ 1.5,\ 2.5,\ 3.5,\ 4.5 \text{ fm} \Big\}.
$$

one gets the energy of the ground state $E^{3/2^-,1/2^-}_{\text{g.s.}} = -38.14$ MeV and of excited state $E^{3/2^-,1/2^-}_{\text{exc.}} = -29.67$ MeV.

Results of calculations described in Sec. 2.1.1. and 2.1.2. are summarized in Fig. 1.

2.1.3. Three channels resonating group calculation for $^7\mathrm{Li}$. –

Fujiwara and Tang [3] have performed recently an extensive calculation in the basis

$$
(^4\mathrm{He} - {}^3\mathrm{H}),\ (^6\mathrm{Li}((1s)^4 (1p)^2,\ I((1p)^2) = 1) - n),
$$
$$
(^6\mathrm{Li}((1s)^4 (1p)^2,\ I((1p)^2) = 0) - n)
$$

which is similar to that used sec. 2.1.1. The nucleon – nucleon interaction employed is purely central, but different than the Volkov potential. It consists of three Gaussian functions. The parameters of these were chosen to yield a satisfactory description of some properties of $^2\mathrm{H}$, $^3\mathrm{H}$ and $^4\mathrm{He}$.

Their results contain several detailed studies on bound states and scattering. In the study of the bound states them found a substantial improvement when other configurations are added to the basic $(^4\mathrm{H} - {}^3\mathrm{H})$. The ground state (L = 1) is influenced mostly by $(^6\mathrm{Li}(I(1p^2) = 1) - n)$ and its binding is increased for 0.68 MeV. The configuration with $I(1p^2) = 0$ influences L = 3 state. Except for insignificant differences this calculation provided for bound states and thresholds equal conclusions as that in sec. 2.1.1.

Additional features in the elastic scattering appeared as the consequences of the coupling other channels $^6$Li + n and $^6$Li + n to $^4$He + $^3$H.

Conclusion coupling low lying configurations is essential        when one has to obtain a quantitative description of the ground and lowest lying states. In Fig. 1 we summarized this effects on the ground state of $^7$Li. Adding lowest cluster configurations of $^7$Li to the dominant ($^4$He $- ^3$H) makes the ground state energy for 0.60 $-$ 0.68 MeV [2] lower than the minimum energy obtainable by the two–cluster structure ($^4$He $- ^3$H) alone. The effect of admixing higher configurations to the low lying spectrum is significant: it eliminates redundant states calculated by uncoupled cluster configurations, it improves position of the first excited state and makes the shape of $^7$Li in that state more  consistent. In the elastic scattering $^4$He + $^3$H, Fujiwara and Tang [3]   found additional features when other channels are coupled to the dominant one.

As the nucleus $^7$Li is the "best" example of the two–cluster structure, expects that in other nuclei the effect of coupling channels will be even more significant.

## 2.3. A three–cluster description of the nucleus $^6$Li

The nucleus $^6$Li is the only nucleus in which one can include all channels by using the basis of three–cluster configurations of different relative position and relative momenta of the clusters $^4$He, p and n. This offers a hope that we can learn more about the effective interaction in the cluster model.

The nucleus $^6$Li is described by a superposition of three–cluster configurations ($^4$He $-$ p $-$ n)$-$ the wave function being written in the GC coordinate representation, the separated clusters described by Slater determinants of harmonic oscillator single particle functions. The first generator coordinate $s_1$ is the distance between the proton and the neutron wells and the second $s_2$ connects the alpha particle well with the midpoint of $s_1$ (Jacobi choice). All harmonic oscillator wells are of the same size ($\beta = m\omega/\hbar$).

As usually we start with a Slater determinant representing nucleons distributed in the lowest of the three wells and coupling the proton and the neutron spins $\sigma_1$ and $\sigma_2$ to the total spin I,$\nu$ :

$$\phi_{I\nu}(x,s_1,s_2) = [\not{A}\{\phi_{4He}(x_{4He},s_{4He})\,\phi(\sigma_1,x_p,s_p)\,\phi(\sigma_2,x_n,s_n)\}]_{I\nu} ,$$

$x = (x_1,x_2,\ldots x_6)$. Projection of this function to the basis of eigenstates of the total angular momentum J,M and parity $\pi$ is performed by applying the operator

$$P^{JM} = \sum_K (2J+1)/(8\pi^2) \int d\Omega\, D^J_{MK}(\Omega)\,R(\Omega) ,$$

so that the basis in which the Harmonian is diagonalized is spanned by functions

$$\psi^{JM\pi}(x) = \sum_{KI\nu} \int ds_1 \int ds_2\, f_{I\nu K}(s_1,s_2)\, P^J_{MK}\, P^\pi\, \phi_{I\nu}(x,s_1,s_2)$$

$$= (2+1) \sum_{KI\nu} \int ds_1 \int ds_2\, f_{I\nu K}(s_1,s_2) \sum_{L_1L_2\ell\mathcal{L}\mathcal{J}\ell\nu'\mu'} \frac{1}{2}(1 + \pi(-)^{L_1+L_2}) \tag{14}$$

$$\phi_{I\nu}(x,L_1L_2\ell\mathcal{L}\mathcal{J}\ell,s_1s_2)(-)^{M+K}\begin{pmatrix} J & I & \mathcal{L} \\ -M & \nu' & \mu \end{pmatrix}\begin{pmatrix} J & I & \mathcal{L} \\ -K & \nu & \mu' \end{pmatrix} B^*(L_1L_2\ell\mathcal{J}\ell\mu's_1s_2)$$

In the last step we expand the function $\phi_{I\nu}(x,s_1,s_2)$ in terms of bispherical

## TABLE 1.

Parameters of interactions used in the study of effective interaction in $^6$Li. The properties of the 2H obtained with these interaction are also listed. The upper values in the columns $E_{gs}$ and Q correspond to BETA = 0.4489 fm, the lower to BETA = 0.53 fm, respectively.

| | gama1 | gama2 | gama3 | V01 | V02 | V03 | $E_{gs}$ (2H) | Q (2H) |
|---|---|---|---|---|---|---|---|---|
| $V_2^c$ (Volkov,1965) | 0.617 | 1.960 | | −60.653 | 61.140 | | −0.23<br>−0.13 | |
| $V^{t1}$ (Hasegava and Nagata,1971) | 0.510 | 4.000 | | − 8.000 | −110.000 | | | |
| $V_2^c + V^{t1}$ | | | | | | | −0.61<br>−0.56 | 0.166<br>0.196 |
| $V^{t2}$ | 0.292 | 4.000 | | − 8.000 | −110.000 | | −2.19 | 0.483 |
| $V_2^c + V^{t2}$ | | | | | | | −2.17 | 0.495 |
| $V^{t3}$ | 0.380 | 4.000 | | − 8.000 | −110.000 | | −2.05 | 0.379 |
| $V_2^c + V^{t3}$ | | | | | | | −2.06 | 0.404 |
| $V_{hn}^c$ (Hasegava and Nagata, 1971) | 0.320 | 2.854 | 6.80 | − 6.00 | −546.00 | 1655. | −1.60 | |
| $V_{hn}^c + V^{t1}$ | | | | | | | −2.03<br>−2.03 | 0.131<br>0.155 |
| $V^{t4}$ (Eikenmaier and Hackenbroick, 1971) | 0.406 | 2.094 | | − 1.18 | −100.94 | | | |
| $V_2^c\ V^{t4}$ | | | | | | | −0.41<br>−0.38 | 0.072<br>0.119 |

## TABLE 2.

Properties of the $^6$Li calculated with three effective interactions in the Table 1.
and for two values of oscillator parameters $\beta$.

| | $M^c$ | $\beta$ [fm$^{-2}$] | $V_2^c$ | $V_2^c+V^{t1}$ | $V_2^c+V^{t2}$ | $V_2^c+V^{t3}$ |
|---|---|---|---|---|---|---|
| $E_{g.s.}$ | | | −28.40 | −28.82 | −30.55 | −30.44 |
| $E(\alpha+d)$ | | | | 0.16 | 0.31 | 0.34 |
| $E(\alpha+p+n)$ | 0.6 | 0.4489 | | 0.77 | 2.50 | 2.39 |
| $E(\alpha)$ | | | −28.05 | −28.05 | −28.05 | −28.05 |
| $E(d)$ | | | | − 0.61 | − 2.19 | − 2.05 |
| $Q(d)$ | | | | 0.166 | 0.483 | 0.379 |
| $E_{g.s.}$ | | | −31.04 | −31.55 | −33.24 | −33.11 |
| $E(\alpha+d)$ | | | | 2.89 | 3.00 | 3.01 |
| $E(\alpha+p+n)$ | 0.5 | 0.4489 | | 3.50 | 5.19 | 5.06 |
| $E(\alpha)$ | | | −28.05 | −28.05 | −28.05 | −28.05 |
| $E(d)$ | | | | − 0.61 | − 2.19 | − 2.05 |
| $Q(d)$ | | | | 0.166 | 0.483 | 0.379 |
| $E_{g.s.}$ | | | −28.93 | −29.37 | −30.98 | −30.95 |
| $E(\alpha+d)$ | | | | 0.01 | 0.01 | 0.09 |
| $E(\alpha+p+n)$ | 0.6 | 0.53 | | 0.57 | 2.18 | 2.15 |
| $E(\alpha)$ | | | −28.80 | −28.80 | −28.80 | −28.80 |
| $E(d)$ | | | | − 0.56 | − 2.17 | − 2.06 |
| $Q(d)$ | | | | 0.196 | 0.494 | 0.404 |
| $E_{g.s.}$ | | | −31.90 | −32.34 | −33.95 | −33.90 |
| $E(\alpha+d)$ | | | | 2.98 | 2.98 | 3.05 |
| $E(\alpha+p+n)$ | 0.5 | 0.53 | | 3.54 | 5.15 | 5.10 |
| $E(\alpha)$ | | | −28.80 | −28.80 | −28.80 | −28.80 |
| $E(d)$ | | | | − 0.56 | − 2.17 | − 2.05 |
| $Q(d)$ | | | | 0.196 | 0.494 | 0.404 |

harmonics $B(L_1 L_2 \ell \mu \hat{s}_1, \hat{s}_2)$ with coefficients $\phi_{l\nu}(x, L_1 L_2 \ell \mathcal{M} s_1, s_2)$, where the orbital momenta $L_1$ and $L_2$ for the relative rotation corresponding to $s_1$ and $s_2$, respectively, couple to the total orbital angular momentum $\mathcal{L}$.

The sums over intercluster angular momenta $L_1$ and $L_2$ are approximated by including only their lowest values relevant for states under consideration. The list of lowest components contributing to the ground state of $^6$Li is $(L_1, L_2, \mathcal{L}, S)$ =

$$(0,0,0,1), \quad (2,0,2,1), \quad (2,0,2,1), \quad (1,1,1,0).$$

At least two terms $L_1 = 0,2$ were necessary in order to include the effect of the tensor nucleon — nucleon interaction. Limiting to $L_2 = 0$ only offers a reasonable description of the ground state energy and both thresholds $\alpha + d$ and $\alpha + p + n$, $L_2 = 0,2$ are necessary for obtaining in addition the quadrupole moment of the $^6$Li. However, the component $(2,0,2,1)$ influences the ground state energy negligable. (See also ref. 16). In this discussion we concentrated to energies and thresholds, so we include only the components $(0,0,0,1)$ and $(2, 0, 2, 1)$.

The microscopic hamiltonian is the same as (4) except that the nucleon — nucleon interaction $V_{ij}^N$ contains the central part $V_{ij}^{NC}$ and the tensor part $V_{ij}^{NT}$, for both we assume to be expressible by a set of appropriate gaussian functions. For the $V_{ij}^{NC}$ again we have chosen the effective potential V2 of Volkov. The strength of the central part is fixed by the binding energy of the $^4$He. The spin—isospin parameter $M^C$ does not influence the energies of $^2$H and $^4$He, this freedom is used later to fix the interfragment energy.

The tensor part $V_{ij}^{NT}$ we used is of the form

$$V_{ij}^{NT} = \Sigma(W_k^T + M_k^T P^M) V_{ok}^T r_{ij}^2 \exp(-\gamma_k^T / 2 \, r_{ij}^2)(3(\sigma_i(r_i - r_j)\sigma_j(r_i - r_j))/r_{ij}^2 - (\sigma_i \cdot \sigma_j)),$$

their parameters are chosen in order to obtain properties of $^2$H. A few choices of the parameters are shown in Table 1. Both choices of Hasegawa and Nagata [16], and of Eikermeier and Hackenbroich [17] produce too small a quadrupole moment. The Volkov central interaction combined with Hasegawa and Nagata tensor parts also do not reproduce satisfactorily the properties of $^2$H. Other combinations we explored contain different long range components.

Properties of $^6$Li are obtained by solving the Hill—Wheeler equation in a space spanned by 25 GC functions (14). The following numerical values obtained with the interactions $V2 + V^{T1}$, $V2 + V^{T2}$, $V2 + V^{T3}$ illustrate the typical general properties of the proposed interactions. Results are shown in Table 2. Interactions which are chosen differ mostly in the range parameters of the tensor interaction.

The dependance of the binding energy on the spin — isospin parameter $M^C$ is strong (about 9 percent, $\sim 2.97$ MeV, for the change of $M^C$ from 0.5 to 0.6) and practically independent on other parameters of the interaction. As the energies $^2$H and $^4$He do not depend on $M^C$, this strong dependance is very convenient as it can be used to fix the interfragment energy.

The dependance of results on the oscillator parameter is rather weak (about 2 percent, 0.7 MeV for a change from 0.4489 fm$^{-2}$ to 0.53 fm$^{-2}$). The binding energy has its maximum at 0.53 fm$^{-2}$. The nucleus $^4$He has the maximum binding at practically equal value of $\beta$.

So let us concentrate to results corresponding to $\beta = 0.53$ fm$^{-2}$ where one can reach energies and thresholds close to experimental values. This is shown in this table:

$ds'_2$

|  | | | $E(^6Li)$ | $E(^4He + {}^2H)$ | $E(^4He+p+n)$ |
|---|---|---|---|---|---|
| Interaction | $V2 + V^{T1}$, | $M^C = 0.495$ | 32 50 | 3.14 | 3.70 |
|  | $V2 + V^{T2}$, | $M^C = 0.550$ | 32 50 | 1.53 | 3.70 |
|  | $V2 + V^{T3}$, | $M^C = 0.550$ | 32.50 | 1.64 | 3.70 |

(The Coulomb interaction is not included in this calculation, therefore the binding of $^4He$ and $^6Li$ are accordingly increased).

Let me summarize our experience with effective interaction in calculations with $^7Li$ and $^6Li$. In both cases we have used the same form of the central part, the effective potential of Volkov V2, the tensor part, which was necessary in calculation of $^6Li$, was also sum of two gaussian terms. It came out that parameters determining energies of $^4He$, $^2H$ and $^6Li$ are almost independent. The oscillator parameter $\beta$ and the strength parameters of the central interaction $V_{ok}^{NS}$, $k = 1,2$, are determined from the binding energy of $^4He$, the parameters of the tensor interaction describe well the nucleus $^2H$, the parameter $M^C$ of the central interaction is left for regulating the energy between clusters.

## 3. MICROSCOPICAL DESCRIPTIONS OF COLLISION OF NUCLEI BASED ON VARIATIONAL METHODS

As there aren't many bound states in light nuclei, the calculation of reaction parameters is of major interest in studying properties of light nuclei. There are two realisations of the microscopic theory of reaction, both based on the variational principle, the resonating group (RG) method and the generator coordinate (GC) method for collision which differ in methods of calculations and approximation related to the calculation. Both methods proved manageable for light systems and already provided many useful results.

There is, however a disturbing feature common to all variational methods: none of the existing methods is immune against false predictions (false resonances).

### 3.1. Variational methods without constraints

The majority of variational calculations are based either on proper functionals being quadratic functions of the trial function(s) a d its (their) derivatives [5,6,7]

$$F[\psi] = \int \mathcal{L}(x, \psi, \psi') \, dx$$

(all formulae in this chapter are written for s—wave for the sake of simplicity), or on improper functionals containing in addition the second derivatives [8,9,10,11,12]. For the first type of functionals there exists the general theory [5], also there are two realisations both with a pair of trial functions, one by Morse and Feshbach [6] and another by Tamm [7]. Both realisations, however, lead to false resonances, the number of which is proportional to the number of square integrable functions introduced to describe the compound system.

All existing proposals of the second type can be derived from the functional

$$F = - <\psi, (H-E)\psi> + k.K + k.tg$$

with the domain

$$\mathcal{D}\{\psi \mid \psi = \varphi(x) + s(x,\vartheta) - K \cdot c(x,\vartheta); \begin{bmatrix} s(x,\vartheta) \\ c(x,\vartheta) \end{bmatrix} = \begin{bmatrix} \cos\vartheta & \sin\vartheta \\ -\sin\vartheta & \cos\vartheta \end{bmatrix} \begin{bmatrix} s(x) \\ c(x) \end{bmatrix}$$

$s(x) \rightarrow \sin kx, \ c(x) \rightarrow \cos kx, \ x \rightarrow \infty : \varphi(x)$ is a square integrable function $\}$,

E is the scattering energy, $k\hbar = \sqrt{2mE}$.

The $\vartheta$ is a nonvariational parameter whose value can be either fixed at the beginning or chosen at the end to avoid false predictions. This projection after variation can reduce the number of false resonances, but it is proven [18] that it is impossible to avoid completely false predictions in none of unconstrained methods.

It is possible, however, to combine two of the proposed methods to check whether calculated parameters are false [19]. For example, the Kohn variational method, which corresponds to the choice tg $\vartheta = 0$, and the inverse Kohn method, which corresponds to tg $\vartheta = \infty$, must have false resonances at different places. If the calculations with both methods provide results satisfactorily close to each other, these results are acceptable. If the results differ, one has to change the basis of the square integrable functions used to express $\varphi(x)$ and repeat the calculations until a satisfactory agreement is reached. This procedure can be tedious and time consuming.

### 3.2. A variational method for scattering with constraints

As correlations induced by the Schroedinger equation only do not restrict properly variational parameters to prevent the occurrence of false predictions, an attempt is made [18] to introduce additional correlations through physical constraints. Two relations are employed with the aim to guarantee the principle of causality:

(i) It is required that the solution for free dynamics $\psi_0(E)$ and the solution for interacting dynamics $\psi(E)$ satisfy the equality

$$< \psi_0(E), \psi_0(E')> \ = \ < \psi(E), \psi(E')> .$$

In order to illustrate the content of this condition let me write it for a well behaved potential, for l = 0 wave

$$\psi_0(x,k) = \sin kx / \cos \eta(k)$$

$\psi(x,k) = \sin(kx + \eta(k))/\cos \eta(k) + \overline{\varphi}(x,k):$

$$\frac{\text{tg } \eta(k) + \text{tg } \eta(k')}{2(k + k')} + \frac{\text{tg } \eta(k) - \text{tg } \eta(k')}{2(k - k')} + <\overline{s}(k), \varphi(k')> + <\overline{\varphi}(k), s(k')> \tag{15}$$
$$+ \text{tg } \eta(k) <\overline{c}(k), \overline{\varphi}(k')> + \text{tg } \eta(k') <\varphi(k), \overline{c}(k')> = 0$$

and for k = k'

$$d \text{ tg } \eta(k)/dk = 2<\varphi(k), \varphi(k)> + 4<\overline{s}(k), \overline{\varphi}(k)> + \text{tg } \eta(k)[1/k + 4<\overline{\varphi}(k), \overline{c}(k)>] . \tag{16}$$

In deriving the last two relations, scattering parameters were treated as distributions. To prove their validity it was necessary to use the Levinson theorem.

The implication of eqs. (15) and (16) is that variational parameters for different energies are not independent, contrary to the treatment in methods without constraints.

(ii) The analyticity of variational parameters by requiring that the variational Jost function satisfy the dispersion relation for the exact solution. Using the same example and writing the wave function as $\psi = \bar\varphi + \bar\tau(k)/k \sin(kx + \bar\eta)$, this requirement becomes

$$\log\left[\tau(\sqrt{E}) + \sigma\tau(\sqrt{E})\right] = -\frac{2}{\pi} P \int_0^\infty \frac{\eta(\sqrt{E'}) + \delta\eta(\sqrt{E'})}{E' - E} dE' + \sum_{i=1}^{N_{BC}} \log\frac{k^2 - k_i^2}{k^2}, \quad (17)$$

where $-k_i^2$ are binding energies. Assuming that $\delta\eta(\sqrt{E}) \neq 0$ only for $E' \epsilon [E - \omega, E + \omega]$, where $\omega$ is small but not zero, the relation (17) becomes

$$\log\frac{\tau(\sqrt{E}) + \delta\tau(\sqrt{E})}{\tau(\sqrt{E})} = -\frac{2}{\pi} P \int_{E-\omega}^{E+\omega} \frac{d\eta(\sqrt{E'})}{E'-E} dE'$$

and, straightforwardly

$$\frac{d\tau(\sqrt{E})}{\tau(\sqrt{E})} = -\frac{2}{\pi}\frac{\omega}{E}\delta\left(\frac{d\eta(\sqrt{E'})}{d\sqrt{E}}\right) + 0(\omega^3). \quad (18)$$

This relation says that $\delta\tau(\sqrt{E}) = 0$ *is not permissible* — all methods previously discussed violate this condition.

The consequence of these results will be illustrated with a functional similar to that of Morse and Feshbach [6]

$$F[u,v] = \left\{\frac{1}{2}\int_0^\infty [u'^2 - v'^2 - k^2(u^2 - v^2)] - V u^2\right\}/v(0)^2$$

with the domain $(u,v)|u(x) = \varphi + \frac{\tau}{k}(\cos\eta \sin kx + \sin\eta\,\bar c(x))$, $\varphi(0) = 0$, $\varphi \epsilon L^2 \cap C^2$. $\bar c(0) = 0$, $\bar c(x) = \cos kx$, $x \to \infty$; $v(x) = \frac{\tau}{k}\sin(kx + \eta)$. The Schrödinger equation follows for arbitrary $\delta\varphi, \delta\eta, \delta\tau$ from the stationary value of $F$,

$$(19)$$

$$\frac{k}{\tau}\sin\eta <\delta\varphi, \bar Hu> - \frac{k}{\tau}<\varphi, \bar Hu>\frac{\delta\tau}{\tau} - \delta\eta\left[\frac{k}{\tau}\cos\eta <\varphi, \bar Hu> + <\sin kx, \bar Hu> = 0\right]$$

In Kohn's method one puts $\delta\tau = 0$ contrary to the condition (18) — this is convenient choice leading to a linear set of equations for variational parameters, but unfortunately the method provides false resonances.

Variational equations for scattering parameters follows from three equations (16), (18) and (19). One calculates $\delta\eta'$ from eq. (16) and eliminates it from eq.(18) and introducing so obtained expression into eq. (19). By putting $Z = \text{ctg}\,\eta$ and $\epsilon = 4\omega/k\pi$ one obtains

$$<\delta\varphi, Hu + \epsilon(1 + Z^2) <\varphi, Hu> u> = 0$$

$$Z <\varphi, Hu> + <\bar s, Hu> - \epsilon <\varphi, \bar Hu>[(1/4/k + <\bar s, c_o>)(Z^2 - 1)/(Z^2 + 1) +$$

$$+ 2Z <c, c_o>/(Z^2 + 1) + <\varphi, c> Z - <\varphi, \bar s>] = 0, \quad (20)$$

$c_o = c - \cos k_x$. These equations have to be solved in the limit $\epsilon \to 0^+$.

*Features of the constrained variational method:*

(1) The equations for variational parameters $Z$ and $\delta\varphi$ are nonlinear. They have to be solved by iteration. Poljšak [19] found an efficient procedure for

solving the system (20) for the most often used approximation $\varphi = \sum_i^{N_c} c_i \varphi_i$, $N_c$ finite; $\varphi_i$ being square integrable functions.

(2) If the potential contains the Coulomb potential $q^2/r$ and a short range potential $W(r)$ one has to define $\eta_1(k)$ and $\tau_1(k)$ by

$$\psi_1(kr) \sim \frac{\tau_1}{k} \sin(kr - 1\pi/2 + \arg \Gamma (l + 1 - iq^2/k) + \frac{q^2}{k} \log 2kr + \eta_1(k)), \quad r \to \infty .$$

Then the dispersion relation for $\tau_1(k)$ and $\eta_1(k)$ are same as before, eq. (17).

(3) For the generalization to elastic scattering of composite nuclei in the GC representation one writes, by denoting the channel $j = (k, J, M, \alpha)$,

$$\varphi_j = \sum_{p,K} c_{pK} P_{MK}^J \phi(r, S_p) ,$$

$$\varphi_{j \, assympt.} = \frac{\tau_j(k)}{k} (\cos \eta_j \hat{\mathcal{F}}_j + \sin \eta_j \mathcal{O}_j)$$

$$= \frac{\tau_j(k)}{k} \sum_{m,M_l} \left\{ \cos \eta_j \int dS \, f_j(S) \, \phi_j(x,S) + \sin \eta_j \int ds \, g_j(S) \, \phi_j(r,S) \right\}$$

$$< lm \, lM_l | JM > < s_1 \nu_1 \, s_2 \nu_2 | lM_l > ,$$

where $f_l$ and $g_l$ satisfy integral equation

$$\int dS \exp[-\beta'(r - S)^2/2] \begin{Bmatrix} f_{lm}(S) \\ g_{lm}(S) \end{Bmatrix} = \frac{1}{r} \begin{Bmatrix} F_l(kr) \\ G_l(kr) \end{Bmatrix} Y_{lm}(\hat{r}) i^l$$

and $F_l$ and $G_l$ are Coulomb functions. The functions $\phi_j(r,S)$ are two– cluster functions like (1).

(4) Generalization to the inelastic scattering of composite nuclei is rather straightforward.

Generalization to reactions when rearrangement channels are open is still unsolved. Dispersion relations are complicated in this case.

(5) Economy: Equations for variational parameters, (20), contain only those matrix elements which are needed in Kohn variational method.


## 5. CONCLUSION

1. During last few years computational technology has developed so that it has become possible to study more precisely the merits and usefulness of the cluster model. There were several extensive calculations by Tokyo, Minneapolis and Ljubljana groups on coupling different two–cluster low–lying channels – calculations with three– cluster configurations and coupling two– and three– cluster configurations. The conclusion from these calculations is that one has to include several low–lying channels (cluster– structures) in order to describe quantitatively properties of light nuclei. This is shown on the nucleus $^7$Li which is consider to be the most perfect two– cluster nucleus.

The effective interaction in the microscopic cluster model is an open problem. It is a difficult problem due to the overcompleteness of the two– (three–) centre basis. However, there is plenty of information from calculations on several systems with the same phenomenological effective nucleon–nucleon interaction.

From calculations on $^6$Li and $^7$Li we come to the conclusion that most low-lying properties can be described by a rather simple effective interaction consisting of the central part of Volkov and a two—term tensor interaction having radial parts of the gaussian form. The parameters of the interaction are chosen so that the strength of the central part reproduces the binding of $^4$He, the Majorana spin-isospin parameter regulates the energy between clusters, and the tensor part takes care about the binding of the deuteron cluster. The simplicity of this scheme will hopefully encourage somebody to think of a model for the effective interaction in the cluster structure model!

2. So far, a feasible and trustworthly variational model for calculating parameters of reactions does not exist. Methods using no additional constraints require sometimes tedious testing. The constrained variational method guaranteeing the conservation of the norm and the analyticity of variational parameters works well for elastic and inelastic scattering. The rather simple form of the dispersion relations in these cases is the consequence of the existence of the system of coupled— channel equations for elastic and inelastic scattering.

It is still not clear how to extend the constrained variational calculation to re-arrangement channels. However, success in this field would extend the feasibility and reliability of description of more complicate systems.

3. The occurrence of three charged particles in the outgoing channels is not rare in reactions of light nuclei in astrophysical conditions. However, there is no useful variational method able to handle this situation.

## REFERENCES

1. M.V. Mihailović and M. Poljšak, Phys. Lett. 66B (1977) 209; Nucl. Phys. A311 (1978) 377
2. R. Beck, R. Krivec and M.V. Mihailović, Nucl. Phys. A363 (1981) 365
3. Y. Fujiwara and Y.C. Tang, preprint, University of Minnesota, 1984
4. M.V. Mihailović and D. Mitić, to be published
5. J. Douglas, Trans. Am. Math. Soc. 50 (1941) 71
6. P.M. Morse and H. Feshbach, Methods of Theoretical Physics (Mc Graw—Hill, New York, 1953), p. 1127
7. N.E. Tamm, Ž.E.T.F. 18 (1948) 337
8. W. Kohn, Phys. Rev. 74 (1948) 1763
9. L. Hulten, Arkiv. Mat. Astron. Fys. 35A (1948) 25; L. Hulten and S. Skarlem, Phys. Rev. 87 (1952) 297
10. R.K. Nesbet and R.S. Oberoi, Phys. Rev. A6 (1972) 1855
11. F.E. Harris and H.H. Michels, Phys. Rev. Lett. 22 (1969) 2036
12. K. Takatsuka and T. Fueno, Phys. Rev. 163 (1967) 964
13. A.B. Volkov, Nucl. Phys. A74 (1965) 33
14. M. Bouten and P. van Leuven, Nucl. Phys. A169 (1971) 407
15. V.I. Kukulin, V.M. Krasnopol'sky, V.T. Voronchev and P.B. Sazonov, Nucl. Phys. A417 (1984) 128
16. A. Hasegawa and S. Nagata, Progr. Theor. Phys. 45 (1971) 1786
17. A. Eikermeier and H.H. Hackenbroich, Nucl. Phys. A169 (1971) 407
18. M. Poljšak, Ph.D.Thesis, Univer. of Ljubljana, 1982; M. Poljšak and M.V. Mihailović, to be published
19. R. Beck, M.V. Mihailović, M. Poljšak, Nucl. Phys. A351 (1981) 295

# NEW METHOD FOR STUDYING CLUSTERING IN NUCLEI

F. Iachello
A.W. Wright Nuclear Structure Laboratory
Yale University
P.O. Box 6666
New Haven, Connecticut 06520

ABSTRACT. A new method for describing cluster configurations in nuclei is discussed. This method makes use of algebraic techniques and introduces bosonic degrees of freedom with $J^P=0^+,1^-$. The method is best suited for applications in heavy nuclei where it has been suggested recently that alpha-clustering may play an important role.

## 1. INTRODUCTION

Clustering phenomena play an important role in the structure of light nuclei. A large number of phenomenological and microscopic calculations have been performed here, some of which are reported at this Conference. Recently, it has been pointed out that several puzzling properties of the light actinide nuclei, with neutron number >126 and proton number >82, appear to indicate that clustering may play an equally important role in these nuclei[1]. Because of the complexity of the problem, very few microscopic clustering calculations have been performed in this region[2]. In order to investigate whether or not clustering plays an important role here, we have developed a semiphenomenological model in which the coupling of the clustering degrees of freedom to the deformation (essential for describing this region) can be treated in a relatively straightforward way. The model has been applied to the study of the Th and Ra isotopes[3]. One important earmark of this new approach is the occurrence of enhanced E1 transitions when the charge to mass ratio of the two clusters is not identical. Large E1 transitions have been observed in $^{18}O$ and $^{218}Ra$[4],[5]. In this article, I will describe briefly this new method, which I will call nuclear vibron model (NVM), and show some applications to heavy nuclei.

## 2. THE NUCLEAR VIBRON MODEL

From a phenomenological point of view, the relative motion of two clusters can be studied by introducing an intercluster potential, $V(r)$, and solving the appropriate Schrödinger equation. An alternative approach, which has been found useful in molecular physics[6], is that of

101

*J. S. Lilley and M. A. Nagarajan (eds.), Clustering Aspects of Nuclear Structure, 101–115.*
© *1985 by D. Reidel Publishing Company.*

treating the clustering degrees of freedom algebraically. Contrary to the case of nuclear deformation, where the collective variable is a quadrupole variable, in the case of clustering, the collective variable is a dipole variable, Fig. 1. The situation encountered in clustering can be treated algebraically by introducing bosonic degrees of freedom

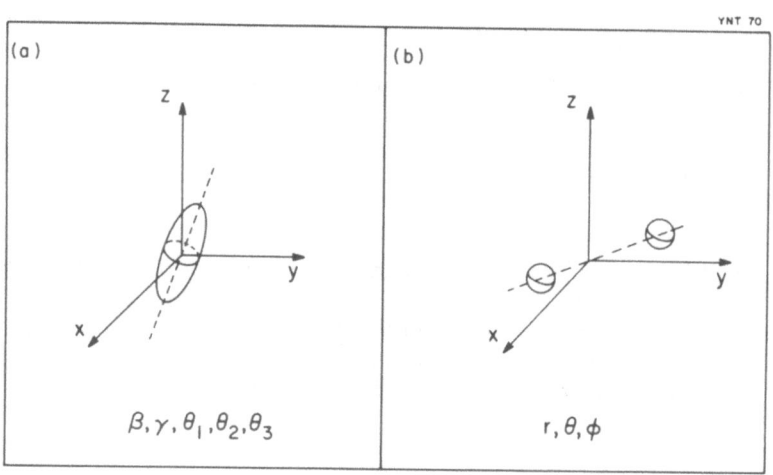

Fig. 1. (a) A deformed body with quadrupole deformation; (b) A molecular cluster configuration.

with angular momenta $J^P = 0^+$ and $1^-$. I will denote the corresponding boson operators by $\sigma^\dagger$ and $\pi_\mu^\dagger$ ($\mu = \pm 1, 0$) (or alternatively $s^{\dagger}$ and $p_\mu^{\dagger}$, $\mu = \pm 1, 0$). The Hamiltonian describing clustering can then be expanded into bilinear products of boson creation and annihilation operators. Retaining the lowest order terms yields,

$$H = E_0 + \varepsilon_\sigma (\sigma^\dagger \cdot \tilde{\sigma}) + \varepsilon_\pi (\pi^\dagger \cdot \tilde{\pi}) + \sum_{L=0,2} \frac{1}{2}(2L+1)^{\frac{1}{2}} c_2 [(\pi^\dagger \times \pi^\dagger)^{(L)} \times (\tilde{\pi} \times \tilde{\pi})^{(L)}]^{(0)}$$

$$+ v[(\pi^\dagger \times \pi^\dagger)^{(0)} \times (\tilde{\sigma} \times \tilde{\sigma})^{(0)} + (\sigma^\dagger \times \sigma^\dagger)^{(0)} \times (\tilde{\pi} \times \tilde{\pi})^{(0)}]^{(0)}$$

$$+ u_1 [(\pi^\dagger \times \sigma^\dagger)^{(1)} \times (\tilde{\pi} \times \tilde{\sigma})^{(1)}] + u_0 [(\sigma^\dagger \times \sigma^\dagger)^{(0)} \times (\tilde{\sigma} \times \tilde{\sigma})^{(0)}]^{(0)} , \qquad (1)$$

where $\tilde{\pi}_\mu = (-)^{1-\mu} \pi_{-\mu}$ and $\tilde{\sigma} = \sigma$. The $\sigma$ and $\pi$ bosons have been called vibrons and hence the name nuclear vibron model (NVM) given to this model. The Hamiltonian, Eq.(1), is diagonalized within the space of fixed number of vibrons, M. Due to this constraint, the number of independent parameters is reduced to four, $\varepsilon_\pi$, $c_0$, $c_2$ and $v$, since the others ($\varepsilon_\sigma$, $u_1$ and $u_0$) can be absorbed in $E_0$.

The form, Eq.(1), is convenient for analyzing the types of spectra that one can obtain. In order to do this, one starts by studying the dynamic symmetries of Eq.(1), i.e. those situations for which the Hamiltonian can be diagonalized in closed form. This study can be done by noting that the Hamiltonian (1) has the group structure of the unitary group in four dimensions, $U(4)$[6]. Breaking $U(4)$ in all possible ways, but retaining the rotation group $O(3)$ as a subgroup, yields two solutions

$$U(4) \rightleftarrows \begin{array}{ll} O(4) \supset O(3) \supset O(2), & \text{(I)} \\ U(3) \supset O(3) \supset O(2). & \text{(II)} \end{array} \qquad (2)$$

The energy spectra corresponding to the two chains in Eq.(2) can be obtained by expanding the Hamiltonian in terms of Casimir invariants and are given by

$$E^{(I)}(N,\omega,L,M_L) = E_0^{(I)} + A\omega(\omega+2) + BL(L+1), \qquad \text{(I)}$$

$$E^{(II)}(N,n_\pi,L,M_L) = E_0^{(II)} + \varepsilon_\pi n_\pi + \alpha_\pi n_\pi(n_\pi+2) + \beta_\pi L(L+1). \quad \text{(II)}$$

$$(3)$$

They are shown in Figs. 2 and 3 respectively. Fig. 2 describes a

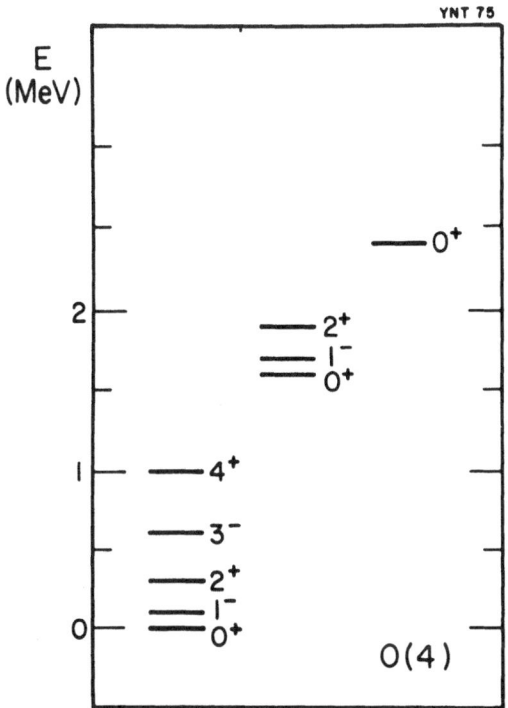

Fig. 2. Typical spectrum of the symmetry I of the nuclear vibron model, M=4, A=-100 keV, B=50 keV.

situation in which the two fragments are well separated, like in a diatomic molecule, and perform vibrations and rotations around an equilibrium configuration. Instead of the quantum number $\omega$, one may use here a vibrational quantum number $v = (M-\omega)/2$. The spectrum of Fig. 2 would arise if one would solve the Schrödinger equation with an intercluster potential, $V(r)$ of the type $V(r) = V_0\{\exp[-2a(r-r_0)] -2\exp[-a(r-r_0)]\}$, (Morse potential). Fig. 3 describes a situation in which one of the

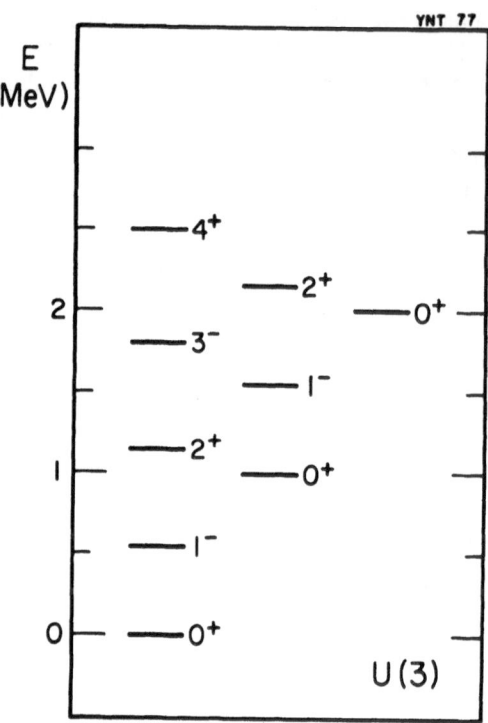

Fig. 3. Typical spectrum of the symmetry II of the nuclear vibron model,
M=4, $\varepsilon_\pi$ =500 keV, $\alpha_\pi$ =0, $\beta_\pi$ =25 keV.

fragments performs (dipole) oscillations relative to the other with
frequency $\varepsilon_\pi/\hbar$. This situation would arise from an intercluster
potential of the type $V(r) = -V_0/\{1+\exp[a(r-r_0)]\}$, (Woods–Saxon potential). Figs. 2 and 3 have been constructed using Eq.(3) together with
the rules that determine the values of the quantum numbers contained in
a given configuration, M. These are:

(I)   The values of $\omega$ are given by $\omega$ = M, M−2, ..., 1 or 0 (M=odd or
      even); the values of L are given by L=$\omega$, $\omega$−1, ..., 1 of 0 and
      $-L \leqslant M_L \leqslant L$, as usual.

(II)  The values of $n_\pi$ are given by $n_\pi$ = M, M−1, ..., 1, 0; the values of
      L are given by L = $n_\pi$, $n_\pi$−2, ..., 1 or 0 ($n_\pi$ = odd of even) and
      $-L \leqslant M_L \leqslant L$.

By varying the parameters $\varepsilon_\pi$, $c_0$, $c_2$ and v in Eq.(1) one can obtain all
possible types of spectra intermediate between the two limits, Eq.(3),
corresponding to solutions of the Schrödinger equation with potentials
intermediate between the Morse and the Woods–Saxon type. From this point
of view, the nuclear vibron model represents a convenient way to solve
the Schrödinger equation with arbitrary potentials.

## 3.   ELECTROMAGNETIC TRANSITION RATES

Electromagnetic transition rates can be calculated algebraically by
introducing appropriate operators and evaluating their matrix elements

between states previously obtained. Retaining only the lowest order
terms, one has transition operators with multipolarity E0, E1, M1 and
E2, whose explicit expression in terms of creation and annihilation
operators is

$$
T^{(E0)} = \alpha_0 (\pi^\dagger \times \tilde{\pi})^{(0)} + \beta_0 (\sigma^\dagger \times \tilde{\sigma})^{(0)},
$$

$$
T^{(E1)} = \alpha_1 (\pi^\dagger \times \tilde{\sigma} + \sigma^\dagger \times \tilde{\pi})^{(1)},
$$

$$
T^{(M1)} = \alpha_1' (\pi^\dagger \times \tilde{\pi})^{(1)},
$$

$$
T^{(E2)} = \alpha_2 (\pi^\dagger \times \tilde{\pi})^{(2)}. \tag{4}
$$

It is important to note here that, since in a clustering configuration
the collective variable is a dipole variable, one expects large electric
dipole (E1) transitions. In order to have a non-vanishing value of $\alpha_1$,
however, one must be in a situation in which the center of mass and
center of charge do not coincide. Thus, for example, $\alpha_1$ vanishes in the
cluster configuration $^{12}$C-$\alpha$ but it is different from zero in $^{14}$C-$\alpha$. A
measure of the collectivity of the E1 transitions can be obtained by
considering the molecular sum rule derived recently by Alhassid, Gai and
Bertsch[7]

$$
S_1 (E1; A_1 + A_2) = \left( \frac{9}{4\pi} \right) \frac{(Z_1 A_2 - Z_2 A_1)^2}{A A_1 A_2} \left( \frac{\hbar^2 e^2}{2m} \right), \tag{5}
$$

where $Z_1$, $Z_2$, $A_1$, $A_2$ are the charges and masses of the two fragments.
The occurrence of large E1 transitions in asymmetric cases, suggested by
the nuclear vibron model[8], has been confirmed recently by a microscopic
calculation of the cluster states in $^{14}$C-$\alpha$ by Assenbaum, Langanke and
Weiguny[9]. Their calculated B(E1; $1_1 \to 0_2$) value exhausts in this
case $\approx$ 40% of the molecular sum rule, Eq.(5).

## 4.    COUPLING OF CLUSTERING AND DEFORMATION

One of the main advantages of the algebraic description of clustering discussed above is that, because of its simplicity, it can be easily
extended to more complex situations. One of these complex situations
occurs when one or both fragments are deformed. I will discuss here the
case in which only one fragment is deformed, Fig. 4. This situation can
be described by treating both the clustering and the deformation degrees
of freedom algebraically. This is done by introducing two sets of
bosons, one $(\sigma, \pi)$ describing the clustering, and the other (s,d)
describing the internal degrees of freedom of one of the fragments. The
latter are assumed to be of the quadrupole type and thus described by
(s,d) bosons (interacting boson model[10]). The corresponding Hamiltonian
can be written as

$$
H = H_{sd} + H_{\sigma\pi} + V_{sd;\sigma\pi}. \tag{6}
$$

The coupling term will contain in general several contributions

$$V_{sd;\sigma\pi} = w_0\big[(d^\dagger \times \tilde{d})^{(0)} \times (\pi^\dagger \times \tilde{\pi})^{(0)}\big]^{(0)} + w_1\big[(d^\dagger \times \tilde{d})^{(1)} \times (\pi^\dagger \times \tilde{\pi})^{(1)}\big]^{(0)}$$
$$+ w_2\big[(d^\dagger \times \tilde{s} + s^\dagger \times \tilde{d} + \chi\ d^\dagger \times \tilde{d})^{(2)} \times (\pi^\dagger \times \tilde{\pi})^{(2)}\big]^{(0)}. \tag{7}$$

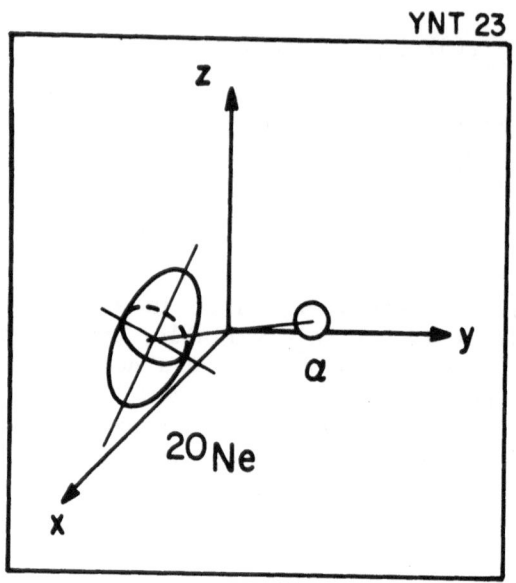

Fig. 4. The clustering configuration $^{20}$Ne-$\alpha$.

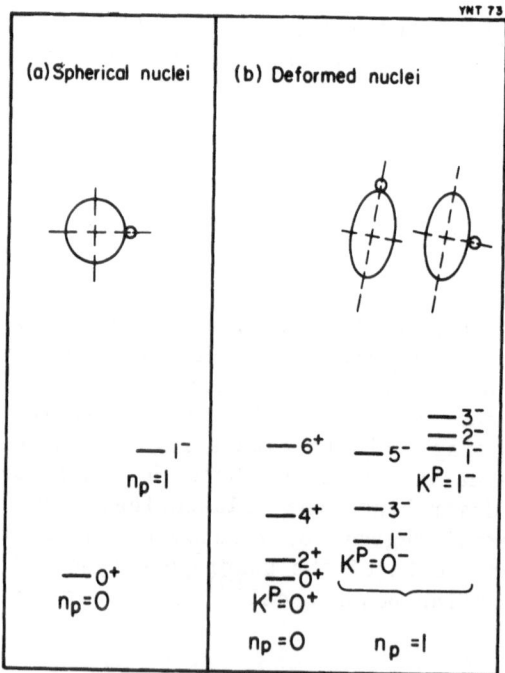

Fig. 5. (a) The spectrum of an $\alpha$-particle oscillating in a spherical nucleus; (b) the spectrum of an $\alpha$-particle oscillating in an axially deformed nucleus. Only low-lying states are shown.

The most important term in the coupling is the last one, since it contains the matrix elements of the quadrupole operator for (s,d) bosons

$$Q_{sd}^{(2)} = (d^\dagger \times \tilde{s} + s^\dagger \times \tilde{d})^{(2)} + \chi \ (d^\dagger \times \tilde{d})^{(2)}, \tag{8}$$

which are usually large. The spectra of Figs. 2 and 3 are changed in the presence of a coupling. Consider, for example, oscillations of a cluster in a deformed nucleus. This case corresponds to a coupling of the limit II of the nuclear vibron model[10], and it can be solved analytically[11]. The results are shown in Fig. 5. Each vibrational multiplet, characterized by values of $n_\pi = 0,1,\ldots$, splits, in deformed nuclei, into bands. These bands correspond to different orientations of the vibrations with respect to the symmetry axis, as shown in the top part of the figure. For example, the state with $n_\pi = 1$ splits into two parts with projection along the symmetry axis $K^P = 0^+,1^-$. The former corresponds to vibrations perpendicular to the symmetry axis.

## 5.    CONFIGURATION MIXING

Another difficulty which one encounters in dealing with clustering in nuclei is that, quite often, two (or more) configurations coexist in the same nucleus. Without entering in a detailed discussion of the character of the different configurations, I mention, as an example, the case of $^{20}$Ne, Fig. 6. Here the ground state band appears to be a shell

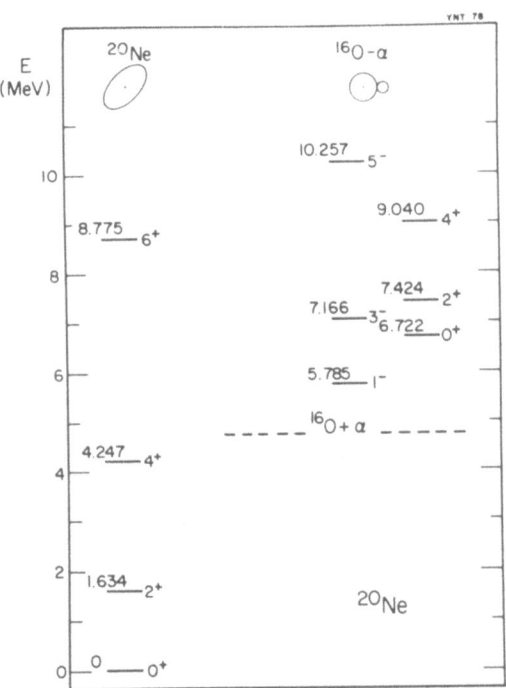

Fig. 6. Coexisting configurations in $^{20}$Ne.

model band, while the excited $K^P = 0^-$ and $K^P = 0^+$ bands appear to be more
related to localized cluster bands[12]. The two types of configuration,
shell model and localized cluster obviously mix. A proper treatment of
configuration mixing is by no means simple. A technique used in the
calculations performed so far has been that of introducing a fixed
number of bosons, $N_T$, and of considering the various configurations as
if they would be composed of M bosons of $(\sigma,\pi)$ type and N bosons of
$(s,d)$ type, with $N_T = N+M$, that is

$$(0\alpha) \qquad\qquad N = N_T \quad, \ M = 0$$

$$(1\alpha) \qquad\qquad N = N_T - 2, \ M = 2 \qquad\qquad\qquad (9)$$

$$(2\alpha) \qquad\qquad N = N_T - 4, \ M = 4$$

$$\ldots\ldots \qquad\qquad\qquad\qquad ,$$

where $(0\alpha)$, $(1\alpha)$, $(2\alpha)$, ... denote configurations with zero, one, two,
... alpha particles. A calculation of configuration mixing in this
scheme requires a knowledge of the energy gaps

$$\Delta_{1\alpha} = E_{1\alpha} - E_{0\alpha},$$

$$\Delta_{2\alpha} = E_{2\alpha} - E_{0\alpha}, \qquad\qquad\qquad\qquad (10)$$

and of the mixing Hamiltonian

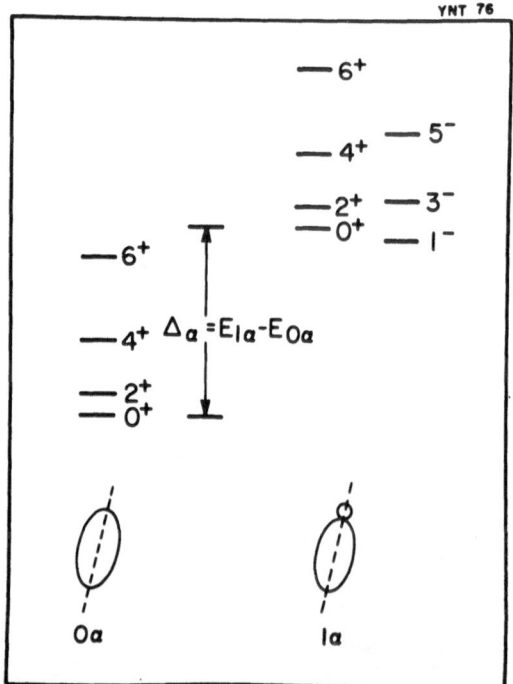

Fig. 7. Mixing of $(0\alpha)$ and $(1\alpha)$ configurations.

$$V_{mix} = (\alpha\ s^\dagger.s^\dagger + \beta\ d^\dagger.d^\dagger)\ (\alpha'\ \tilde{\sigma}.\tilde{\sigma} + \beta'\ \tilde{\pi}.\tilde{\pi})$$
$$+ (\alpha'\ \sigma^\dagger.\sigma^\dagger + \beta'\ \pi^\dagger.\pi^\dagger)\ (\alpha\ \tilde{s}.\tilde{s} + \beta\ \tilde{d}.\tilde{d}). \qquad (11)$$

The effect of the mixing is to displace further the bands. In particular, the mixing between (0α) and (1α) configurations displace the positive parity bands relative to the negative parity bands since the latter do not appear in the (0α) configuration, Fig. 7.

6.    MICROSCOPIC INTERPRETATION

Much of the discussion presented in the previous sections has been based on an algebraic treatment of the collective degrees of freedom. In a more fundamental description, the collective degrees of freedom must be derived from a microscopic theory. In this derivation, one may follow two routes:

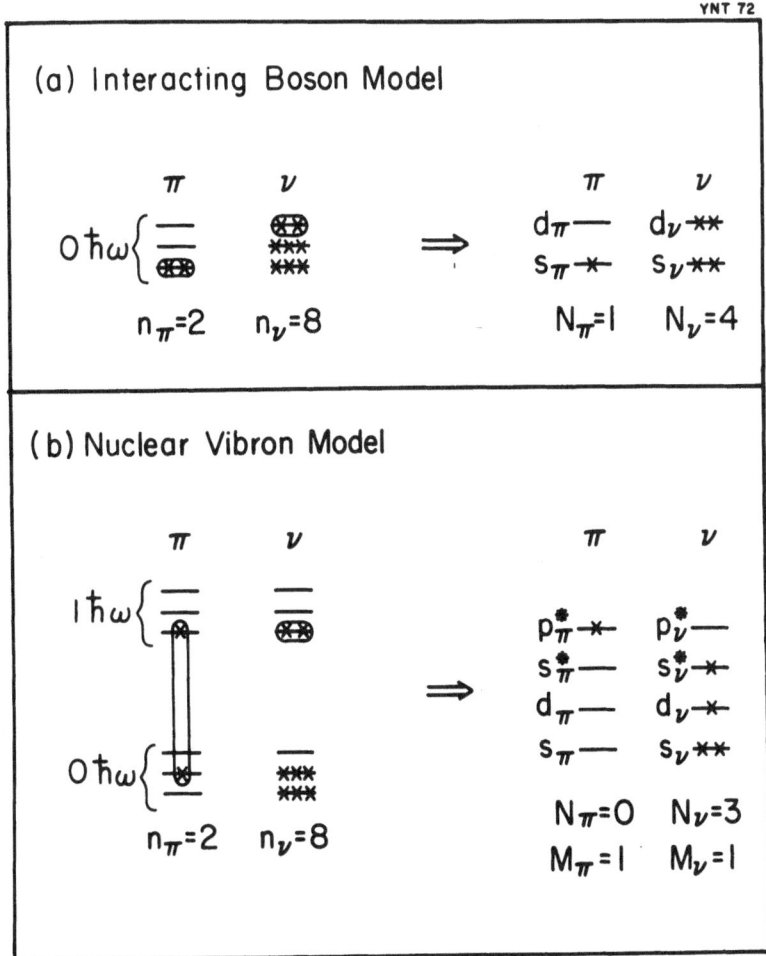

Fig. 8. Comparison between the microscopic interpretation of the basic building blocks of (a) the interacting boson model; (b) the nuclear vibron model. In this figure the $(\sigma,\pi)$ bosons are denoted by $(s^*,p^*)$.

(i)  start from the shell model and construct localized clusters by
     superimposing several shell model configurations. It is interesting
     to compare the situation here with that encountered in the micro-
     scopic treatment of the quadrupole collective degree of freedom. In
     the latter case, the collective features appear to be generated by
     correlations in the valence shell ($0\hbar\omega$) and thus one requires a
     horizontal truncation of the shell model space. In the case of
     clustering, the collective features appear to be generated by
     correlations involving several shells ($n\hbar\omega$) and thus one requires a
     vertical truncation of the shell model space[13]. This situation is
     depicted in Fig. 8. A consequence of this observation is that,
     although the technique described in Sect. 5 is by far the simplest
     to deal with configuration mixing, it may actually not be the most
     appropriate. In fact, while in the case of the interacting boson
     model, one can associate coupled pairs to the $J^P = 0^+, 2^+$ bosons and
     thus choose N as the number of valence pairs, in the case of the
     nuclear vibron model, this may not be possible, since the vibrons
     are closer to particle-hole rather than particle-particle pairs.
     The choice of their number, M, is then not related to the number of
     valence particles and it must be considered a free parameter.

(ii) A second route for approaching this problem may be provided by the
     resonating group (RGM) or generator coordinate methods (GCM). A
     calculation of this type is, for example, that of Ref.[99]. Here, the
     parameters appearing in the vibron Hamiltonian, Eq. (1), or in the
     mixing Hamiltonian, Eq. (11), could in principle be derived by

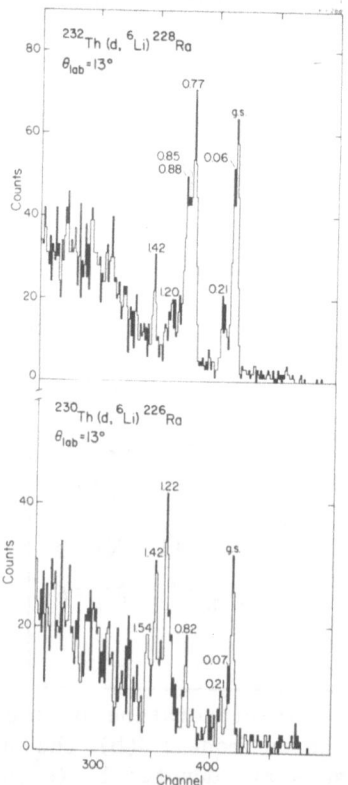

Fig. 9. Spectra of $^{226}$Ra and $^{228}$Ra
obtained in a (d, $^6$Li) reaction
at $\theta_{lab} = 13°$.

using a technique similar to that used in the microscopic derivation of the parameters of the interacting boson model[14], i.e. by
equating matrix elements of the appropriate operators in the algebraic and microscopic space.

Further developments of these techniques, especially for applications to heavy nuclei, would be of utmost importance for gaining insight
into the structure of the model.

## 7.    CLUSTERING IN HEAVY NUCLEI

Several features of heavy nuclei appear to indicate that clustering
may play here a role as important as that played in light nuclei. Two of
these are:
(i)  Recent (d,$^6$Li) experiments[15,16] show a large population of excited
     $0^+$ states, as shown in Fig. 9.
(ii) Low-lying negative parity states are observed with small hindrance
     factor (HF) in α-decay, Fig. 10; furthermore, the El matrix ele-
     ments connecting negative and positive parity states are large[5].

In view of these experimental observations, Daley[3] has performed
detailed calculations of the $_{88}$Ra and $_{90}$Th isotopes using the model
described above. His results are summarized in Fig. 11. The calculations
appear to describe the observed features, including the behaviour of the
relative α-decay widths and transition probabilities.

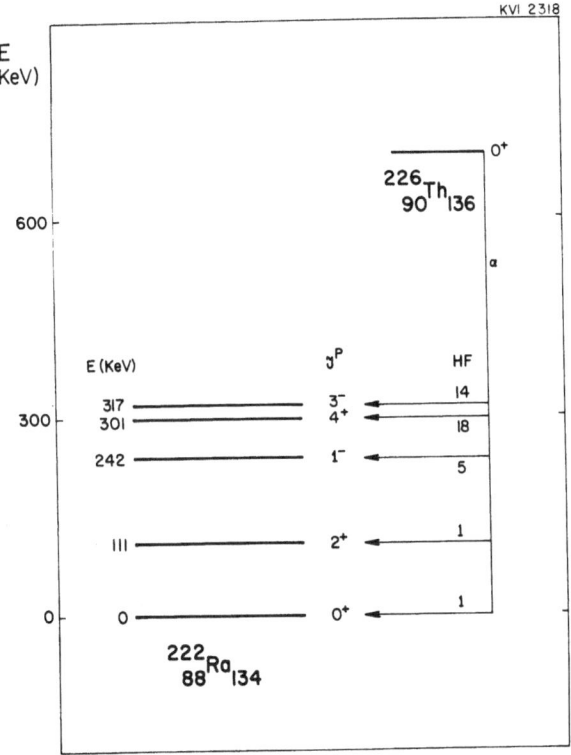

Fig. 10. The experimental spectrum of $^{222}_{88}$Ra$_{134}$.

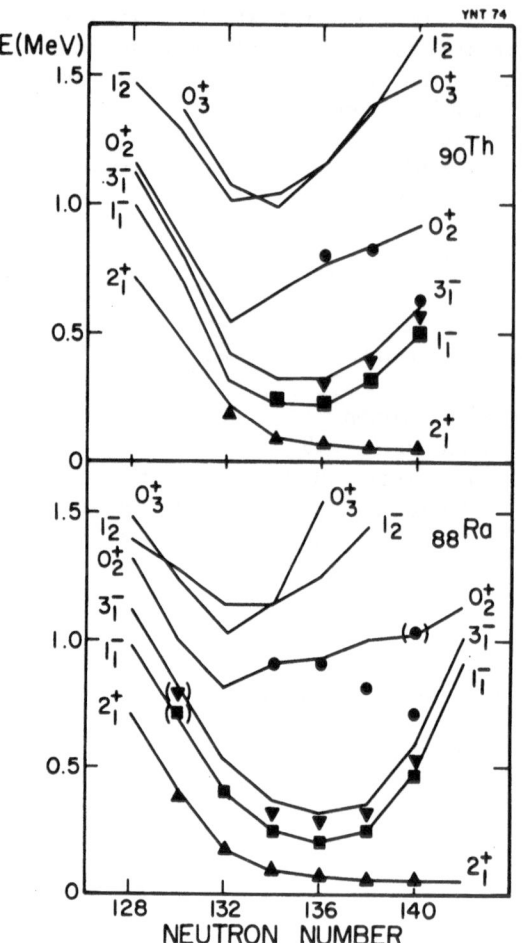

Fig. 11. Energy levels of $_{88}$Ra and $_{90}$Th as calculated by Daley[3].

Two types of calculations were attempted: one with a rigid cluster structure (O(4) limit of the vibron model) and one with a soft structure (U(3) limit of the vibron model). The latter, shown in Fig. 11, appears to describe the data more accurately. The comparison between experimental and theoretical spectra is shown in more detail in Fig. 12. The calculations of Daley have been extended recently by Daley and Gai[17] to include high spin states.

The structure of the light actinide nuclei has also been discussed in terms of stable octupole deformations[18]. An important point which deserves further attention is to what extent the two interpretations are different. It is clear, for example, that a cluster configuration, when expanded in terms of multipoles, contains at the same time a dipole, a quadrupole and an octupole, Fig. 13. Thus, the interpretation in terms of octupole deformations may, in fact, be rather similar to that in terms of clustering. Moreover, in order to obtain large E1 transitions, octupole deformations alone are not sufficient. One must, in addition, displace the center of mass from the center of charge by moving protons away from neutrons. Again, this effectively gives rise to clustering.

Another point which deserves further attention is to what extent

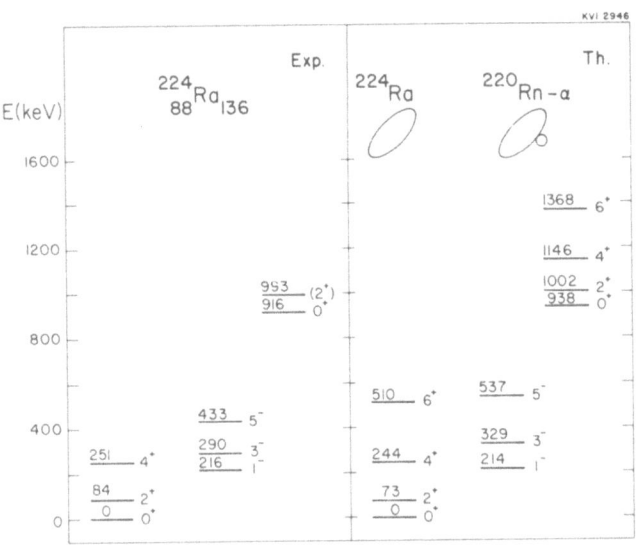

Fig. 12. Comparison between experimental and theoretical energy levels in $^{224}_{88}Ra_{136}$.

clusters larger than α-particles play a role. This could occur in several ways. For example, by having multicluster configurations of the type α-A-α, ..., or by having configurations of the type A-$^{12}$C, A-$^{16}$O, ..., where A denotes the core nucleus. Configurations of the latter type are particularly intriguing in view of the recent observation of decay of $^{223}$Ra into Pb+C[19].

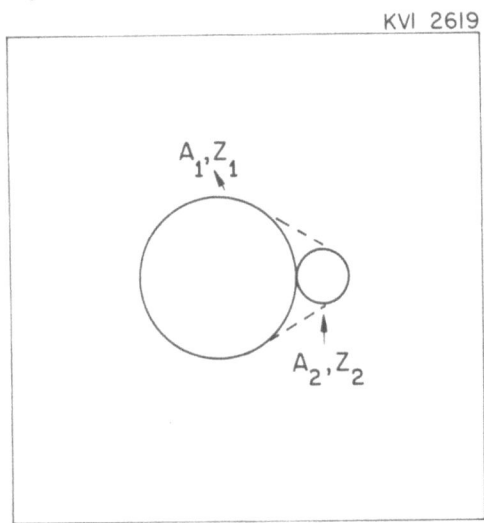

Fig. 13. A cluster configuration $(A_1,Z_1) - (A_2,Z_2)$.

# 8.   APPLICATIONS TO LIGHT NUCLEI

Although the vibron model is best suited for applications to heavy nuclei, it may nonetheless be applied to light nuclei, where it may help elucidating some aspects of clustering phenomena well known in this region. Three applications have been discussed so far:

(i)   Levels in $^{18}O$. Gai et al[4] recently have measured E1 transition rates between members of the $^{14}C$-$\alpha$ cluster band and found considerable enhancement of these transitions. The measured $B(E1;1_1^- \rightarrow 0_2^+) = 2.8 \times 10^{-2}$ W.u. exhausts a significant fraction of the molecular sum rule ($\approx 13\%$). This measurement appears to be, so far, the clearest evidence for the occurrence of a collective dipole mode.

(ii)  Resonances in $\alpha$-scattering. Cseh[20] has analyzed, using the vibron model, resonances in $\alpha$-scattering and found some evidence for the occurrence of molecular bands, although this evidence is not completely conclusive. In this analysis, coupling to the deformation and configuration mixing was not included and the number of vibrons M was taken as a free parameter.

(iii) Resonances in $^{12}C$-$^{12}C$ scattering. Erb and Bromley[21] have classified the resonances observed in $^{12}C$-$^{12}C$ scattering in terms of a rigid molecular-like structure (O(4) limit of the vibron model). Here again, the number of vibrons M has been taken as a free parameter.

# 9.   CONCLUSIONS

I have presented in this article a new technique to deal phenomenologically with cluster configurations in nuclei. The advantage of this technique is that it can take into account in a straightforward way several physical effects, such as the coupling of clustering to deformation and the mixing of configurations.

In order to determine the usefulness of this scheme, detailed applications of it to both light and heavy nuclei should be performed. Furthermore, and most importantly, a microscopic interpretation of the building blocks of the model should be given by expanding the suggestions presented in Sect. 5. For example, a generator coordinate method calculation of $^{212}Po$ seen as $^{208}Pb+\alpha$ would be most welcome. The microscopic interpretation may shed some light on a proper treatment of some aspects of the model which are at present not well focused, as, for example, the choice of the vibron number H and the treatment of configuration mixing.

Another important question, which is, to some extent, independent from any detailed description in terms of one model or another, is whether or not clustering configurations play an important role in heavy nuclei. The experimental evidence accumulated so far in the light actinide nuclei is very suggestive, and it indicates that this possibility must be taken seriously. Further experimental studies of this interesting region are needed in order to make conclusive statements.

Finally, the model discussed above stresses the importance of collective dipole transitions. Two cases ($^{18}O$ and $^{218}Ra$) have been found so far. It would be interesting to see whether or not other cases exist.

## 10.  ACKNOWLEDGEMENTS

I wish to thank H. Daley for performing the calculations in $_{88}Ra$ and $_{90}Th$, D.A. Bromley and M. Gai for searching and discovering dipole collectivity in $^{18}O$ and $^{218}Ra$, A. Weiguny for interesting discussions on the role of El transitions in light nuclei, J. Cseh for his interest in the vibron model and R.H. Siemssen for his hospitality at the Kernfysisch Versneller Instituut where this article was completed. This work was performed in part under the Department of Energy Contract No. DE-AC 02-76 ER 03074.

## REFERENCES

1.   F. Iachello and A.D. Jackson, Phys.Lett. 108B (1982) 151.
2.   I. Tonozuka and A. Arima, Nucl.Phys. A323 (1979) 45.
3.   H. Daley and F. Iachello, Phys.Lett. 131B (1983) 281.
4.   M. Gai, M. Ruscev, A.C. Hayes, J.F. Ennis, R. Keddy, E.C.Schloemer, S.M. Sternbenz and D.A. Bromley, Phys.Rev.Lett. 50 (1983) 239.
5.   M. Gai, J.F. Ennis, M. Ruscev, E.C. Schloemer, B. Shivakumar, S.M. Sternbenz, N. Tsoupas and D.A. Bromley, Phys.Rev.Lett. 51 (1983) 646.
6.   F. Iachello, Chem.Phys.Lett. 78 (1981) 581; F. Iachello and R.D. Levine, J.Chem.Phys. 77 (1982) 3046.
7.   Y. Alhassid, M. Gai and G.F. Bertsch, Phys.Rev.Lett. 49 (1982) 1482.
8.   F. Iachello, Phys.Rev. C23 (1981) 2778.
9.   H.J. Assenbaum, K. Langanke and A. Weiguny, These Proceedings.
10.  A. Arima and F. Iachello, Ann.Phys. (N.Y.) 99 (1976) 253; A. Arima and F. Iachello, Ann.Phys. (N.Y.) 111 (1978) 201; A. Arima and F. Iachello, Ann.Phys. (N.Y.) 123 (1979) 468.
11.  H. Daley and F. Iachello, to be published.
12.  A. Arima, Proc.Int.Conf. on Nucl. Physics, Munich (1973), J. de Boer and H.J. Mang eds., Vol. 2, p. 184.
13.  M. Kamimura, T. Matsuse and K. Takada, Progr.Theor.Phys. 47 (1972) 1537.
14.  T. Otsuka, A. Arima and F. Iachello, Nucl.Phys. A309 (1978) 1.
15.  J. Jänecke, F.D. Becchetti, D. Overway, J.D. Croissart and R.L.Spross, Phys.Rev. C23 (1981) 101.
16.  A.M. van der Berg, N. Blasi, R.H. Siemssen, W.A. Sterrenburg and Z. Sujkowski, Nucl.Phys. A422 (1984) 45.
17.  H.J. Daley and M. Gai, These Proceedings.
18.  R.K. Sheline and G.A. Leander, Phys.Rev.Lett. 51 (1983) 359 and references therein.
19.  H.J. Rose and G.A. Jones, Nature 307 (1984) 245.
20.  J. Cseh, Phys.Rev. C27 (1983) 2991.
21.  K. Erb and D.A. Bromley, Phys.Rev. C23 (1981) 2781.

# FUSION - QM RESONANCES

# INTERACTIONS BETWEEN 'LIGHT' HEAVY IONS*

U. Mosel
Institut für Theoretische Physik
Universität Giessen
D-6300 Giessen, West Germany

ABSTRACT. New developments in the theoretical descriptions of heavy ion interactions are discussed. In particular, it is emphasized that polarization effects are important for a description of molecular-type reactions and have to be included in any theory. It is then shown that even in strong coupling situations the DWBA is expected to work if the proper "dressed" potentials are used and thus does not constitute a weak coupling approximation. Crucial for a deeper understanding of nuclear molecules are obviously the new measurements of alignment and γ-decay in molecular resonances. Calculations of these quantities are discussed in the light of the latest experiments.

## 1. INTRODUCTION

Since the pioneering experiments of Bromley et al.[1]) in which correlated resonances were observed in many different reaction channels of $^{12}C+^{12}C$ it has been speculated that these resonances might be indications of molecular configurations formed during the scattering process. Since then structures, although often wider, have also been observed at higher energies in $^{12}C+^{12}C$ (ref. 2) as well as in a number of other systems[3]).

Many theoretical descriptions of the observed cross sections, at least for the broader structures, have been proposed. These descriptions often rely on different variants of direct reaction theories, but all of them describe the observed cross sections - mostly excitation functions - with similar success. To this question if and how much one can actually learn from direct reaction model studies I will address myself in the first part of this talk. In the next two parts I will then try to review our present knowledge about two crucial types of experiments, i.e. measurements of the alignment and of the γ-decay between resonances. All along through this talk I will in particular point out why these types of experiments can possibly bridge the gap between reaction theory descriptions of the observed structures on one side and descriptions in terms of the structure of the combined system on the other.

*Work supported by BMFT and GSI Darmstadt.

119

*J. S. Lilley and M. A. Nagarajan (eds.), Clustering Aspects of Nuclear Structure, 119–131.*
© 1985 by D. Reidel Publishing Company.

## 2.  POLARIZATION EFFECTS

In this chapter I will summarize our present understanding why so many
seemingly quite different reaction theories seem to describe the measured excitation functions all equally well. A dramatic example is $^{12}$C+
$^{12}$C where Austern-Blair[4]), DWBA[5]), Coupled Channel with weak[6]) and Coupled Channel Calculations with strong interactions[7]) have all given
comparable fits to the data. At first sight this "success" implies that
nothing can be learned from such calculations.

However, the key to a deeper understanding of this situation lies
in the observation that these different reaction models all use different diagonal and coupling potentials. Thus part of the interactions
that are, for example, treated explicitly in the CC calculations are included in the potentials used in the DWBA descriptions[8]). That such polarization effects are indeed important is most dramatically illustrated
by the case of $^{16}$O+$^{16}$O where correlated resonant structures have been
observed both in the $3_1^-$ as well as in the mismatched $0_2^+$ channel[9]). This
correlation can be understood only by a CC-calculation that couples
these two channels so strongly that their effective potentials lie essentially on top of each other (see fig. 1b) so that both channel wavefunctions resonate at the same energy. Therefore, a DWBA calculation with a

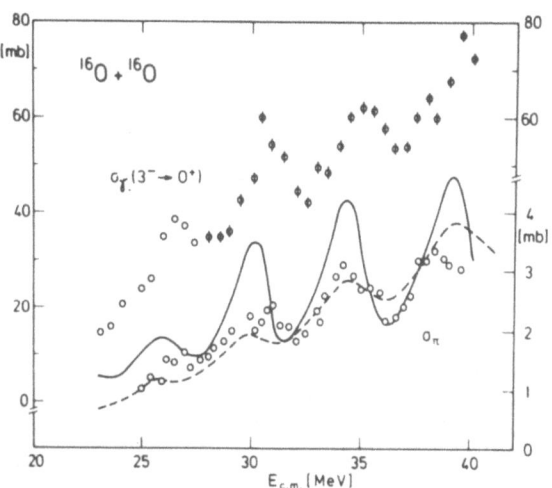

Figure 1a. Experimental and calculated cross sections for inelastic $3^-$ and $0_2^+$ cross sections in $^{16}$O+$^{16}$O (from ref. 16).

similar shift in the exit channel would have been equally successful.
The theoretical description of these polarization phenomena, that -
macroscopically speaking - include all the dynamic deformations during
the collision, starts from Feshbach's projection formalism[11]). I will
first give the relevant expression that relate large-space to small-
space coupled channel calculations and then those that make the connection with DWBA calculations.

If we denote the operator projecting on the explicitly treated

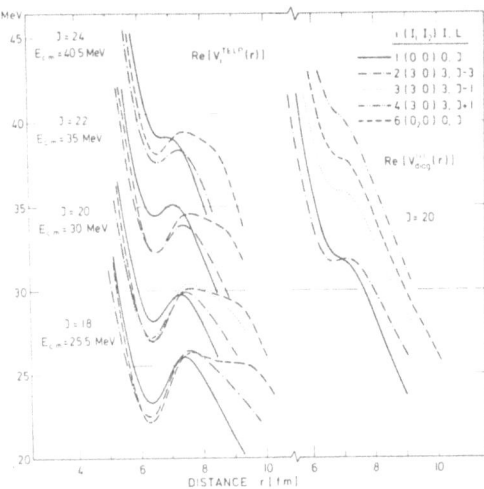

Figure 1b. Trivially equivalent local potentials that contain the CC effects (left) and diagonal potentials (right) for the calculation whose results are shown in fig. 1a. The different channels are indicated in the upper right hand corner (from ref. 10).

states in the CC calculation by P (subchannels denoted by $P_i$) and all the other channels that are left out, mostly because of technical reasons like e.g. computer-time restrictions, by Q(P+Q=1) then one obtains the projected equations:

$$(E-H_{PP}) P\Psi = H_{PQ} Q\Psi \tag{1}$$

$$(E-H_{QQ}) Q\Psi = H_{QP} P\Psi \ . \tag{2}$$

Solving the second equation for $Q\Psi$ and inserting into the first gives the well known expression:

$$(E-H_{PP}-H_{PQ} \frac{1}{E-H_{QQ}} H_{QP}) \ P\Psi = 0 \ . \tag{3}$$

The third term in parentheses on the lhs. is the polarization potential.
     The problem here lies in evaluating the propagator $(E-HQQ)^{-1}$. In order to study the polarization effects we have started our CC calculations in a large space (P+Q), assumed to be "nature". We can then evaluate the Green function (denoted by $G_{kl}^J; k,l \epsilon Q$) in the Q space exactly[12]). Using this Green function the polarization potential is given by:

$$\Delta V_{i,j}^J (r,r') = \sum_{kl} V_{ik}^J (r) G_{kl}^J (r,r') V_{lj}^J (r')$$

$$i,j \ \epsilon \ P; \ k,l \ \epsilon \ Q \tag{4}$$

where $V_{ik}$ are the matrix elements of the "bare" coupling interaction.

Note that $\Delta V$ is J- and E-dependent as well as nonlocal. The $\Delta V$'s that are obtained from such calculations are typically surface-peaked and largest for the grazing partial waves. The T-matrix finally is given by:

$$T_{ji} = T^0 \delta_{ij} + <\phi_j^{(-)} | (H+\Delta V)_{ji} | P_i \psi^{(+)}> \tag{5}$$

where i,j stand for $P_i, P_j$ and $\phi_j^{(-)}$ fulfills the "optical model" equation:

$$\left[E - (H+\Delta V)_{ji}\right] \phi_j^{(-)} = 0 \quad . \tag{6}$$

It is now clear that different CC spaces will lead to different interactions - but, of course, identical cross sections - if one starts from the same bare Hamiltonian. What is even more interesting is that the T-matrix can exactly be brought into the form of a DWBA amplitude, the latter thus being applicable to any coupling, no matter how strong it is[13]. Thus the DWBA, which is widely believed to be a weak coupling limit of the CC method, is really a much more general method, at least in its common use in which couplings and potentials are adjusted to some set of experimental data. This will now be shown.

Consider a transition between the entrance channel 0 and a final subchannel f which alone form the subspace P. The exact T-matrix is then (see eq.5):

$$T_{f0} = <\phi_f^{(-)} \left[H+\overline{\Delta V}\right]_{f0} \psi_0^{(+)}> \tag{5a}$$

with

$$\left[E - (H+\overline{\Delta V})_{ff}\right] \phi_f^{(-)} = 0 \tag{6a}$$

and

$$\left[E - (H+\Delta V)_{00}\right] \psi_0^{(+)} = 0 \quad . \tag{6b}$$

The polarization potentials $\Delta V$ and $\overline{\Delta V}$ contain the couplings to all states in Q+f and Q, respectively.

If now $(H+\Delta V)_{00}$ is adjusted such as to give a good description of elastic scattering in an optical model calculation then $\psi_0$ is an optical model wavefunction for the ingoing wave and $T_{f0}$ in eq. 5a has exactly the form of a DWBA amplitude. Since this statement does not rely on any weak coupling limit, following Satchler[8] I prefer to call this method from now on the DW method.

There are 2 points noteworthy: first, in the DW method the potentials in the ingoing and outgoing channel are not the same (see eqs. 6) with the biggest differences to be expected for strongly coupled inelastic excitations. Second, the coupling interaction is connected to the potential in the excited state and not in the ground state. Again for strongly coupled states this invalidates the standard prescription to simply use the derivative of the elastic channel potential for the coupling. In other words: observation of different coupling strengths

for heavy - and light-ion induced reactions would reflect such polarization effects. The same type of argument was also applied to two-step DW
calculations in ref. 13 and Tanimura has very recently shown that indeed in general an n-step DW calculation is completely equivalent to a
CC calculation, if the proper potentials are used[14]). Success in fitting
data with a DWBA code thus does not imply the existence of weak coupling.

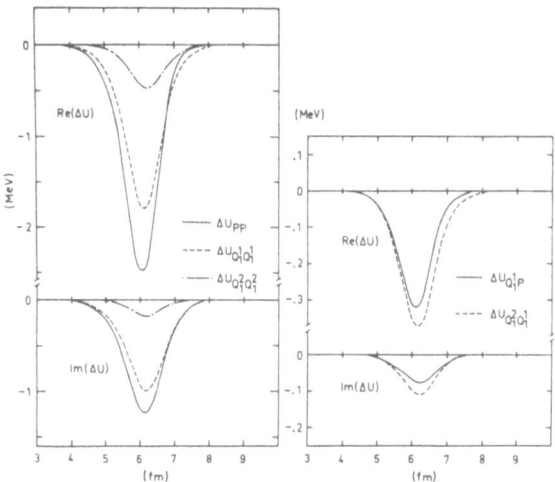

Figure 2. Polarization potentials for diagonal (left) and off-
diagonal (right) terms for different channels (solid: $(0^+, 0^+)$;
dashed: $(2^+, 0^+)$, dot-dashed: $(2^+, 2^+)$ in $^{12}C + ^{12}C$ scattering
(from ref. 13).

Fig. 2 gives an indication of what the polarization potentials look
like, calculated here for $^{12}C + ^{12}C$; the curves on the left give the diagonal polarization potentials for the elastic (solid curve), the aligned
substate of the single $2^+$ excitation channel (dashed) and for the mutual
$2^+$ excitation channel (dash-dotted), all for the grazing partial wave
$J=18$ at $E_{cm}=30$ MeV. All terms are surface-peaked, as one would have
naively expected. The effective coupling strength (on the right), to be
used in the DW calculation, acquires an imaginary part and its absolute
value is actually decreased by about 10 % compared to the bare value.
If a nuclear molecule with considerable overlap of the two nuclei
is to be reached, polarization effects have to be very large since such
states correspond to strongly distorted nuclear configurations. Indeed,
Chandra and myself have given arguments some years ago[15,16]) that a
molecule in $^{12}C + ^{12}C$ could be connected with a well developed shape iso-
mer in $^{24}Mg$ that is directly reached from the entrance channel by a
diabatic process. It is well known from many dynamical studies[17,18])
that such diabatic processes play a dominant role in the initial stages
of heavy ion collisions. The possible connection between "molecular
bands" with rotational states in the combined system has also been investigated by the Lund group[19]) and by Noto et al. in a paper contributed to
this conference[20]).
Finally, also the parity dependence of optical potentials extracted

from data on $^{16}O + ^{20}Ne$ by Kondo and Robson et al.[21] might have its origin in polarization effects since the sign of the parity dependence that is empirically needed seems to be opposite to that given by microscopic theories.

## 3. ALIGNMENT AND γ-DECAY OF $^{12}C + ^{12}C$

If indeed the observed structures (molecules) are connected with (maybe highly deformed) states of the combined nucleus then two experimental signatures are immediately apparent. First, the rotation of the dumbbell-like molecule should lead to a particular partition and orientation of angular momenta and second, there should be quantum transitions between different rotational states of the molecule. Experiments have been performed recently that address both of these questions and I will review them here together with some relevant new calculations.

In a classical picture one expects that formation of a molecular state corresponds to a sticking situation[16]. Sticking should lead to a transfer of angular momentum from the relative motion ($l_i$) into rotation of the combined system whose magnitude is given by (for symmetrical systems):

$$\Delta l = \frac{2}{7} l_i .$$

Typically, for $^{12}C + ^{12}C$ at an energy of $E_{cm} \approx 25$ MeV, $l = 14-16$ is dominant so that one expects an angular momentum transfer of $4\hbar$. Second, one expects that this angular momentum is parallel to the incoming relative angular momentum $\vec{l}_i$.

The big problem then is, how to actually observe these features. If, and only if, these are preserved on separation then one expects that the mutual $2^+$ excitation carries the signature of the molecule and that the angular momenta of the $^{12}C$ nuclei should be parallel to each other and to $\vec{l}_i$.

The probability for preservation of the molecular signatures during the separation can be estimated from the corresponding decay widths. These show[2] that e.g. for the resonance at $\approx 25$ MeV the decay into the single $2^+$ state is two times as probable as that into the mutual $2^+$ channel and that for the resonance at $\approx 19$ MeV the latter decay is essentially forbidden.

Measurements of the spin-alignment in such reactions as $^{12}C + ^{12}C$ and others have recently been performed[22-25]. The alignment in the single excitation $2^+$ channel in $^{12}C + ^{12}C$ is given in fig. 3. Here the alignment is defined by:

$$P_{zz} = \frac{1}{I(2I-1)} \left[ \sum_M 3M^2 P_M - I(I+1) \right]$$

where $P_M$ is the probability for population of the magnetic substate M. At the higher energies the measured alignment on the average does not exceed the average value of 1/3, corresponding to equal occupation of all magnetic substates. Most of the theoretical predictions of align-

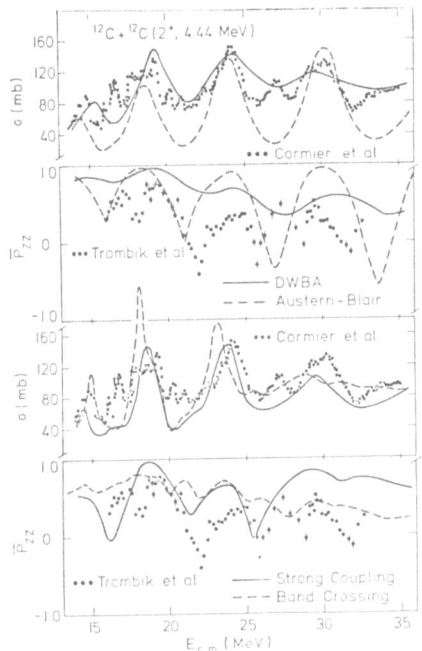

Figure 3. Comparison of experimental cross section and alignments with different model calculations (ref. 26).

ment[26]) give only an approximate description of the alignment data
whereas the cross sections are usually described much better. The theoretical calculations usually give a too strong alignment due to a strong
preference for the aligned configuration that is favored by energy-angu-
lar momentum matching. It is interesting to note that the calculations[27,
28]) do much better for the alignment in systems such as $^{16}O+^{16}O$ $(3^-)$
(ref. 29) where the inelastic coupling is smaller. This could indicate
that the polarization terms have not yet been properly treated for the
strongly coupled $^{12}C+^{12}C$ system.

For the mutual $2^+$ excitation channel, the experimental results change
dramatically; the alignment is now bigger and clearly correlated with
the gross structure in the cross sections (see fig. 4). Even more exciting is the result of a recent $\gamma$-$\gamma$-correlation experiment at the Darm-
stadt-Heidelberg crystal ball that shows that the $(|M_1|,|M_2|) = (2,2)$
component dominates (with more than 50 % probability) in the regions of
peaks in the cross section[24,25]. Since any other orientation would correspond to a large mismatch of $4\hbar$, it is very probable that the two
angular momenta are aligned parallel to each other and to the incoming
angular momentum with about 50 % probability in the $(2^+,2^+)$ channel[25].

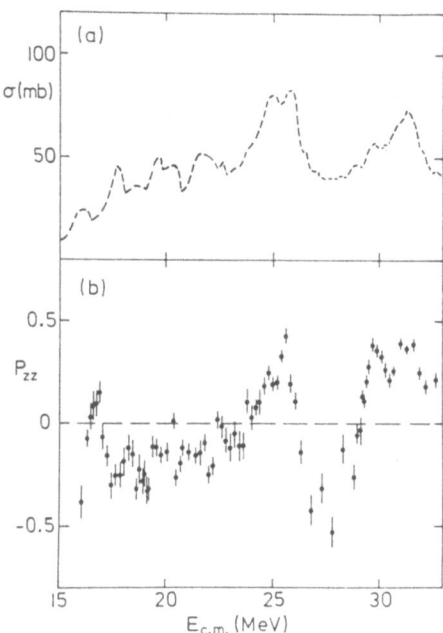

Figure 4. Cross section and alignment for the $(2^+, 2^+)$ mutual
excitation channel in $^{12}C + {^{12}}C$ (from ref. 25).

It is thus tempting to associate this fraction of this channel with a
molecular configuration (the other components are largest off resonance). Thus these alignment measurements seem to support the simple
expectations that one has for a rigidy rotating sticking complex. The
calculations, however, still do not give a perfect picture of the measured alignments, in particular the sharp minima off resonance are not
reproduced but the overall magnitude is described well[28].

Such a rotating nucleus should also be able to emit γ-radiation.
It has, therefore, been a problem that the two first experiments[30,31] to
look for this radiation have both failed to see any γ-decay. Estimates
based on the Bohr-Mottelson formula for a rotating nucleus with a qua-
drupole moment as determined in ref. 15 give $(\Gamma_\gamma / \Gamma) \sim 10^{-5}$ whereas the
experimental upper limits were $8 \cdot 10^{-6}$. Newer experiments[32] have pushed
this upper limit down to $6 \cdot 10^{-7}$.

A possible explanation for this failure to see the γ-transition is
that the theoretical estimate was probably too simplistic because it
assumed an unchanged intrinsic structure at the two rotational states.
If, however, polarization effects are large, as expected on the basis
of all the proceeding discussions, then this assumption is probably not
justified.

A first effort was made some years ago to incorporate this effect in a DW calculation[33]. Tanimura and myself have recently taken up those calculations again and have extended them to include coupled channel effects.

Even if the initial and the final channel are well specified there are still a number of possibilities for the intermediate resonance states that are illustrated in fig. 5. In view of the earlier discussion

INTERNAL CONTRIBUTIONS

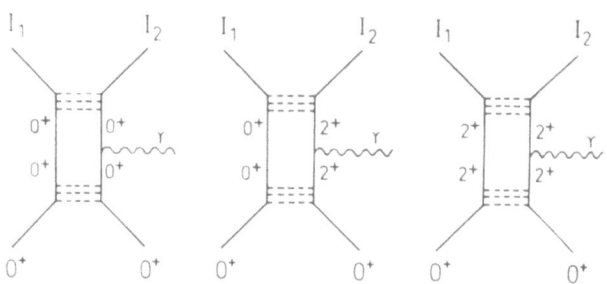

Figure 5. Intermediate resonance states in $^{12}C + ^{12}C$ $\gamma$-emission.

only the graph on the right, in which a mutual $2^+$ excitation is formed in the intermediate state, is the one to associate with a molecular state. The bremsstrahlung cross section is then given for electric quadrupole radiation by[34]:

$$\frac{d\sigma}{d\Omega dq} \sim \sum_\mu |<f|r^2 Y_{2\mu}|i>|^2$$

where $|i>$ and $(f)$ are the initial and final scattering states, here obtained from a coupled channel calculation. Tanimura has spent a large effort to find optimal parameters for the CC calculation at the initial and final energy, reproducing simultaneously the elastic and single $2^+$ excitation angular distributions and the integrated mutual $(2^+, 2^+)$ cross section[35]. The wavefunctions obtained from those calculations can thus be used to calculate the transition matrix element.

Fig. 6 shows the calculated cross section for different final channels as a function of final energy, assuming a fixed bombarding energy $E_{c.m.}$ = 25.3 MeV. It is seen that the cross section exhibits a fairly sharp resonance - like structure at $E_f \approx 18.7$ MeV and that the values for the $(0^+, 0^+)$ and the $(0^+, 2^+)$ final states are of similar magnitude. The $(2^+, 2^+)$ final channel is here suppressed because $\Gamma_{2^+2^+}$ is very small at $E_{c.m.} \approx 19$ MeV (ref. 2).

The partial decay width $\Gamma_\gamma/\Gamma$ that can be extracted from this calculated cross sections is:

$$(\Gamma_\gamma/\Gamma)_{theor} = 5 \cdot 10^{-7}$$

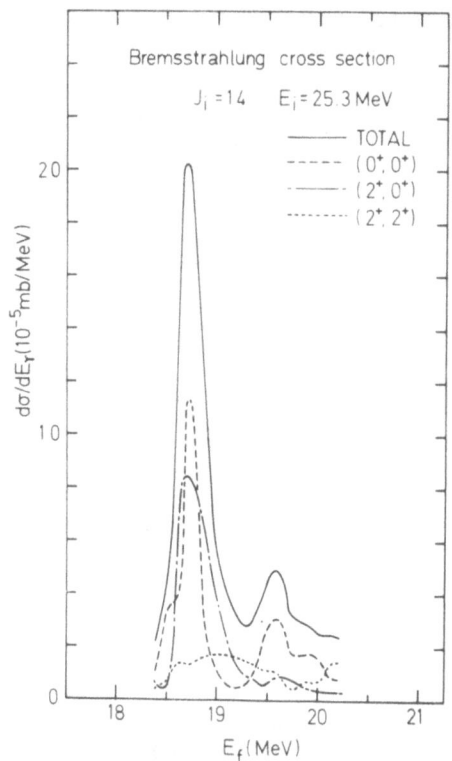

Figure 6. Calculated $\gamma$-emission cross section for different final channels as function of final energy.

and thus just at the upper limit of the new experiments[32]).

In the calculated cross sections one has coherent contributions from all the intermediate states shown in fig. 5 as well as from external bremsstrahlung. In fig. 7 we decompose the cross section into its intermediate state contributions. Here the very interesting result emerges that only those two intermediate states play any significant role in which either one or both Carbon nuclei are excited; the mutual $(2^+, 2^+)$ state contributes about 50 % to the total yield.

This is an extremely interesting result and in line with the experimentally determined M-substate populations[24]) (see above). It suggests that these experiments should be repeated at the higher bombarding energy of $E_{c.m.} \approx 30$ MeV so that the final state, following $\gamma$-decay, could decay into the $(2^+, 2^+)$ channel ($\Gamma_{2^+2^+} = 0.45$ at $E_{c.m.} = 25$ MeV vs. $\leq 0.15$ at 19 MeV). This would eliminate the additional decay factor associated with the second vertex in fig. 5.

In nonsymmetrical systems such as $^{12}C+^{16}O$ one expects dipole radiation as the dominant decay mode[36]). It is thus gratifying to see in a contribution to this conference that such experiments are indeed presently undertaken at Yale[37]). The quadrupole transition strengths for

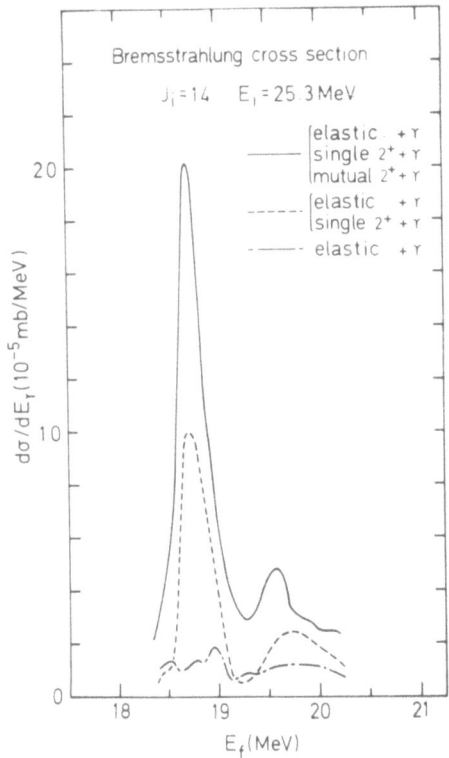

**Figure 7.** Decomposition of the γ-emission cross section into intermediate state contributions.

this system have also been calculated in a contribution to this conference[38]) and show similar magnitudes as the values quoted above.


SUMMARY AND CONCLUSION

Although more than 20 years have passed since the notion of nuclear molecular states was proposed there is still no definitive experiment proving or disproving their existence. I feel strongly that part of this difficulty has to be attributed to the fact that most measurements have only dealt with cross sections trying to find resonances and gross structures. Since these resonances are, however, often strongly fragmented and since there is a gradual transition from simple shape resonances to molecular states such experiments can only be suggestive, but not definitive. The new generation of highly sophisticated experiments, however, that try to determine alignment, maybe also polarization, and the γ-decay open up a critical period for "molecular physics". The first results obtained so far are encouraging as they seem to show the signatures of sticking, i.e. of a rigidly rotating combined system.

Theoretical descriptions are still mainly based on direct reaction theories. As such they cannot incorporate the properties of the combined system and have to rely on, often ambiguous, fits to selected data. At present, nevertheless the γ-decay calculations look quite promising whereas similar quality alignment calculations are not available. It would thus be very desirable to have calculations that describe as many details of the cross sections as possible and give simultaneously the γ-decay widths and the alignments.

ACKNOWLEDGEMENT

The calculations reported here were mostly performed by O. Tanimura and R. Wolf; they deserve most of the credit for these developments. I also gratefully acknowledge many stimulating discussions with V. Metag.

REFERENCES

[1]) D.A. Bromley et al., Phys. Rev. Lett. 4 (1960) 365

[2]) T.M. Cormier et al., Phys. Rev. Lett. 40 (1978) 924

[3]) Resonances in Heavy Ion Reactions, ed. K. Eberhard, Lecture Notes in Physics 156 (Springer, Heidelberg 1982)

[4]) R.L. Phillips et al., Phys. Rev. Lett. 42 (1979) 566

[5]) O. Tanimura and U. Mosel, Phys. Lett. 105B (1931) 334

[6]) Y. Kondo et al., Phys. Rev. C19 (1979) 1356

[7]) O. Tanimura and T. Tazawa, Phys. Rep. 61 (1980) 253

[8]) G.R. Satchler, Direct Nuclear Reactions, Clarendon Press, Oxford, 1983

[9]) W.S. Freeman et al., Phys. Rev. Lett. 45 (1980) 1479

[10]) O. Tanimura and U. Mosel, Phys. Rev. C24 (1981) 321

[11]) H. Feshbach, Ann. Phys. (NY) 19 (1962) 287

[12]) R. Wolf et al., Z. Physik 305 (1982) 179
     R. Wolf et al., Nucl. Phys. A414 (1984) 162

[13]) O. Tanimura et al., Phys. Lett. 132B (1983) 249

[14]) O. Tanimura, to be published

[15]) H. Chandra and U. Mosel, Nucl. Phys, A298 (1978) 151

[16]) U. Mosel, in ref. 3, p. 358

[17]) D. Glas and U. Mosel, Nucl. Phys. A298 (1978) 151

[18]) W. Cassing et al., Z. Physik A314 (1983) 309

[19]) I. Ragnarsson et al., Phys. Script. 24 (1981) 215

[20]) H. Noto, contribution to this conference

[21]) Y. Kondo et al., Nucl. Phys. A410 (1983) 289
Y. Kondo et al., contributed papers to this conference

[22]) S.J. Willett et al., Phys. Rev. C28 (1983) 1986

[23]) W. Trombik et al., Phys. Lett. 135B (1984) 271

[24]) D. Konnerth et al., Proc. XV Masurian Summer School on Nuclear
Physics, Mikolajki, 1983, to be published in Nukleonika

[25]) W. Trautmann, Brookhaven preprint, 1984, to be published in Comments
on Nuclear and Particle Physics

[26]) O. Tanimura and U. Mosel, Phys. Lett. 114B (1982) 7

[27]) Y. Abe, private communication

[28]) O. Tanimura, private communication

[29]) N. Kato et al., Phys. Lett. 120B (1983) 314

[30]) R.L. McGrath et al., Phys. Rev. C24 (1981) 2374

[31]) V. Metag et al., Phys. Rev. C25 (1982) 1486

[32]) V. Metag, in: Nuclear Physics with Heavy Ions,
P. Braun-Munzinger ed., (Harwood Academic Publ., Chur 1984), p. 391

[33]) J.S. Blair and H. Sherif, University of Washington, Nuclear Physics
Laboratory, Annual Report 1981

[34]) K. Alder et al., Rev. Mod. Phys. 28 (1956) 432

[35]) O. Tanimura et al., to be published

[36]) Y. Alhassid et al., Phys. Rev. Lett. 49 (1982) 1482

[37]) M. Ruscev et al., contribution to this conference

[38]) K. Kato et al., contribution to this conference

# RECENT PROGRESS IN THE STUDY OF HEAVY-ION RESONANCES

R. R. Betts
Argonne National Laboratory,
Argonne, Illinois 60439 USA

ABSTRACT.   Recent results from the study of resonances in heavy-ion reactions are reviewed.  Emphasis is placed on the range of occurrence of this phenomenon and on the constraints placed by the data for the heaviest systems on various model interpretations.  Experiments which probe specific spectroscopic properties of the observed resonances will be discussed, together with the relationship of these results to nuclear structure models.

## 1. INTRODUCTION

The discussion of heavy-ion resonances forms a natural part of a conference devoted to clustering aspects of nuclear structure and reactions, the nuclear molecule being the archetype of this feature of nuclear structure.  In the six years since the last conference in this series, there has been tremendous activity in this area of research, spurred in part by the wider availability of precision heavy ion beams.  Evidence of this activity is found in the two conferences wholly devoted to the subject of heavy-ion resonances [1,2] held in the last four years, and the talks on heavy-ion resonances which have formed important parts of numerous other meetings on heavy-ion reactions and nuclear structure.

One of the most exciting developments in this field has been the realization that heavy-ion resonances, previously thought to be confined to collisions between a few p-shell nuclei, are in fact a much more widespread phenomenon.  The establishment of the phenomenology of the occurrence of these resonances is an area of major interest, and the systematics of resonance appearance are proving to provide important clues as to the underlying nuclear structure and also to place stringent constraints on possible model interpretations.

In parallel with the study of even heavier systems, important new results have emerged from studies of the systematics of spins, energies and specific spectroscopic properties of the resonances. Although these results do not yet allow us to uniquely define the degrees of freedom responsible for the resonance spectra they do

133

*J. S. Lilley and M. A. Nagarajan (eds.), Clustering Aspects of Nuclear Structure, 133–151.*

provide the first steps in this direction, and it may well be that
the final unravelling of this puzzle will clarify the relationships
between seemingly dissimilar approaches to models of nuclear structure.

## 2.  SYSTEMATICS OF OCCURRENCE

For many years, following the observation of resonances in the scat-
tering and reactions of $^{12}C$ + $^{12}C$, $^{12}C$ + $^{16}O$ and $^{16}O$ + $^{16}O$, it was
assumed that the occurrence of resonances was unique to these three
systems.   It was felt that the increased number of open reaction
channels for heavier systems would result in a strong damping of any
resonant states and thus render them effectively unobservable.   This
belief was shaken by the observation of a strong backward rise in
the elastic scattering of $^{16}O$ + $^{28}Si$ [3].    Similar results were
subsequently obtained for a number of neighboring systems [4-9].
The data for $^{16}O$ + $^{28}Si$ elastic scattering at a bombarding energy of
55 MeV  are  shown  in  Fig. 1.    The  backward  angle  cross-section

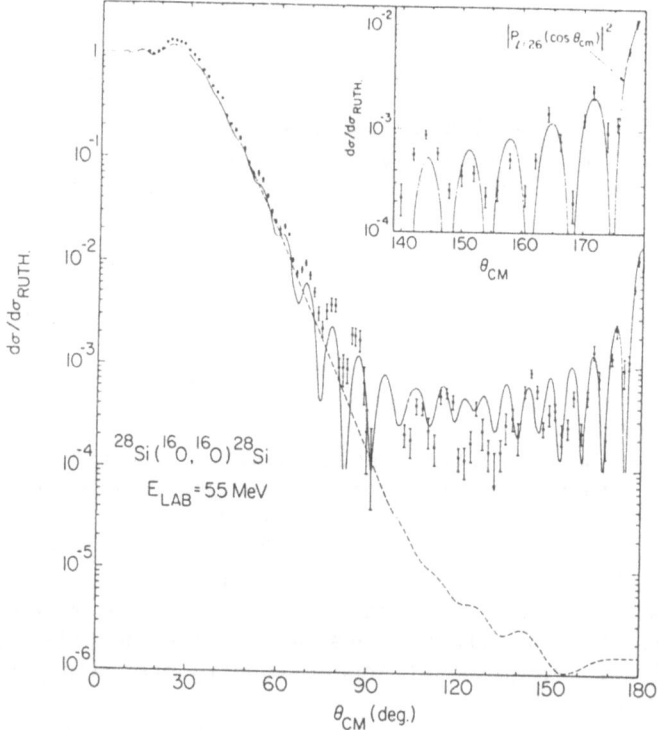

Figure 1.   $^{28}Si(^{16}O,^{16}O)$ elastic scattering angular distribution.

oscillates rapidly with angle and is reasonably well described by a
pure Legendre polynomial squared dependence, indicating the domin-
ance of at most a few partial waves, and therefore suggestive of

resonance formation. This suggestion was confirmed by excitation function measurements [5] of the large angle elastic and inelastic scattering for this system which showed a number of broad structures each of which was fragmented into much narrower structures. These excitation functions were subjected to an extensive analysis [10] which showed that although there were quite strong correlations between the elastic and inelastic yields, statistical fluctuations could not be ruled out as the origin of the observed narrow structures.

These results were extended to an even heavier system with the observation [11,12] of similar resonance behavior for $^{28}$Si + $^{28}$Si, and the study of this and neighboring systems has provided a focus of recent experimental efforts. An elastic scattering angular distribution for the $^{28}$Si + $^{28}$Si system measured at a bombarding energy of 120 MeV – twice the Coulomb barrier – is shown in Fig. 2. The

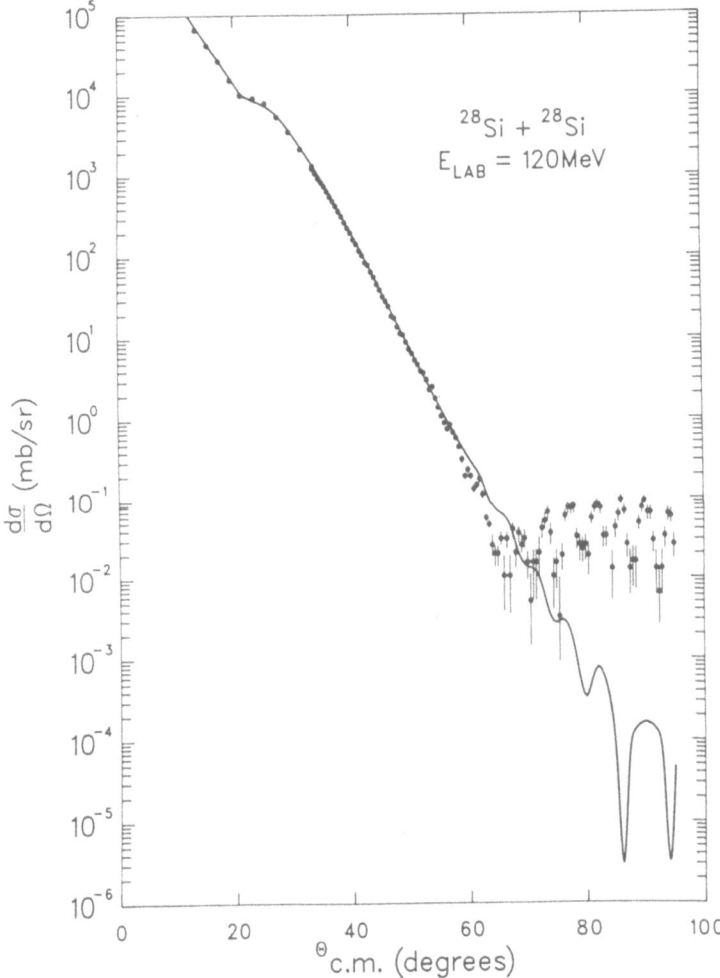

Figure 2.    $^{28}$Si($^{28}$Si, $^{28}$Si) elastic scattering angular distribution.

data forward of $\Theta_{cm} \approx 60°$ are characteristic of elastic scattering under the influence of Coulomb repulsion with strong absorption for small impact parameters - the curve drawn through the data is the result of an optical model calculation with such a strongly absorbing potential. For larger angles, as was observed in the $^{16}O + ^{28}Si$ case, the data show rapid angular oscillations with a period equal to that of a squared Legendre polynomial of order 40, equal to the grazing angular momentum.

A spectrum of elastically and inelastically scattered particles for the $^{28}Si + ^{28}Si$ system, measured [13] at large angles, is shown in Fig. 3. This indicates that the elastic scattering channel is,

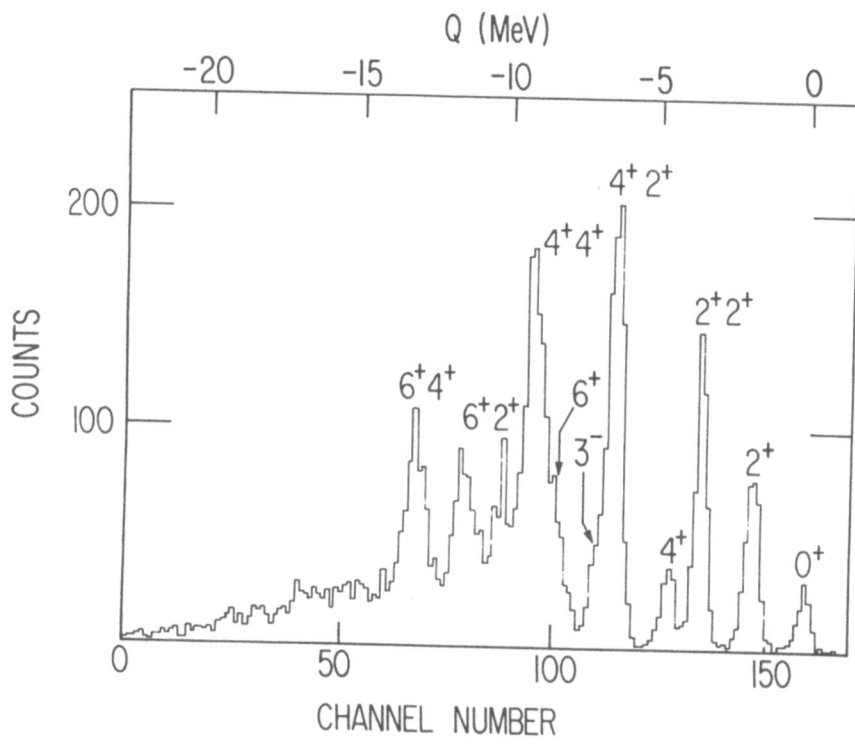

Figure 3. Q-value spectrum for $^{28}Si + ^{28}Si$ large-angle scattering.

so to speak, only the tip of the iceberg and that the large angle yields are dominated by transitions to quite negative Q-values and to combinations of inelastic excitations with large spins. These Q-values and fragment spins are similar to those that would be obtained from a simple sticking model of the collision process - suggesting that the observed large angle cross-sections result from the formation and decay of a long-lived complex which scissions at quite large deformations. The long time scale of the process leading to the large angle yields is underlined by the results of excitation function measurements [14] for large angle elastic and inelastic scattering channels shown in Fig. 4, which display narrow correlated structures of width $\Gamma_{cm} \simeq 150$ keV. These data have been subjected

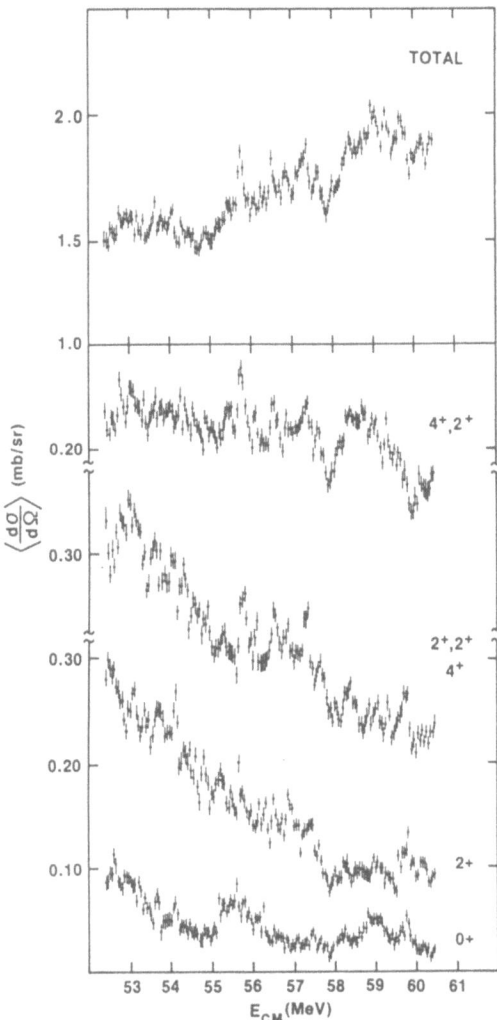

Figure 4.   Excitation functions for $^{28}$Si + $^{28}$Si large-angle elastic and inelastic scattering.

to a detailed statistical analysis and the narrow structures shown [15] to be inconsistent with statistical fluctuations at better than the 99% confidence level.   On the basis of these data and analyses it can therefore be concluded that the narrow structures are more or less isolated resonances with spins of order 40 ℏ lying at excitation energies of 70 MeV or so in the compound nucleus.

Similar data have been obtained [16,17] for a number of systems close to $^{28}$Si + $^{28}$Si.   To date detailed excitation functions have been measured for the $^{24}$Mg + $^{24}$Mg, $^{24}$Mg + $^{26}$Mg, $^{28}$Si + $^{30}$Si, $^{30}$Si + $^{30}$Si and $^{32}$S + $^{32}$S systems and rather more fragmentary data obtained for $^{24}$Mg + $^{28}$Si, $^{32}$S + $^{24}$Mg and $^{40}$Ca + $^{40}$Ca. A number of interesting systematics emerge from these data. Figure 5 shows the

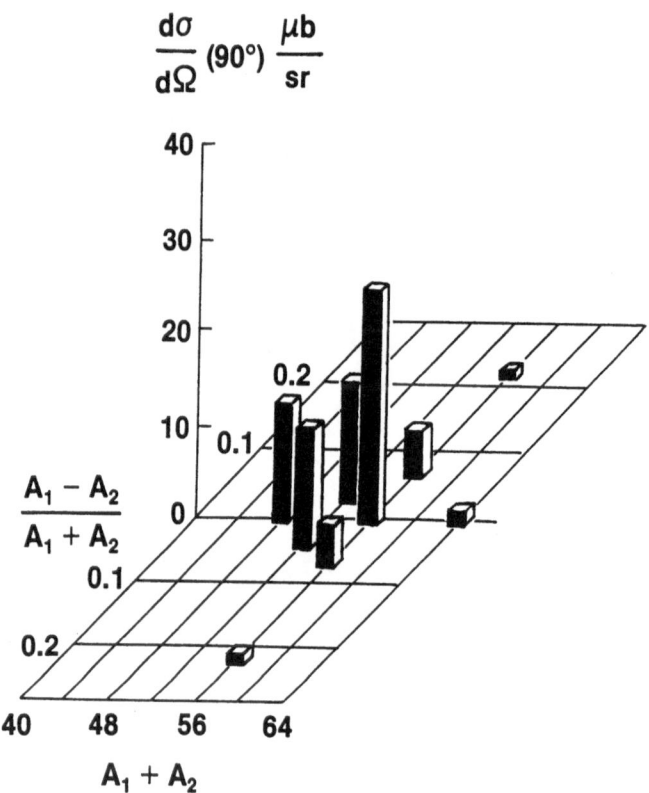

Figure 5. Systematics of $\Theta_{cm}$ = 90° elastic scattering for various α-particle nuclei.

average $\Theta_{cm}$ = 90° cross-sections at energies close to twice the Coulomb barrier for the elastic scattering of $^{24}$Mg, $^{28}$Si and $^{32}$S from each other, plotted versus combined mass ($A_1 + A_2$) and mass-asymmetry ($A_1-A_2$)/($A_1+A_2$). The experimental cross-sections for the symmetric systems have been reduced by a factor of four to account for the effects of symmetrization. All of these systems show evidence for the formation of resonances and the variations in these cross-sections therefore presumably represent a real nuclear structure effect in the formation and decay of the resonance states. The data show a maximum for $^{28}$Si + $^{28}$Si with a decline for both increasing mass and mass-asymmetry which would tend to suggest the importance of shell effects centered near N=Z=28 and a favoring of decays into symmetric fragments. At present, however, this incomplete picture can only lead to speculation and a wider knowledge of these systematics is clearly desirable.

Another strong dependence is observed as a function of the addition of neutrons to the colliding ions.   This effect is known from the studies of lighter systems but is perhaps appears most dramatically in the comparison of data [17] for the $^{28}$Si + $^{28}$Si, $^{28}$Si + $^{30}$Si and $^{30}$Si + $^{30}$Si systems shown in Fig. 6.   The total large-angle

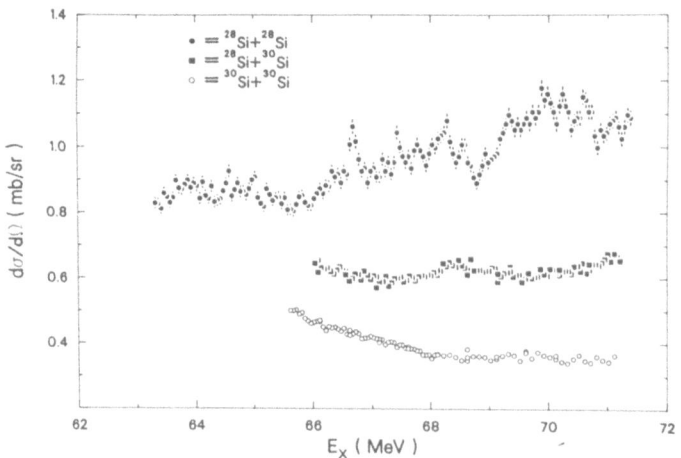

Figure 6.   Excitation functions for $^{28}$Si + $^{28}$Si, $^{28}$Si + $^{30}$Si and $^{30}$Si + $^{30}$Si large-angle scattering.

elastic, inelastic and transfer cross-sections are shown plotted as a function of compound nucleus excitation energy.   Associated with the addition of neutrons we see not only a disappearance of the narrow resonances which appear so prominently in the $^{28}$Si + $^{28}$Si data but also a decrease in the total large angle cross-section of roughly a factor of two for each pair of added neutrons.   These data therefore also strongly suggest the importance of shell and/or alpha particle clustering effects in the determination of resonance occurrence and in the determination of the overall features of the orbiting cross-sections.

## 3.   SURFACE TRANSPARENCY IN HEAVY-ION SCATTERING

Many attempts [18-22] have been made to describe the energy-dependent structure observed in heavy-ion collisions in terms of resonances in the ion-ion scattering potential and the coupling of resonances in the elastic channel to inelastic excitations or other more complex degrees of freedom.   An apparently essential ingredient in these calculations is that the absorptive part of the potential be weak in the surface region of the colliding nuclei in order that the standing wave resonances formed in the real potential for grazing partial waves not become too broad.   Calculations of angular distributions and excitation functions within this framework have

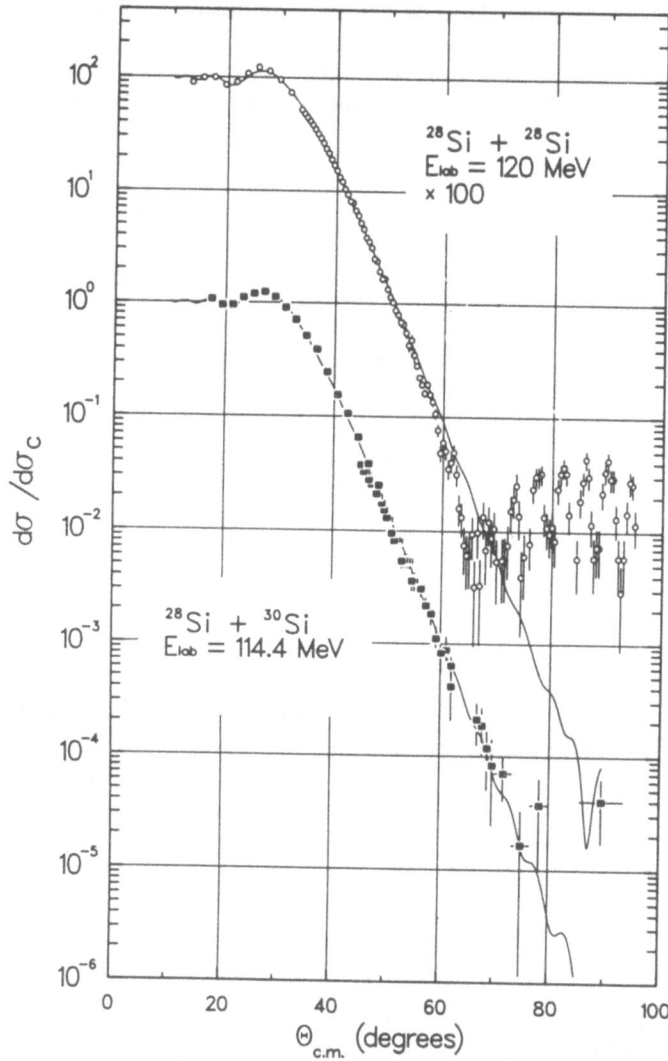

Figure 7.    Comparison of $^{28}$Si + $^{28}$Si and $^{28}$Si + $^{30}$Si elastic
scattering.

been reported for a number of systems and have, in general, proved
quite successful in reproducing the qualitative features of the
experimental observations.    Within the framework of this approach,
the damping of resonances for non alpha-particle systems is understood to arise as a result of increased absorption in the surface
due to coupling of the entrance channel to the increased number of
reaction channels coming from the addition of valence particles.
     For the heavier systems, as mentioned earlier, the forward
angle elastic scattering data seem to be consistent with strong
absorption in the surface whereas resonance behavior is still

observed at back angles. To further elucidate this point, a comparison has been made of elastic scattering angular distributions [23] for $^{28}$Si + $^{28}$Si at 120 MeV and $^{28}$Si + $^{30}$Si at 114.4 MeV as shown in Fig. 7. Within the strong absorption diffraction theory due to Frahn [24], these two angular distributions should have identical shapes and thus differences between the two could be taken to signify differences in the overall features of the absorptive potential, the real potential being expected to remain essentially unchanged. We see that, for angles less than 60°, the two angular distributions are in fact identical, whereas at large angles the measured cross-sections differ by at least an order of magnitude. The curves shown with the data are optical model calculations with the same strongly absorbing potential and are in good agreement with the measurements for $d\sigma/d\sigma_c \geqslant 10^{-3}$. Calculations [25] with surface transparent potentials, while they do agree qualitatively with the large angle cross-sections for $^{28}$Si + $^{28}$Si and do reproduce the general features of the excitation functions, have large oscillations at forward angles which are manifestly not present in the data. This analysis suggests, therefore, that the observed differences in the large angle cross-sections for these two systems arise from differences in detail in the absorption rather than some overall feature such as surface transparency, and that the resonance component may be better described by coupling to some specific degree of freedom which has its origin in the structure of the composite system.

Alternatively, it may be that more complex optical model parameterizations can provide better agreement with these experimental results and thus resolve this inconsistency. A particularly interesting new suggestion along these lines has come from the results of a detailed phase-shift analysis of $^{12}$C + $^{12}$C elastic scattering [26]. The absorption coefficients ($n_\ell$) extracted from the measured angular distributions are shown plotted versus bombarding energy in Fig 8. The dashed curve shows the absorption coefficients calculated with a standard Woods-Saxon surface-transparent potential [27], whereas the solid curves were calculated with a potential which has an additional attractive component outside the normal real potential. The latter calculation does, indeed, give a somewhat better description of the gross behavior of the experimental phases. It is tempting to associate this secondary minimum, added in a purely ad hoc fashion, with the second minimum observed in deformed shell model calculations [28] of the potential energy surface for $^{24}$Mg. The theoretical justification for associating a feature of an adiabatic calculation with the optical potential is, however, lacking at the present time.


4.  FISSION AND SHAPE ISOMERS

Many of the features observed in the large angle cross-sections for the heavier systems such as complete energy damping and fragment spins are characteristic of the decay of a fully equilibrated

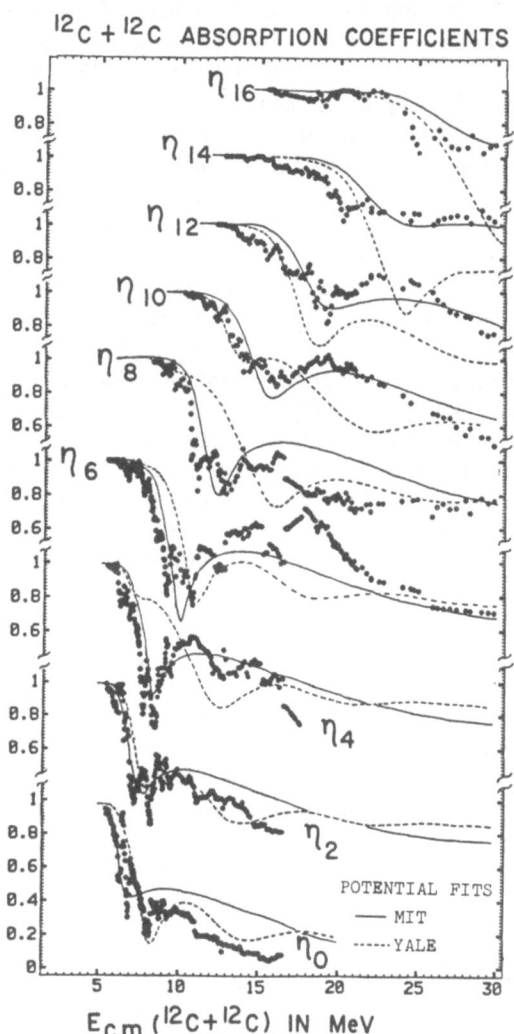

Figure 8. Absorption coefficients deduced from a phase shift analysis of $^{12}C + ^{12}C$ elastic scattering.

system. It is therefore worthwhile to ask the question, to what extent are the overall features of these long time-scale processes describable as due to fission or statistical decay of the compound nucleus. In particular, the strong isotopic dependence of the large angle cross-sections for $^{28}Si + ^{28}Si$, $^{28}Si + ^{30}Si$ and $^{30}Si + ^{30}Si$ is rather well correlated with the expected increase of the compound nucleus fission barrier with increasing N/Z. This possibility has been investigated by measurements of the cross-sections for fission-like fragments produced in $^{16}O + ^{40}Ca$ and $^{16}O + ^{44}Ca$ reactions. Preliminary results [29] of these measurements are shown as a function of bombarding energy in Fig 9. The observed "fission" yield for the $^{44}$Ca target lies approximately a factor of two lower

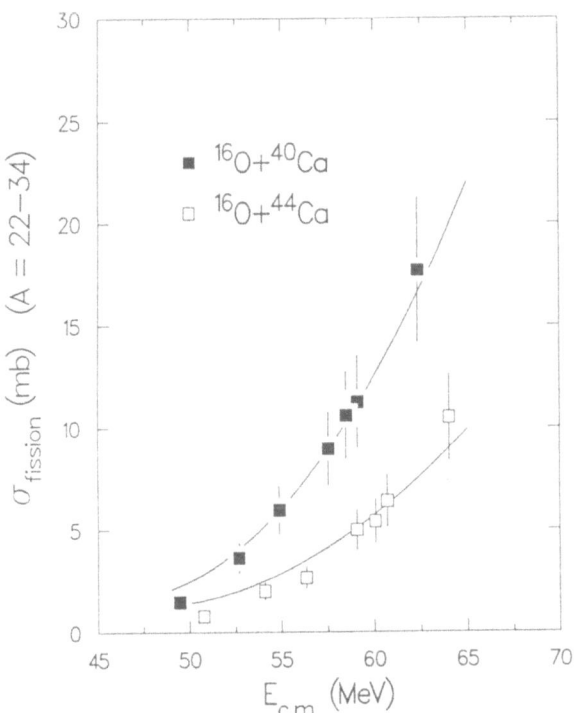

Figure 9. Energy dependence of "fission" cross-sections for the $^{16}$O + $^{40}$Ca and $^{16}$O + $^{44}$Ca systems.

than that for $^{40}$Ca over the whole energy range. The relative magnitude and overall energy dependence of these cross-sections are well reproduced by statistical model calculations which reproduce the measured total fusion cross-sections and we therefore conclude that the production of these symmetric masses is strongly influenced by the compound nucleus fission barrier.

Given the importance of a fission or fission-like mechanism in the large angle scattering processes, it may then be hypothesized that the narrow resonances correspond to fissioning shape isomeric states which lie in secondary minima in the compound nucleus potential energy surface. That such "superdeformed" shape isomers may exist is supported by the results of Strutinsky-type calculations of potential energy surfaces for nuclei in the mass regions of interest [30]. An example of such a calculation is shown in Fig. 10 where the potential energy surface of $^{48}$Cr at spin 32 is shown plotted versus β and γ deformations. This calculation shows clearly a secondary minimum for large prolate deformations - the vibrational states in which could then be associated with resonances observed

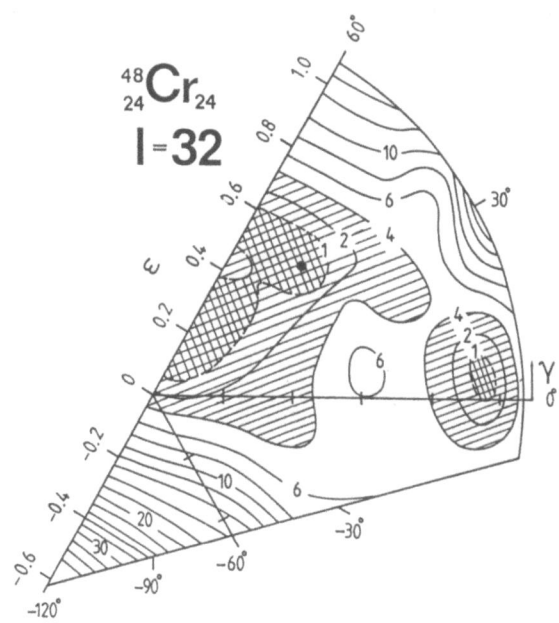

Figure 10. Potential energy surface for $^{48}$Cr at spin 32 calculated in the Nilsson Strutinsky prescription.

[16] in $^{24}$Mg + $^{24}$Mg scattering. Similar calculations have been performed for a number of nuclei in this mass region and show not only the appearance of superdeformed minima in $^{48}$Cr and $^{56}$Ni but also the disappearance of these minima in $^{58}$Ni and $^{60}$Ni – as observed experimentally with the vanishing of resonances in the $^{28}$Si + $^{30}$Si and $^{30}$Si + $^{30}$Si systems.

It is clear that this hypothesis can qualitatively account for many of the features of the observed resonances and an outstanding experimental challenge is to pursue these studies to even heavier systems where similar calculations again predict [31] the occurrence of such semi-stable superdeformed shapes.

## 5. RESONANCE SPECTROSCOPY

Obviously, one of the ultimate goals in the study of heavy ion resonances is to obtain a complete experimental picture of resonance energies, spins and other spectroscopic properties that can then be matched against the predicted resonance spectra, decay patterns, etc. This is clearly a most difficult task and despite intensive study for only a few of the lighter systems has any semblance of order emerged. The best studied of all systems in which resonances are seen is $^{12}$C + $^{12}$C and, as is illustrated in the upper portion of Fig. 11, the low-lying spectrum of resonances is by now quite well characterized [32]. An attempt to understand these data in terms of

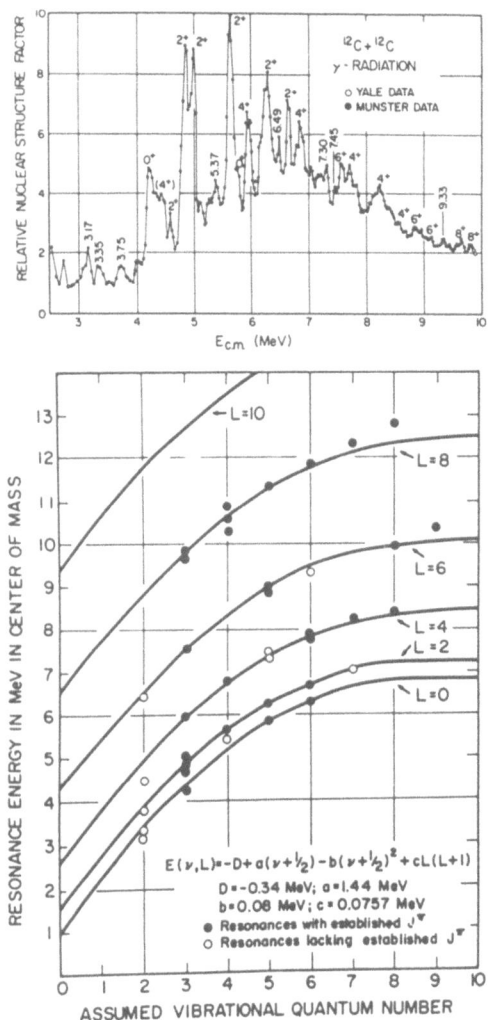

Figure 11.    Resonances in the $^{12}C$ + $^{12}C$ system near the Coulomb
barrier (upper) together with O(4) fit to the experimental spectrum
(lower).

a rotation-vibration spectrum [33,34] was reasonably successful in
accounting for the observed resonances, as has been a more recent
analysis in the framework of the interacting boson approximation
[35], shown in the lower section of Fig. 11.  These two approaches
differ in the nature of the vibrational mode which is coupled to the
rotation of the system, the former assuming the usual quadrupole
phonons familiar from the geometric model whereas the IBA analysis
assumes a dipole [36] or p-boson.  To date, there has been no
resolution of this fundamental difference although the search for
enhanced quadrupole and dipole electromagnetic transactions is a

subject of quite some activity.  As a comment on this, Fig. 12 shows
the  spectrum  of  single  particle  levels  in  an  axially  deformed

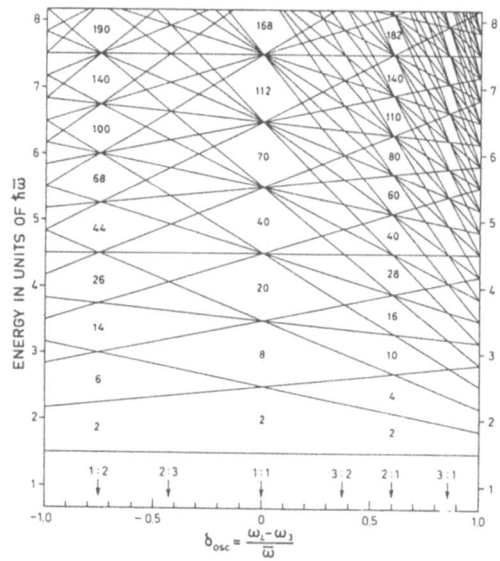

Figure 12.    Single particle energy levels in an axially deformed
harmonic oscillator potential.

harmonic oscillator potential [33].   This simplified figure clearly
shows the large shell gaps which appear at large prolate deformations ($\delta = 0.6$) which in more detailed calculations are responsible
for the "superdeformed" shape isomers.  It is important to note that
whereas the spherical shells ($\delta = 0.0$) contain only levels of a
single parity, the superdeformed shells contain equal mixtures of
levels with both parities.   It might therefore be expected that at
large  deformations  the  residual  interactions  might  lead  to  the
existence of collective dipole and octupole modes in addition to the
more familiar quadrupole vibrational modes.
       Although the measurement of energies and spins forms an essen-
tial step in the spectroscopy of heavy ion resonances, the above
discussion shows that, in general, such information is insufficient.   What is required in addition is a knowledge of the patterns
of decay strengths of each of the resonances.   Only then can the
familial relationships between resonances inherent in the theoretical models be established.   The search for electromagnetic transitions between resonances has already been mentioned.  Evidence has
been found [38,39] for "collective" E1 transitions in both light
($^{18}O$) and heavy ($^{218}Ra$) nuclei, but not yet in a truly heavy ion
resonance situation.   Considerable effort has been expended [40] in
the search for molecular γ-ray transitions and the results thus far,
although only upper limits, indicate that the full collective E2

strength is far from concentrated in a single transition between two resonances.

Some more positive systematics have emerged from a detailed analysis of $\alpha$-decay widths of $^{12}C + ^{12}C$ resonances. The $\alpha$-decay widths for a number of $J^{\pi} = 12^+$ resonances in $^{24}Mg$ have been measured [41] to excited states of $^{20}Ne$. The widths for decay to the excited 8 particle - 4 hole band were found to be systematically much larger than those to the ground and other low-lying bands. It is concluded on the basis of these results that this enhancement reflects the microscopic structure of the $^{12}C + ^{12}C$ resonances as multi-particle multi-hole "quartet" states which then places the resonance states on an equal footing with much better understood excitations throughout the s-d shell. On the theoretical side, Strutinsky-type calculations for $^{24}Mg$ and $^{20}Ne$ again show the presence of "superdeformed" minima in the potential energy surface and the states in these minima have been associated with specific multi-particle multi-hole configurations [28]. What is now required are calculations of the spectrum of levels in the superdeformed minima, together with estimates of their particle decay widths.

For the heavier systems, the spectroscopy is in a much more primitive state. A detailed study [42] has been made of the decay widths of the three narrow J=34 resonances observed near $E_{c.m.} = 46$ MeV in the $^{24}Mg + ^{24}Mg$ system. Excitation functions for the elastic scattering and two ground-state transfer channels are shown in Fig. 13. The analysis of these data in terms of single isolated resonances indicates that the reduced widths for decay into $^{24}Mg + ^{24}Mg$ are an order of magnitude larger than those for decay into other mass channels. The sum of the widths observed in the elastic and inelastic $^{24}Mg + ^{24}Mg$ channels accounts for some 30% of the observed total width whereas the widths into other mass channels total at most a few percent. These results then indicate a mass-symmetric configuration for the resonant states, although the absolute values of the reduced widths shows that this is far from a pure molecular configuration. The inability of these measurements to account for the observed total width of the resonances suggest a significant spreading width which should manifest itself in the appearance of these resonances in compound nuclear decay channels such as $\alpha$-emission.


5.  RELATIONSHIP TO OTHER WORK

There is now considerable evidence that heavy-ion resonances are a widespread phenomenon which have their roots in the nuclear structure of the composite system. It is of interest to speculate on the relationship of these observations with other results from nuclear structure and atomic physics.

An obvious connection exists between the theoretically predicted high spin fissioning shape isomers and the well known and studied fissioning shape isomers in actinide nuclei. This latter class of excitation has been extensively studied and the expected

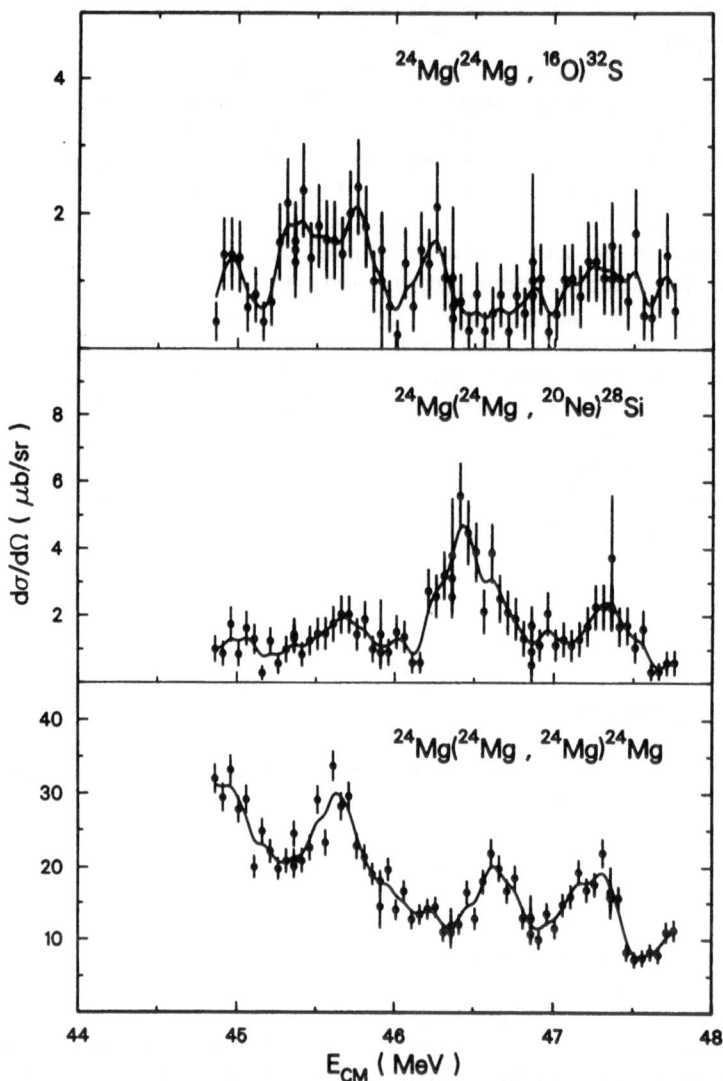

Figure 13. Comparison of excitation functions for three ground
state transitions in the $^{24}$Mg + $^{24}$Mg system.

extreme deformation of these isomers has recently been verified
experimentally [43]. Such superdeformed states have also been
predicted to occur in a number of other mass regions. In the mass
150 region where the fission decay channel is suppressed, evidence
has been found that such states not only exist, but have lifetimes
comparable to those for γ-ray or n-emission [44,45]. From the γ-ray
work, it has been possible to deduce the moment of inertia of the
emitting nucleus and the value obtained is in agreement with that
predicted by the deformed shell model calculations. Of the known
nuclear species, it then appears that superdeformed or molecule-like
configurations are ubiquitous and represent a quite general feature
of nuclear structure.

For much heavier systems, there is evidence from the study of fission fragment angular distributions that extended nuclear configurations have been formed [46] with lifetimes intermediate between those of the compound nucleus and the rotation time of the system. This quasi-fission process is believed to be the precursor of deep inelastic scattering and the study of the dependence of this process on the neutron and proton number of both the composite system and of the final fragments is of quite some interest.

Finally, in the heaviest nuclear species it is possible to produce, studies of positron spectra [47,48] have revealed the unexpected existence of narrow lines. Such a spectrum for the U + Cm system is shown in Fig. 14. From the narrow width of the observed

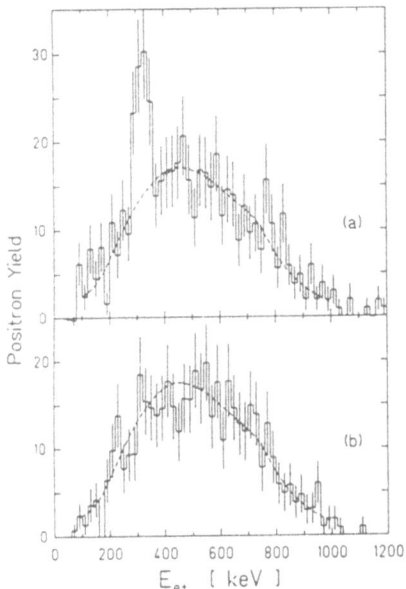

Figure 14. Positron spectra observed in coincidence with scattered ions in the U + Cm system.

structures it has been concluded that a long-lived nuclear system has been formed even though the bombarding energy is considerably below that at which nuclear interactions should take place. The possibility exists again, therefore, that a long range interaction has led to the formation of a long-lived extended nuclear system which would be quite analogous to our current picture of the resonances seen in much lighter systems.

The connection between the above observations and the resonances in nuclei with A < 56 is perhaps tenuous, nevertheless the attempt to link these results is clearly a fascinating challenge.

## 6. ACKNOWLEDGMENTS

The unpublished results presented here are the result of a collaboration between ANL, Pennsylvania and BNL. In particular, Surender
Saini, Stephen Sanders and Flemming Videbaek have made large contributions. I am indebted to those who sent reprints, preprints and
unpublished results. I apologize for sins of omission. This work
was performed under the auspices of the Office of High Energy and
Nuclear Physics, Division of Nuclear Physics, U. S. Department of
Energy under contract number W-31-109-ENG-38.

## 7. REFERENCES

1. Proceedings of International Conference on Resonant Behavior
   on Heavy Ion Systems, June 1980, Greece, ed. G. Vourvopoulos
   (National Printing Office of Greece, Athens, 1980).
2. Proceedings of Workshop on Resonances in Heavy Ion Reactions,
   October 1981, Bad Honnef, West Germany, ed. K. A. Eberhard
   (Springer, Berlin, 1982).
3. P. Braun-Munzinger et al., Phys. Rev. Letters **38**, 944 (1977).
4. J. Barrette et al., Phys. Rev. Letters **40**, 445 (1978).
5. M. R. Clover et al., Phys. Rev. Letters **40**, 1008 (1978).
6. M. Paul et al., Phys. Rev. Letters **40**, 1310 (1978).
7. C. K. Gelbke et al., Phys. Rev. Letters **41**, 1778 (1978).
8. J. C. Peng et al., Phys. Letters **80B**, 35 (1978).
9. S. M. Lee et al., Phys. Rev. Letters **42**, 429 (1979).
10. J. Barrette et al., Phys. Rev. **C20**, 1759 (1979).
11. R. R. Betts, S. B. DiCenzo and J. F. Petersen, Phys. Rev.
    Letters, **43**, 253 (1979).
12. R. R. Betts, S. B. DiCenzo and J. F. Petersen, Phys. Letters
    **100B**, 117 (1981).
13. R. R. Betts et al., Phys. Rev. Letters **46**, 313 (1981).
14. R. R. Betts, B. B. Back, and B. G. Glagola, Phys. Rev. Letters
    **47**, 23 (1981).
15. S. Saini and R. R. Betts, Phys. Rev. **C29**, 1769 (1984).
16. R. W. Zurmühle et al., Phys. Letters **129B**, 384 (1983).
17. Unpublished data from an ANL-Pennsylvania-BNL collaboration.
18. B. Imanishi, Nucl. Phys. **A125**, 33 (1968).
19. W. Scheid, W. Greiner and R. Lemmer, Phys. Rev. Letters **25**,
    176 (1970).
20. A. Gobbi et al., Phys. Rev. **C7**, 30 (1973).
21. D. Hahn, W. Scheid and J. Y. Park in Proceedings of Workshop
    on Resonances in Heavy Ion Reactions loc. cit p. 337.
22. Y. Abe et al., in Proceedings of International Conference on
    Nuclear Physics with Heavy Ions April 1983, Stony Brook, USA
    ed. P. Braun-Munzinger (Harwood, New York 1984) p. 369.
23. F. Videbaek, et al., (to be published).
24. W. E. Frahn, Ann Phys. (N.Y.) **72**, 524 (1972).
25. A. Thiel, W. Greiner and W. Scheid, Phys. Rev. **C29**, 864 (1984)
    and private communication.

26.  E. R. Cosman, C. E. Ordonez and R. J. Ledoux in <u>Proceedings of International Conference on Fusion Reactions Below the Coulomb Barrier,</u> June 1984, Cambridge, USA, ed. S. Steadman (Springer, Berlin), in press.

27.  W. Reilly et al., Nuovo Cimento **13A**, 897 (1973).

28.  G. Leander and S. E. Larsson, Nucl. Phys. **A239**, 93 (1975).

29.  S. J. Sanders et al., (to be published).

30.  T. Bengtsson, M. Faber, M. Ploszajczak, I. Ragnarsson and S. Åberg, Lund MPh-84/01 preprint and private communication.

31.  W. Nazarewicz et al., Lund MPh-84/08 preprint.

32.  K. A. Erb et al., Phys. Rev. **C22**, 507 (1980).

33.  N. Cindro et al., Phys. Rev. Letters **39**, 1135 (1977).

34.  N. Cindro and W. Greiner, J. Phys. **G9**, L175 (1983).

35.  K. A. Erb and D. A. Bromley, Phys. Rev. **C23**, 2781 (1981).

36.  F. Iachello, Phys. Rev. **C23**, 2778 (1981).

37.  Aa. Bohr and B. Mottelson in <u>Nuclear Structure</u> (Benjamin, Reading Mass 1975) v. II, p. 592.

38.  M. Gai et al., Phys. Rev. Letters **50**, 239 (1983).

39.  M. Gai et al., Phys. Rev. Letters **51**, 646 (1983).

40.  V. Metag et al., in <u>Proceedings of International Conference with Heavy Ions</u> loc. cit. p. 391.

41.  R. J. Ledoux et al., Phys. Rev. C (in press).

42.  S. Saini et al., (to be published).

43.  C. E. Bemis, Jr. et al., Phys. Rev. Letters **43**, 1854 (1979).

44.  B. M. Nyako et al., Phys. Rev. Letters **52**, 507 (1984).

45.  W. Kühn et al., Phys. Rev. Letters **51**, 1858 (1983).

46.  B. B. Back et al., Phys. Rev. Letters **50**, 818 (1983).

47.  J. Schweppe et al., Phys. Rev. Letters **51**, 2261 (1983).

48.  M. Clemente et al., Phys. Letters **137B**, 41 (1984).

POLARIZATION STUDIES

# CLUSTERING ASPECTS OF POLARIZED BEAM STUDIES

R.C. Johnson
Department of Physics
University of Surrey
Guildford
Surrey
GU2 5XH
England

ABSTRACT. Recent analyses of experiments with low energy polarized $^6$Li snd $^7$Li beams are reviewed. It is emphasized that elastic scattering polarization observables show a much stronger sensitivity to details of the projectile structure than do unpolarized cross-sections. The way that the measured vector and tensor analysing powers for Li scattering from a range of heavier targets can be understood in terms of simple cluster models for the ground and low lying excited states of the projectiles is described.

Although the simple cluster picture can explain the main features of the Li elastic scattering data systematic deviations do exist which it is difficult to understand in this way. The evidence for a novel type of spin-dependence in heavy-ion scattering is discussed. The quite different physics involved in the spin-dependence of light and heavy-ion interactions is emphasised.

Recent work on (d,$\alpha$) reactions with polarized deuterons is also reviewed and discussed from the point of view of nuclear structure information.

## 1. INTRODUCTION

For reasons which have mainly to do with atomic shell structure the development of techniques for producing and accelerating beams of polarized ions has been confined to nucleons, deuterons, tritons, $^3$He, $^6$Li and $^7$Li, and $^{23}$Na. It is of considerable interest to this Conference that this list includes examples of nuclei for which a very simple picture of non-overlapping clusters is a good starting point. As I shall explain, the experiments which have been carried out so far with polarized $^6$Li and $^7$Li by the Heidelberg-Marburg Group [1] show a consistency with the simple cluster picture. Detailed features of the cluster structure show themselves in a definite way in the data, which so far have been confined to energies near the Coulomb barrier. There

155

J. S. Lilley and M. A. Nagarajan (eds.), Clustering Aspects of Nuclear Structure, 155–177.
© 1985 by D. Reidel Publishing Company.

is a hint that deviations from the simple picture are important at higher energies.

Much of my talk will be about the elastic scattering of $^6$Li and $^7$Li by various targets. Unless otherwise stated I use the term "polarized" to refer to any spin state with non-random occupation of magnetic sub-states. It is remarkable that elastic scattering with polarized beams of these nuclei display considerable sensitivity to projectile structure. The differential cross-section for unpolarized projectiles of the energies of the Heidelberg-Marburg experiments ($E \leqslant 20$ MeV) definitely do not show such sensitivity.

## 2. DESCRIPTION OF POLARIZED BEAM EXPERIMENTS

The experimental results I will discuss involve particles with spin $I > \frac{1}{2}$ in their ground states: $^6$Li has $I = 1$ and $^7$Li has $I = \frac{3}{2}$. For spin-$\frac{1}{2}$ particles incident on an unpolarized target the analyzing power is specified by a vector $\underline{A}$ which points in the direction of $\underline{n}$, the normal to the reaction plane. I will use the notation which is more commonly used in this field in which the real quantity $iT_{11}(\theta)$ which is simply proportional to the magnitude of $\underline{A}$ is quoted.

The quantity $iT_{11}$ is called the vector analyzing power. It is often expressed in terms of the left-right asymmetry for a process induced by a polarized beam, but it can also be expressed in terms of the change in intensity observed when all the beam particles are in a definite magnetic sub-state with respect to the direction $\underline{n}$. For example

$$iT_{11}(\theta) = (\sigma(N_{\frac{1}{2}} = \tfrac{1}{2}) - \sigma_0)/\sqrt{2}\sigma_0 = A(\theta)/\sqrt{2} \ , \tag{1}$$

where $\sigma_0$ is the cross-section for an unpolarized beam.

For $I = 1$ the complete specification of the analyzing power of a reaction requires further quantities which can be conveniently organised in terms of tensors of higher rank than 1. Referred to the conventional right-handed Madison co-ordinate system, with z-axis along the incident beam direction and y-axis along $\underline{n}$, these are usually denoted by the real quantities $T_{20}(\theta)$, $T_{21}(\theta)$ and $T_{22}(\theta)$. In general these provide three, independent, angular distributions in addition to $iT_{11}(\theta)$ and $\sigma_0(\theta)$. The population of magnetic sub-states in the incident beam can be chosen (I refer to papers by experimentalists of how this is actually done) so that these quantities are simply obtained from differences in measured intensities; for example

$$iT_{11}(\theta) = \tfrac{1}{2} \sqrt{3}(\sigma(N_1 = \tfrac{2}{3}, N_0 = \tfrac{1}{3}) - \sigma_0)/\sigma_0 \ , \tag{2}$$

$$T_{20}(\theta) = \sqrt{2}(\sigma(N_1 = N_{-1} = \tfrac{1}{2}, N_0 = 0) - \sigma_0)/\sigma_0 \ . \tag{3}$$

The beam involved in the expression for $T_{20}$ is referred to as "purely aligned". In the case when the alignment axis is $\underline{n}$ this tensor analyzing component is referred to as $^TT_{20}$ by the Heidelberg-Marburg Group to distinguish it from the corresponding component in the Madison co-ordinate system. Of course, components in different co-ordinate systems are related; for example

$$^{T}T_{20} = -\tfrac{1}{2} T_{20} - \tfrac{1}{2} \sqrt{6}\, T_{22} \ . \tag{4}$$

In heavy-ion scattering a co-ordinate system of particular usefulness is one in which the y-axis is as in the Madison system, but the z-axis is perpendicular to the apsidal vector, $\underline{R}_0(\theta)$ to the turning point of the classical Coulomb orbit i.e. perpendicular to the momentum transfer in elastic scattering (see Fig. 1). Tensor components in this system are denoted by barred quantities, $\overline{T}_{20}$, etc. [2]. The point about using different co-ordinate systems is that different dynamical assumptions often show up as a characteristic relation between components which is most simply expressed in a particular co-ordinate system

For $I = \tfrac{3}{2}$ similar definitions of vector and $2^{nd}$ rank tensor analyzing powers apply and in particular

$$T_{20}(\theta) = (\sigma(N_{\frac{3}{2}} = N_{-\frac{3}{2}} = \tfrac{1}{2}) - \sigma_0)/\sigma_0 \ . \tag{5}$$

In addition now we have the possibility of $3^{rd}$ rank tensors. As yet very little is known experimentally about these observables.

An important rule-of-thumb for elastic scatterring is that in a situation where spin-dependent forces are perturbative, which is often the case, an observable of a given rank is most sensitive to terms in the Hamiltonian of the same rank in the spin operator; hence, $iT_{11}(\theta)$ is dominated by $\underline{L}\cdot\underline{I}$ potentials, $T_{20}(\theta)$ by second rank tensor spin forces, etc.. The experimentalist can choose which parts of the Hamiltonian he wishes to emphasise by suitably choosing the population of the magnetic

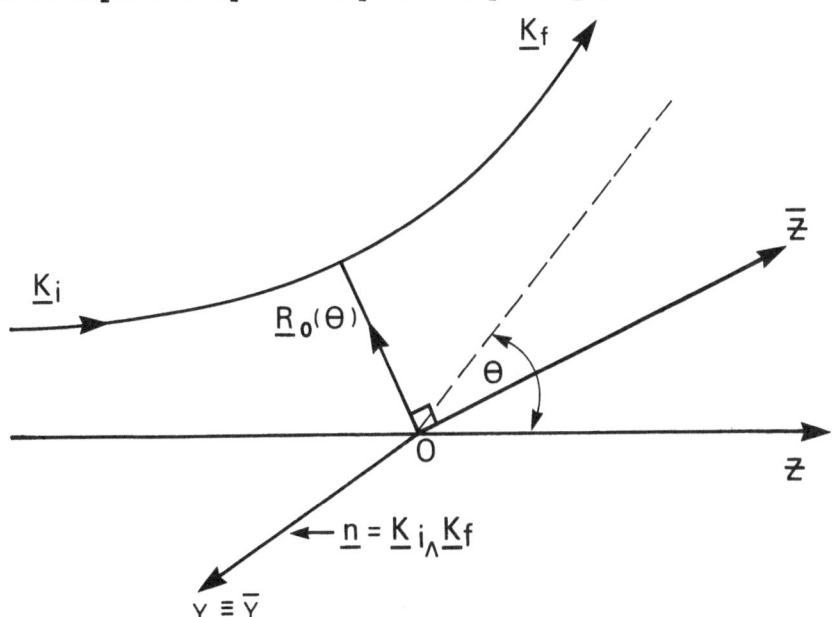

Figure 1. Madison Convention and "barred" co-ordinate systems.

sub-states of the incident beam. It is remarkable that modern experimental techniques have brought this type of approach within our powers.

## 3.   TRANSFER REACTIONS

I first want to discuss transfer reactions with polarized beams. This topic is likely to become even more important in the future when modern polarized sources become available.

Polarization phenomena in transfer reactions with light ions are known to show a strong sensitivity to the structure of the initial and final states involved. The classic example is the $(d,p)$ reaction where the vector analyzing powers for two final states with the same neutron orbital angular momentum $\ell_n$, but different $j_n(= \ell_n \pm \frac{1}{2})$ have opposite sign. Tensor analyzing powers are extremely sensitive to the amplitude and phase of the deuteron D-state component. These results are well established experimentally and can be understood in a direct reaction framework [3].

Of relevance to this Conference are recent studies of $(d,\alpha)$ reactions with vector and tensor polarized deuterons at Birmingham and TUNL. These studies [4-9] are providing important tests of the amplitude and phases of deuteron-like cluster components in shell-model wavefunctions as well as giving the clearest evidence to date for the direct observation of effects arising from a D-state component of the alpha-particle.

In the simplest one-step direct reaction model, the $A(d,\alpha)B$ reaction amplitude involves two vertices: $d + (np) \rightarrow \alpha$ and the associated overlap $\langle d,(np)|\alpha\rangle$, and $A \rightarrow (np) + B$ and its overlap integral $\langle B(np)|A\rangle$, where $(np)$ denotes a cluster with the quantum numbers of the deuteron. The possible angular momentum structure of these overlaps is severely restricted by the spins of the nuclei involved. For example, the $\langle d,(np)|\alpha\rangle$ overlap has two components in which the deuteron clusters have relative angular momentum $\ell = 0$ and $\ell = 2$. The latter component arises mainly from non-spherical components in the $\alpha$-particle wavefunction induced by the tensor force in the nucleon-nucleon interactions. There is considerable interest in checking the predictions of 4-nucleon calculations of these components. In the case of a $0^+ \rightarrow 1^+$ transition the $\langle A|(np), B\rangle$ overlap also contains $L = 0$ and $L = 2$ components in general. Polarization studies are important, because for example, in the absence of a deuteron-nucleus spin-orbit potential, the tensor analyzing powers in a $0^+ \rightarrow 1^+$ transition are predicted to be zero unless $L = 2$ admixtures are present in at least one of the overlaps.

An important consideration here is that at the low energies studied so far (9 MeV $< E_d <$ 16 MeV) the only property of the $\alpha$-particle D-state which enters the tensor analyzing powers is the number $D_2(d,\alpha)$ which is closely related to the asymptotic D/S ratio of the $\langle d,(np)|\alpha\rangle$ overlap. This number is not related in any simple way to the D-state probability, but it is nevertheless an important property of the $\alpha$-particle, as is the analogous quantity for the deuteron [10,11]. All the transitions studied so far involve this same quantity $D_2$, whereas of course they cover a range of completely unrelated $\langle A|(np),B \rangle$ overlaps. The relative phases of the components of the latter and the sign and magnitude of $D_2(d,\alpha)$ have a crucial influence on the tensor analyzing power data. The Birmingham–Surrey study [8] showed that the $^{40}Ca(d,\alpha)^{38}K$ data at 12·3 MeV could be fitted with $D_2 = 0$, but only if the relative amplitude of $L = 2$ and $L = 0$ (np) clusters in $^{40}Ca$ had the opposite sign to that predicted by shell–model calculation [12]. Consistency with the shell–model calculations could be obtained provided $D_2(d,\alpha)$ was given a value close to $- 0·31$ $fm^2$. This value is close to that obtained from other [5,6,9] $(d,\alpha)$ studies as well as microscopic calculations of $\alpha$–particle wavefunctions [4,6].

The reactions $(d,^6Li)$, $(^6Li,d)$, $(^6Li,\alpha)$, $(^7Li,\alpha)$ hold out similar promise with regard to the structure of d and $\alpha$–cluster states as well as the $\langle d,\alpha|^6Li \rangle$ and $\langle \alpha,t|^7Li \rangle$ vertices. The deuteron–nucleus spin–orbit force can by itself generate tensor analyzing powers in a transfer reaction and this can cause problems. The vector analyzing power can often give an important guide here as well as confirming the dominant orbital angular momentum transfer involved. The $^6Li$ and $^7Li$ spin–orbit forces are probably less important, and we can learn here from the insights gained from elastic scattering which I will discuss later. A more serious problem is the importance of multi–step processes. In some cases these can be ignored with some confidence, but the subject has been surrounded with considerable controversy and a lot of open questions remain [13]. The $(d,\alpha)$ studies I have highlighted certainly ignore these contributions and obtain a consistent account of the polarization data. It is probably significant that the latter involve ratios of intensities so that any renormalization of direction reaction amplitudes by higher–order processes can possibly be ignored.

Polarization phenomena in heavy ion transfer reactions have also aroused some interest, although these studies have often been confined to measurements of polarization in the final state in correlation measurements of decay products. In the case of vector polarization observables the sensitivity of their sign and magnitude to the spin of the final state in a direct reaction model has already been mentioned. Since much of the heavy–ion data does not involve transitions to discrete final states (although see the talk by Rae at this Conference), it is clear that in so far as a direct reaction mode is important, the observed polarization will be the result of an average over quantities whose signs are a strong function of the structure of the nuclei involved as well as reaction mechanism questions such as the relative contribution from near–side and far–side orbits. These points have been well brought out by Bond in a contribution to this conference [14].


## 4. ELASTIC SCATTERING OF $^6Li$ AND $^7Li$

My discussion will be confined to elastic scattering as being of most relevance to this Conference, although experiments with polarized Li beams, and very recently $^{23}Na$, have included fusion, inelastic scattering, and transfer reactions as well.

The vector and tensor analyzing powers involve different aspects of the structure of $^6$Li and $^7$Li and it is therefore convenient to discuss them separately.

### 4(i)  Vector Analyzing Power

Double folding models based on the M3Y effective nucleon-nucleon interaction have been very successful in providing an understanding of optical potentials for nucleons and heavy ions [15,16]. There have been serious problems [16,17,18] in the case of $^6$Li and $^7$Li for which folding models tend to over-estimate the depth of the real potential by a factor of 2. If we interpret this in a simple minded way as a renormalization of the effective interaction we would expect double folding models to overestimate the $^6$Li and $^7$Li spin-orbit forces. The latter are, of course, of most relevance to the vector analyzing power, $iT_{11}(\theta)$.

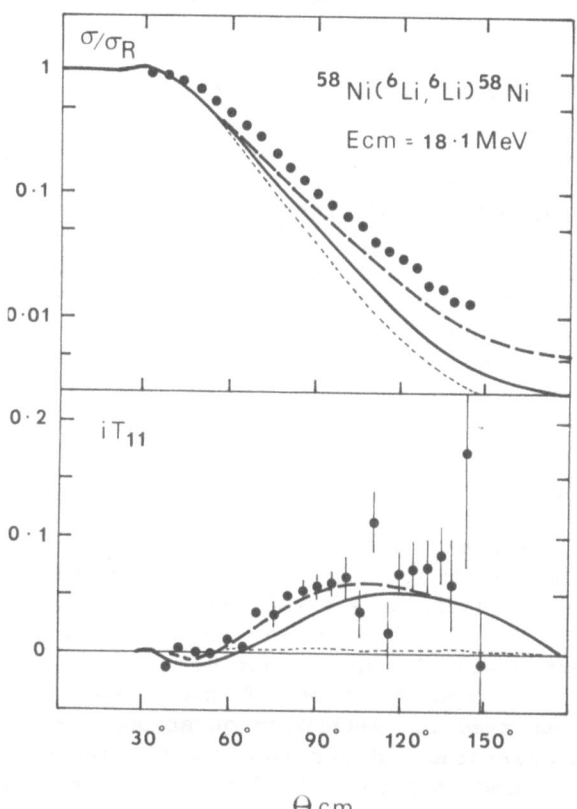

Figure 2. Data [43] and calculations [19] for the differential cross-section (ratio to Rutherford) and vector analyzing power for $^6$Li + $^{58}$Ni elastic scattering at $E_{cm}$ = 18.1 MeV. The short-dashed, solid and dashed curves are the results of one-channel, two-channel, and four-channel calculations, respectively, including the ground and excited states shown in Fig. 4.

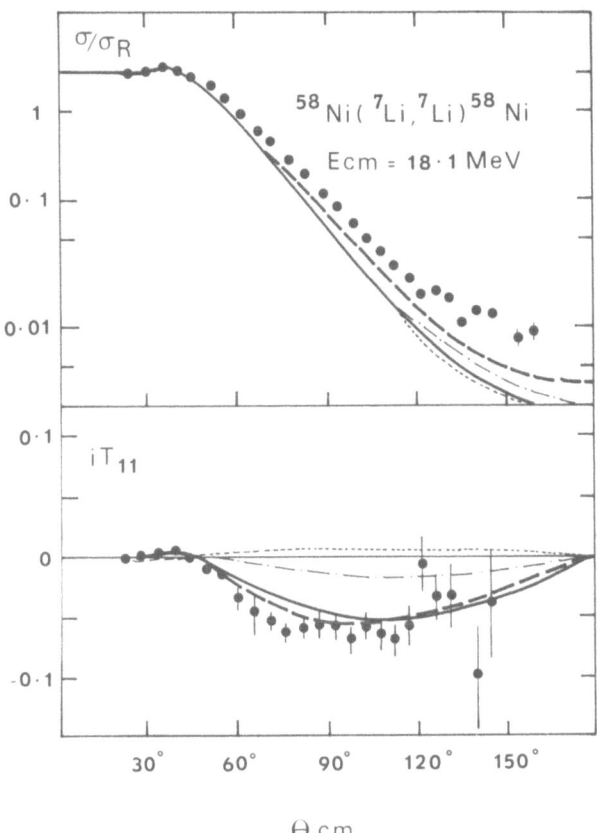

Figure 3. Data [44] and calculations [19] for the differential cross-section (ratio to Rutherford) and vector analyzing power for $^7Li + ^{58}Ni$ elastic scattering at $E_{cm} = 18.1$ MeV. The short–dashed and dash–dotted curves are the results of one–channel calculations neglecting and including the tensor contribution from the ground state matrix element, eq. (7), respectively. The solid and dashed curves are the results of two channel and four channel calculations, respectively, including the ground and excited states shown in Fig. 4.

What is the experimental situation? Figs. 2 and 3 show the differential cross-section and $iT_{11}(\theta)$ for $^6Li$ and $^7Li$ elastic scattering from $^{58}Ni$ at $E_{cm} = 18.1$ MeV. The cross-sections show a typical Fresnel pattern and are almost identical for the 2 isotopes, indicating a similarity between the central parts of their optical potentials. The vector analyzing powers are quite different. They have similar, rather small, magnitudes but opposite signs. Although the magnitudes are a small fraction of their maximum possible values, folding model spin–orbit potentials based on any reasonable model for the ground states of $^6Li$ and $^7Li$ give real spin–orbit forces which have the same sign and a similar magnitude for the two isotopes and which generate $iT_{11}(\theta)$ which are considerably smaller than the data, as well as giving the same sign for $^6Li$ and $^7Li$. An example [19] of this type of result is shown by the short-dashed curves in Figs. 2 and 3, to be discussed in more detail later. Thus the simple renormalization which does for the central potentials cannot possibly work for the analyzing powers.

An explanation of the situation and a excellent consistent account of the data without invoking any freely adjustable parameters has been given independently by both the Surrey group [19] and the Tokyo-Kyushu group [20]. They point out that a mechanism involving excitations of the projectile provides a natural explanation of the sign difference and a quantitative account of the observed magnitudes.

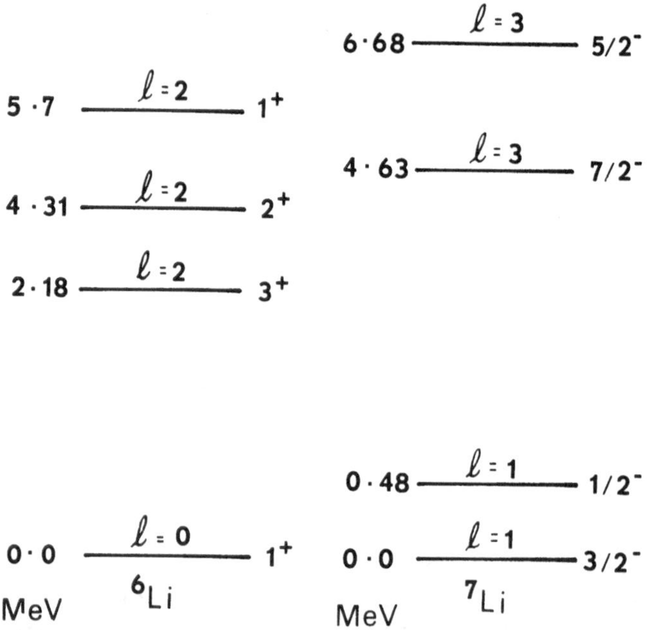

$6 \cdot 68 \underline{\quad \ell = 3 \quad} 5/2^-$

$5 \cdot 7 \underline{\quad \ell = 2 \quad} 1^+$

$4 \cdot 63 \underline{\quad \ell = 3 \quad} 7/2^-$

$4 \cdot 31 \underline{\quad \ell = 2 \quad} 2^+$

$2 \cdot 18 \underline{\quad \ell = 2 \quad} 3^+$

$0 \cdot 48 \underline{\quad \ell = 1 \quad} 1/2^-$

$0 \cdot 0 \underline{\quad \ell = 0 \quad} 1^+$    $0 \cdot 0 \underline{\quad \ell = 1 \quad} 3/2^-$

MeV    $^6Li$        MeV    $^7Li$

Figure 4. Low-lying levels of $^6Li$ taken into account in the projectile excitation calculations of refs. [19] and [20]. The projectile excitation effects are dominated by the $3^+$ state in $^6Li$ and the $\frac{1}{2}$ - state in $^7Li$.

Both $^6Li$ and $^7Li$ are known to have a well-developed deuteron structure. The low-lying T = 0 states of $^6Li$ can be thought of as $\alpha + d$ with various states of intercluster angular momentum coupled to the spin of the deuteron, and similarly for $^7Li$ in terms of $(\alpha + t)$ clusters. The relevant states are shown in Fig. 4 with their intercluster angular momentum assignments. At the energies near the Coulomb barrier involved in the Heidelberg experiments the most important contributions arise from configurations in which the inter-cluster separations are large so that explicit antisymmetrization can be ignored.

The Hamiltonian is assumed to be

$$H = H_6 + K_R + V_{\alpha A}(\underline{R} + \tfrac{1}{3}\,\underline{r}) + V_{dA}(\underline{R} - \tfrac{2}{3}\,\underline{r}) \; , \qquad (6)$$

where I have taken the $^6$Li case for definiteness. In eq. (6), $H_6$ is the internal Hamiltonian for $^6$Li and $K_R$ is the $^6$Li-target relative kinetic energy operator. The potentials $V_{\alpha A}$ and $V_{dA}$ are complex optical potentials chosen to describe $\alpha$ and d scattering from the target A ($\equiv$ $^{58}$Ni) at an appropriate-energy and which therefore implicitly include effects due to excitations of $^{58}$Ni. The co-ordinates $\underline{R}$ and $\underline{r}$ describe the $^6$Li-$^{58}$Ni and d-$\alpha$ separations, respectively.

A key feature of the mechanism invoked in refs. 19 and 20, is that even if the cluster-target potentials are purely central they can excite the cluster states through matrix elements of the form

$$<(\ell s)IM_I|V_{\alpha A} + V_{\alpha A}|(\ell' s)I'M'_I> \; ,$$

$$\qquad (7)$$

where $\ell, \ell'$ are intercluster angular momenta, I, I' are possible spins of $^6$Li and s = 1 is the deuteron spin (s = $\tfrac{1}{2}$ in the $^7$Li case). These matrix elements are large and when the coupled-channel equations associated with eqs. (6) and (7) are solved they generate the observed cross-sections and analyzing powers, including those for the excitations of the first excited state of $^7$Li, without the use of effective charges.

The results of coupled-channels calculations, taken from ref. 19, are shown in Figs. 2 and 3. The agreement with experiment is excellent. The single channel calculation, in which the vector analyzing powers are generated by the cluster-$^{58}$Ni spin-orbit potentials, are shown by the short-dashed curves. As already mentioned, this calculation gives essentially zero $iT_{11}(\theta)$, although the main features of $\sigma(\theta)$ are given by this single-folding model. Thus, the picture that emerges from this work is that the effective spin-orbit interaction is almost entirely of dynamical origin and has very little to do with the spin-dependence of the nucleon-nucleon interaction. This situation should be contrasted with the case of the nucleon and deuteron spin-orbit potentials which are understood microscopically, at least qualitatively, as arising from the spin-dependence of the nucleon-nucleon interaction.

A very nice feature of the work of ref. 19 is that they provide a simple way of understanding the effective spin-orbit force induced by projectile excitation and why it has a different sign for $^6$Li and $^7$Li at the energy of the Heidelberg experiment. Fig. 5 shows the angular momentum coupling involved in a 2-step contribution to elastic $^6$Li scattering via the first excited 3$^+$ state in which $\ell$ = 2 and therefore $\underline{\ell}$ and the deuteron spin $\underline{s}$ are parallel. Following the earlier discussion the deuteron-target spin-orbit interactions is ignored so that $\underline{s}$ is a constant of the motion. For $^6$Li incident with orbital angular momentum L relative to the target, its total angular momentum is $\underline{J} = \underline{L} + \underline{I}$. The Fig. 5(a) shows that if J = L + I the orbital angular momentum of $^6$Li$^*$(3$^+$) relative to the target must be L' = L - 2 in the semi-classical limit, whereas L' = L + 2 if J = L - I (Fig. 5(b)). There exists, therefore, a kinematic connection between J and L'.

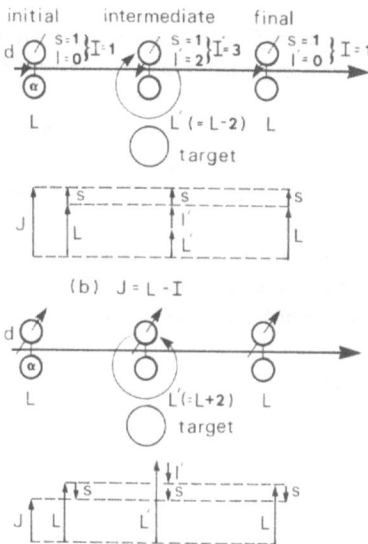

**Figure 5.** Schematic representation of ⁶Li scattering with projectile excitation in the intermediate state. The symbols are defined in the text. Parts (a) and (b) are for partial waves with J = L + I and J = L - I, respectively

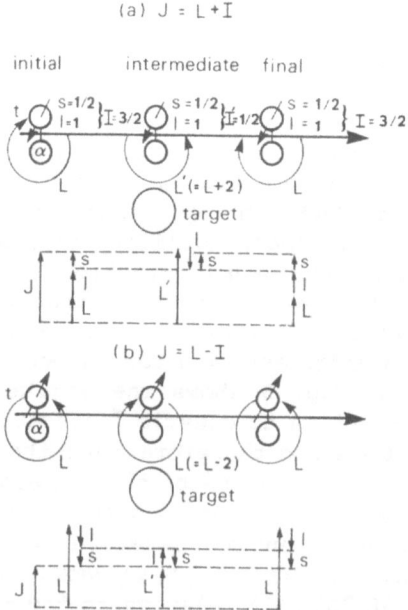

**Figure 6.** Schematic representation of ⁷Li scattering with projectile excitation in the intermediate state. See caption to Fig. 5.

Dynamical effects arise because the intermediate channel has a negative Q value and hence, in grazing partial waves, channels with $L' = L - 2$ will be strongly favoured over channels with $L' = L + 2$ because of barrier effects. This in turn means that coupling to the inelastic channel is most effective if $J = L + I$ and least effective if $J = L - I$, i.e. the effect of the coupling back on the elastic channel depends on J for a given L and therefore is equivalent to a spin—orbit interaction (with additional central and tensor contributions).

In the case of $^7$Li a similar argument shows that the correspondence between J and L' is opposite to the $^6$Li case because $\ell$ and $\underline{s}$ (the triton spin) are antiparallel in the $\frac{1}{2}^-$ excited state and this can explain the different sign for the induced spin—orbit force in this case (see Fig. 6).

An appropriate quantum mechanical version of these ideas has been worked out in detail in refs. 19 and 20 (see also [21]). Expression for the effective spin—orbit, tensor and central interaction associated with projectile excitation have been given. Unfortunately these interactions are very non—local and are therefore difficult to use as a guide to phenomenology, but they do clearly express the way that the induced spin—dependence depends on the quantum numbers of the spectrum of the projectile, including the circumstances in which there is complete cancellation between the contribution from different states. Further work along these lines would be desirable.

Vector analyzing powers have also been measured for $^6$Li and $^7$Li scattering from other targets. Some earlier calculations for $^6$Li scattering from $^{28}$Si and $^{16}$O had reported good agreement with predictions of spin—orbit forces obtained from folding—models, but more recent work by Windham, et al., [22] has shown that these calculations contained inaccuracies and that in fact, consistent with the results for $^{58}$Ni and $^4$He targets, the projectile excitation mechanism makes a crucial contribution to the vector analyzing power. The folding model spin—orbit potential cannot explain the data.

The situation for a $^{12}$C target at the same incident energy appears to be somewhat different. In 2 contributions to this conference [23,24] Kamimura, Sakuragi, Yahiro and Tanifuji have made a rather complete set of calculations for the observables in $^7$Li scattering by $^{12}$C for comparison with data from Heidelberg at 21.1 MeV. The data exhibit much more angular structure than the $^{58}$Ni case. Kamimura et al., find that at angles forward of 90° the calculations agree quite well with the cross—section and polarization data and did not change substantially when excitations of both the projectile and the 1st excited 2$^+$ state of $^{12}$C were included. At large angles, however, there are big discrepancies between the calculations whether calculated in a single or double—folding framework. Clearly for this lighter target and lower Coulomb barrier the reaction mechanism is much more complicated than in the $^{58}$Ni case. Unfortunately the data do not extend into the angular region of maximum sensitivity.

The influence of projectile excitation on the spin—dependence of $^{19}$F scattering by $^{28}$Si at 60 MeV is presented [25] in contribution No. 69, within the framework of a t + $^{16}$O cluster model of $^{19}$F. In this case the authors find that although the projectile excitation mechanism is clearly important for a detailed account of the $^{19}$F spin–orbit force, there is considerable cancellation between the contributions from multistep processes via the $\frac{5^+}{2}$ (0·20 MeV) and $\frac{3^+}{2}$ (1·55 MeV) spin–orbit doublet. A factor of two rather than an order of magnitude increase over the folding model predictions for $iT_{11}(\theta)$ is obtained in this case. Similar results have been obtained by Windham [26]. There are unfortunately no polarized $^{19}$F measurements to compare with these predictions, although there is indirect evidence from studies of transfer reactions with unpolarized beams. A strong $^{19}$F spin–orbit force for $^{19}$F–$^{28}$Si scattering at 60 MeV has been suggested as an explanation of a strong j–dependence of the differential cross–section in $^{28}$Si($^{19}$F,$^{16}$O)$^{31}$P reaction [27]. The spin–orbit force required in the transfer reaction appears to be much stronger than that induced by projectile excitation, although a final verdict on this must await the result of calculations of the direct influence of multistep processes via excited states of $^{19}$F to the transfer reaction, rather than simply the induced spin–orbit effect in the elastic channel.

Further evidence for the importance of dynamically induced spin–orbit effects for heavy–ions has been obtained from the analysis of spin–flip measurements in inelastic scattering. In this case also folding models fail to fit spin–flip probabilities by 2 or 3 orders of magnitude, and contributions from multistep processes involving transfer reactions and projectile excitation must be invoked [28]. The situation is reminiscent of that found in very low energy deuteron elastic scattering from heavy targets [29].

## 4(ii) Tensor Analyzing Powers

A first understanding of the main features of the tensor analyzing powers in $^6$Li and $^7$Li scattering from $^{58}$Ni was provided by the "shape effect" picture of the Heidelberg–Marburg group [30]. $^7$Li has a large negative spectroscopic quadrupole moment ($Q^{charge} = -3.4 \pm 0.6$ e.fm$^2$). In a purely aligned beam of $^7$Li ions only the sub–states M = $\pm \frac{3}{2}$ are populated and with equal probability. In such a beam the quadrupole moment will also be aligned. If M refers to n, the normal to the reaction plane, and if Q < 0 it is clear (see Fig. 7) that the grazing impact parameter (when the densities of the projectile and target just overlap) will be smaller for a randomly orientated beam than for a beam aligned along n. This means in turn that the $\frac{1}{4}$ – point in the Fresnel pattern which describes the cross–section will be at a smaller angle for an aligned beam than for an unpolarized beam. It is easily deduced that $\sigma$(aligned) $\leqslant \sigma$(unpolarized) in the way shown in Fig. 7, and hence also that $^TT_{20}(\theta)$ has a similar shape (see eq. (5)). The same reasoning for $^6$Li, which has a very small quadrupole moment predicts $^TT_{20}(\theta)$ to be very small.

The experimental situation is shown in Fig. 8. The qualitative features agree with the above argument, except that for $^6$Li, $^TT_{20}(\theta)$ is definitely positive at $E_{cm} = 18 \cdot 1$ MeV whereas the $^6$Li quadrupole moment is negative. I shall return to this question later.

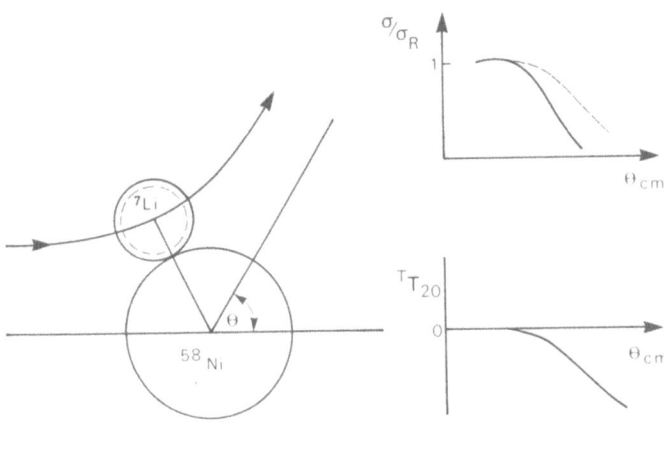

$^T T_{20} = (\sigma \text{(aligned)} - \sigma_0) / \sigma_0$

———————  aligned

— — — —  unpolarized

**Figure 7.** "Shape effect" picture for tensor analyzing powers in $^7$Li scattering.

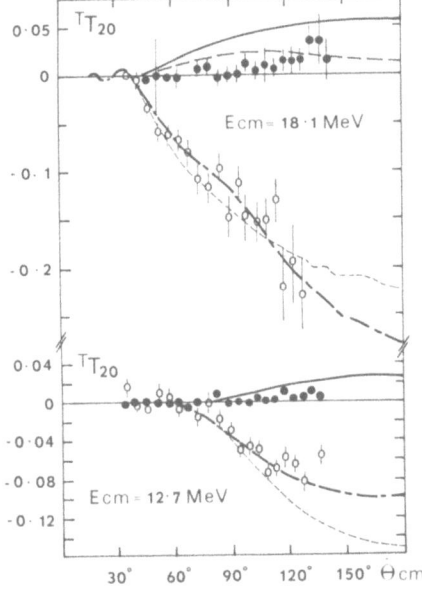

**Figure 8.** The tensor analyzing power $^T T_{20}$ for beams aligned perpendicular to the scattering plane in $^6$Li and $^7$Li scattering from $^{58}$Ni at $E_{cm} = 18 \cdot 1$ MeV (upper Fig.) and at $12 \cdot 7$ MeV (lower Fig.). The data for $^6$Li (solid circles) are from refs. [43,45] and for $^7$Li (open circles) are from refs. [30,46]. The calculations are from ref. [19]. The solid and dashed curves are the results for $^6$Li scattering, and the short dashed and dash-dotted curves for $^7$Li scattering, in the one-channel and full multichannel calculations, respectively, including the states shown in Fig. 4.

The same "shape effect" picture can be used to make statements about the other components of the tensor analysing power. If it is assumed that the differential cross-sections for different alignment axes are proportional to the distance between the surfaces of the two nuclei (assumed to have uniform density) at the turning point of the classical Coulomb orbit it can easily be deduced that all three components of the analyzing power in the Madison co-ordinate system are related to ${}^T T_{20}(\theta)$ by [30]

$$T_{20}(\theta) = (1 - 3\sin^2 \tfrac{1}{2} \theta)^T T_{20}(\theta) \ , \tag{8a}$$

$$T_{21}(\theta) = - \sqrt{\tfrac{3}{2}} \sin\theta \ {}^T T_{20}(\theta) \ , \tag{8b}$$

$$T_{22}(\theta) = - \sqrt{\tfrac{3}{2}} \cos^2 \tfrac{1}{2} \theta \ {}^T T_{20}(\theta) \ . \tag{8c}$$

It is shown in ref. [30] that these relationships are satisfied extremely well by the data. A completely equivalent way of expressing the relationships (8a–c) is obtained by using the barred co-ordinate system described in section 2 and shown in Fig. 1. In terms of these components the "shape effect" relations (8a–c) can be expressed as (see refs. [31] and [33])

$$\bar{T}_{21}(\theta) = 0 \ , \tag{9a}$$

$$\bar{T}_{22}^{(\theta)} = - (\tfrac{3}{2})^{\frac{1}{2}} \bar{T}_{20}^{(\theta)} \ . \tag{9b}$$

These relations are scattering angle independent and much simpler than (8a–c).

A comparison between the predictions of (9a) and (9b) with the data [30] at 18·1 MeV is shown in Figs. 9 and 10, which are taken from ref. [32]. The relations (9a,b) are very well satisfied, although there are definite discrepancies which I will return to later. For a ${}^{12}C$ target at 21 MeV there are even larger deviations from the relations (8a,b) at large angles [24].

The derivation of the "shape effect" relations (8) or (9) is based on somewhat dubious assumptions. A more rigorous approach is through the concept of a tensor potential.

When a diagonal matrix element of the type (7), or its double-folding model analogue, is evaluated using a model for the ${}^7Li$ ground state which correctly accounts for its quadrupole moment a tensor potential of the form

$$V_T(R)T_R \ , \tag{10}$$

where

$$T_R = (\underline{I}\cdot\underline{R}/R)^2 - \tfrac{1}{3} \underline{I}^2 \ , \tag{11}$$

is automatically generated. This potential displays a dependence on the alignment of the spin with respect to centre-of-mass position R to be expected for a potential describing the scattering of a deformed object.

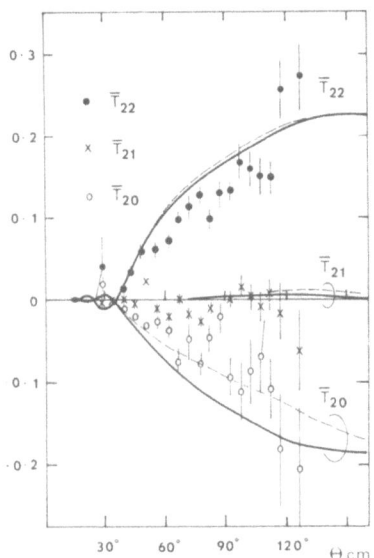

Figure 9. Data [30] and calculations [32] for tensor analyzing power components for $^{7}$Li scattering from $^{58}$Ni at 18·1 MeV in the "barred" co-ordinate system. The solid curves are one-channel folding model predictions with a $T_R$ potential discussed in the text. The dash curves include a $T_p$ potential as in the text.

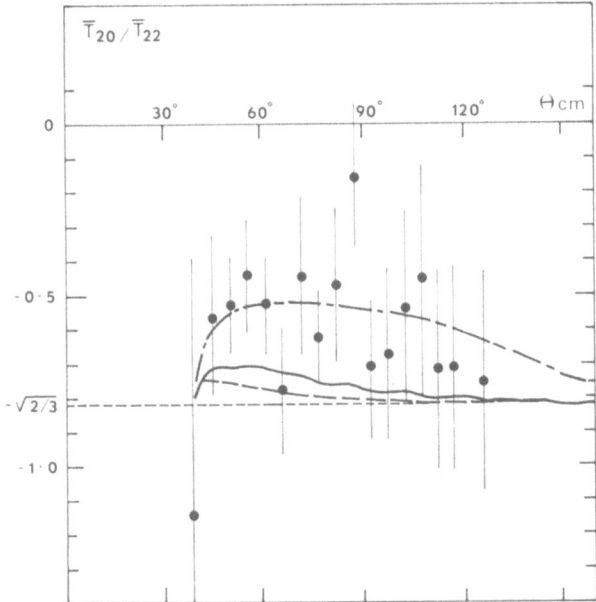

Figure 10. Data [30] and calculations [32] for the ratio $\bar{T}_{20}/\bar{T}_{22}$ for "barred" tensor analyzing power components for $^{7}$Li + $^{58}$Ni scattering at 18·1 MeV. The solid curves and dashed curves and one channel and multichannel calculations, respectively, and the dash-dotted curves include $T_p$ tensor potential as discussed in the text.

When the interactions involved are Coulombic the form factor $V_T(R)$ is proportional to the charge quadrupole moment and $R^{-3}$. At very low energies this relation has been used to measure [34] $Q^{charge}$. Diagonal and off-diagonal terms of this type are familiar from Coulomb excitation studies.

In the current context ($E \simeq 20$ MeV on a Ni target) strong interaction contributions to $V_T(R)$ are extremely important. In the surface region, which is of most relevance, calculation in the $^7$Li case shows that $V_T(R)$ has a sign which is determined by the quadrupole moment but the radial dependence has, of course, the short range characteristic of the densities of the two nuclei. Detailed calculations using the simple cluster model wavefunction for $^7$Li described earlier give excellent agreement with experiment [19,20]. A comparison in terms of the barred components of the tensor analyzing power is shown in Figs. 9 and 10 (solid curves). Note that the relations (9a,b) are very accurately satisfied by the full quantum mechanical calculations.

At first sight the measured results for $^6$Li are puzzling. As already mentioned, the quadrupole moments of $^6$Li and $^7$Li have the same sign but $^T T_{20}$ has opposite sign at the same incident energy. One explanation for this discrepancy within the cluster model framework has been given in refs. [35] and [19]. These authors point out that it is natural to assume that the tensor force in $V_{np}$ will generate a positive quadrupole moment in the deuteron cluster in $^6$Li similar to that of a free deuteron. The observed small negative quadrupole moment of $^6$Li ($Q^{charge} = -0.064$ e.fm$^2$) is then understood as arising from a cancellation between this contribution and a negative contribution generated from an $\ell = 2$ component in the relative motion of the d and $\alpha$ clusters. They estimated this component to have an amplitude of magnitude 0.08. When a $^6$Li wavefunction with this structure is used to calculate the $T_R$ form factor it is found [35,19] that the contribution from the deuteron cluster dominates at large distances, because, for a given position of the $^6$Li centre-of-mass the deuteron can overlap more effectively than the $\alpha$-particle with the target. Hence at large impact parameters $^6$Li presents a tensor potential characteristic of a positive quadrupole moment.

The quantitative situation is shown in Fig. 8, which is taken from ref. [19]. Projectile excitation effects also play a role here, but the main difference between $^6$Li and $^7$Li arises as described above. In this example, therefore, experiments with aligned beams probe the quadrupole distribution of $^6$Li in more detail than can be summarised by simply quoting its quadrupole moment.

It should be emphasised that the $^6$Li model used in ref. [35] had no real dynamical basis: the relative admixtures of $\ell = 0$ and $\ell = 2$ in the intercluster motion were simply adjusted to give the correct $^6$Li quadrupole moment when combined with that of the deuteron cluster. It is therefore most encouraging that a consistent dynamical model which gives some support for these ideas has recently been developed. In a contribution to this Conference Merchant and Rowley [36] have shown that the tensor force between the $\alpha$ and d clusters induced by the deuteron D-state gives rise to an $\ell = 2$ admixture of the correct sign and similar magnitude to that obtained in ref. [35]. The remaining discrepancies can be interpreted as evidence for a difference between the properties of the deuteron cluster and a free deuteron. The deuteron cluster is expected to be smaller (see e.g. [37]) and to have a different quadrupole moment [38]. Unfortunately, a consistent RGM microscopic calculation of $^6$Li is extremely difficult, basically because of the radically different function space required to describe an alpha particle and a deuteron. (See e.g. [39] for a calculation which does include a two-body tensor force and hence does at least produce a non-spherical deuteron cluster.).

The Merchant-Rowley model uses a d-$\alpha$ tensor force generated by the nucleon-$\alpha$ central interaction and the deuteron D-state. It should be noted that such a model produces tensor analyzing powers in d + $\alpha$ elastic scattering which differ substantially from experiment [40].

## 5. DEVIATIONS FROM THE "SHAPE EFFECT" PICTURE

We have seen that the data show definite deviations from the "shape effect" picture in the sense that (a) the sign of $^T T_{20}$ is not simply determined by the Q of the projectile; (b) there are deviations from (8a,b) or equivalently (9a,b). An advantage of working in terms of the tensor potential is that it allows an understanding of the significance of these deviations. We have already described a situation of type (a) (the $^6$Li case). Further important insight into (b) is provided by Nishioka and Johnson [31,32].

For heavy-ion scattering (although not for deuteron scattering) the influence of the $\underline{L} \cdot \underline{I}$ forces on tensor analyzing powers can be ignored and the tensor force treated perturbatively to a very good approximation. According to ref. [2] the tensor analyzing powers can then be written as

$$\sigma_0(\theta)\bar{T}_{2q}(\theta) = 2\mathrm{Re}(\bar{M}_{00}(\theta)\bar{M}^*_{2q}(\theta)) \; , \tag{12}$$

where $\sigma_0(\theta)$ is the unpolarized differential cross-section and the $\bar{M}_{2q}(\theta)$ are multipole components of the scattering amplitude. The k = 2 components are given to first order by

$$\bar{M}_{2q} = \langle \chi^{(-)}_{\underline{K}_f} | V_{2q} | \chi^{(+)}_{\underline{K}_i} \rangle \tag{13}$$

where the $|\chi^{\pm}\rangle$ are distorted waves associated with the central parts of the interaction and $V_{2q}$ is that part of the tensor interaction which depends on $\underline{R}$ (i.e. $V_T(R)Y_{2q}(\hat{R})$). Parity conservation, reciprocity and rotational invariance require $\bar{M}_{21} = 0$ and hence the condition (9a)

$$\bar{T}_{21} = 0 \; , \tag{14}$$

is seen to be a perfectly general and essentially model independent property of the tensor analyzing powers for heavy-ions.

If it is further assumed that the distorted waves in eq. (13) cause the integral to be dominated by the region of the classical turning point (Fig. 1) $\underline{R}_0$, then the angular arguments of $V_{2q}(\underline{R}_0)$ are $\theta_0 = \pi/2$

and $\Theta_0 = 0$ or $\pi$ in the barred co-ordinate system. Nishioka and Johnson refer to this as the "turning-point" model. Then, since for these angles $Y_{22} = -(\frac{3}{2})^{\frac{1}{2}} Y_{20}$, we have

$$\bar{T}_{22} = -(\frac{3}{2})\bar{T}_{20} \tag{15}$$

which is the second of the "shape-effect" relations (9b).

A quantum-mechanical derivation of the "turning-point" model can be given for a wide class of short range tensor forces of the type defined in eqs. (10) and (11), and specific calculations show that eqs. (14) and (15) are very accurately satisfied by quantum mechanical calculations for heavy ions.

The "turning-point" model allows us to make predictions for other types of tensor force which are allowed by general considerations. Satchler denoted these as $T_R$, $T_p$, and $T_L$ and they are all second rank tensors in the spin-operator $\underline{I}$. The $T_R$-type has already been defined in eqs. (10) and (11). $T_p$ and $T_L$ are defined analogously with $\underline{R}$ replaced by the centre-of-mass momentum operator $\underline{P}$ and angular momentum operator $\underline{L} = \underline{R} \wedge \underline{P}$, respectively, and involve the combinations $(\underline{I} \cdot \underline{P})^2$ and $(\underline{I} \cdot \underline{L})^2$. When the operators $\underline{P}$ and $L$ are replaced in eq. (13) by values appropriate to the classical turning point it is found that

$$\bar{T}_{20} \neq 0, \qquad \bar{T}_{22} = 0 , \tag{16}$$

for a pure $T_p$ interaction, but

$$\bar{T}_{22} = +\sqrt{\frac{3}{2}} \, \bar{T}_{20} , \tag{17}$$

for a pure $T_L$ interaction. Eq. (14) is, of course, satisfied in all cases.

The turning-point model, therefore, predicts very different relations between $\bar{T}_{20}$ and $\bar{T}_{22}$ for the 3 types of tensor interaction. These relations can be used to identify the effects of the $T_p$ or $T_L$ interactions which are otherwise hidden by the existence of the dominant $T_R$-interaction. Quantum mechanical calculations show that in fact the relations (16) and (17) are satisfied accurately by the appropriate forces.

Nishioka and Johnson [31,32] have used these ideas to provide evidence for a $T_p$ interaction in $^7Li + ^{58}Ni$ scattering at $E_{cm} = 18 \cdot 1$ MeV. The experimental ratio $\bar{T}_{20}/\bar{T}_{22}$ is shown in Fig. 10 and displays a definite deviation from the ratio $-(\frac{2}{3})^{1/2}$ expected for a pure $T_R$ interaction. Projectile excitation effects (dashed-curves in Fig. 10) modify $\bar{T}_{20}$ and $\bar{T}_{22}$ and give an improved fit to the data, but preserve the ratio appropriate to a $T_R$ interaction. Target excitation effects do not appear to give rise to any relevant spin-dependent effects although they do, of course, contribute strongly to the central interactions through the central parts of the cluster target potentials. Any tensor force effects from target excitation and transfer reactions can be shown on theoretical grounds to be predominantly of the $T_R$ type and satisfy eqs. (9a,b).

The dashed curves in Fig. 9 and the dashed-dotted curves in Fig. 10

show the predictions for $^7$Li + $^{58}$Ni of the cluster folding-model described in Section 4 with an additional $T_p$ interaction with form-factor

$$V_{T_p}(R) = V_0 (1 + \exp[(R - R_0)/a])^{-1}$$

in which $V_0$ = 2·0 MeV, $R_0$ = 1·3 $((7)^{1/3} + (58)^{1/3})$ fm , and a = 0·6 fm. The $T_p$ interaction affects only $\bar{T}_{20}$ as the "turning-point" model predicts and improves the fit to the data. The $\bar{T}_{20}/\bar{T}_{22}$ ratio is also well fitted, although the experimental error bars are large. A similar analysis for the $^{12}$C data, which show [24] even stronger deviations from the "shape-effect" relations have not yet been carried out.

We have seen that interactions of the $T_R$ type arise naturally for a deformed projectile. How can $T_p$ and $T_L$ interactions arise? Terms of the type are generated in double-folding models when the full complexity of the nucleon-nucleon interaction is taken into account. For a spin-zero target they can be thought of as arising through the $p^2$ dependence of the strength of the nucleon-nucleus spin-orbit potential. As such, they give rise to effects which are much smaller then the $T_p$ term suggested by the $^7$Li data and quoted above.

A force of the $T_p$ type can be generated by Pauli blocking effects in the scattering of a deformed composite projectile as first shown for deuterons by Ioannides and Johnson [38]. The overlap of the Fermi seas of the projectile and target depends on the relative orientation of the projectile's spin and its momentum for a deformed projectile. The $T_p$ interaction used to generate the calculations in Figs. 9 and 10 is in fact similar to that calculated by Ioannides and Johnson for high energy deuteron scattering by a finite nucleus. Unfortunately, it is very difficult to separate out the effects of a $T_p$ force in the case of deuterons because of the indistinguishability of $T_R$ and $T_p$ effects at low energies [41], and because of the large second order effects of the L·I potential at intermediate energies [42]. The latter effects are very small for heavy ions, so that heavy-ions provide a good opportunity to see this novel spin dependence which reflects in a very explicit way the rôle of the Pauli principle in the scattering of composite nuclei. It would be very interesting to have accurate data for a range of targets and energies and involving all three components of the tensor analyzing powers of $^6$Li, $^7$Li and $^{23}$Na. For the latter, calculations show that the "turning-point" predictions are satisfied to very high accuracy, so that discrimination between the different types of tensor force should be excellent. There is also a need for theoretical estimates of the $T_p$ potential for heavy ions.

## 6. CONCLUDING REMARKS

My emphasis in this talk has been on elastic scattering experiments with polarized $^6$Li and $^7$Li beams. Simple cluster model ideas of the ground states and low-lying excited states of these nuclei have been shown to be extremely useful in providing a framework within which the existing data can be quantitatively understood. The spin-dependence of heavy-ion scattering has been shown to have a very different origin from that of light-ions. The spin-dependence of heavy-ion scattering is mainly of dynamical origin and has very little to do with the spin-dependence of the nucleon-nucleon interaction. In fact the closest analogy is perhaps the study of the spin-dependence of the nucleon-nucleon interaction itself and what it tells us about nucleon structure, e.g. the long-range nucleon-nucleon tensor force is a manifestation of the non-spherical meson cloud around a nucleon.

Studies of tensor analyzing powers reveal new aspects of the non-spherical components of the projectile. Studies of this type are, of course, well known from Coulomb excitation and other studies; but with polarized beams the effects can be seen even in elastic scattering and when nuclear forces are important new features beyond simply the quadrupole moment may be involved. Most of the experiments so far have involved energies near the Coulomb barrier where penetrations are small. At higher energies many of the models described here will break down; but the existence of a body of data at lower energies (the Heidelberg-Marburg data) whose interpretation is rather well understood will be of immense importance in the understanding of the new data and will hopefully become available soon from Daresbury, Florida and Heidelberg.

## 7. ACKNOWLEDGEMENTS

I am very grateful to my colleagues H. Nishioka, J.A. Tostevin, G. Windham and F.D. Santos for the many discussions we have had about the physics discussed here and for showing me their results prior to publication.

## REFERENCES

1. For a recent account of the polarized heavy ion source at Heidelberg see D. Kramer, K. Becker, K. Blatt, R. Caplar, D. Fick, H. Gemmeke, W. Haeberli, H. Jansch, O. Karban, I. Koenig, L. Luh, K.-H. Mobius, V. Necas, W. Ott, M. Tanaka, G. Tungate, I.M. Turkiewicz, A. Weller, and E. Steffens, Nucl. Instr. and Meth. **220** (1984) 123; an extensive list of references to the experimental results relevant to this talk can be found in ref. 19 below.

2. D.J. Hooton and R.C. Johnson, Nucl. Phys. **A175** (1971) 583; R.C. Johnson, Nucl. Phys. **A293** (1977) 92.

3. L.D. Knutson and W. Haeberli, in <u>Prog. Part. and Nucl. Physics</u> (Pergamon Press 1980) Vol. **3**, p.127.

4. F.D. Santos, S.A. Tonsfeldt, T.B. Clegg, E.J. Ludwig, Y. Tagishi and J.F. Wilkerson, Phys. Rev. **C25** (1982) 3243.

5. J.A. Tostevin, Phys. Rev. **C28** (1983) 961.

6. F.D. Santos, Prog. Theor. Phys. **70** (1983) 1679.

7. F.D.Santos and A.M. Eiro, Portugaliae Physica **15** (1984) 65.

8. J.A. Tostevin, J.M. Nelson, O. Karban, A.K. Basak and S. Roman, preprint (July 1984).

9. B.C. Karp, E.J. Ludwig, W.J. Thompson and F.D. Santos, preprint (July 1984).

10. R.C. Johnson and F.D. Santos, Part. Nucl.**2** (1971) 285.

11. T.E.O. Ericson and M. Rosa-Clot Nucl. Phys. **A405** (1983) 497; T.E.O. Ericson, Nucl. Phys. **A416** (1984) 281c.

12. W. Chung and B.H. Wildenthal (unpublished).

13. W.T. Pinkston, Comm. Nucl. Part. Phys. **12** (1984) 133.

14. P.D. Bond, Contribution No. 94.

15. G. Bertsch, J. Borysowicz, H. McManus and W.G. Love, Nucl. Phys. **A284** (1977) 399.

16. G.R. Satchler and W.G. Love, Phys. Repts. **55C** (1979) 183.

17. M.F. Steeden, J. Coopersmith, S.J. Cartwright, N.M. Clarke and R.J. Griffiths, J. Phys. **G6** (1980) 501.

18. G.R. Satchler, Phys. Lett. **83B** (1979) 287.

19. H. Nishioka, J.A. Tostevin, R.C. Johnson and K.-I. Kubo, Nucl. Phys. **A415** (1984) 230.

20. H. Ohnishi, M. Tanifuji, M. Kamimura, Y. Sakaguchi and M. Yahiro, Nucl. Phys. **A415** (1984) 271.

21. R.C. Johnson, Lecture Notes of 1983 RCNP-Kikuchi Summer School (Edited by H. Ogata, M. Fujiwara and A. Shimizu, Osaka University, Osaka 1983) p.194.

22. G. Windham, H. Nishioka, J.A. Tostevin and R.C. Johnson, Phys. Letts. **138B** (1984) 253.

23. Y. Sakuragi, M. Kamimura, M. Yahiro and M. Tanifuji, Contribution No. 30.

24. M. Kamimura, Y. Sakuragi, M. Yahiro and M. Tanifuji, Contribution No. 31.

25. S. Ohkubo and M. Kamimura, Contribution No. 69.

26. G. Windham, private communication.

27. S. Kubono, D. Dehuhard, D.A. Lewis, T.K. Li, J.L. Artz, D.J. Weber, P.J. Ellis and A. Dudek-Ellis Phys. Rev. Lett. **38** (1977) 817.

28. Q.K.K. Liu, P.J. Ellis and S. Chakravarti, Contribution No. 118.

29. J.A. Tostevin and R.C. Johnson, Phys. Lett. **124B** (1983) 135.

30. Z. Moroz, P. Zupranski,R. Bottger, P. Egelhof, K.-H. Mobius, G. Tungate, E. Steffens, W. Dreves, I. Koenig and D. Fick, Nucl. Phys. **A138** (1982) 294 and references therein.

31. H. Nishioka and R.C. Johnson, Contribution No. 3.

32. H. Nishioka and R.C. Johnson, Contribution No. 4.

33. D. Fick, 'Polarization in Heavy Ion Reactions', Lectures at Oak Ridge National Laboratory, June 1982 (Edited by B. Shivakumar and A.W. Wright, ORNL/TM-8816)

34. P. Egelhof, W. Dreves, K.-H. Mobius, E. Steffens, G. Tungate, G. Zupranski, D. Fick, R.Bottger and F. Roesel, Phys. Rev. Letts. **44** (1980) 1380.

35. H. Nishioka, J.A. Tostevin and R.C. Johnson, Phys. Lett. **124B** (1983) 17.

36. A.C. Merchant and N. Rowley, Contribution No. 108.

37. R. Beck, F. Dickman and A.T. Kruppa, Contribution No. 50.

38. A.A. Ioannides and R.C. Johnson, Phys. Rev. **C17** (1978) 1331 and references therein.

39. H.M. Hofmann and T. Mertelueier, Contribution No. 106.

40. R. Frick, H. Clement, G. Graw, P. Shiemenz and N. Seichert, Phys. Rev. Lett. **44** (1980) 14.

41. R.P. Goddard, Nucl. Phys. **A291** (1977) 13.

42. E.J. Stephenson, C.C. Foster, P. Schwandt and D.A. Goldberg, Nucl. Phys. **A359** (1981) 316.

43. K. Rusek, Z. Moroz, R. Caplar, P. Egelhof, K.-H. Mobius, E. Steffens, I. Koenig, A. Weller and D. Fick, Nucl. Phys. **A407** (1983) 208.

44. G. Tungate, R. Bottger, P. Egelhof, K.-H. Mobius, Z. Moroz, E.Steffens, W. Dreves, I. Koenig and D. Fick, Phys. Lett. **98B** (1981) 347.

45. W. Dreves, D. Zupranski, P. Egelhof, D. Kassen, E. Steffens, W. Weiss and D. Fick, Phys. Lett. 78B (1978) 36.

46. P. Zupranski, W. Dreves, P. Egelhof, K.-H. Mobius, E. Steffens, G. Tungate and D. Fick, Phys. Lett. 91B (1980) 358.

BREAK-UP, FRAGMENTATION AND FISSION

# ELASTIC AND INELASTIC SCATTERING AND BREAKUP OF CLUSTER-LIKE NUCLEI

M. Kamimura, M. Yahiro*, Y. Iseri, M. Nakano† and Y. Sakuragi

Department of Physics, Kyushu University, Fukuoka 812, Japan
*Shimonoseki University of Fisheries, Shimonoseki 759-65, Japan
†Division of Physics, U.O.E.H., Kitakyushu 807, Japan

ABSTRACT. Extensive analyses based on a three-body coupled-channel approach are reported for scattering and breakup of deuteron, $^3$He, $^{6,7}$Li and $^{12}$C nuclei. Microscopic cluster-model wave functions are employed for the bound states and resonant and non-resonant breakup states of the projectile nuclei. As well as the bound states of the projectile, the breakup continuum states are treated in the framework of the method of coupled discretized continuum channels developed by the present authors. The following experimental data are reproduced by the coupled-channel calculations: Coincidence cross sections for the elastic breakup (d,pn) and ($^3$He,dp) processes and differential cross sections for the elastic and inelastic scattering and breakup of $^{6,7}$Li and $^{12}$C. Dynamical polarization potentials induced by projectile breakup and by inelastic excitation are exactly calculated and a puzzle for the potential due to the projectile breakup is solved.

## 1. INTRODUCTION AND THE MODEL

As is well known, the microscopic cluster model has been successful in systematic studies of the structure of light nuclei [1-2]. One of the most important reasons of the success of the cluster model may be stated as follows: On account of the total antisymmetrization intervening between the constituent clusters, the microscopic cluster model is able to describe not only cluster-like states but also most of shell-like states simultaneously; loosely coupling solutions of the relative motion between the clusters stand for the cluster-like states, while compactly coupling solutions stand for the shell-like states (in the latter solutions, the clusters melt due to the antisymmetrization). Wave functions so-obtained by the microscopic cluster model have explained a variety of experimental data including electron scattering form factors.

It is therefore of interest to apply those precise wave functions to the study of reaction mechanism, especially to the study of (peripheral) heavy-ion reactions in which the light nuclei concerned are used

Presented by M. Kamimura

J. S. Lilley and M. A. Nagarajan (eds.), Clustering Aspects of Nuclear Structure, 181–197.
© 1985 by D. Reidel Publishing Company.

as projectiles. In contrast to the case where the light nuclei are
targets and are bombarded by light ions such as protons, in the case
where the nuclei are projectiles, an absolutely important role can be
played by channel couplings between the elastic channel and excited
channels and, especially, between the excited channels. This is
because potentials of those couplings are comparable, in magnitude, to
the elastic-channel distorting potential in the peripheral region which
is important for the scattering; in the internal region, the former
potentials are much smaller than the latter but the region is not
important at all in the peripheral reactions due to the presence of the
strongly absorptive potential.

Thus, analyses of scattering and breakup of light heavy-ions can
give a good way to test the precise knowledge of both the diagonal and
transition densities of the light nuclei in the surface region where the
clusterization may be of special importance (note that the latter
densities include the transitions between the excited states). Since
most of light nuclei are rather easy to breakup into the constituent
clusters, it is particularly important to take into consideration the
resonant and non-resonant breakup of the projectile in the field of the
target nucleus as well as the excitation of the projectile into its bound
excited states.

## 1.1. Coupled-Channel Approach Based on Cluster-Model Wave Functions

The purpose of the present paper is to study elastic and inelastic
scattering of light nuclei and breakup of the projectiles from the viewpoint of the application of the precise cluster-model wave functions
which describe the bound and breakup states of the projectiles. One of
the most sophisticated approaches to the study in which those projectile
states are explicitly taken into account may be to perform coupled-
channel (CC) calculations in which all the CC form factors are derived
by doubly folding an effective nucleon-nucleon interaction into the
distribution of the diagonal and transition densities of the projectile
and the target nucleus.

This approach may be outlined as follows: We expand the total wave
function of the colliding system in terms of the internal states of the
projectile, $\phi_i(a)$, and those of the target, $\phi_j(A)$ :

$$\Psi = \sum_{ij} \phi_i^{(a)} \phi_j^{(A)} \chi_{ij}(R) , \tag{1.1}$$

where $\chi_{ij}(R)$ describes the relative motion between the projectile in a
state "i" and the target in a state "j". As mentioned above, we take
into account the projectile-breakup continuum. Because of limitations
in actual calculations, however, we discretize the continuum states and
include them in the set of $\{\phi_i(a)\}$ which are composed of discrete wave
functions; this discretization method based on a three-body model will
be sketched in section 2.1. A set of CC equations for $\chi$'s may be given
by

$$\left[ -\frac{\hbar^2}{2\mu} \nabla^2 + V_{ij,ij}(R) - (E-\epsilon_{ij}) \right] \chi_{ij}(R) = -\sum_{i'j'} V_{ij,i'j'}(R) \chi_{i'j'}(R). \tag{1.2}$$

Here, the form factors $V(\mathbb{R})$ may be calculated by folding an effective nucleon-nucleon interaction $v_{NN}$ as

$$V_{ij,i'j'}(\mathbb{R}) = (\phi_i^{(a)} \phi_j^{(A)} | v_{NN} | \phi_{i'}^{(a)} \phi_{j'}^{(A)}),\qquad(1.3)$$

where the round brackets denote integration over the internal coordinates of the two nuclei. For the moment we ignore the effects of antisymmetrization between two nuclei (many-particle exchange may not be significant in peripheral heavy-ion reactions, and the one-nucleon knock-on exchange effect can be approximately included [3] in $v_{NN}$).

## 1.2. Double-Folding Model for Elastic Scattering

Let $i,j=0$ denote the ground state of the colliding nuclei ($\epsilon_{ij}=0$). Then $\chi_{00}$ describes the relative motion of the two nuclei while they both remain in their ground states. If $\chi_{00}$ is projected out from the CC equations (1.2), an exactly equivalent equation for $\chi_{00}$ is obtained in the form [3]

$$\left[ -\frac{\hbar^2}{2\mu}\nabla^2 + U_{eff} - E \right] \chi_{00} = 0 \qquad(1.4)$$

Here, $U_{eff}$ is an effective potential which may be expressed [4] as

$$U_{eff} = V_{00,00} + \sum_{iji'j'} V_{00,ij} \left(\frac{1}{E-H+i\epsilon}\right)_{ij,i'j'} V_{i'j',00} \qquad(1.5)$$

$$= U_F + \Delta U, \text{ say,}$$

where the first term $U_F(V_{00,00})$ is real and is the so-called double folded potential. The remaining term $\Delta U$, which we refer to a dynamical polarization (DP) potential, is induced by the coupling of the elastic channel to all the other channels. This term certainly generates the the imaginary part and also contributes to the real part.

The double folding (DF) model of Satchler and Love [3] employs $U_F$ alone for the real part (at least in the surface region which is important for the elastic heavy-ion scattering) and assumes that the DP potential $\Delta U$ due to all the excited channels contributes to the imaginary part only. Approximate estimations of $\Delta U$ to the second order in $V$ [5] suggested that Re $\Delta U \ll$ Im $\Delta U$. The real part $U_F$ has usually been calculated with the use of an effective interaction, called the M3Y interaction [6], based upon a G-matrix derived from the Reid nucleon-nucleon potential. The knock-on single-nucleon exchange effect has been approximately added to the interaction in the $\delta$-function form. In actual analysis of experimental data, $U_{eff}$ has been assumed

$$U_{eff}(R) = N_R U_F(R) + i W(R) \qquad(1.6)$$

or

$$U_{eff}(R) = N_R U_F(R) + i N_I U_F(R). \qquad(1.7)$$

where $N_R$ is a renormalization factor for $U_F$, and the Woods-Saxon shape has usually been assumed for $W(R)$. $N_R$ and $W(R)$ (or $N_R$ and $N_I$) are optimized so as to fit the observed differential cross section of elastic scattering. If the optimum value of $N_R$ remains close to $N_R=1.0$, we consider that the use of $U_F$ alone for the real part is successful, namely the DF model is successful for the scattering. The DF model with $N_R \approx 1.0$ has been shown [3] to reproduce the observed elastic cross sections for a large number of systems at bombarding energies $E_{lab}/A \approx 5 \sim 10$ MeV.

## 1.3. Failure of Double-Folding Model for Scattering of $^{6,7}$Li and $^9$Be and a Puzzle for Projectile Breakup Effect

Exceptions of the success have been reported by various authors [3,7] in the case of the elastic scattering of $^6$Li, $^7$Li and $^9$Be in a wide range of the bombarding energies. The folded potential $U_F$ in this case must be reduced in strength by a factor of about two; namely, an "unhappiness" value of $N_R \sim 0.5$ is required; otherwise, the position of the diffraction peaks of the observed angular distribution are not reproduced even though $W(R)$ is widely varied.

It was often stated that this strong renormalization of the real part may be attributed to effects of breakup of the projectiles because those nuclei have outstandingly small threshold energy of the breakup into contituent clusters. Actually, Thompson and Nagarajan [8] and Sakuragi, Yahiro and Kamimura [9,10] showed that, in the case of $^6$Li scattering, inclusion of the projectile-breakup channels leads to a good fit to the observed angular distribution without any renormalization of the real part. The same was also shown in the case of $^7$Li scattering [11, 12].

The following discussion given by Satchler and Love [3] is to be reminded of: In light of approximate estimations [5] of the DP potential $\Delta U$ induced by the excitation of collective states, coupling of the breakup channels is naturally expected to induce the imaginary part mainly and contribute to the real part attractively but almost negligibly. The observed effect is ,however, strongly repulsive ($N_R \sim 0.5$); in strength, it amounts to about half of $U_F$, at least in the surface region which is important for the scattering.

Therefore, it was a puzzle in ref. [3] whether inclusion of the breakup effect really induces such a large, repulsive real potential. The puzzle was beautifully solved by Sakuragi, Yahiro and Kamimura [10] in their precise CC analysis of the $^6$Li projectile breakup effect. They discovered that the DP potential due to the projectile breakup is of completely opposite type compared with the potential due to the excitation of collective states.

## 1.4. Three-Body Coupled-Channel Approach to Projectile Breakup

The purpose of the present paper is firstly to show successful applications of the microscopic cluster-model wave functions of $^6$Li, $^7$Li and $^{12}$C to elastic scattering of the nuclei and inelastic excitation and breakup of the projectile. Excitation of the target is not considered explicitly. The CC form factors (1.3) are calculatd with the use of those wave functions and the CC equations (1.2) are solved.

Secondly, we shall show that the projectile breakup process is well described by the method of coupled discretized continuum channels (CDCC) which the present authors have developed [10,13,14]. This method is found to provide a hopeful approach to nuclear reaction mechanisms on the basis of three-body (three-cluster) model. To demonstrate the usefulness of the CDCC in the analysis of actual breakup cross sections, we shall report its successful application to coincidence cross sections for elastic breakup (d,pn) and ($^3$He,dp) processes. An important role of the CC effect due to the breakup channels will be exhibited in the comparison of the result of CDCC with that of prior-form DWBA.

The third purpose of the present work is to discuss the DP potentials induced by projectile breakup states, by well-developed cluster-like excited states and by shell-like excited states. The potentials are exactly calculated in terms of the wave-function-equivalent local potential with the use of the CC wave functions solved and the CC form factors. A schematic CC calculation is performed to simulate various types of the CC effects, and a systematic understanding of the DP potentials will be given.

In the present paper, we are interested in the scattering of deuteron, $^3$He, $^6$Li, $^7$Li and $^{12}$C. The first two have only non-resonant breakup continuum as their excited states. The $^6$Li($^7$Li) nucleus has the well developed cluster-like ground state and the same resonant and non-resonant excited states that are strongly coupled to the ground state. The $^{12}$C nucleus has the shell-like ground and excited states as well as well-developed cluster-like resonant states and non-resonant three-alpha breakup states. We can then enjoy various types of excitation mechanisms in the scattering of those nuclei.

In section 2, the method of CDCC is briefly sketched and its successful application is demonstrated in the breakup of deuteron and $^3$He projectiles. In section 3, scattering and breakup of $^6$Li and $^7$Li are extensively analyzed. Scattering of $^{12}$C is studied in section 4. A systematics of the DP potential is discussed in section 5. Concluding remarks are given in section 6.


## 2. DEUTERON AND $^3$He PROJECTILE BREAKUP

### 2.1. Method of CDCC

In order to study projectile breakup processes on the basis of the three-body model, the present authors have developed the method of coupled discretized continuum channels(CDCC) [10,13,14] (related references are therein); an extensive test of the validity of this method was done in ref. [14]. We briefly sketch the method taking the case of deuteron projectile as an example.

The following prescription was recommended in ref. [14]: The continuum of deuteron breakup states is truncated as follows. The relative angular and linear momenta, $\ell$ and $k$ respectively, are restricted to $\ell=0$ and 2 and $0 \leq k \leq 1$ fm$^{-1}$. The k-continuum for each $\ell$ is

discretized into eight bins with an interval $\Delta k=1/8$ fm$^{-1}$.                    The
exact proton-neutron wave function is averaged within each bin.  The
averaged wave function is assumed to be the wave function of the discretized deuteron breakup state corresponding to that bin.  We refer to the
set of the averaged wave functions as $\{\Phi_i(a)\}$.  As for the target, we
employ $\Phi_0(A)$ alone and ignore the others ;   we are interested in
elastic breakup of deuteron projectile.  The form factors $V_{i0,i'0}$ of Eq.
(1.3) may then be obtained by folding the proton- and neutron-target
optical potentials into the diagonal and transition densities of the
proton-neutron pair (for the case of the deuteron and $^3$He breakup, we
employ the single-folded form factors).

        By solving the CC equations (1.2), we obtain S-matrix elements for
the elastic scattering and for the excitation of the discretized breakup
states.  Continuous S-matrix elements for the breakup into the continuum
states may be given, as a function of k, by smoothly interpolating the
discrete S-matrix elements with respect to k [14].  Convergence of the
elastic and breakup S-matrix elements was examined and verified in ref.
[14] with respect to increasing $\ell_{max}$ and $k_{max}$ and narrowing $\Delta k$.  The
triple differential cross section can be calculated from the smoothed
S-matrix elements [15].

## 2.2. Breakup of Deuteron and $^3$He Projectiles

        The purpose of the present sub section is to demonstrate that the
method of CDCC is a powerful tool to investigate the deuteron and $^3$He
breakup processes and suggest that the method is suited for the cases of
heavier projectiles, too.  Detailed discussions of the breakup of deuteron
and $^3$He themselves will be given by de Meijer in this conference.

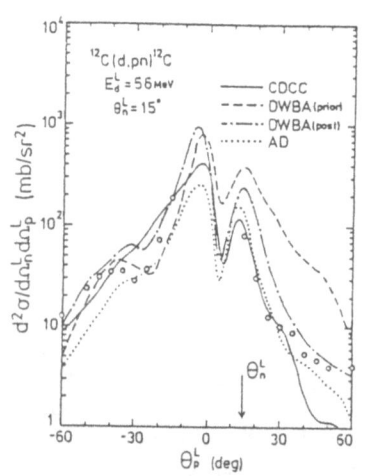

        Figure 1 shows the coincidence cross
section of the elastic breakup $^{12}$C(d,pn)
reaction at $E_d$=56 MeV.  Experimental data are
given by Matsuoka et al. [16].  Calculated
results are given by four kinds of models;
i) the prior-form DWBA calculation by Matsuoka
et al. [16], ii) the adiabatic treatment of
deuteron breakup by Amakawa and Austern [17],
iii) the post-form DWBA calculation by Baur,
Shyam, Rösel and Trautmann [18], and iv) the
CDCC calculation by Yahiro, Iseri, Kamimura
and Nakano [15].  Three calculations ii)to iv)
give almost the same quality of fitting to the
experimental data, while the prior-form DWBA
much overestimates the data at $\theta_p^L > 0$.

        The CDCC work made clear the following:
the transition from the incident channel to
the individual breakup channels can be described by first order Born approximation, but
the breakup cross section is strongly affected
both by the multi-step processes among the
breakup channels themselves and by the choice
of the exit-channel distorting potentials.

Fig. 1.  Coincidence cross
section of the elastic
breakup $^{12}$C(d,pn) reaction.
Data are from ref.[16].

The latter two factors in the exit channels work coherently for the break-up with the low p-n momentum $k$ ($\theta_p^L > 0$) but nearly cancel out for the break-up with the intermediate $k$ ($\theta_p^L < 0$). This is precisely the reason why the prior-form DWBA is not successful for the coincidence cross section for $\theta_p^L > 0$ but is successful for $\theta_p^L < 0$.

The fact that the result given by the adiabatic treatment (AD) is quite close to that by CDCC for $\theta_p^L > 0$ is reasonable because AD is consider-ed to be a good approximation for the low-k breakup. Validity of AD has been discussed, for example, in refs. [14,15,17,19-22]. Among the discussions, it is to be specially noted [22] that application of AD to the polarization phenomena should be done very carefully, since the vector analyzing power is drastically affected by Q-values of the strongly coupled channels; in AD, Q-values of the breakup channels are automatically degenerated at the ground-state value (or some values dependent on $\ell$).

The post-form DWBA calculation [23] and the CDCC one [24] are also compared to each other in Fig. 2 for the coincidence cross sections of the elastic breakup $^{28}$Si($^3$He,dp) reaction together with the experimental data [23]. Both calculations reproduce the data satisfactorily. The small difference between post-form DWBA and CDCC in the results for (d,pn) and ($^3$He,dp) might be due to the fact that the final-state interaction between p and n (d and p) is weak. One way of exami-ning the validity of the post-form DWBA calculation for (d,pn) would be to check the deviation of the DWBA T-matrix $\langle \chi_p \chi_n | V_{pn} | \phi_d \chi_d \rangle$ from the CCBA one $\langle \chi_p \chi_n | V_{pn} | \phi_d \chi_d + \Psi(d*) \rangle$. Here $\phi_d \chi_d$ is the incident-channel wave function and $\Psi(d*)$ refers to the breakup-channel component in the CDCC calculation. The contribution of $\Psi(d*)$ is neglected in post-form DWBA. The validity of the DWBA was examinied in a limiting case [20] of the neutron being trapped by a target, specially for $^{58}$Ni(d,p) $^{59}$Ni(g.s.) at $E_d$=80 MeV, and it was found (Fig. 3) that the contribution of $\Psi(d*)$ to the (d,p) cross section is important for $\theta_p \gtrsim 30$ and even exceeds the contribution of $\phi_d \chi_d$ for $\theta_p \gtrsim 60$.

The post-form DWBA will be hard to work for the case of the breakup of such projec-tiles as have resonant excited states which couple strongly to the ground state ($^{6,7}$Li, $^{12}$C, $^{20}$Ne, etc.); the presence of the strong final-state interaction will be severe for the DWBA. On the other hand, CDCC is quite suited for including the resonant breakup

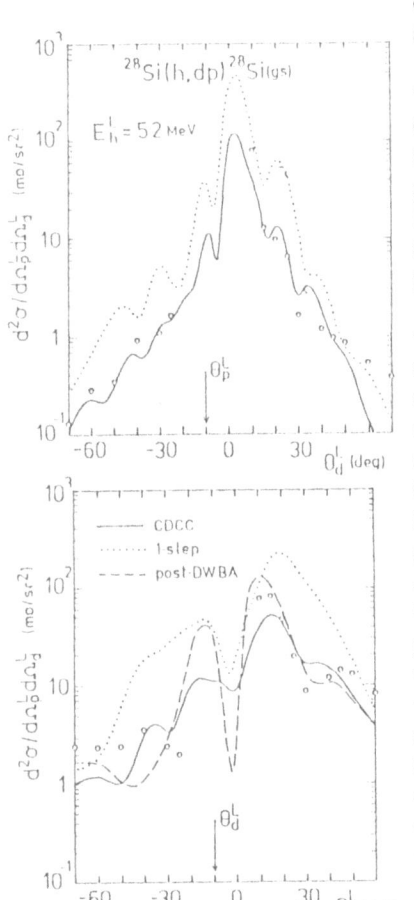

Fig. 2. Coincidence cross sections for the elastic breakup $^{28}$Si($^3$He,dp)[23,24].

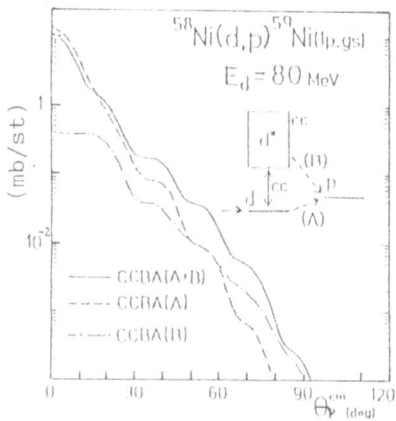

[10]; a successful application to the $^6$Li breakup cross sections for the resonance states will be reported in section 3. In connection with Fig. 3, a similar CDCC study of ($^6$Li,d) reactions will be of special interest, since no work has been performed to examine the contribution of the $^6$Li breakup (resonant and non-resonant) states to the α-transfer cross sections.

Fig. 3. Contribution of the deuteron breakup states to the (d,p) cross section [20].

# 3.  ELASTIC SCATTERING AND BREAKUP OF $^6$Li AND $^7$Li

## 3.1. Dynamical Polarization Potential Due to $^6$Li Breakup

As mentioned in section 1, exception of the success of the double-folding (DF) model with $N_R \approx 1.0$ was the scattering of $^{6,7}$Li and $^9$Be [3,7]. In Fig. 4, a result of the DF calculation for the scattering of $^6$Li+$^{28}$Si

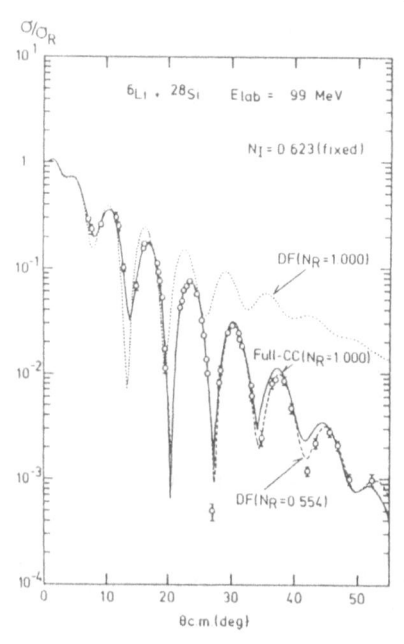

at 99 MeV in shown in the dashed line. The double folded real potential $U_F$ must be reduced in strength by muliplying a renormalization factor $N_R$=0.54. The dotted line shows a DF calculation with $N_R$=1.0 and the same $N_I$ of the best-fit DF calculation. The large difference between both results is caused by the difference of the real part only. If $N_I$ is increased, the dotted line comes down but the oscillation becomes out of phase. The same is seen in Fig. 5 for $^6$Li+$^{40}$Ca at 156 MeV.

Thompson and Nagarajan [8] showed that, in the framework of the adiabatic treatment (AD) of the breakup degree of freedom, inclusion of the $^6$Li breakup effect gives rise to a good fit to the observed data for the scattering of $^6$Li from $^{12}$C, $^{40}$Ca and $^{208}$Pb without any renormalization of the real part; the case of $^6$Li+$^{40}$Ca is shown in Fig. 5. Their AD is based on a three-body treatment of the α+d+target system, and the $^6$Li-target interaction is generated by the α-target and d-target optical potentials. The interaction is not directly related to the double-folded $^6$Li-target potential $U_F$, the result of AD suggested that the remarkable renormalization of $U_F$ might closely related to the breakup of the projectile.

Fig. 4. DF and CDCC calculations of $^6$Li+$^{28}$Si [10]. The data are taken from [25].

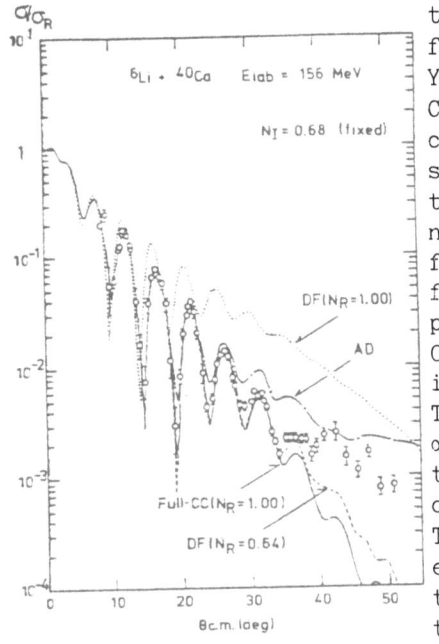

Fig. 5.  AD [8] and DF and CDCC [10] calculations for $^6$Li+$^{40}$Ca.  The data are from [26].

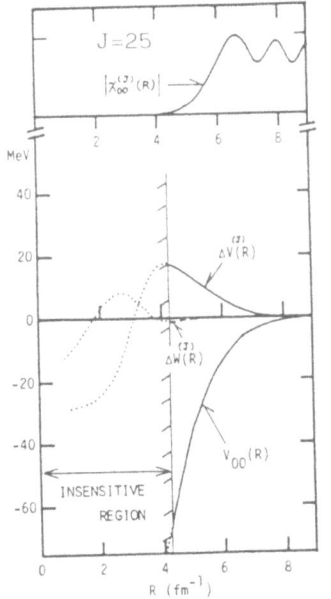

Fig. 6.  The DP potential due to $^6$Li breakup [10]

The direct relation between the projectile breakup effect and the renormalization factor $N_R$ for $U_F$ was studied by Sakuragi, Yahiro and Kamimura [9,10] in the microscopic CDCC framework.  They composed the bound and continuum $^6$Li wave functions with a microscopic $\alpha$-d cluster model taking into account the total antisymmetrization between the six nucleons.  The wave functions were successfully examined by the electron-scattering form factors and by the $\alpha$-d scattering phase shifts.  The $\alpha$-d continuum states with $0<k<1.5$ fm$^{-1}$ ($\varepsilon_{max}=35$ MeV) are discretized into four bins for each of $\ell=0$, 1 and 2.  The lowest bin of $\ell=2$ stands for the D-wave $\alpha$-d resonance states.  The wave functions of the ground and discretized breakup states of $^6$Li compose a set of $\{\phi_i^{(a)}\}$ in Eq. (1.1).  Target-excited states were ignored but effects of the states were considered to be taken into account in the imaginary part of the CC potentials.  The complex CC potentials, were assumed to be obtained by multiplying a complex factor $N_R+iN_I$ to all $V_{ij,ij'}(R)$'s in Eq. (1.3); but $N_R=1.0$ is kept in the CC calculations, $N_I$ being the only one adjustable parameter.

In Fig. 4, the result of the CDCC calculation [10] for $^6$Li+$^{28}$Si at 99 MeV is shown by the solid line which fits the data very well; here the same $N_I$ of the best-fit DF was employed.  For $^6$Li+$^{40}$Ca at 156 MeV, the same is seen in Fig. 5.  Since the result of CDCC is quite close to that of the best-fit DF, one may make the conjecture that the dynamical polarization (DP) potential induced by the breakup channels contributes strongly and repulsively to the real part only, at least in the surface region which is important for the scattering.

In ref. [10] the DP potential due to the $^6$Li breakup was exactly calculated in the form of the wave-function-equivalent (trivially-equivalent) local potential for each J, the total angular momentum; use is made of the CC wave functions solved and the breakup form factors.  In the case of the scattering of $^6$Li+$^{40}$Ca at 99 MeV, the real part $\Delta V^{(J)}$ and the imaginary part $\Delta W^{(J)}$ of the DP potential are illustrated in Fig. 6; here a grazing angular momentum J=25

is chosen and J-dependence was found to be small.  It is clearly seen in
Fig. 6 that the exact calculation of the DP potential verifies the aforementioned conjecture on the property of the potential; in the surface
region, the real part is as strongly repulsive as is required to understand the remarkable renormalization mentioned before, while the imaginary
part is negligibly small.  This feature of the DP potential due to $^6$Li
breakup is found to hold for a wide range of the bombarding energy; see
ref. [27] for $^6$Li+$^{28}$Si at E($^6$Li)=46~154 MeV.

The character of the DP potential shown in Fig. 6 is newly discovered
and is completely opposite to that of the DP potential induced by the
excitation of the $2^+(3^-)$ collective state; the imaginary part of the
latter potential is considerably strong, whereas the real part is very
small [5].  This difference is found [28] to arise from the presence of
the large imaginary part ($N_I$~0.6) of the coupling potentials between the
$^6$Li incident channel and the breakup channels (c.f. section 5).

## 3.2. $^6$Li Breakup Cross Sections

The CDCC calculations of $^6$Li scattering give simultaneously the
cross sections for the resonance states and discretized non-resonance
states.  Figure 7 shows the scattering of $^6$Li+$^6$Li at E($^6$Li)=156 MeV; the
cross section for the $3^+$ resonance at 2.19 MeV was measured not by the
projectile breakup but by the inelastic excitation of the target $^6$Li.  In
the CDCC calculation ($N_R$=1.0) [30], the parameter $N_I$ was so chosen that
the elastic-scattering data was reproduced.  Thus,  the calculated
inelastic cross section for the $3^+$ resonance was absolutely predicted,
and the agreement is quite good for $\theta_{c.m.}$<40:

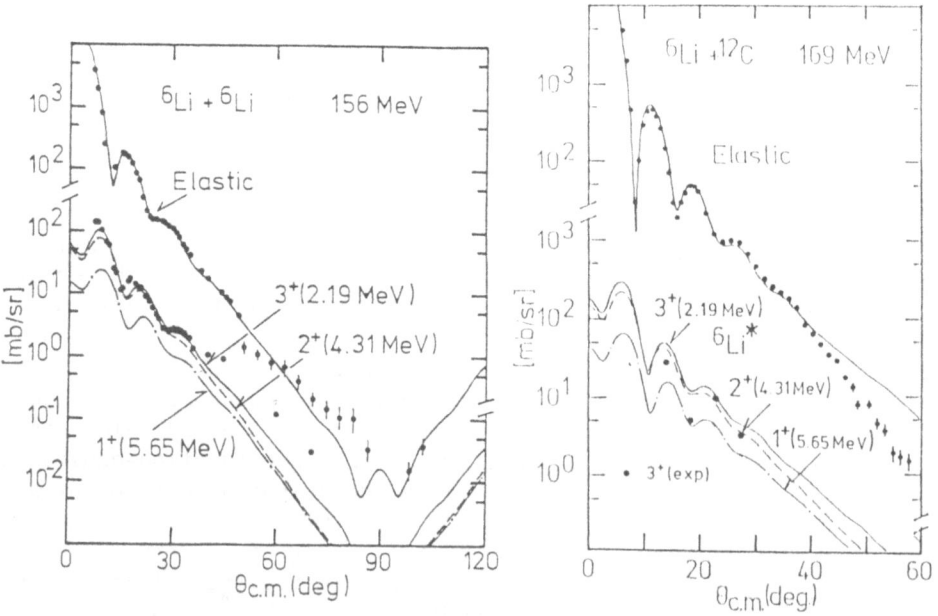

Fig. 7.  CDCC calculation of $^6$Li        Fig. 8.  CDCC calculation of $^6$Li+$^{12}$C
+$^6$Li [30].  The data are from [29].     [30].  The data are from [31,32].

In Fig. 8, a CDCC calculation predicts [30] for $^6Li+^{12}C$ at $E(^6Li)=$ 169 MeV, the elastic breakup cross sections for the $3^+$, $2^+$ and $1^+$ resonances of $^6Li$ at $\epsilon_x=2.19$, 4.31 and 5.65 MeV, respectively; the value of $N_I=0.466$ was determined so as to fit the elastic scattering data [31]. An $\alpha$-d coincidence experiment for the elastic breakup $^{12}C(Li,d)$ reaction has recently been performed at $E(^6Li)=178$ MeV [32]. Differential cross sections of $^6Li*$ were derived for the $^6Li$ continuum states in $0<\epsilon_x<1.0$ MeV and in $1.0<\epsilon_x<4.0$ MeV. The cross section for the former may be considered to be for the $3^+$ resonant state      ; the experimental result is given in Fig. 8 (though the bombarding energy is slightly different). It is striking that the CDCC prediction which was given before the experimental result came out agrees with the data satisfactorily.

### 3.3. $^7Li$ Scattering

Effect of the $^7Li$ projectile breakup has been studied by Nagarajan, Thompson and Johnson [11] based on the same adiabatic treatment as used in the $^6Li$ case [8] and by Sakuragi, Yahiro and Kamimura [12] a based on CDCC. They found that the $^7Li$ breakup states play a very important role in reproducing the elastic scattering data satisfactorily. Figure 9 by CDCC [12] illustrates individual roles of the $3/2^-$(g.s.) and $1/2^-$ bound states, the $7/2^-$ and $5/2^-$ resonant states and $1/2^-$ to $7/2^-$ non-resonant breakup states. The effect of the $7/2^-$ resonance is prominent.

Scattering of polarized $^6Li$ and $^7Li$ from $^{58}Ni$ at $E_{lab}\sim20$ MeV and the related polarization phenomena were precisely studied in CC methods, for example, in refs. [22,34] and is reviewed by Johnson in this conference (further references may be seen in his talk).

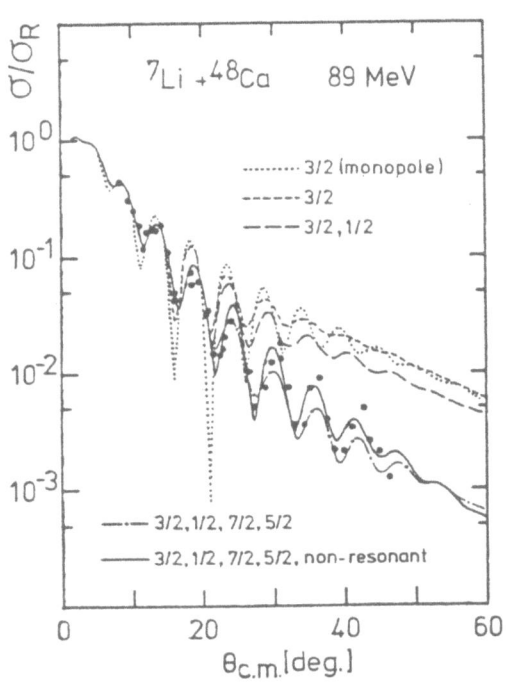

Fig. 9. CDCC calculation of $^7Li+^{48}Ca$. The data are from [33].

## 4. $^{12}C\rightarrow3\alpha$ BREAKUP EFFECT ON THE SCATTERING OF $^{12}C$

The double-folding (DF) model with $N_R\approx1.0$ has been successful for the scattering of $^{12}C$ at $E(^{12}C)/12\lesssim10$ MeV [3]. For higher energies of $E(^{12}C)/12\gtrsim20$ MeV, however, a strong renormalization, $N_R\approx0.6\sim0.84$, of the folded real potential has been found to be necessary to reproduce the observed cross sections [35,37]. This renormalization must be due to the excitation of $^{12}C$, which is of interest from the viewpoint of the excitation of soft heavy ions in the field of the target nucleus. In the present sections we first discuss coupled-channel (CC) effects of the

$^{12}$C excited states including breakup states [36,37] in the high-energy scattering of $^{12}$C+$^{12}$C, $^{13}$C+$^{12}$C and $^{16}$O+$^{12}$C. Secondly, we discuss the inelastic scattering of $^6$Li from $^{12}$C.

As for the $^{12}$C wave functions $\{\Phi_I^{(a)}\}$ in Eq. (1.1), the CC work of Sakuragi, Fukushima and Kamimura [36,37] employed the totally antisymmetrized 3α resonating group wave functions, $\Phi_{IM}^{(C)} = \mathcal{A}[\phi_\alpha\phi_\alpha\phi_\alpha u_{IM}]$. Fukushima and Kamimura [42,43] calculated $\Phi_{IM}^{(C)}$ for I=0 to 4 by means of diagonalizing a twelve-body total Hamiltonian in a subspace spanned by a set of $L^2$-normalizable basis functions for the relative wave functions $u_{IM}$. The wave functions of the $0_1^+$, $2_1^+$, $4_1^+$, $0_2^+$, $2_2^+$, $1_1^-$ and $3_1^-$ reproduced well their energies, electron-scattering form factors, B(Eλ) strengths and α-decay widths. The $0_2^+$, $2_2^+$ and $1_1^-$ states are well-developed clustering states. On the other hand, the $0_1^+$, $2_1^+$, $4_1^+$ and $3_1^-$ states are of shell-like character and are represented as compactly coupled solutions of the relative motion between the totally antisymmetrized 3α clusters.

Among the $^{12}$C states, refs. [36,37] employed the $0_1^+$, $2_1^+$, $0_2^+$, $2_2^+$ and $3_1^-$ states. Non-resonant 3α breakup continuum states were described in terms of a finite number of discrete states which were obtained simultaneously by the aforementioned diagonalization. Among the discretized non-resonant states, the lowest two for each I =$0^+$ and $2^+$ were taken up; coupling of the other members are negligible. The nuclear part of the diagonal and coupling potentials of the $^{12}$C+A system (A= the ground state of $^{12}$C, $^{13}$C and $^{16}$O) is assumed to be given by Eq. (1.3) by folding the M3Y interaction [6] together with the zero-range exchange term [3]. The double-folded potentials are multiplied by a complex factor $N_R$+i$N_I$. In the CC calculations shown in the following, we persist in taking $N_R$=1.0 and leave $N_I$ as the only one adjustable parameter. The excitation of $^{13}$C and $^{16}$O and the simultaneous excitation in $^{12}$C+$^{12}$C were ignored.

In Fig. 10, the results of the CC calculations [36,37] are given in the solid lines: (a) $^{12}$C+$^{12}$C at E($^{12}$C)=300 MeV, (b) $^{16}$O+$^{12}$C at E($^{16}$O)=168, 216 and 311 MeV and (c) $^{13}$C+$^{12}$C at E($^{13}$C)=260 MeV. For these five E's, E($^{12}$C)/12=25, 11, 14, 19 and 20 MeV, respectively if $^{12}$C is regarded as the projectile in (b) and (c). A good agreement with the observed differential cross sections is obtained; the inelastc cross sections for the $2_1^+$ state of $^{12}$C in (a) and (b) at 168 MeV are simultaneously reproduced.

The dotted lines in Fig. 10 are given by the $0_1^+$ one-channel calculation with the use of the same value of $N_I$ as in the CC calculation. The large difference between the solid and dotted lines shows a strong CC effect of the whole excited channels considered here. Roles of the individual channels may be understood by comparing the four kinds of lines in Fig. 10. $2_1^+$ channel shows the strongest coupling to the incident channel, but the well-developed clustering states ($0_2^+$ and $2_2^+$) and the non-resonant breakup states play an important role, especially in reproducing the enhancement at backward angles.

The $0_1^+$-one-channel calculaton (DF) in which the optimum value of $N_R$ and the optimum imaginary potential within the Woods-Saxon shape were searched [36,37] gave $N_R$=0.66 for $^{12}$C+$^{12}$C, $N_R$=0.98, 0.92 and 0.84 for $^{16}$O +$^{12}$C at 168, 216 and 311 MeV, respectively and $N_R$=0.76 for $^{13}$C+$^{12}$C (the lower in Fig. 10c); the imaginary potential becomes much stronger than that used in the full CC calculation for the $0_1^+$ channel. Since the optimum value of $N_R$ is very close to 1.0 for the scattering of $^{16}$O+$^{12}$C at 168 and

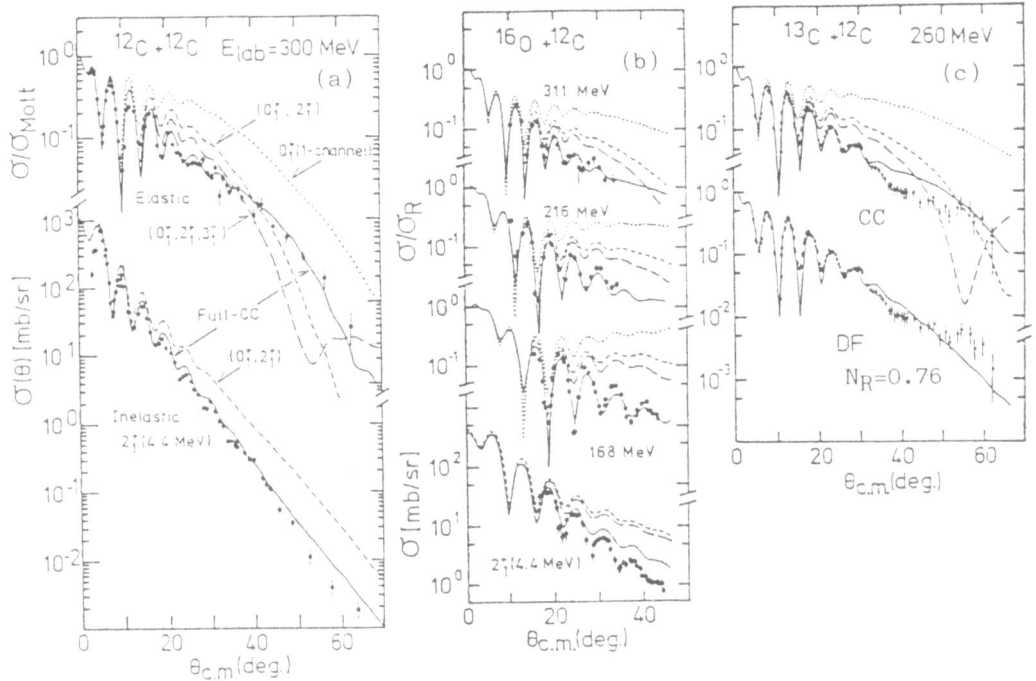

Fig. 10. Angular distributions of the scattering of (a) $^{12}$C+$^{12}$C,
(b) $^{16}$O+$^{12}$C and (c) $^{13}$C+$^{12}$C. The solid lines are given by the (full)
CC calculation [36,37]. The dotted, short-dashed and long-dashed
lines are given by the $0^+_1$ one-channel, the $0^+_1$-$2^+_1$ two-channel and the
$0^+_1$-$2^+_1$-$3^-_1$ three-channel calculations respectively. The data are from
(a) [35], (b) [38] and (c) [39]. The lower half of (c) shows a DF
calculation with $N_R$=0.76.

216 MeV, one may consider that the DF model still work in this energy
region, $E(^{12}C)/12$=11~14 MeV, and the CC effect of the excited states can
be replaced by renormalizing the imaginary part of the optical potential.
Significant deviation of $N_R$ from 1.0, however, is seen for the other cases
for which $E(^{12}C)/12$=20~25 MeV. The CC effect in the cases is considered
to induce a sizable real, repulsive potential as well as the imaginary
part.

In order to see the CC effect explicitly, the DP potential induced
by the excited channels was calculated in the form of the wave-function-
equivalent (trivially equivalent) local potential in the same manner as
in ref. [10]. Let $\Delta U^{(J)}(R)$ denote the DP potential, which depends on the
total angular momentum J. $\Delta U^{(J)}(R)$ are illustrated in (a) and (b) Fig.
11 for the case of $^{16}$O+$^{12}$C at 168 and 311 MeV; a grazing angular momentum
is taken for J, but the J-dependence of $\Delta U$ is small in the surface region
which is important for the scattering. At 168 MeV, Re$\Delta U$ is repulsive but
small as expected before (i.e. $N_R$=0.98 in the DF calculation). On the
other hand, at 311 MeV, Re$\Delta U$ becomes, indeed, more strongly repulsive as
was expected (i.e. $N_R$=0.84). In both cases, the considerable large Im$\Delta U$
is really generated.

Figure 11c shows that the well-developed clustering states ($0^+_1$ and

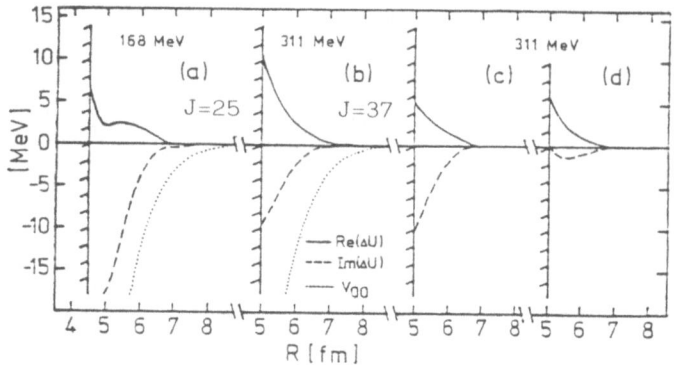

Fig. 11.  DP potential $\Delta U^{(J)}(R)$ due to the excited states of $^{12}C$ in $^{16}O+^{12}C$, (a) at $E(^{16}O)=168$ MeV and (b) at 311 MeV.  $V_{00}$ is the real part of the folded potential of the incident channel.  (c) Contribution to $\Delta U^{(J)}(R)$ of the $2^+_1$ and $3^-_1$ states and (d) of the $0^+_2$ and $2^+_2$ and non-resonant breakup states.  All are taken from [57].

$2^+_1$) plus the non-resonant breakup states contribute to the real part (repulsive) but weakly to the imaginary part.  This feature is consistent with that of the DP potential due to the breakup of $^6Li$ which has resonant and non-resonant breakup states, too [10].  On the other hand, the $2^+$ and $3^-$ states of $^{12}C$ sizably induce both of real (repulsive) and imaginary potentials at $E(^{12}C)/12=20$ MeV; this feature has not been seen in the previous studies [5] of the DP potential due to the excitation of collective $2^+$ and $3^-$ states in other nuclei.

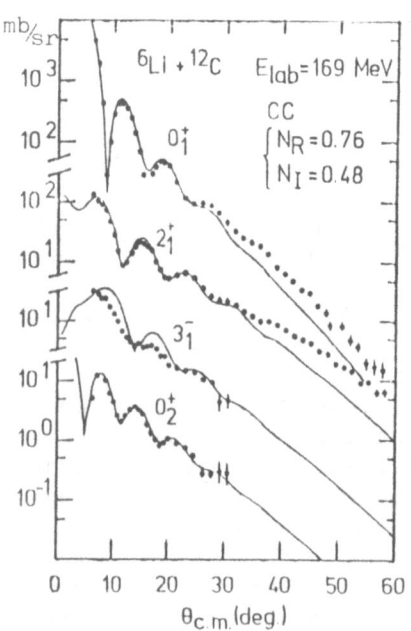

The inelastic scattering of $^6Li$ from $^{12}C$ at $E(^6Li)=169$ MeV was studied in the CC framework by Sakuragi and Kamimura [41]. In this case, they ignored the $^6Li$ breakup channels but renormalized the real part of the CC potentials of the $^{12}C$ states following the results of section 3.  The case of $N_R=0.76$ and $N_I=0.48$ produced the elastic cross section in Fig. 12.  The inelastic cross sections were then absolutely predicted.  The calculation gave a beautiful fit to the observed $2^+_1$, $3^-_1$ and $0^+_2$ cross sections.

Fig. 12.  CC calculation of the inelastic scattering of $^6Li$ from $^{12}C$ ($0^+_1$, $2^+_1$, $3^-_1$, $0^+_2$) at 169 MeV [41]. The data from [31].

## 5.   A SYSTEMATICS OF DYNAMICAL POLARIZATION POTENTIALS

In this section, we report a schematic but systematic study of the dynamical polarization (DP) potentials by Sakuragi, Yahiro and Kamimura [40]. They considered $^6$Li scattering from $^{28}$Si at 99MeV and took a two-state CC model in which the ground state of $^6$Li couples to the D-wave resonant state only.  The same CC framework as in section 3 was employed.

The double-folded form factors, Eq. (1.3), were multiplied by a complex factor $N_R+iN_I$.  The $N_R$ and $N_I$ were fixed to $N_R=1.0$ and $N_I=0.6$ for the diagonal and reorientation potentials (c.f. section 3), since the $N_I$ dependence is rather weak.  On the other hand, $N_R$ and $N_I$ for the coupling potential was varied widely, and the DP potential due to the excited channel was calculated for several combinations of $N_R$ and $N_I$ and is shown in Fig. 13 for R>5 fm the region of which is important for the scattering.

The case of "A" at $N_R=1.0$ and $N_I=0.6$ stands for the actual DP potential due to the resonant breakup of $^6$Li projectile (see Fig. 6).  The case of "B" at $N_R=1.0$ and $N_I=0.0$~0.3 simulates the DP potential due to the $^{12}$C excitation (see fig. 11) in the $^{16}$O+$^{12}$C scattering.  The case "C" corresponds to the well known type of DP potential due to the excitation of collective states [5]; it is to be noted that this potential is only a limit case of the wide possibility of the DP potentials.  From Fig.13, it is understood that the magnitude of the imaginary part of the coupling (transition) potential affects decisively on the DP potential; it causes even a sign change in the imaginary part of the induced DP potential.

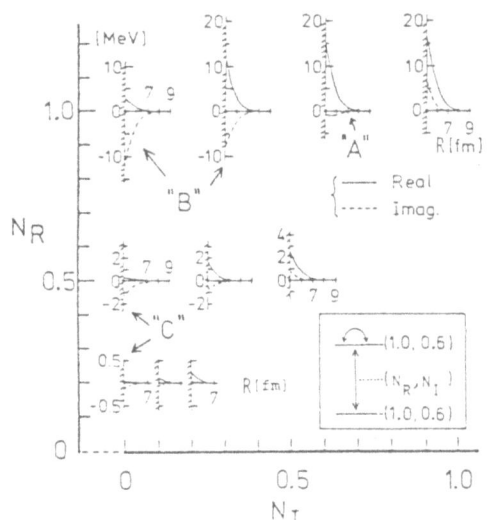

Fig. 13.  Systematics of DP potentials given in a two-state CC model [40] with respect to the strength of the coupling potential, $N_R$ and $N_I$. Calculation is made for $^6$Li+$^{28}$Si at 99 MeV.

## 6.   CONCLUDING REMARKS

We have proposed a microscopic coupled-channel (CC) approach to the scattering and breakup of soft heavy ions, and we have examined successfully the approach in the case of the $^6$Li, $^7$Li and $^{12}$C projectiles.  The wave functions of the ground and excited states of the projectiles were described by the microscopic cluster model and were well checked by various **experimental** data such as electron-scattering form factors.  The breakup continuum states of the projectiles were also taken into account explicitly in the framework of the method of coupled discretized continuum channels.  The real part of the diagonal and coupling potentials was

derived by folding the M3Y interaction into the projectile and target
states concerned.  The geometry of the imaginary part of the potentials
was assumed to be the same as the real part, at least in the surface
region which is important for the reactions.  The strength factor of the
imaginary part, $N_I$, is the only one adjustable parameter.

A lot of elastic scattering cross sections were beautifully reproduced by the CC approach.  Some inelastic and breakup cross sections were
absolutely predicted, after the optimum $N_I$ was determined through the
elastic scattering, and the predicted cross sections gave a good fit to
the data.  Those analyses made clear that the couplings of the projectile
breakup channels are very strong and absolutely contribute to the elastic
scattering cross sections.  The imaginary part of the coupling (transition)
potentials was found to play an essentially important role in the CC
mechanism.  A future study is expected on the origin of the imaginary part
of the coupling potentials as well as that of the diagonal ones.

Explicit calculation of the dynamical polarization (DP) potential
induced by the breakup channels explained the reason of the failure of
the double-folding model for the scattering of the soft projectiles.  The
DP potential is found to be of completely different character compared
with that due to the excitation of collective states.  A systematic
study of the origin of a variety types of DP potentials is in progress
and is expected to reveal precise CC mechanisms among the various excited
states in light nuclei, such as shell-like states, cluster-like states
and non-resonant breakup states.

Since the microscopic cluster model have produced a lot of precise
wave functions of light nuclei, further extensive applications of the
microscopic CC approach based on such wave functions will much more
intimately and successfully combine the cluster-structure study with the
heavy-ion reaction study.

REFERENCES
[1] K. Ikeda et al., Prog. Theor. Phys. Suppl. 52 (1972), 68 (1980).
[2] K. Wildermuth and Y.C. Tang, A Unified Theory of the Nucleus
    (Viewig, Bansweig, 1977).
[3] G. R. Satchler and W. G. Love, Phys. Report 55 (1979) 183.
[4] H. Feshbach, Ann. Phys. (N.Y.) 19 (1967) 287.
[5] N. Vinh Mau, Phys. Letters 71B (1977) 5.
    W.G. Love, T. Terasawa and G.R. Satchler, Nucl.Phys A291(1977) 183.
    B. Sinha, Phys. Rev. Letters 42 (1979) 690.
    K.-I. Kubo and P.E. Hodgson, Nucl. Phys. A366 (1981) 320.
[6] G. Bertsch, J. Borysowicz, H. McManaus and W.G. Love, Nucl.
    Phys. A284 (1977) 399.
[7] C.W. Glover, R.I. Cutler and K.W. Kemper, Nucl.Phys.A341(1980) 137.
    G.R. Satchler, Phys. Letters 83B (1979) 284.
[8] I.J. Thompson and M.A. Nagarajan, Phys. Letters 106B (1981) 163.
[9] Y. Sakuragi, M. Yahiro and M. Kamimura, Prog. Theor. Phys. 68
    (1982) 322.
[10] Y. Sakuragi, M. Yahiro and M. Kamimura, Prog. Theor. Phys. 70
    (1983) 1047.

[11] M.A. Nagarajan, I.J. Thompson and R.C. Johnson, Nucl. Phys. A385 (1982) 525.
[12] Y. Sakuragi, M. Yahiro and M. Kamimura, Proc. 1983 RCNP Inter. Symp. on Light ion reaction mechanism (Contributed papers),1983,Osaka,p.84.
[13] M. Yahiro and M. Kamimura, Prog. Theor. Phys. 65(1981)2046, 2051.
[14] M. Yahiro, M. Nakano, Y. Iseri and M. Kamimura, Prog. Theor. Phys. 67 (1982) 1467.
[15] M. Yahiro, Y. Iseri, M. Kamimura and M. Nakano, Phys. Letters 141B (1984) 19.
[16] N. Matsuoka et al., Nucl. Phys. A391 (1982) 357.
[17] H. Amakawa and N. Austern, Aust. J. Phys. 36 (1983) 633.
[18] G. Baur, R. Shyam, F. Rösel and D. Trautmann, Phys.Rev.C28(1983)946.
[19] H. Amakawa, S. Yamaji, A. Mori and K. Yazaki, Phys.Lett. 82B(1979)13.
[20] Y. Iseri, M. Yahiro and M. Nakano, Prog. Theor. Phys. 69 (1983)1038.
[21] M. Yahiro, Proc. 1983 RCNP Inter. Symp. on Light ion reaction mechanism, Osaka, 1983.
[22] H. Ohnishi, M. Tanifuji, M. Kamimura, Y. Sakuragi and M. Yahiro Nucl. Phys. A415 (1984) 271.
[23] R. J. de Meijer and R. Kamermans, Rev. Mod. Phys. to be published.
[24] Y. Iseri, M. Yahiro, M. Kamimura and M. Kawai, a contribution to this conference and private communication.
[25] P. Schwandt et al., INCF Annual report, 1979, p. 93.
[26] Z. Majka, H.J. Gils and H. Rebel, Z. Phys. A288 (1978) 139.
[27] Y. Sakuragi, M. Yahiro and M. Kamimura, a contribution to this conf.
[28] Y. Sakuragi, M. Kamimura and M. Yahiro, a contribution to the 1984 INS-RIKEN Inter. Symp. on Heavy ion physics, 1984, Tokyo.
[29] S. Micek et al., private communication.
[30] Y. Sakuragi, M. Yahiro and M. Kamimura, a contribution to the 1984 INS-RIKEN Intern. Symp. on Heavy ion physics, 1984, Tokyo.
[31] K. Katori et al., private communication.
[32] K. Katori et al., private communication.
[33] M.F. Steeden, et al., J. Phys. G. Nucl. Phys. 6 (1980) 501.
[34] N. Nishioka, J.A. Tostevin, R.C. Johnson and K.-I. Kubo, Nucl. Phys. A415 (1984) 230.
[35] H.G. Bohlen et al., Z. Phys. A308 (1982) 121.
[36] Y. Sakuragi, Y. Fukushima and M. Kamimura, Proc. 1983 RCNP Intern. Symp. on Light ion reaction mechanism, 1983, Osaka.
[37] Y. Sakuragi and M. Kamimura, preprint submitted to Phys. Letters.
[38] J.C. Hieberl and G.T. Garbey, Phys. Rev. 135 (1964) B346; M.E. Brandan and A. Menchaca-Rocha, Phys. Rev. C23 (1981) 1272.
[39] H.G. Bohlen et al., Annual Report 1983, HMI, Berlin, p. 38.
[40] Y. Sakuragi, M. Yahiro and M. Kamimura, a contribution to the 1984 INS-RIKEN Intern. Symp. on Heavy ion physics, 1984, Tokyo.
[41] M. Kamimura and Y. Sakuragi, Proc. the IPCR Symp. on Nuclear physics at 10~100 MeV/u, November, 1983, Tokyo.
[42] Y. Fukushima and M. Kamimura, Proc. Intern. Conf. on Nuclear structure, 1977, Tokyo, p. 225.
[43] M. Kamimura, Nucl. Phys. A351 (1981) 456.

# DIRECT AND SEQUENTIAL BREAKUP OF CLUSTER-LIKE NUCLEI

A.C. Shotter,
Department of Physics,
James Clerk Maxwell Buildings,
The King's Buildings,
Mayfield Road,
EDINBURGH. U.K.

ABSTRACT. The interaction of light ion projectiles with target nuclei is characterised by ejectiles travelling in the forward direction with velocities similar to that of the incident projectiles. If such ejectiles are produced by projectile breakup, then experiments measuring simultaneously the kinematic characteristics of two or more fragments can provide valuable information concerning the types of breakup processes.

## 1. INTRODUCTION

The breakup reaction mechanism for heavy ion reactions is associated with those events where, due to the interaction with the target nucleus the projectile breaks up into two or more fragments. It is possible to further classify the breakup reaction into sequential breakup and direct breakup. These two categories are schematically illustrated in fig. 1, by time-separation diagrams for the projectile and target.

For sequential breakup the incoming projectile, denoted by Pg.s. (projectile in ground state), interacts with the target nucleus denoted by Tg.s.. As a result of the interaction the projectile is excited. The excited projectile (p*) then moves away from the target nucleus and, if the excitation energy is sufficient, it decays into the two particle channel F1 + F2. Normally there is little, if any, final state inter- action between the fragments and the target. Sequential breakup may be subdivided as either a) quasielastic, where the target is unexcited (T*=Tg.s.), or b) inelastic, where the target is left in an excited state. In direct breakup the interaction of the target field across the projectile is sufficient to break the projectile into two fragments F1, F2 directly without the projectile going through any intermediate state. If such a breakup process occurs in the peripheral region between projectile and target, then it is possible that there is little final state interaction for the fragments F1, F2 with the target. In this case the target will probably remain unexcited and the event is there- fore called quasielastic direct breakup. However, it is most likely that the target will be excited either by $i$) target-projectile interaction in the initial state or $ii$) target-fragment interaction in the final

199

*J. S. Lilley and M. A. Nagarajan (eds.), Clustering Aspects of Nuclear Structure, 199–213.*
© *1985 by D. Reidel Publishing Company.*

state. It is possible that both *i)* and *ii)* operate simultaneously as
illustrated in fig. 1. This breakup process is identified as inelastic
direct breakup.

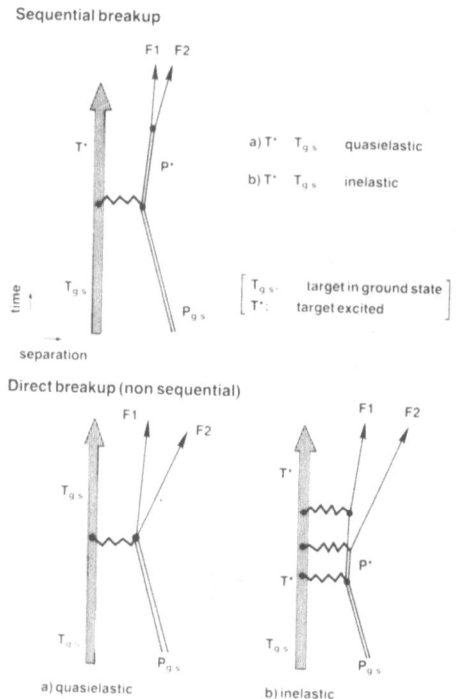

Fig. 1. Types of projectile breakup.

Since the direct breakup process involves a rapid disintegration
of the projectile, it is sometimes referred to as non-sequential breakup
to distinguish it from the delayed breakup associated with sequential
breakup.

Various types of experiments have been undertaken to investigate
breakup reactions; a brief summary of the main types is given below.

1.1. Single arm inclusive experiments.

Many experimental investigations have been conducted for various pro-
jectile/target/energy combinations, where the aim is to measure inclusive
spectra for various ejectiles.

1.2. Two arm coincidence experiments.

The aim of these experiments is to investigate those reaction events
where two fragments are emitted from the breakup of the projectile.

## 1.3. Fragment γ-ray coincidence experiments.

If breakup occurs and one of the fragments is captured by the target, the target will become highly excited. The target will de-excite by neutron emission and finally γ-emission. The γ-emission is characteristic of the residual nucleus after neutron evaporation. Measuring the yield of the escaped fragment in coincidence with γ-rays therefore provides valuable information concerning this breakup-fusion process.

## 1.4. Target residual recoil experiments.

For those events where breakup fusion occurs the residual nucleus will recoil with a velocity appropriate to the momentum absorbed. By ranging the products in a series of thin foils and subsequently γ-counting these foils, it is possible to get information concerning the energy and type of fused breakup fragment.

## 1.5. Multifragment experiments.

For some light ion projectiles of intermediate energy ~ 10-20 MeV/A breakup into more than two fragments becomes significant. To obtain information concerning these processes several detectors are needed to record the fragments.

Each of these types of experiments have yielded valuable information concerning different features of the breakup process. However, due to lack of space this paper will be concerned mainly with experiments in categories 1.1, 1.2, and 1.5, although passing reference will be made to some experiments associated with 1.3 and 1.4.

## 2.    INCLUSIVE EXPERIMENTS.

If a particular nuclear reaction is dominated by quasielastic breakup of either the direct or sequential type, then we would expect that the fragments would emerge with a velocity similar to that of the projectile. Even if one of the fragments fuses with the target the remaining fragment, if there is no strong final state interaction between it and the target, will emerge with a velocity similar to the projectile. In other words for such events $E_f = (m_f/m_p)E_p$, for forward scattering angles, where f refers to fragment and p to projectile. Figure 2 shows $\alpha$ and t spectra for $\theta = 15^o$ and $30^o$ for the reaction $^{120}Sn(^7Li,x)$ for 70 MeV $^7Li$ (1). All the spectra show bell-shaped distributions that have maxima corresponding to the ejectile travelling with the projectile velocity. (In fig. 2-6, the arrows mark the energy where the ejectile is travelling with this velocity.) The yield of these distributions increases strongly with decreasing angle. For the $^4He$ spectra, at $30^o$, it seems that the bell-shaped distribution sits on a lower energy component which is probably in part due to the evaporative tail of a compound nucleus reaction.

Figure 3 shows the spectra for light ion ejectiles from the reaction $^{208}Pb(^7Li,x)$ $E_{Li} = 70$ MeV (1). The spectra for emission of $^4He, ^3H, ^2H$ and p are again dominated a broad bump associated with ejectiles moving

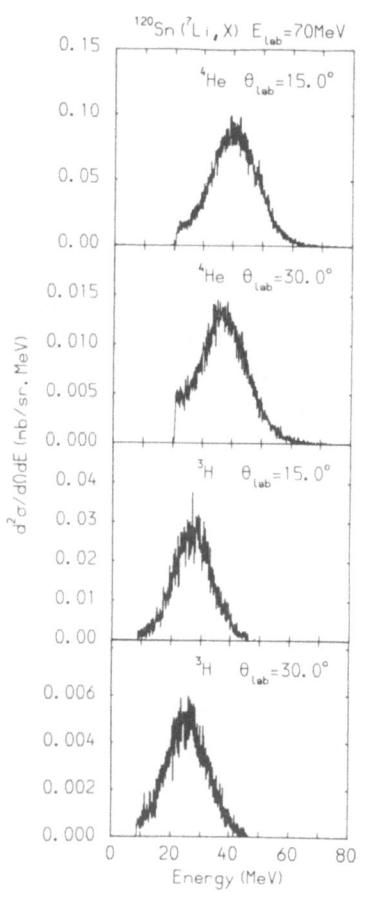

Fig. 2.

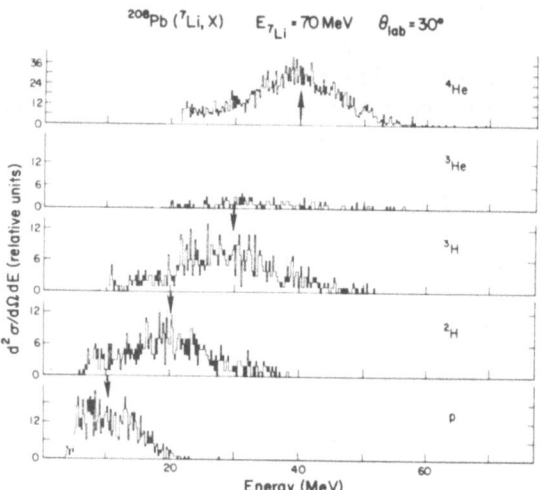

Fig. 3.

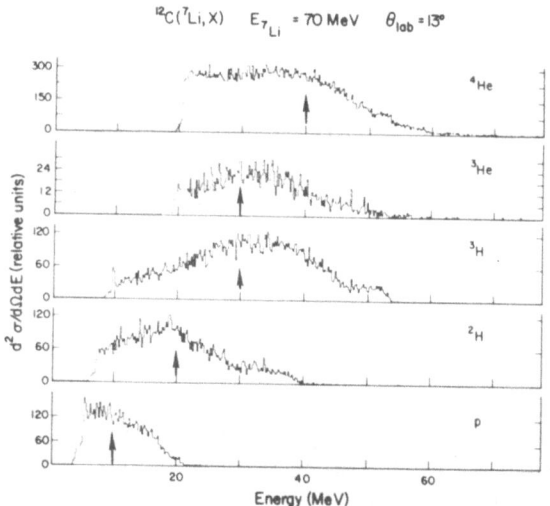

Fig. 4.

with beam velocity. The low yield of $^3$He compared to $^3$H is probably due to the high Q for $^7$Li → $^3$He + n+$^3$H (23.2 MeV) compared to $^7$Li → $^4$He+$^3$H (2.4 MeV). These ejectile bumps are also seen for light targets as can be seen in fig. 4, which shows similar ejectile spectra as fig. 3, but for a $^{12}$C target (1). These spectra will obviously be influenced by disintegration of the target as well as the projectile. It is expected that such a target disintegration would yield a large flux of α particles below beam velocity, and this probably accounts for the large background on which the velocity peak appears to sit.

For a heavier projectile, fig. 5 shows $\alpha$ ejectile spectra for the reaction $^{208}Pb(^{12}C,x)$ $E_{^{12}C}$ = 187 MeV at several angles (2). At the more backward angles the spectra are dominated by the exponential tail that is probably associated with the compound nucleus component. At the most forward angle it can be seen that a beam velocity component begins to strongly influence the overall spectrum shape. For $^{8}Be$ ejectiles this component is even more dramatically evident as can be seen in fig. 6; for $^{8}Be$ the compound nucleus component would be expected to be small.

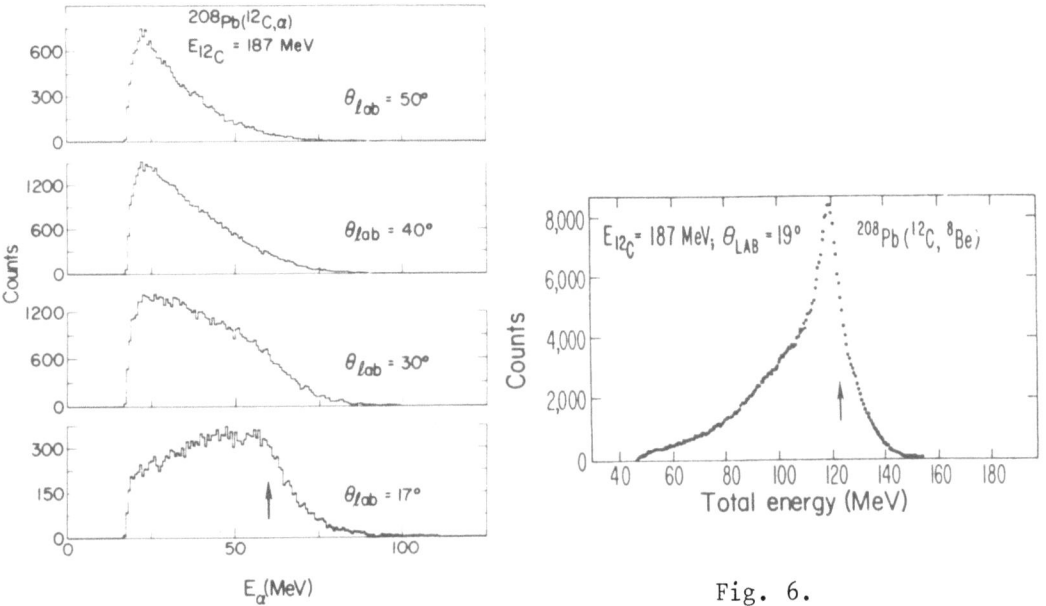

Fig. 5.

Fig. 6.

For even heavier projectiles such as $^{16}O$ and $^{20}Ne$ the ejectile spectra show similar characteristics. Thus for $^{16}O$ projectiles at 315 MeV (3), scattered from a Pb target the ejectile spectra are dominated by a bump that is near beam velocity, although for the lighter ejectiles the maximum of the spectra falls below the energy corresponding to beam velocity. A similar situation arises for $^{20}Ne$ projectiles for example $^{40}Ca(^{20}Ne,x)$ $E_{ca}$ = 149, 262 MeV (4).

By way of summary fig. 7 shows the magnitude of the ratio of ejectile velocity to projectile velocity for different reaction conditions of projectile, target and ejectile type. It can be seen that for a wide variety of situations the condition $E_f = (m_f/m_p)E_p$ holds reasonably well. The breakup reaction mechanism can certainly give rise to these peaks, however it is possible that other reaction mechanisms could also contribute. If, for example, the projectile transfers part of its mass to the target nucleus then Brink's semiclassical rules (5) for optimum transfer also give $E_f = (m_f/m_p)E_p$. Brink's rules only apply when the mass transfered is small compared to target and projectile masses. A recent calculation (6) that attempts to modify the rules for large mass transfers shows that for this situation $E_f < (m_f/m_p)E_p$.

If quasielastic breakup is a significant source of beam velocity ejectiles then we should expect a significant number of events where all the fragments associated with the projectile breakup escape from the target nucleus. Therefore coincidence experiments which aim to measure the characteristics of at least two emitted fragments are of great importance.

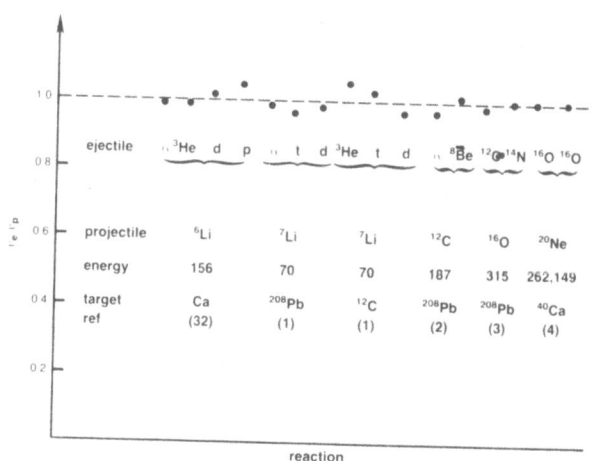

Fig. 7.   Ratio of ejectile to projectile velocity for various reactions.

## 3.   TWO ARM COINCIDENCE EXPERIMENTS.

There are many published reports of experiments of this type and therefore some selection is necessary. In this section some specific experiments associated with quasielastic breakup will be discussed and in section 5 some two arm coincidence experiments concerning inelastic breakup will be considered.

## 3.1. Quasielastic breakup of $^7$Li.

When a projectile breaks into two fragments then, provided the target does not rapidly disintegrate, it is possible to completely specify the reaction kinematically if the emission energies and angles of the fragments are measured. The angle between the two fragments is related to their relative centre of mass energy $\varepsilon_r$. For a given separation angle $\theta$ between the fragment detectors the minimum $\varepsilon_r$ that can be observed is given by $\varepsilon_r \sim \theta^2 m_1 m_2 E^*/(m_1+m_2)^2$, for small $\theta$ where $m_1$ and $m_2$ are the masses of the two fragments of centre of mass energy $E^*$. For $^7$Li the most probable breakup mode is the $\alpha + t$ channel, $Q = 2.47$ MeV. A convenient way to display the experimental results is to plot the measured energies of $\alpha, t$ pairs on an $E_\alpha$, $E_t$ diagram as shown schematically in fig. 8. Different diagonal grouping of events would correspond to different target excitation energies, starting from zero target excitation $Q_0$, 1st excited state $Q_1$, etc. For a given diagonal, the distribution of events along it would be determined by the relative centre of mass energy distribution of the projectile fragments.

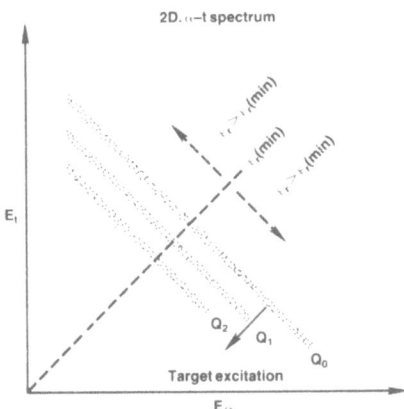

Fig. 8. Schematic distribution of $^7$Li projectile breakup into $\alpha$ and t fragments, energies $E_\alpha$ and $E_t$.

A diagonal distribution for a particular target state can be projected onto either of the energy axes. Such a projection for the reaction $^{120}$Sn($^7$Li,$\alpha$+t)$^{120}$Sn$_{g.s.}$, $E_{Li}$ = 70 MeV is shown in fig. 9b,c,d for various ejectile angles (7). In this experiment two detectors were placed in a line perpendicular to the reaction plane with an opening angle of 5°. The two narrow peaks in each of the projected spectra correspond to the two kinematic solutions where the $^7$Li projectile is excited to the 4.63 MeV state before breakup. The results from a Monte Carlo calculation that simulates such a breakup process and for the same experimental geometry is shown in fig. 9a. It can be seen that no events due to sequential breakup would be expected between 29 MeV and 48 MeV. Therefore the experimental events between these limits must be due to processes other than sequential breakup (8). This new process appears to be dominant at 11.5°. At 11.5° the closest distance of approach between the $^7$Li and $^{120}$Sn nuclei is 17 fm which is outside the nuclear range between the nuclei. This strongly suggests that this new breakup process, at least for 11.5°, originates from the difference between the forces for target, $\alpha$-cluster and target, t-cluster systems.

In normal Coulomb excitation calculations the excited nucleus does not break up, and it is assumed that the Coulomb trajectory follows the Rutherford path (9). For a situation where the projectile does break up the normal calculation is still assumed to be valid even when the association time of the two nuclei during the breakup process is not much greater than the collision time (10). For the breakup situation considered here the association time is ~ $8 \times 10^{-22}$s (determined from $\Delta E \Delta t = \hbar$, assuming $\Delta E \approx 0.8$ MeV) and the collision time is ~ $4 \times 10^{-23}$s.

Since the ground state of $^7$Li is a $\frac{3}{2}^-$ state, the most significant Coulomb multipole term leading to breakup will be E1, corresponding to an $\alpha$-t relative motion of $\ell$ = 0. The cross section for Coulomb breakup to continuum states of energies between $\varepsilon_r$ to $\varepsilon_r$ + $d\varepsilon_r$ may therefore be written

$$d\sigma_{E1} = \left(\frac{Ze}{\hbar v}\right)^2 B(\varepsilon_r)df_{E1}(\theta,\xi)d\varepsilon_r$$

where Z = target charge, v = projectile velocity, and $df_{E1}(\theta,\xi)$ is the usual Coulomb excitation function (9). The reduced transition probability $B(\varepsilon_r)$ is a function of the relative $\alpha$-t energy $\varepsilon_r$, and may be determined by its relationship to the photodisintegration cross section of $^7$Li (i.e. $^7$Li$(\gamma,t)\alpha$) or $\alpha$-t capture cross section (i.e. $\alpha(t,\gamma)^7$Li) by

$$B(\varepsilon_r) = \frac{9}{16\pi^3} \frac{c}{\omega} \sigma_{\gamma,^7Li}(E_\gamma)$$

or

$$B(\varepsilon_r) = \frac{9}{16\pi^3} c\hbar\left(\frac{6}{7}\right) \frac{mc^2}{(E_\gamma)^3} \varepsilon_r \sigma_{\alpha t,\gamma}(\varepsilon_r)$$

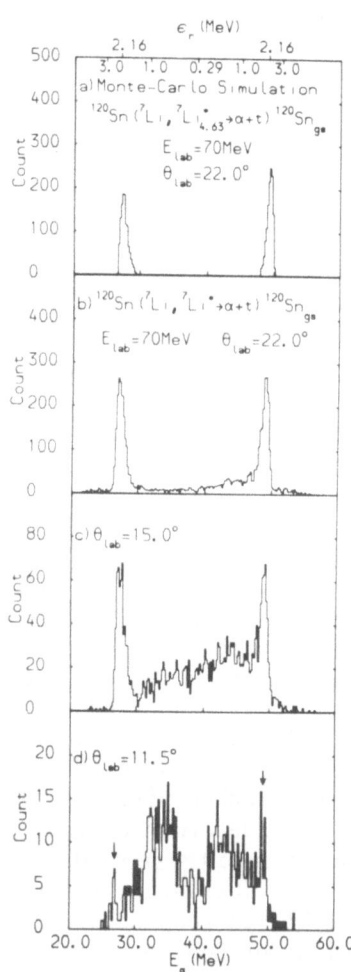

Fig. 9. Projected $E_\alpha$ spectra for various angles.

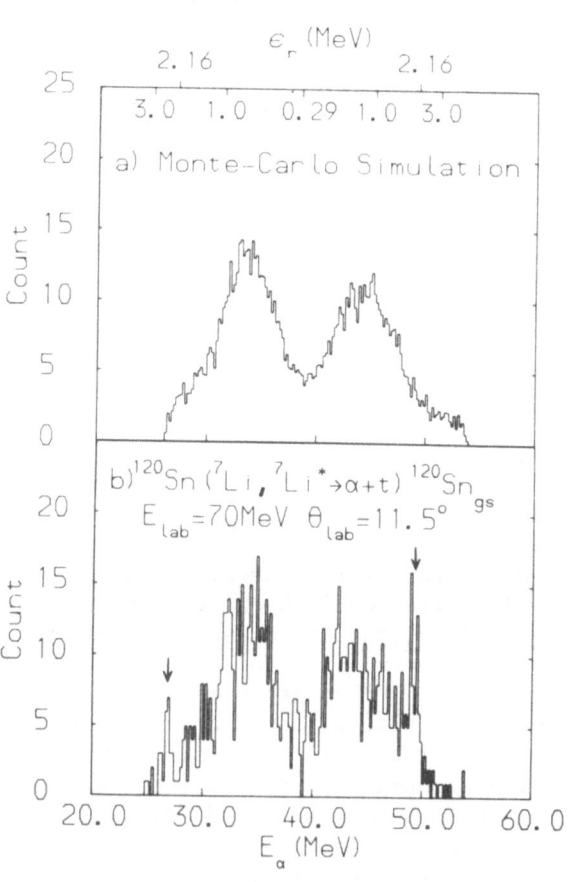

Fig. 10. Simulated and experimental projected $E_\alpha$ spectra.

In fact for the $\varepsilon_r$ energies relevant here, more data exists for the $\alpha(t,\gamma)^7Li$ reaction than for $^7Li(\gamma,t)\alpha$. If the latest fusion data of Ottewell (11) is used to determine $B(\varepsilon_r)$ the resulting $d\sigma_{E1}$ can be used in a Monte Carlo simulation to determine the projected $E_\alpha$ spectra for a particular experimental configuration. The spectrum simulated in this way to correspond to the same experimental situation as the 11.5° data (fig. 9d) is shown in fig. 10. The simulated spectrum compares very favourably both in magnitude and shape to the experimental spectrum. Therefore this indicates that the Coulomb force plays an important role in the direct breakup of $^7Li$. This also may be true for $^6Li$ projectiles (12).

## 3.2. Quasielastic breakup for projectiles heavier than $^7Li$.

Since the Coulomb force plays a dominant role in breakup of $^7Li$ at forward angles it is natural to enquire whether this is true of other projectiles. An example of $^{12}C$ breakup into the $\alpha + {}^8Be$ channel is shown in fig. 11. (2). In this figure the projected spectrum is of a similar nature to that in fig. 9 except that the quasielastic diagonal distribution corresponding to $Q_0$ is projected onto the $^8Be$ energy axis. Most of the events appear to originate from excitation of the projectile to the 7.6 MeV and 9.6 MeV states with little evidence of any direct breakup characterised by a broad distribution of events. A similar situation occurs for $^{16}O \rightarrow {}^{12}C + \alpha$ breakup. An example of a projected $^{12}C$ spectrum for quasielastic breakup is shown in fig. 12 (13).

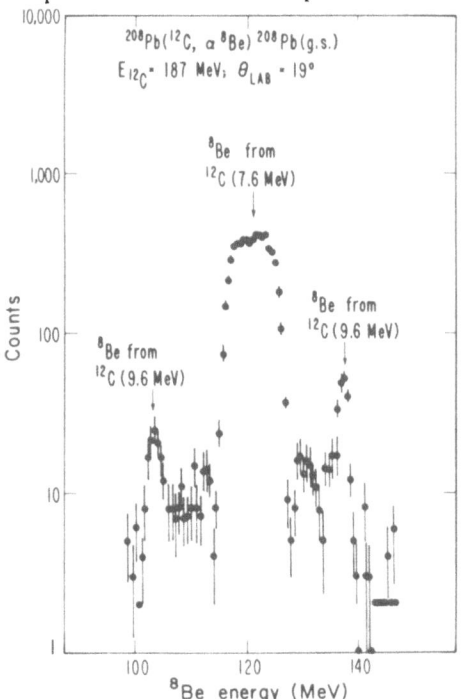

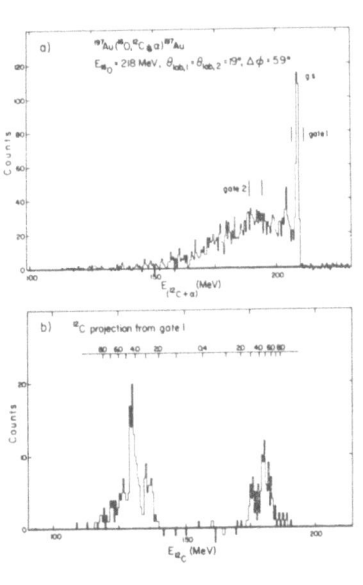

Fig. 11.  Projected $^8Be$ spectrum.

Fig. 12.  Projected $^{12}C$ spectrum.

The peaks in this projected spectrum can be identified with states in
$^{16}$O. Again there is no evidence of a continuous distribution expected
for direct breakup. A similar situation occurs for $^{20}$Ne → $^{16}$O + α
breakup. Most of the quasielastic breakup events can be identified with
discrete states in $^{20}$Ne with little evidence of direct breakup (14, 15).

There are probably two reasons why direct Coulomb breakup is
observed in $^{7}$Li but not in projectiles like $^{12}$C, $^{16}$O, $^{20}$Ne. Table 1
lists for the four projectiles the lowest Q value for particle breakup
and the value of $(Z_1/M_1 - Z_2/M_2)$ for the charge Z and mass M of the two
fragments 1 and 2. It can be seen that the breakup threshold for $^{7}$Li is
much lower than for other projectiles, and since Coulomb excitation
strongly depends upon the transition energy this lower threshold
obviously favours $^{7}$Li. For photodisintegration of a nucleus into two
clusters $(Z_1, M_1)$ and $(Z_2, M_2)$ the interaction Hamiltonian is proportional
to the quantity

$$ r^L Y_{LM} (\frac{Z_1}{M_1^L} + (-1)^L \frac{Z_2}{M_2^L}), $$

where r is the separation between the clusters and L the multipole order
of the transition (16). The lowest multipole order E1 will result in
the greatest transition probability, and this is proportional to $(Z_1/M_1 -
Z_2/M_2)$. As can be seen from table 1, only $^{7}$Li has a finite value for
this quantity. For the other projectiles the lowest multipole for disintegration will be E2, and therefore for this reason their transition
probabilities also will be significantly smaller than for $^{7}$Li.

Table 1.

| Reaction | Q(MeV) | $Z_1/M_1 - Z_2/M_2$ |
|---|---|---|
| $^{7}$Li → α + t | 2.47 | 0.17 |
| $^{12}$C → α + $^{8}$Be | 7.36 | 0 |
| $^{16}$O → α + $^{12}$C | 7.16 | 0 |
| $^{20}$Ne → α + $^{16}$O | 4.72 | 0 |

The situation discussed above is for forward scattering angles where
the Coulomb force is expected to dominate the reaction mechanism. However for scattering angles near grazing, both the Coulomb and nuclear
forces will influence the breakup process. For angles greater than
grazing, the nuclear force will probably dominate and gives rise to a
strong sequential component in the backward scattering angles for $^{7}$Li
breakup, fig. 9. Does the nuclear force lead to direct breakup? The
answer is probably yes, but it will be difficult to identify this component in the experimental data due to strong final state interactions
of the fragments with the target. If a light projectile is considered
as two clusters, one being an α, then their relative motion when they

come under the influence of a target field is determined by
$(F_\alpha/A_\alpha - F_i/A_i)$, where $A_\alpha$, $A_i$ are the masses of the two clusters and $F_\alpha$,
$F_i$ are the target forces on them. Figure 13 shows this quantity plotted
for $^{16}O$ and $^7Li$ projectiles as a function of the projectile-target
separation (17).

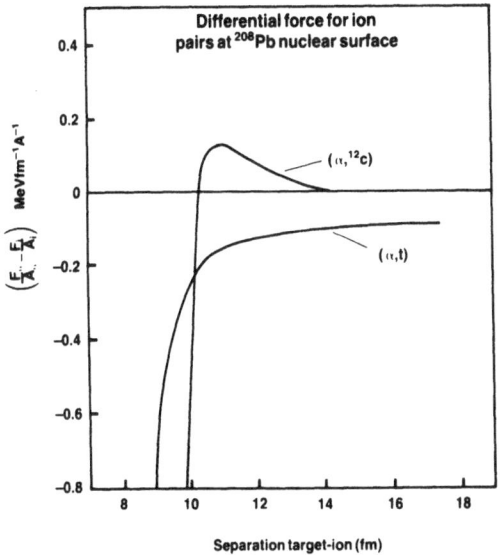

Fig. 13. Differential          Fig. 14. Breakup probability.
         cluster force.

If higher order effects are ignored, breakup can only occur when
this differential force is finite, and for $^7Li$ it is finite for distances
outside the nuclear force range. This again just means that Coulomb
breakup is possible for $^7Li$. For $^{16}O$ the differential force only becomes finite inside the nuclear field radius. This means that if direct
breakup occurs the fragments can be further deflected due to final state
interactions (f.s.i.). These f.s.i. will tend to increase the opening
angle between the fragments, and so any experimental reconstruction of
the event would infer that the relative energy between the two fragments was higher than it was at the point of breakup. Such events
could get lost amongst other events that originate from sequential
breakup, where the projectile has been excited to an energy where the
state density is high. The fragments can also lose energy to the target,
in which case the breakup event will be associated with the inelastic
breakup mode rather than the quasielastic mode.

4.    RELATIONSHIP BETWEEN INCLUSIVE AND COINCIDENCE EXPERIMENTS FOR
      QUASIELASTIC BREAKUP.

As pointed out in the previous section, the fragments originating from
the direct breakup of a projectile not only scatter but they can lose
energy to the target, and even be fused with it. It is therefore expected that the number of inclusive events for one type of fragment will be
greater than the number of such fragments in coincidence with another

fragment. Takada et. al. indeed found that for the quasielastic breakup of $^{20}$Ne into $^{16}$O + $\alpha$ at 260 MeV, in the forward direction, only about 15% of the inclusive yield could be accounted for by the total yield from the coincident component (18). To account for this they proposed the breakup-fusion model where one of the fragments after breakup fuses with the target. The transition amplitude for the coincident fragment yield has the form $\langle \chi_1^- \phi_1 \chi_2^- \phi_2 | V_p | \chi_p^+ \phi_p \rangle$ yield where the wavefunction $\chi_p^+$ for the centre of mass motion of the projectile, internal wavefunction $\phi_p$, is determined by the relevant optical potential; similar notation also applies to the fragments 1 and 2. The inclusive cross section for fragment 2, where fragment 1 fuses, is determined by the quantity $\langle \chi_1^+ | W_1 | \chi_1^+ \rangle$, where the centre of mass motion wavefunction $\chi_1^+$ is now determined not by the ordinary optical model equation, but one which has a source due to the breakup process, i.e. $(\nabla^2 - U - E)\chi_1^+ = \rho_1$, where U = optical model potential for fragment 1 (imaginary part $W_1$), and $\rho_1$ is the source of fragments originating from breakup (19). Using this model Udagawa et. al. have been able to account for the difference between the inclusive and coincidence yields for some reactions. Recently they have extended this model to $(\alpha,p)$ reactions (20) and are able to account for the high energy component of the p spectra.

One of the interesting results of these calculations is the behaviour of the source function with respect to projectile target separation. The magnitude of this function peaks inside the nuclear surface by about 2fm. It is interesting to note that the differential target force per nucleon for the projectile clusters also peaks inside the surface by about the same amount. El Lithi also has arrived at a similar conclusion via a classical trajectory calculation for $^7$Li breakup from $^{208}$Pb at 70 MeV (21). In this calculation the real parts of the optical potentials for the $\alpha$ and t fragments are used to determine the $\alpha$ and t motions. The yield of $^7$Li breakup as a function of target projectile separation peaks inside the nuclear peripheral region by several fm (fig. 14).

If one of the breakup fragments fuses with the target then considerable energy and angular momentum will be transfered to it. Experiments that measure gamma rays in coincidence with fast fragments can give valuable information concerning the partial fusion process which is followed by an (xn) reaction (22,23,24). Experiments that measure the energy of the recoiling nucleus following partial fusion also provide valuable information (25).

A series of elegant experiments that involve a determination of neutron multiplicity for different emission fragments clearly shows that there are two sources of neutrons (26). These are a cold source, that originates from a sequential process where the residual fragments exchange little energy with the target, and a hot source, where one of the fragments has fused with the target.

So far discussion has centered on the fusion of one fragment with the target. However as mentioned in Section 2 the inclusive quasielastic peak could also be thought of as arising from a transfer process. Although it would appear that these two mechanisms are distinct, since one involves a single step and the other a two-step mechanism for single fragment production, this is by no means clear. Indeed because the two reaction mechanisms evolve on a similar time scale the models associated with these mechanisms may be equivalent ways of describing the same phenomena.

## 5.    INELASTIC BREAKUP PROCESSES.

The processes discussed in the previous section mostly refer to coincident projectile fragment events corresponding to quasielastic scattering where the target is left in its ground state.  For those inelastic breakup events where the target is excited, the total energy of coincident fragments will be lower than for quasielastic events (fig. 8).  Such events could arise from the projectile inelastically scattering with the target before escaping in an excited state to sequentially breakup later. They also could arise from a direct breakup of the projectile into fragments which inelastically scatter in the target nuclear field.  For this latter situation one might expect that the f.s.i. of the fragments would destroy any correlation between them associated with the initial breakup dynamics.  If this simple picture is true, then one might expect that the yield of a fragment at a particular energy would be independent of the energy of the complementary breakup fragment.  In this situation the coincident fragment spectrum would be proportional to the inclusive spectrum for that fragment.  The first data to show such an effect came from the Birmingham University group (27).  Figure 15 shows some of their data.

$$^{14}N + {}^{58}Ni \quad 148 \text{ MeV}$$

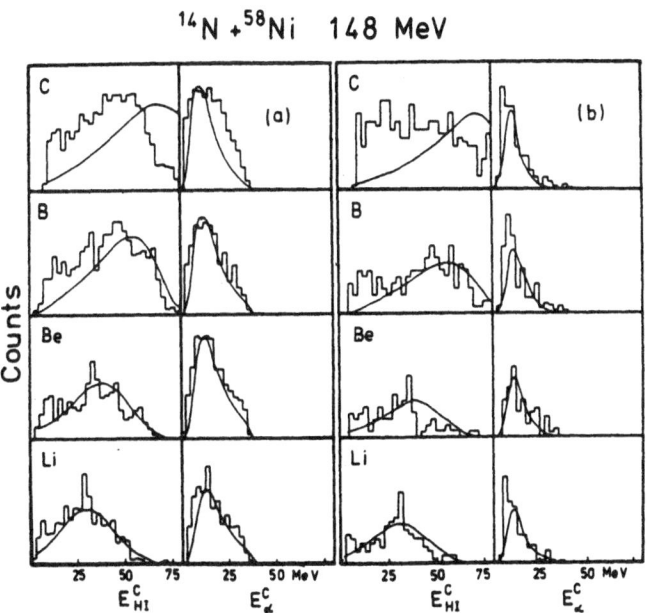

Fig. 15.   Projected coincidence spectra (a)$\theta_{HI}$=13°, $\theta_{\alpha}$= 26° and (b)$\theta_{HI}$=13°, $\theta_{\alpha}$= 51° solid lines:  shapes of inclusive  spectra.

It is clear that the coincidence data spectra do indeed follow the inclusive data. In fact they were able to show that for two fragments 1 and 2

$$\frac{d^4\sigma}{d\Omega_1 dE_1 d\Omega_2 dE_2} = K \frac{d^2\sigma}{d\Omega_1 dE_1} \frac{d^2\sigma}{d\Omega_2 dE_2} ,$$

where K is a constant which is independent of fragment energy and angle. However a similar, but more recent, study of the same reaction at higher projectile energy shows that the value of K can vary under certain circumstances, especially when a heavy fragment is detected at a large scattering angle (28). At this stage it is difficult to speculate what these results mean, but clearly the reaction mechanism is more complex than implied by the above equation. In addition this experiment shows the division between inelastic direct breakup and inelastic sequential breakup. When the angle of separation between the two fragment detectors is not too large then a localisation of coincident events in a $E_1 \times E_2$ plot can be associated with projectile excited states. For a large separation between the detectors no such localisation is apparent.

6.    VARIATION OF BREAKUP MODE WITH PROJECTILE ENERGY.

As the projectile energy increases, then for breakup events that occur in the peripheral region of the target, the chance that one or both fragments fuses with the target will decrease with energy. In other words the chance that breakup fusions occur decreases, and fragment multiplicity increases, with increasing energy. This is now evident from several types of experiments. Thus Murphy et. al (29) used a streamer chamber to study the fragment multiplicity for breakup of 260 MeV $^{16}$O. They deduce a significant increase in fragment multiplicity over the projectile energy range 8 to 16 MeV/A. An interpretation of these results is that a breakup fragment produced at the target surface can no longer fuse with the target because its angular momentum exceeds the critical value for fusion (30). For light ion projectiles this critical limit occurs at roughly 15 MeV/A (30). Similar conclusions have been reached in other types of experiments using different projectiles (2,31).

7.    SUMMARY

The spectra of ejectiles from reactions with light-ion projectiles show a distinctive maximum near beam velocity for forward scattering angles. Such a spectrum feature can be attributed to projectile breakup. Coincidence experiments which measure the kinematic relationship between breakup fragments can yield direct information concerning the types of breakup mechanisms. Breakup reactions can be classified as either quasielastic or inelastic depending on the energy transfered by the projectile to the target. For $^7$Li projectile the quasielastic breakup can proceed either directly or sequentially through the excited states of the projectile. At forward angles the Coulomb force between projectile and target is mainly responsible for direct breakup. For other projectiles such as $^{12}$C, $^{16}$O, $^{20}$Ne breakup proceeds primarily through the projectile excited states; direct breakup is more difficult to identify.

For the inelastic breakup mode there appears to be a loss of correlation between the breakup fragments. This is thought to arise due to the strong final state interactions of the fragments with the target.

The breakup fusion model has been used to associate quasielastic inclusive and coincident fragment yields. The chance that any one breakup fragment fuses with the target decreases with projectile energy.

ACKNOWLEDGEMENTS.

I would like to sincerely thank my past and present colleagues with whom I have had the pleasure to work with on the topics discussed in this paper: A.N. Bice, J. Cerny, N.E. Sanderson, M.N. Nagarajan, I.J. Thompson, D. Branford, A. El Lithi, T. Davinson and V. Rapp.

REFERENCES

1.   Data (Unpublished): Edinburgh Nuclear Physics Group.
2.   A.N. Bice et.al. Nucl. Phys. A390 (1982) 161.
3.   C.K. Gelbke et.al. Phys. Lett. 42C (1978) 312.
4.   T. Udagawa et.al. Phys. Rev. 20C (1979) 1949.
5.   D.M. Brink, Phys. Lett. 40B (1972) 37.
6.   B.G. Harvey & M.J. Murphy, Phys. Lett. 130B (1983) 373.
7.   A.C. Shotter et.al. Submitted for publication.
8.   A.C. Shotter, Phys. Rev. Lett. 46 (1981) 12.
9.   K. Alder & A. Winther, Electromagnetic excitation, 1975.  IBSN 07204 02883.
10.  H.A. Weidenmuller & A. Winther, Ann. Phys. 66 (1971) 218.
11.  D.F. Ottewell, Ph.D. Thesis 1976, Univ. Brit. Columbia.
12.  H.Gemmeke et.al. Phys. Lett. 96B (1980) 47.
13.  A.N. Bice et.al. LBL 13366 report (1982) 107.
14.  H. Homeyer et.al. Phys. Rev. 26C (1982) 1335.
15.  T. Shimoda et.al. Proc. Int. Conf. Light-ion reaction mechanisms. RCNP (1983) 645.
16.  S.A. Moszkowski, Alpha-Beta- and Gamma-ray Spectroscopy. North Holland (1965).
17.  R.A. Broglia & A. Winther, Heavy ion reactions (1981) 114 ISBN. 0-8053-1302-8(v.1).
18.  E. Takada et.al. Phys. Rev. 23C (1981) 772.
19.  T. Udagawa et.al. Phys. Lett. 118B (1982) 45.
20.  T. Udagawa et.al. Phys. Lett. 135B (1984) 333.
21.  A. El Lithi; private communication.
22.  K. Siwek-Wilczynska et.al. Nucl. Phys. A330 (1979) 150.
23.  C.M. Castaneda et.al. Phys. Rev. 16C (1977) 1437.
24.  H. Utsunomiya Phys. Rev. 28C (1983) 1975.
25.  D.Parker, J.Asher, T.Conlon and I.Naqib, A.E.R.E.rep.R-11217(1984)
26.  H. Homeyer et.al.  Z. Phys. A314 (1983) 143.
27.  R.K. Bhowmik et.al. Phys. Rev. Lett. 43 (1979) 619.
28.  T. Fukuda et.al. Phys. Rev. 27C (1983) 2029.
29.  M.J. Murphy et.al. Phys. Lett. 120B (1983) 75.
30.  K. Siwek-Wilczynska et.al. Phys. Rev. Lett. 43 (1979) 1599.
31.  R.L. Robinson et.al. Phys. Rev. 27C (1983) 3006.
32.  B. Neumann et.al. Z. Physik A296 (1980) 113.

BREAKUP PHENOMENA IN NUCLEAR COLLISION PROCESSES WITH $^3$He AND $^4$He
PROJECTILES

R.J. de Meijer
Kernfysisch Versneller Instituut, Rijksuniversiteit Groningen
9747 AA Groningen, the Netherlands
and
R. Kamermans
Robert J. van de Graaff Laboratorium, Rijksuniversiteit
Utrecht
3508 TA Utrecht, the Netherlands

ABSTRACT. A review will be given of the experimental data, obtained with
$^3$He and $\alpha$-beams, that have been taken to identify the main processes
contributing to the continuum part in particle spectra. Evidence will be
presented that the continuum predominantly originates from quasi-free
breakup processes in which one cluster in the projectile behaves as
spectator, whereas the remaining cluster interacts with the target
nucleus.

1. INTRODUCTION

    Clustering phenomena in nuclei have been of large interest in last
two decades. This aspect of the nucleus was experimentally frequently
investigated via transfer and knockout reactions. In this talk it will
be pointed out that the continuum part of particle spectra to a large
extent originates from processes which clearly demonstrate the presence
of clusters in nuclei.
    Energy spectra of particles emitted in nuclear reactions at 10-40
MeV/amu show a characteristic pattern: at high ejectile energies one
observes the transitions to well isolated states in the residual
nucleus, whereas at the low-energy side of the spectrum contributions
from (pre-)equilibrium emission are present. A large part of the cross
section, however, goes into the so-called continuum region. In this
region one can expect contributions via transitions to the many overlapping states or to broad structures present at those excitation
energies in the residual nucleus. Reactions with three or more particles
in the final state, however, will also give rise to a continuous energy
distribution in kinematically incomplete measurements.
    Over the last years the growing interest in this continuum is
mainly motivated by the increase of the cross section to this region by
going to higher incident energies and heavier projectiles. In addition,
a better understanding of the underlying mechanisms will also facilitate
the study of collective phenomena as giant resonances and of deeply-

215

*J. S. Lilley and M. A. Nagarajan (eds.), Clustering Aspects of Nuclear Structure, 215–227.*
© *1985 by D. Reidel Publishing Company.*

bound hole states or highly-unbound particle states, since these nuclear structure studies involve the subtraction of this continuum "background".

This review deals primarily with direct projectile breakup, a fast process that leads to at least three bodies in the final state. This process has been observed for deuterons, $^3$He, $^4$He and heavier ions in interactions with nuclei and is probably a phenomenon that occurs in all reactions between target nuclei and all energetic complex projectiles. The interaction of the experiments with deuterons of Helmholz et al.[1] have led Serber[2] to develop a simple model that still forms the basis for the present understanding of this type of reactions. Deuteron breakup was studied for many targets at various bombarding energies[3].

Compared to the case of deuteron breakup little was known on the breakup properties of $^3$He and $^4$He until quite recently. The investigations on $^3$He induced direct breakup started with an investigation of the properties of the continuum region in inclusive ($^3$He,d) reactions by Matsuoka et al.[4]. Studies on the breakup of the α-particle were started by Budzanowski et al.[5] and Wu et al.[6]. Since then detailed work by several groups have revealed different reaction processes contributing to the continuum region. The experimental data and their interpretation for $^3$He and $^4$He will be the subject of this talk. For a more extensive

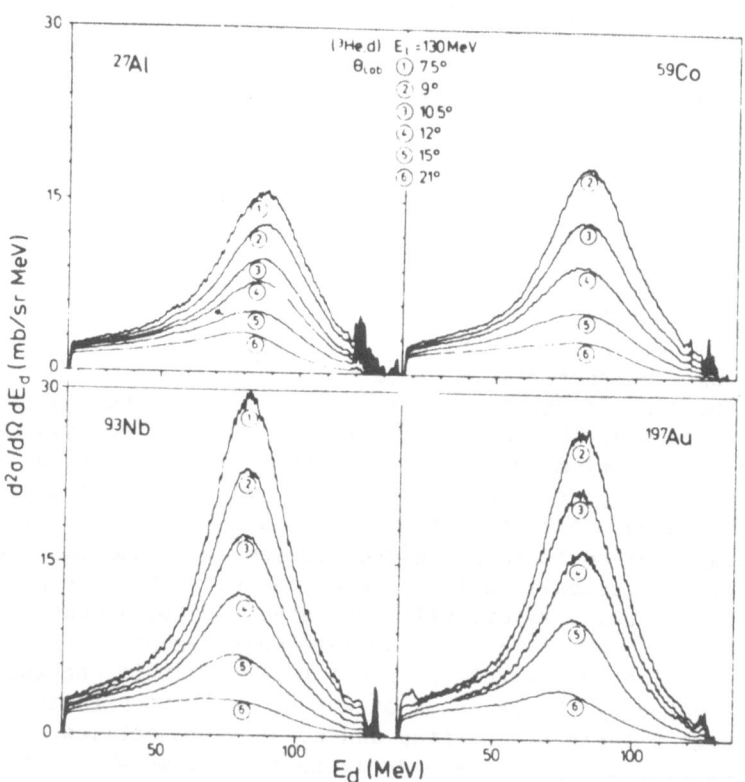

Figure 1. Examples of spectra from the ($^3$He,d) reaction at $E_{3\text{He}}$ = 130 MeV. Taken from ref.[8].

review of both experimental and theoretical investigations of breakup studies with these projectiles we refer to ref.[7].

The use of He projectiles has the advantage that the structure of these projectiles is still rather simple; therefore the number of reaction processes is limited. $^3$He is the lightest stable projectile for which a binary partition can lead to two charged particles. This partition has a large spectroscopic factor. Further $^3$He has no well-defined excited states, contrary to $^4$He where several excited states are known. The large binding energy difference between $^3$He and $^4$He (5.5 and 20 MeV, respectively) allows the investigation of the role that the projectile binding energy plays in these processes. These similarities and differences result in features that are common to $^3$He and $^4$He induced reactions and to aspects in which the results for the two projectiles differ. We will first focus on the common features and than discuss the differences.

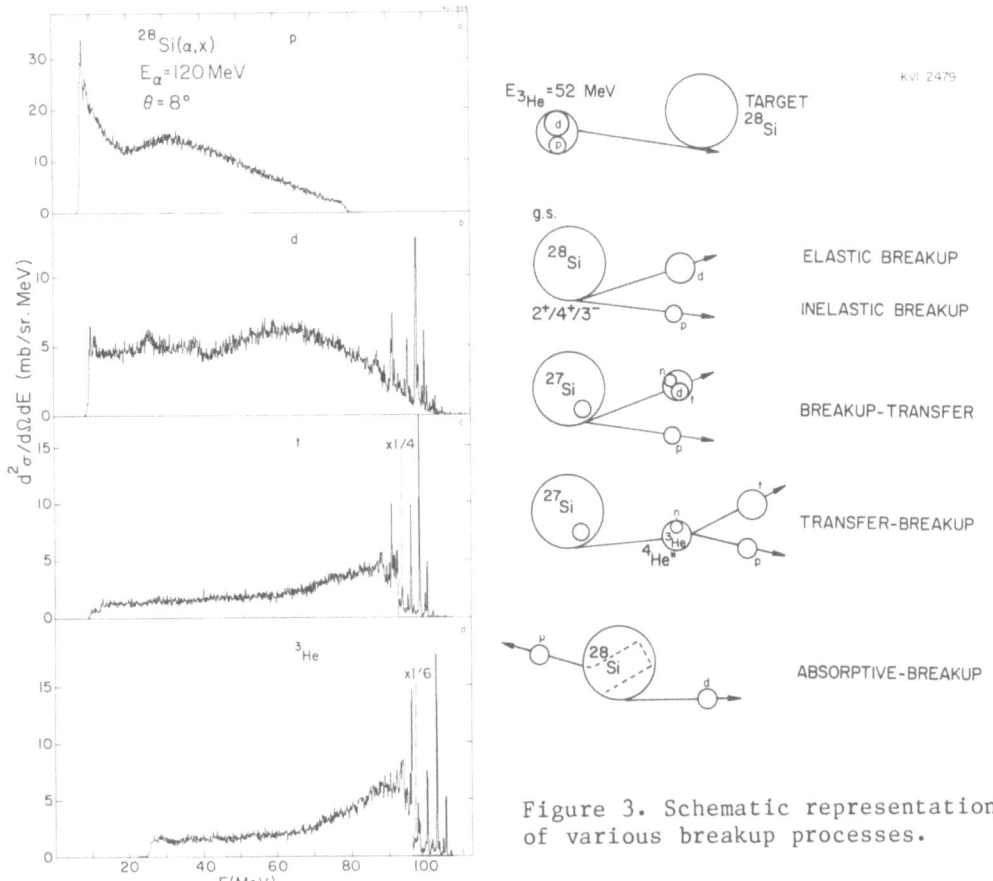

Figure 3. Schematic representation of various breakup processes.

Figure 2. Inclusive p, d, t and $^3$He spectra from the reaction $^{28}$Si($\alpha$,x) at $E_\alpha$ = 120 MeV and $\theta$ = 8°.

## 2. COMMON FEATURES IN $^3$He AND $^4$He INDUCED REACTIONS

### A. Inclusive measurements

Pronounced, bell-shaped enhancements (bumps) occur at forward angles in the continuum part of $^3$He induced p, d and t spectra. Fig. 1 presents as an example ($^3$He,d) spectra at several angles at $E_{3_{He}}$ = 130 MeV[8]. As shown in fig. 2 bumps have also been observed in inclusive p, d, t, and $^3$He spectra from α-induced reactions[9] (see also ref.[6]). In all these spectra the bumps occur at an energy which correponds to the beam velocity (beam velocity energy). The relative importance of the bump decreases rapidly with increasing detection angle. In addition to the bump one observes a component at the low-energy side. This component is present at all angles and has at more backward angles the shape of an exponential tail (tail). These components have been studied over a wide range of targets and incident energies (see ref.[7]).

The occurence of a bump at beam velocity energy suggests a fast process in which the observed particle remains practically undisturbed. This particle was in this process a spectator. Consequently the interaction has occured between the other constituent of the projectile (the participant) and the target nucleus. The inclusive measurements show only a clear signal of the spectator. To establish the characteristics of the participant-target interaction correlation experiments are essential.

### B. Coincidence measurements

Detailed coincidence studies between charged particles have been reported for both $^3$He and α-induced reactions (for details see ref.[7]). The ($^3$He,pd) coincidences, especially at 52 MeV, have played an important role in the understanding of the various processes[10]. For this reason mainly these data will be used as examples in this section.

From all these studies a rather simple picture emerges: the participant-target interaction corresponds to the one known from "conventional" nuclear physics like elastic scattering, inelastic scattering, particle transfer and absorption (capture). This allows a schematic classification of the breakup processes as in e.g. given in fig. 3 for 52 MeV $^3$He on $^{28}$Si[11]. According to this scheme elastic breakup is the process in which the projectile breaks up into two constituents, while the target nucleus remains in its ground state. If instead the target nucleus is left in an excited state the process is named inelastic breakup. Absorptive breakup is the process in which the participant is absorbed by the target nucleus. In fig. 3 this process is presented for a proton participant absorbed by $^{28}$Si; the compound system $^{29}$P decays subsequently by proton emission. In the $^3$He induced reactions also p-t, p-$^3$He and d-d coincidences have been observed. These coincidences were explained by the breakup-transfer process. In this process the participant picks up a nucleon from the target. This process is kinematically distinguishable from the transfer-breakup process in which the $^3$He projectile as an entity picks up a nucleon and forms an unstable state of $^4$He or $^4$Li which subsequently decays.

The identification of elastic breakup, inelastic breakup and absorptive breakup is illustrated in fig. 4. This two-dimensional $E_p$ versus $E_d$ spectrum[12] is obtained with deuterons detected at forward angles ($\theta_d = -10°$) and protons at backward angles ($\theta_p = -145°$). Such a geometry favours processes in which the deuteron is spectator. In fig. 4 one observes loci corresponding to transitions to states in the residual nucleus $^{28}$Si. At relatively low proton energies, there is an intense concentration of events which do not belong to a particular locus. These

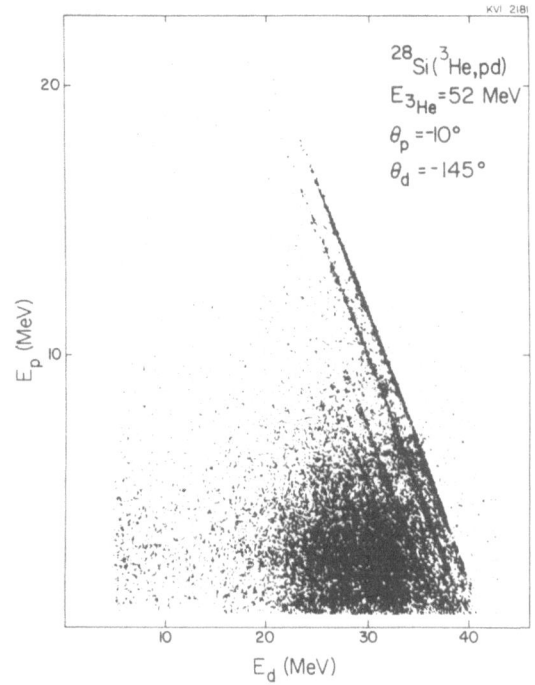

Figure 4. Two-dimensional $E_p$ versus $E_d$ spectrum of the reaction $^{28}$Si($^3$He,dp) at $\theta_d = -10°$ and $\theta_p = -145°$.

events are concentrated around the beam velocity energy of the deuteron indicating a process with a deuteron spectator. The corresponding proton spectra, shown in fig. 5, give exponentially shaped spectra, characteristic for a proton that is statistically emitted. At more forward proton angles the intensity of the loci increases relative to the intense concentration of events and at very forward angles the ground-state transition dominates the spectrum. The events in fig. 4 therefore may be catagorized in the above defined classes:

(i)  The events belonging to the ground-state locus. All particles are left in their ground-state and the process is called elastic breakup.

(ii)  The events on the other loci. In these reactions only the target is excited and the process is called inelastic breakup.

(iii) The intense concentration of events resulting from absorptive breakup.

A more complicated situation occurs when the two coincident particles do not fully account for the projectile. This is illustrated in the ($^3$He,pp) reaction. Fig. 6 shows at the left hand side the p-p coinci-

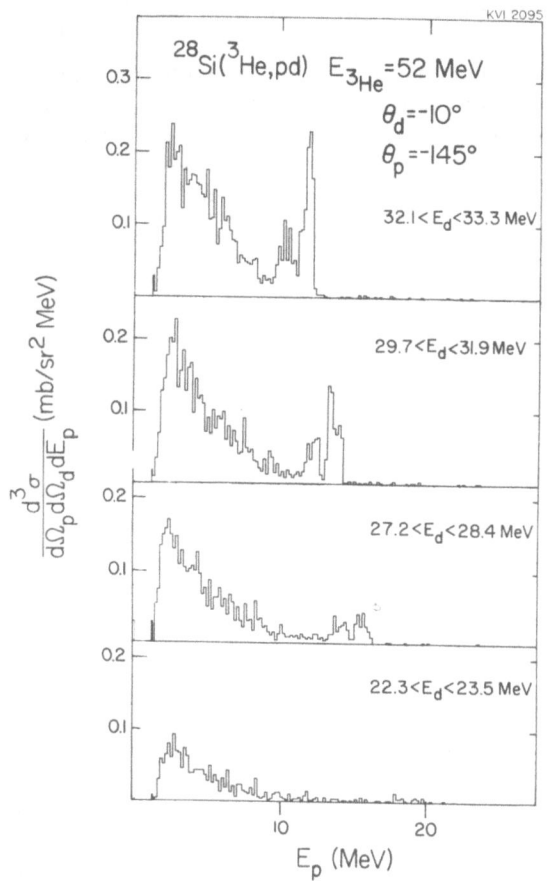

Figure 5. Projected proton spectra for the events shown in Fig. 4 selected for various deuteron energy intervals.

dence data with $\theta_{p1}$ = -10° and $\theta_{p2}$ = +10°. In part a of the figure a number of loci can be seen, corresponding to the ($^3$He,$^2$He) reaction (see refs. [7,13]) and a concentration of events centred around proton beam velocity. This indicates that at this angle an appreciable number of proton pairs occur which are travelling with approximately beam velocity (two-proton spectators). The projections on the two axes have been made for $E_{p1}$ +$E_{p2}$ ≤45 MeV in order to exclude a large contribution from the ($^3$He,$^2$He) events on the loci. The projected spectra (part b and c) have the same shape as the inclusive spectrum: a bump and a tail. Since there is no clear concentration of events at low proton energies it means that the tail in one spectrum is mainly coincident with the bump in the other spectrum. This implies coincidences between a spectator proton and a statistically emitted proton (one-proton spectator breakup). At more backward angles the two-dimensional spectrum looks quite different as can be seen in the right hand part of fig. 6 where $\theta_{p2}$ has been moved to 70°. In part a the loci have disappeared and a concentration of events shows up as a band at low $E_{p2}$. The projection on the $E_{p1}$ axis still results in a spectrum that shows a bump and a tail, whereas the projection on the $E_{p2}$ axis results in a pure tail spectrum. So these events indicate processes with one-proton spectator (bump-tail coinci-

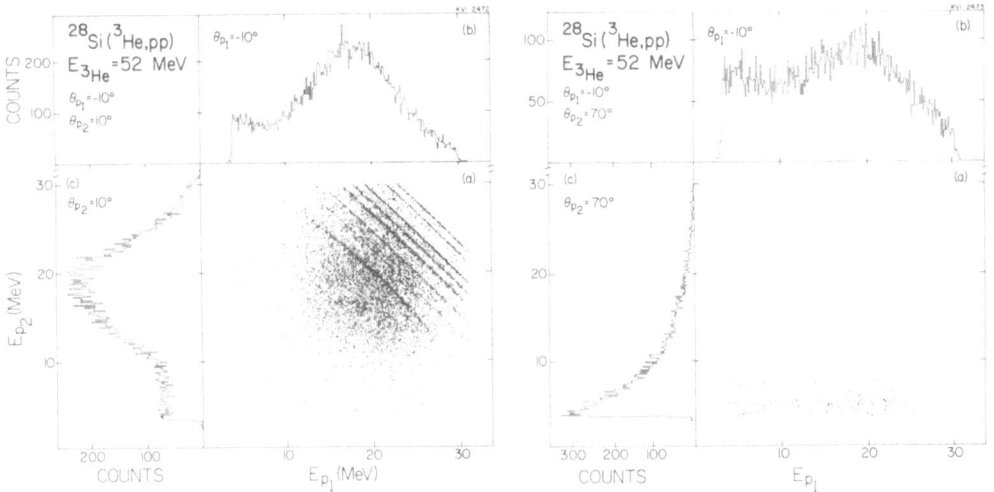

Figure 6. Two dimensional $E_{p2}$ versus $E_{p1}$ spectra for the reaction $^{28}Si(^{3}He,pp)$ at $\theta_{p1} = -10°$ and $E_{3He} = 52$ MeV and the corresponding projections on the axes. The left hand side presents data for $\theta_{p2} = +10°$, the right hand side for $\theta_{p2} = +70°$.

dence) and no-proton spectator coincidences (tail-tail). Similar processes in $\alpha$-induced reactions have been observed by Koontz[14].

The reaction mechanisms involved in these two-, one- and no-proton spectator processes is not yet as well determined as in the processes discussed before. This is due to the fact that not all constituents of the projectile have been detected (kinematically incomplete measurement). In the case of $^{3}He$ the neutron is not detected and hence for the two-proton spectator case the individual contributions of elastic, inelastic and absorptive breakup remain unknown. In the one-proton spectator case one proton is spectator and the second proton arises from the absorption of the deuteron, either in its ground state or in its unbound state, $d$, (absorptive breakup). The two statistically emitted protons in the no-proton spectator case are either due absorption of the full projectile ($^{3}He$) or to the process with a neutron spectator and the absorption of $^{2}He$ (absorptive breakup). Aarts et al.[10] have shown that the coincidence spectra at each angle might be unravelled into contributions from the three processes.

## 3.  THE DIFFERENCES BETWEEN $^{3}He$ AND $^{4}He$ INDUCED REACTIONS

So far we have discussed those processes which lead to common features in the reactions with $^{3}He$ and $\alpha$-projectiles. Especially at lower $\alpha$-energies, however, there exist some distinct differences. In contrast to e.g. the inclusive $(^{3}He,d)$ spectra at $E_{3He} = 52$ MeV, the inclusive $(\alpha,t)$ and $(\alpha,^{3}He)$ spectra do not show pronounced enhancements around beam velocity energy[15]. Also coincidence measurements at $\theta_{1} = -10°$ and $\theta_{2} = +10°$ show hardly any cross section for projectile

breakup[15]. The absence of spectator bumps in the $(\alpha,t)$ and $(\alpha,{}^3He)$
spectra can be understood by considering the three-body phase space
available for this type of reaction.

The three-particle phase space for projectile breakup in the $(d,p)$,
$({}^3He,d)$ and $(\alpha,t)$ reactions have been presented schematically in fig. 7.
From this figure one sees that in the case of the deuteron the bump will
occur at $\frac{1}{2}E_b$ and in the case of $({}^3He,d)$ at $2/3E_b$. In both cases the
phase space is still rather flat. However in the case of the $(\alpha,t)$
reaction at $E_\alpha$ = 65 MeV the spectator energy of $\frac{3}{4}E_b$ lies outside the
available phase space, which explains why hardly any projectile breakup
can be expected at this beam energy. So it is because of the larger
ejectile/projectile mass ratio and the much larger binding energy of the
$\alpha$-projectile that triton spectators are absent in the $(\alpha,t)$ reaction at
$E_\alpha$ = 65 MeV. From these phase space arguments one expects the onset of

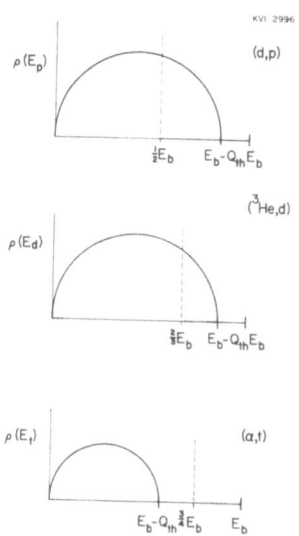

Figure 7. Schematic representation
of phase space distributions,
$\rho(E)$, and beam velocity energies
for the reactions $(d,p)$, $({}^3He,d)$
and $(\alpha,t)$.

the spectator processes in the energy range $80 < E_\alpha < 100$ MeV.

But also at higher $\alpha$-energies marked differences occur. Contrary to
the bell shaped enhancements observed in the $({}^3He,d)$ and $({}^3He,p)$
reactions which are predominantly due to spectator particles the bumps
in the $(\alpha,t)$ and $(\alpha,{}^3He)$ spectra almost exclusively arise from particle
unbound states populated in the single-nucleon transfer reactions.[9].
Since ${}^{28}Si$ is a self-conjugate nucleus the $(\alpha,{}^3He)$ and $(\alpha,t)$ reactions
lead to mirror nuclei. Therefore it is not surprising that the ${}^3He$ and t
spectra in fig. 2 are about the same. The magnitude of the high-energy
side of the continuum and of the discrete transitions is in the ${}^3He$
spectrum about 1.5 times larger than in the corresponding t-spectrum. By
shifting the triton energy spectrum over the Coulomb displacement energy
and subtracting the two spectra results in a difference energy spectrum
as shown in fig. 8. This figure shows that only a bell-shaped bump and
discrete states remain. Together with the angular distribution for this

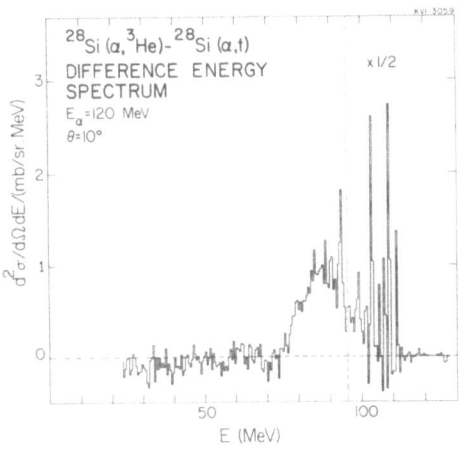

Figure 8. Difference energy spectrum obtained by subtracting from the $(\alpha, {}^3He)$ spectrum the $(\alpha, t)$ spectrum, shifted over the Coulomb displacement energy.

difference bump and the results of $(\alpha, tp)$ coincidence measurements this bump is considered as evidence for transtions to highly-unbound particle states.

In the t-p coincidence measurements at $E_\alpha = 120$ MeV and $\theta_t = -10°$ a process is observed which is strongly forward peaked[9,16]. The corresponding process in ${}^3He$ induced reactions should be observed in d-p coincidences; no evidence for such a process is found in the $({}^3He, dp)$ spectra. Fig. 9 presents the data for four energy intervals in the triton spectrum. The correlation for the high-energy bites presented in part b and c are rather isotropic. The correlations in fig. 9a show on top of a flat part (isotropic component) a forward peaked anisotropic component. Such components have also been observed by Koontz [14] in the coincidences between protons, deuterons and tritons and low-energetic protons and alphas.

The intensity ratio over the isotropic component in the t-p correlation has, as shown in fig. 10a, a maximum for the Q-value interval $-65 < Q < -40$ MeV. Moreover the proton triton relative-energy spectrum shows a broad peak at energies corresponding to $21 < E_x < 26$ MeV in ${}^4He$. In this excitation-energy region several broad states are known and hence this result is not fairly conclusive. From these observations the following processes are proposed for the anisotropic component:

(i) the reaction $(\alpha, \alpha^*)$ in which $\alpha^*$ represents a state in the range $21 < E_x < 26$ MeV.

(ii) quasi-free inelastic breakup of the $\alpha$-particle in which the proton is spectator and the participant triton has an inelastic scattering with the target nucleus. In this process the target nucleus is excited predominantly to $15 < E_x < 40$ MeV.

A comparison with p-t correlations with $\theta_p = -10°$ and d-d correlation with $\theta_d = -10°$ show that the data are certainly consistent with the quasi-free process. The fact that the maximum intensity ratio for the d-d correlation has a maximum for a different Q-value range ($-36 < Q < -26$ MeV) makes the $(\alpha, \alpha^*)$ reaction as responsible mechanism for all anisotropic components rather unlikely. The energy difference between participant deuteron ($E_d \sim 35$ MeV) and triton ($E_t \sim 70$ MeV) could explain

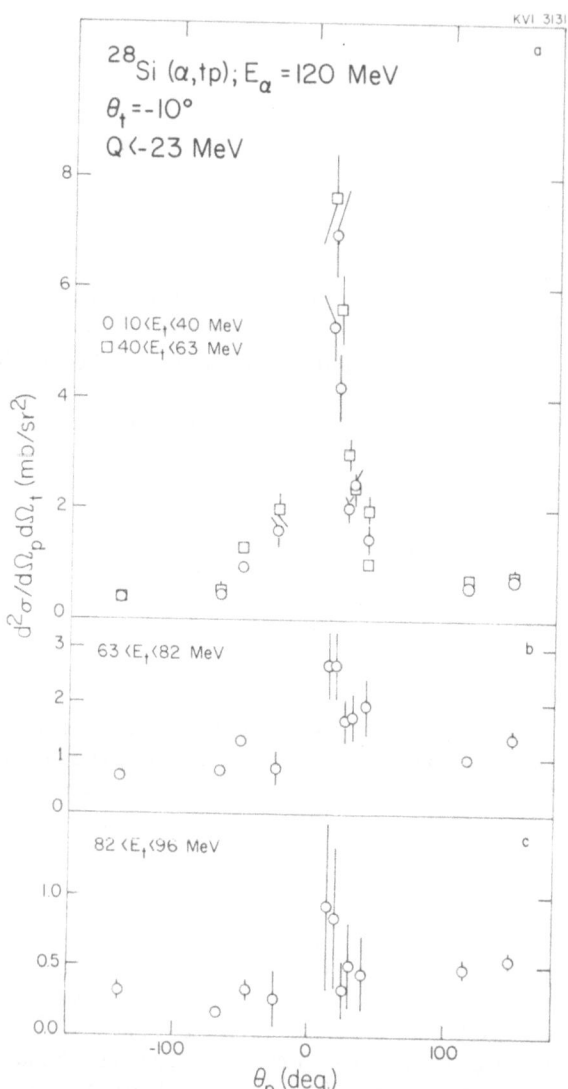

Figure 9. Angular correlation for t-p coincidences of the reaction $^{28}$Si($\alpha$,tp) at $\theta_t = -10°$ and Q < -23 MeV for various triton-energy intervals.

why the target nucleus is less excited in the deuteron scattering than in the triton scattering. Although those results seem to favour the quasi-free process one should keep in mind that the branching ratio for p-t decay is strongly favoured over d-d decay of excited states. Therefore it can not yet be ruled out that the sequential decay mode still shows up stronger in the p-t and t-p coincidences than the quasi-free process, whereas the d-d data are predominantly due to the quasi-free process.

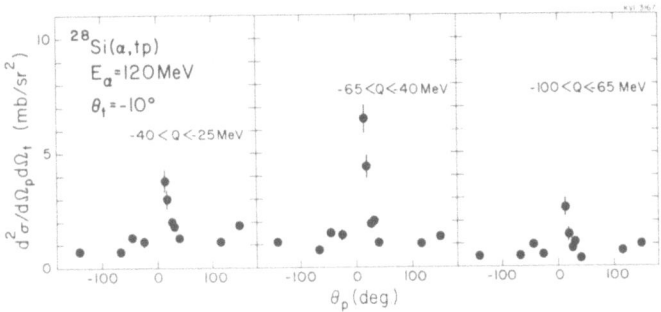

Figure 10. Angular correlation for t-p coincidence of the reaction
$^{28}$Si($\alpha$,tp) at $\theta_t$ = -10° for various Q-value intervals.

## 4. DECOMPOSITION OF THE INCLUSIVE CROSS SECTIONS BY COINCIDENCE MEASUREMENTS

From the angular correlations a comparison can be made with the cross sections observed in the inclusive spectra, by integrating the various processes over angle and energy. For the case of 52 MeV $^3$He on $^{12}$C, $^{28}$Si and $^{58}$Ni the results are presented in fig. 11. The results for ($^3$He,d) show that except for $^{12}$C, absorptive breakup is the dominant

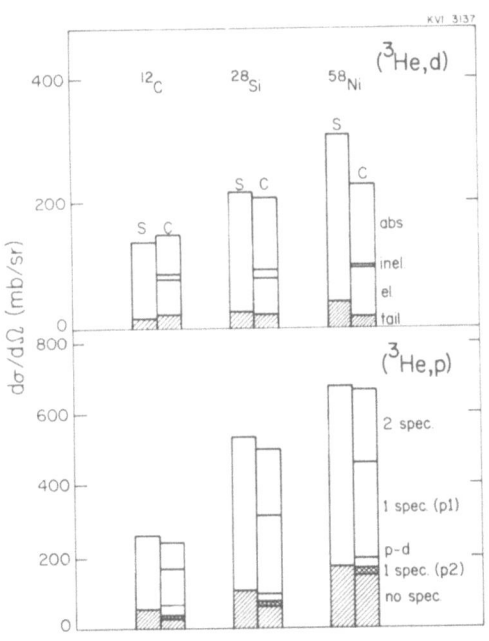

Figure 11. Comparison ($^3$He,d) and ($^3$He,p) inclusive cross sections (S) with those obtained in particle-particle coincidences (C).

reaction process. The comparison in the top part of fig. 11 also shows that for $^{12}$C and $^{28}$Si the singles deuteron yield for both the tail and the bump part is accounted for by the p-d coincidences processes. For $^{58}$Ni about 70% of the singles deuteron yield is explained this way. No significant coincidence yields were observed between deuterons and charged particles other than protons and calculations show that for $^{58}$Ni the missing strength can be attributed to the neutron decay channel in the absorptive breakup process[10].

The same procedure has been applied to reconstruct the inclusive ($^3$He,p) cross sections from various coincidence data. The results presented in the bottom part of fig. 11 indicate that the inclusive proton yield is well reproduced by the coincidence data and that no appreciable cross section is going to an unobserved neutron channel.

## 5.  CONCLUSIONS

In this paper the experimental data on breakup phenomena obtained with $^3$He and $^4$He projectiles have been reviewed. Because inclusive experiments usually can give only global information about the breakup processes, exclusive experiments are essential for the detailed understanding of the reaction mechanisms. This requires a knowledge of three-body kinematics since a certain detector geometry strongly selects the observable processes, even to the extent that processes with a small total cross section might give the impression to be dominant. This might lead to a misinterpretation of the data. Examples of such a selection are coincidence measurements at small relative angles which preferentially give information on the sequential-breakup process. It was shown that in order to unravel the broad structures observed in inclusive experiments the highly energetic light particles have to be detected in coincidence over a wide dynamic energy range and, although it may sound paradoxically, with a good energy resolution. This last requirement is necessary in order to separate the breakup contributions from all other reaction processes such as transitions to closely spaced discrete states.

In inclusive experiments a bump was found to be centered roughly around the beam-velocity energy. The cross sections were found to be strongly forwardly peaked and to be proportional to $A^{1/3}$, indicating that breakup is a peripheral process. Detailed angular correlation measurements revealed that elastic breakup, inelastic breakup and absorptive breakup contribute to the breakup bump. Elastic and inelastic breakup are the processes in which the projectile dissociates into a proton and a deuteron, while leaving the target nucleus in its ground state and an excited state, respectively. Elastic breakup was found to be about an order of magnitude stronger than inelastic breakup but the two processes together account only for a minor part of the bump cross section. The major part arises from absorptive breakup in which the deuteron does not interact with the target nucleus and the proton is captured. Absorptive breakup is the dominant channel at least for $^3$He-projectiles. Especially the measurements of $^3$He induced triton spectra have shown the existence of the breakup-transfer process. A common feature of all the breakup processes is the observation of a spectator particle. The reactions differ by the interaction of the participant and the target nucleus. These interactions resemble those responsible for processes in conventional nuclear reactions.

In addition to the reactions mentioned above, the p-p coincidence data show two-proton, one-proton and no-proton spectator processes for both projectiles, but the details of the reaction mechanisms contributing to these processes are not yet well determined.

The main differences between the $^3$He and $\alpha$-induced breakup reactions are (i) the absence of direct breakup processes in the $(\alpha,t)$ and $(\alpha,^3He)$ channels at $E_\alpha$=65 MeV, (ii) the fact that at higher energies $(E_\alpha$=120 MeV) the bump in the $(\alpha,t)$ and $(\alpha,^3He)$ spectra was found to be exclusively due to sequential particle decay of states populated in conventional transfer reactions, and (iii) the strongly forward peaked process in the $\alpha$-induced t-p coicidences, which contribute to the inclusive triton tail spectrum

The authors wish to thank Drs. E.H.L. Aarts, P.D. Bond and H. Koeslag for their helpful and stimulating discussions. This work is part of the research program of the "Stichting voor Fundamenteel Onderzoek der Materie" (FOM), with financial support of the "Nederlandse Organisatie voor Zuiver-Wetenschappelijk Onderzoek" (ZWO).

## References

1) A.C. Helmholz, E.M. McMillan and D.C. Sewell, Phys.Rev. 72(1947)1003.

2) R. Serber, Phys.Rev. 72(1947)1008.

3) G. Baur and D. Trautmann, Phys.Rep. 25C(1976)293.

4) N. Matsuoka, et al., Nucl.Phys. A311(1978)173.

5) A. Budzanowski, et al., Phys.Rev.Lett. 41(1978)635.

6) J.R. Wu, C.C. Chang and H.D. Holmgren, Phys.Rev.Lett. 40(1978)1013.

7) R.J. de Meijer and R. Kamermans, Rev.Mod.Phys. to be published

8) A. Djaloeis et al., Phys.Rev. C27(1983)2389.

9) H.J. Koeslag et al., KVI Annual Report 1983, p26, and private communication.

10) E.H.L. Aarts, R.A.R.L. Malfliet, R.J. de Meijer and S.Y. van der Werf, Nucl.Phys. in press.

11) R.J. de Meijer, Proc. XX Winter School on Nuclear Physics, Bormio, Italy, 1982.

12) E.H.L. Aarts, R.K. Bhowmik, R.J. de Meijer and S.Y. van der Werf, Phys.Lett. 102B(1981)307.

13) J. van Driel, R. Kamermans, R.J. de Meijer and A.E.L. Dieperink, Nucl.Phys. A342(1980)1.

14) R.W. Koontz, Thesis University of Maryland, 1980 (unpublished).

15) R.J. de Meijer, E.H.L. Aarts, M.B. Greenfield and W.A. Sterrenburg, Nucl.Phys. A402(1983)15.

16) H. Koeslag et al., submitted to Phys.Lett.

# CLUSTER TRANSFER IN HEAVY ION REACTIONS

# MICROSCOPIC FORM FACTORS FOR CLUSTER TRANSFER REACTIONS

R. G. Lovas
Institute of Nuclear Research
Debrecen
P. O. Box 51
H-4001, Hungary

ABSTRACT. The nuclear-structure aspects of the distorted-wave Born approximation (DWBA) model of direct cluster transfer reactions are reviewed. Results for the ingredients of the DWBA form factor are presented for the $\alpha+t$, $\alpha+d$ and $\alpha+^{16}O$ subsystems. The construction of realistic form factors for practical DWBA calculations is discussed. It is concluded that the use of realistic form factors is mainly important for the determination of the absolute spectroscopic factors.

## 1. INTRODUCTION

The study of direct cluster transfer reactions provides information on the cluster structure of the participating nuclei. In the distorted-wave Born approximation (DWBA) model for the reaction $A(a,b)B$ ($a=b+x$, $B=A+x$) and its inverse, this information is contained in some wave-function overlaps characterizing the $a=b+x$ and $B=A+x$ decompositions. The DWBA model is too rough [1] and the cross sections are not sensitive enough to allow us to determine these overlaps by a mere comparison with experiment. It is, however, feasible to extract the norm square of the overlap for $B=A+x$, the spectroscopic factor (SF) of the $B=A+x$ decomposition, provided that the shape of this overlap as well as the shape and magnitude of the other one are known. In determining these quantities, the analyses mostly rely on naive local-potential pictures. The validity of such a picture is highly questionable [2]. It is therefore important to calculate these quantities from reliable microscopic models.

    In this review I report on some recent applications of the cluster model to the calculation of the nuclear-structure quantities involved in the DWBA description of cluster transfer reactions. To define these quantities, I shall first recall the DWBA formulae in schematic terms (sect. 2). Then I shall review some microscopic methods for the calculation of these functions (sect. 3). Next I present some results of such calculations on $^7Li=\alpha+t$, $^6Li=\alpha+d$ and $^{20}Ne=\alpha+^{16}O$, and show their applications to transfer reactions (sects. 4 and 5). Then I shall discuss the local-potential picture and make a proposal for practical applications of the results (sect. 6). Finally, I draw some general conclusions (sect. 7).

*J. S. Lilley and M. A. Nagarajan (eds.), Clustering Aspects of Nuclear Structure, 231–244.*
© *1985 by D. Reidel Publishing Company.*

## 2. CLUSTER-MODEL ASPECTS OF THE DWBA

The exactly antisymmetrized DWBA transition amplitude of the reaction
A(a,b)B is given in the post form by [3]

$$T = \binom{B+b}{b}^{\frac{1}{2}} < \chi_{Bb}^{-} \Phi_B \Phi_b | V_{Bb} - v_{Bb} | A_{Aa} \{ \Phi_A \Phi_a \chi_{Aa}^{+} \} >, \tag{1}$$

where $V_{Bb}$ and $v_{Bb}$ are the exact and the optical potential acting between
B and b, $\chi$ are the distorted waves, $\Phi$ are internal wave functions, and
the interfragment antisymmetrizer is $A_{Aa} = [A!a!/(A+a)!]^{\frac{1}{2}} \Sigma (-)^P P$. In the
conventional model the only explicit departure from exact antisymmetry
is that it neglects the exchanges between A and b. It also introduces
the approximation $V_{Bb} - v_{Bb} \approx V_{bx}$, so that

$$T = \int d\vec{r}_{Bb} \int d\vec{r}_{Aa} \chi_{Bb}^{-}(\vec{r}_{Bb}) f(\vec{r}_{Bb}, \vec{r}_{Aa}) \chi_{Aa}^{+}(\vec{r}_{Aa}), \tag{2}$$

where the form factor $f(\vec{r}_{Bb}, \vec{r}_{Aa}) \equiv F(\vec{r}_{Ax}, \vec{r}_{bx})$ is defined as

$$F(\vec{r}, \vec{r}') = \binom{a}{x}^{\frac{1}{2}} \binom{B}{x}^{\frac{1}{2}} < \Phi_B \Phi_b \delta(\vec{r} - \vec{r}_{Ax}) | V_{bx} | \Phi_A \Phi_a \delta(\vec{r}' - \vec{r}_{bx}) >. \tag{3}$$

The assumption that x is transferred in a well-defined state $\Phi_x$ is in-
troduced by insertion of the projector $\Sigma_{A'b'} | \Phi_{A'} \Phi_{b'} \Phi_x > < \Phi_{A'} \Phi_{b'} \Phi_x |$ in
front of $V_{bx}$. This leads to

$$F(\vec{r}, \vec{r}') = U_{B(Ax)}^{*}(\vec{r}) W_{a(bx)}(\vec{r}'), \tag{4}$$

where the "overlap" or reduced-width amplitude $U_B$ and the "potential
overlap" $W_a$ are defined as

$$U_{B(Ax)}(\vec{r}) = \binom{B}{x}^{\frac{1}{2}} < \Phi_A \Phi_x \delta(\vec{r} - \vec{r}_{Ax}) | \Phi_B >, \tag{5}$$

$$W_{a(bx)}(\vec{r}) = \binom{a}{x}^{\frac{1}{2}} < \Phi_b \Phi_x \delta(\vec{r} - \vec{r}_{bx}) | V_{bx} | \Phi_a >. \tag{6}$$

In $U_B$ the internal motion $\Phi_A \Phi_x$ is "projected out", and thus it
characterizes the Ax intercluster motion in $\Phi_B$. Indeed, $U_B$ is usually
assumed to be proportional to an eigenstate in a local potential $v_{Ax}$. In
the usual analyses $V_{bx}$ is also replaced by a local potential, $v_{bx}$, in
which case $W_a$ reduces to

$$W_{a(bx)}(\vec{r}) \simeq v_{bx}(r) U_{a(bx)}(\vec{r}), \tag{7}$$

and the two overlaps U satisfy the two equations

$$[T + v(r)] U(\vec{r}) = EU(\vec{r}). \tag{8a}$$

The norm squares of the two U are the spectroscopic factors

$$S = <U|U>. \tag{9}$$

In a pure two-cluster model the wave function is

$$\Phi = A_{12}\{\Phi_1\Phi_2\psi(\vec{r}_{12})\} = \int d\vec{r}\psi(\vec{r})\Phi_{\vec{r}}, \tag{10a}$$

with

$$\Phi_{\vec{r}} = A_{12}\{\Phi_1\Phi_2\delta(\vec{r}-\vec{r}_{12})\}. \tag{10b}$$

Then the overlap U may be expressed through $\psi$ as $U=N\psi$, where the kernel $N(\vec{r},\vec{r}')$ of the integral operator N is defined by

$$N(\vec{r},\vec{r}') = <\Phi_{\vec{r}}|\Phi_{\vec{r}'}>. \tag{11}$$

The functions $\psi$ and U are normalized as $<\psi|N|\psi>=<U|N^{-1}|U>=1$, where $N^{-1}$ is defined as $N^{-1}N=NN^{-1}=\Lambda$, with $\Lambda$ being the projector onto the subspace on which N is invertible. It is then obvious that the function

$$\phi \equiv N^{\frac{1}{2}}\psi =N^{-\frac{1}{2}}U \tag{12}$$

is normalized in the usual sense: $<\phi|\phi>=1$. Therefore, it is $\phi$, and not $\psi$ or U, that can be considered the wave function of relative motion.

These considerations bear upon the case of mixed configurations, too. It has been conjectured by Fliessbach [2] that, from among the three functions $\psi$, $\phi$ and U, only $\phi$ can be approximated by a solution of a s.p. Schrödinger equation with a local potential:

$$[T+v'(r)]\phi(\vec{r}) = E\phi(\vec{r}). \tag{8b}$$

Horiuchi's recent findings [4] corroborate this assumption, and the use of (8a) does not seem justifiable in general.

We note that Fliessbach attempted to save the local-potential picture for transfer reactions by deriving a transition amplitude, in which U is replaced by $\phi$ [5]. Since, however, there seems to be no satisfactory derivation of such a model [6,7], we retain the conventional DWBA. Thus our main concern will be to show results for $W_a$ and $U_B$.

## 3. CALCULATION OF THE OVERLAPS AND POTENTIAL OVERLAPS

The computation of U and W is easier in structure models that use harmonic oscillator (h.o.) functions. The h.o. shell model has been used to calculate U [8] and schematic cluster models have been used to calculate U [9,10] as well as W [11,10]. The unphysical tails of such functions are sometimes corrected phenomenologically [12,11]. Here, however, we are primarily concerned with cluster models free of such deficiencies as we wish to study the intercluster motion as determined by the nucleon-nucleon (NN) force. Since the overlaps U also enter into the description of $\alpha$ decay, knock-out reactions etc., they were calculated for quite a few cases long ago [13]. However, as far as I know, microscopic calculations for W have only appeared recently [11,14-16,10].

### 3.1. Exact methods

In the resonating-group method (RGM), the cluster-model wave function consists of terms of the type of (10) with $\Phi_i$ being h.o. Slater determinants from which the c.m. zero-point motion has been removed [17]. In the framework of the generator-coordinate method (GCM) the basis elements are

$$\Phi(\vec{s}) = [(\nu_1\beta_1+\nu_2\beta_2)/(4\pi)]^{\frac{3}{4}} \int d\vec{s} A_{12}\{\Psi_1(\vec{s}_1)\Psi_2(\vec{s}_2)\}, \tag{13}$$

where $\Psi_i(\vec{s}_i)$ is an h.o. Slater determinant centred around $\vec{s}_i$, $\nu_i$ is the number of nucleons, $\beta_i$ is the h.o. parameter, $\beta_i=m\omega_i/\hbar$, of cluster i, $\vec{S}=(\nu_1\vec{s}_1+\nu_2\vec{s}_2)/(\nu_1+\nu_2)$, $\vec{s}=\vec{s}_1-\vec{s}_2$. Thus, in calculating U or W, we have to deal with terms like

$$<\Phi_1'\Phi_2'\delta(\vec{r}-\vec{r}_1'_2)|0|\Phi_{\vec{r}}> \tag{14a}$$

or

$$\Omega(\vec{r},\vec{s}) \equiv <\Phi_1\Phi_2'\delta(\vec{r}-\vec{r}_1'_2)|0|\Phi(\vec{s})>, \tag{14b}$$

with $0=1$ or $V_{12}$. The calculation of (14) requires all the armoury of the calculation of the RGM and GCM kernels [17].

For a $\beta_1=\beta_2$ model of the $^7$Li=$\alpha$+t system, we have shown [15] that a direct evaluation of (14b) by means of Jacobi coordinates is feasible both for $0=1$ and $V_{\alpha t}$. The calculation of U for $^6$Li=$\alpha$+d in a model in which $\beta_1\neq\beta_2$ and both take several values [18] has also been performed with this method. The evaluation of (14) can, however, be formulated more generally by a transformation of the integration variables whereby both the bra and the ket are expressed in terms of Slater determinants. E.g., (14b) can be written as

$$\Omega(\vec{r},\vec{s}) = (2\pi)^{-\frac{9}{2}}(\pi\gamma)^{\frac{3}{4}}\int d\vec{q}\ \exp(-i\vec{q}\vec{r}+q^2/2\gamma)\times$$
$$\times \int d\vec{s}'\exp(i\vec{q}\vec{s}')\omega(\vec{s}',\vec{s}), \tag{15}$$

with $\gamma=\nu_1'\nu_2'\beta_1\beta_2/(\nu_1'\beta_1+\nu_2'\beta_2)$ and

$$\omega(\vec{s}',\vec{s}) = [(\nu_1'\beta_1'+\nu_2'\beta_2')(\nu_1\beta_1+\nu_2\beta_2)/16\pi^2]^{\frac{3}{4}}\times$$
$$\times \int d\vec{s}'<\Psi_1'(\vec{s}_1')\Psi_2'(\vec{s}_2')|0|A_{12}\{\Psi_1(\vec{s}_1)\Psi_2(\vec{s}_2)\}>. \tag{16}$$

For the evaluation of $<|>$ in (16) the well-known technique of handling Slater determinants [19] may be used.

## 3.2. Approximate methods

To evaluate (14) is much easier for $0=1$ than for $0=V_{12}$. One can, however, reduce the calculation of W to that of U with the following trick, called the "equation-of-motion method" (EMM) [15]. Whether the model wave function $\Phi$ is given by (10a) or contains additional terms, it satisfies the Schrödinger equation, $H\Phi=\epsilon\Phi$, in the subspace spanned by $\Phi_{\vec{r}}$. Thus $<\Phi_{\vec{r}}|H-E|\Phi>=0$ holds, which implies

$$W = (E-T)U \qquad (E = \epsilon-<\Phi_1|H_1|\Phi_1>-<\Phi_2|H_2|\Phi_2>), \tag{17}$$

so long as $\Phi_i$ are exact solutions of the Schrödinger equations

$$H_i\Phi_i = <\Phi_i|H_i|\Phi_i>\Phi_i \qquad (i=1,2). \qquad (18)$$

Since $\Phi_i$ are usually not exact, eq. (17) is an approximation.

Having two clusters of equal $\beta$, the computation of U may be further simplified by the use of the orthogonality-condition model (OCM) [20]. According to this model, $\phi$ of eq. (12) satisfies

$$\Lambda(T+v)\phi = E\phi, \qquad (19)$$

where v is an appropriate local potential. Then U and W may be calculated by inverting (12), $U=N^{\frac{1}{2}}\phi$, and by using (17). We note that, since the derivation of (19) makes use of (18), W of the OCM is in fact an approximation to that of the EMM.

To illustrate the performance of the approximate methods, I quote some results for the $^7$Li=$\alpha$+t system [15,16]. To establish the correspondence between the NN and OCM potentials, we used the method of Friedrich and Canto [21]. This consists in optimizing the parameters of v by fitting the energy surface calculated with the OCM to that calculated microscopically. (The energy surface is the expectation value of the Hamiltonian of two clusters, with their respective h.o. potentials pinned down in space, as a function of their distance.)

The OCM overlap is so close to the exact one that their difference could not be resolved in a figure of reasonable size. The radial part w(r) of the microscopic W(r) generated by the Volkov 2 force [22] is compared with its EMM and OCM approximants in fig. 1. We see that both approximations are excellent. From results for the positive-energy region [4,23] we infer that the OCM is likely to be an excellent substitute for the underlying microscopic model for a wide range of A+$\alpha$ systems. Its use is only restricted by the limitations of the microscopic models themselves.

The OCM, as a method for generating U, was also tested by Subbotin et al. [24] They found some disagreement with the RGM presumably for two reasons: they wrongly identified the solution of (19) with $\psi$ instead

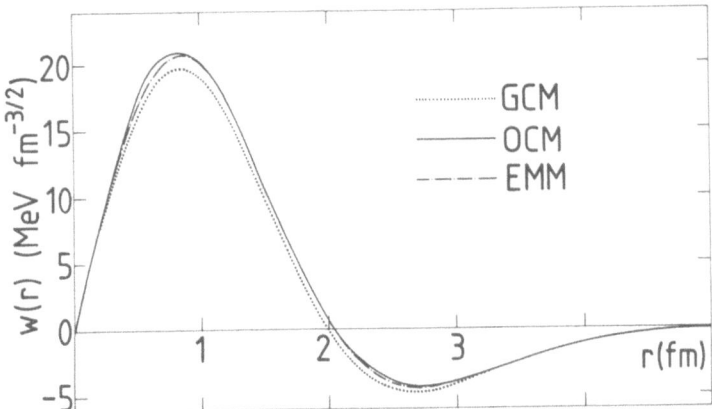

Figure 1. The exact GCM potential overlap and its EMM and OCM approximants for $\alpha$+t.

of φ, and adopted the direct (double-folding) intercluster potential
term for v instead of optimizing it.

## 4. MODEL DEPENDENCE OF THE MICROSCOPIC CALCULATIONS

Calculations have been performed for the $^7Li=\alpha+t$, $^6Li=\alpha+d$ and $^{20}Ne=\alpha+^{16}O$
systems. Each system was assumed to be in a pure two-cluster state. For
$^{6,7}Li$ such an assumption seems to be well-founded [25-27,18], and for
the g.s. band of $^{20}Ne$, too, it describes a major component of the wave
function [28]. For $^7Li$, U and W turned out to be insensitive to the
choice of β, which encouraged us to use single-β models both for $^7Li$
and $^{20}Ne$. However, to allow for the intricate behaviour of the deuteron,
for $^6Li$ we used the dynamical-distortion model of Beck, Dickmann and

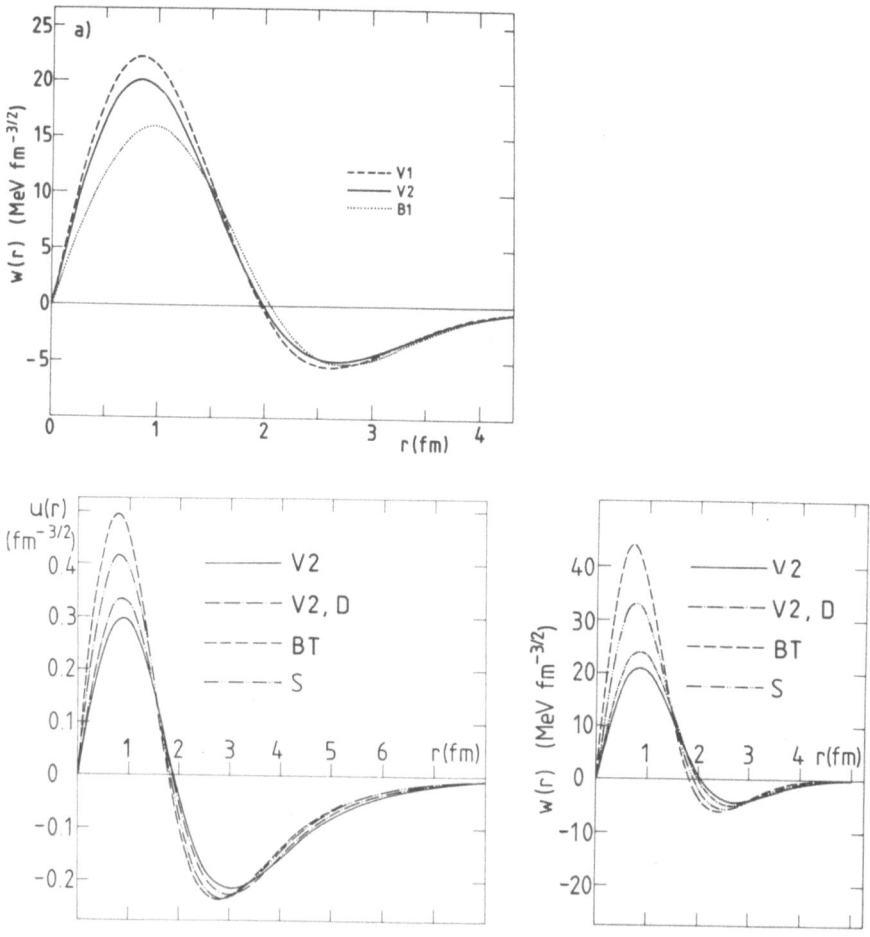

Figure 2. Overlaps u(r) and potential overlaps w(r) for α+t with various
forces. The curve V2,D is the result of an OCM calculation with v chosen
to be the direct intercluster term produced by the V2 force.

Kruppa [18]. We used forces 1 and 2 of Volkov (V1, V2) [22], the force of Brink and Boeker (B1) [29], of Brown and Tang (BT) [30] and of Saito (S) [31], each slightly adjusted to set the cluster separation energy exact as required by transfer calculations. Some of the calculations were performed with the GCM, some others with the OCM.

We tested the dependence of U and W on the NN force for the case of α+t [14,16]. All forces tested reproduce [14,16] the measured radii and moments of $^7$Li. Fig. 2 shows the radial factors u(r) and w(r) calculated with various NN forces and OCM prescriptions. We see that the dependence of u and w on the NN force is non-negligible, with the two limits represented by the "old" forces due to Brown and Tang (BT) and to Brink and Boeker (B1). Since the relative wave function φ in Tang's recent, most realistic calculation [32,26] supports our calculation with V2, the V2 results are to be considered the most realistic.

Kajino, Matsuse and Arima have found that the dynamical distortions of the clusters have no appreciable effect on the u(r) of $^7$Li=α+t [27]. Using V2, we have made similar tests for both u and w of $^6$Li=α+d [33]. In the GCM wave function that we used [18] the size parameters β of both α and d have been taken as GC's, in addition to the distance s.The wave functions of the free α and d, inserted in the bra of (5) and (6), were also GCM wave functions, with the discretized values of the GC's β taken to be the same as those used in $^6$Li. In fig. 3 plotted is also u(r) of a single-β model of the two clusterizations {α+d, $^5$He+p} [34]. We see that, while the $^5$He+p clusterization is unimportant, the cluster dynamical distortion does have a significant effect.

In fig. 3 the potential overlap of Singh and Jain [11] is also shown, along with the overlap and potential overlap of Cook [10]. They are based on a schematic cluster-model wave function and the M3Y force

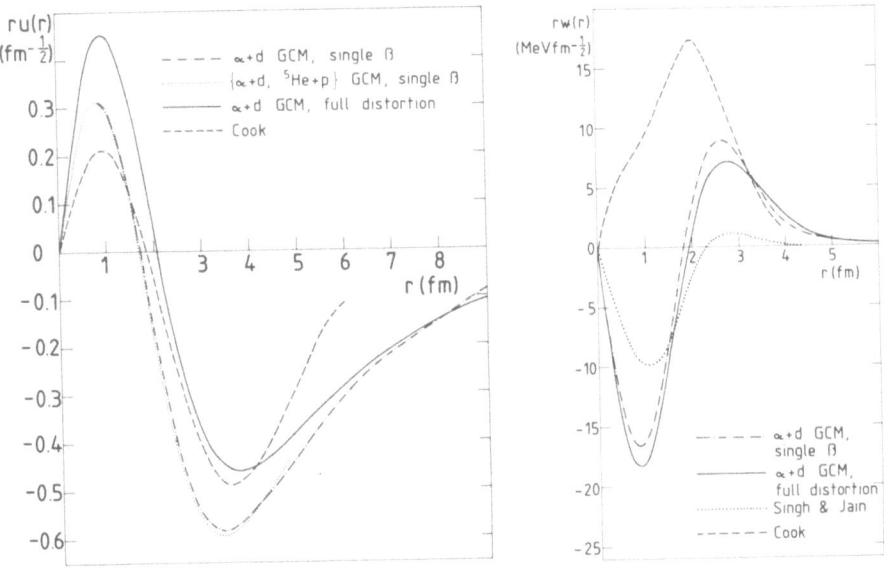

Figure 3. Overlaps ru(r) and potential overlaps rw(r) for α+d in various models.

[35]. Cook states [10] that the w(r) of Singh and Jain is incorrect and
he corrects the errors. However, our earlier criticism [14] on Singh and
Jain's work applies to Cook's work as well: the wave function and the
force are inconsistent, and hence so are u(r) and w(r). Indeed, eq. (17)
applied to Cook's u(r) implies that a consistent w(r) must have a node,
while Cook's w(r) has none. Moreover, Cook has dropped the tail correction of Singh and Jain, and thus his overlap has a Gaussian tail. It is
also extremely surprising that in a rigid single-clusterization model
his α+d SF is as low as 0.49. Thus these results cannot be realistic.

## 5. MICROSCOPIC FORM FACTORS AND THE DWBA

Having realistic overlaps and potential overlaps, one has to see how
they compare with the phenomenological ones, what changes they cause in
the DWBA cross section and whether they help to understand any data.

The microscopic overlaps, potential overlaps and the cross sections
that involve them contain the SF's (9), whose values are given in table
I. It is noteworthy that the distortion reduces the $^6$Li SF substantially.
The $^{20}$Ne SF's differ from earlier values [36,37] mainly because of our
readjustment of the potential, for each state, to the correct energy.

In fig. 4 the microscopic (V2, [14,33]) and phenomenological (eq.
(7), [38]) α+t and α+d potential overlaps and their DWBA predictions for
the $^{12}$C($^7$Li,t)$^{16}$O and $^{16}$O($^6$Li,d)$^{20}$Ne cross sections are compared. The
microscopic and phenomenological w(r) differ drastically, but this difference is reduced in the forward-angle cross section. For ($^7$Li,t) the
behaviour of the cross sections may be understood by assuming that the
main contribution to σ(∿0°) comes from the region 3 fm<r<4.5 fm, whereas
in the ($^6$Li,d) example contributions must also come from r<3 fm. Such a
behaviour, however, must be sensitive to the target, energy and the
(highly ambiguous) optical potentials.

Fig. 5 shows microscopic (OCM with V2, [39]) and phenomenological
[40] overlaps for the g.s. of $^{20}$Ne=α+$^{16}$O, together with sample DWBA calculations. The difference in the cross section comes mainly from the SF
of the microscopic overlap. In agreement with earlier findings [37,24],
the forward-angle angular distribution is almost unaffected.

We see that the dramatic discrepancies between the phenomenological
and microscopic quantities tend to appear in the DWBA somewhat damped.
Yet there are some indications that the microscopic form factors may
help to understand the cluster transfer data. First we mention the old
work of Kubo, Nemoto and Bandō [37], who partially explained a target
dependence of the ($^7$Li,t) reaction by employing microscopic models for

Table I. Theoretical α spectroscopic factors produced by the V2 force

| State | $^7$Li,$\frac{3}{2}^-$ | | $^6$Li,$1^-$ | | $^{20}$Ne,$0^+$ | $^{20}$Ne,$2^+$ | $^{20}$Ne,$4^+$ |
|-------|------------|------|------------------|------------------|-----------------|-----------------|-----------------|
| Model | α+t | α+d {α+d,$^5$He+p} | α+d, distorted | $^{16}$O+α | $^{16}$O+α | $^{16}$O+α |
| S | 1.153 | 1.073 | 1.041 | 0.858 | 0.548 | 0.546 | 0.509 |

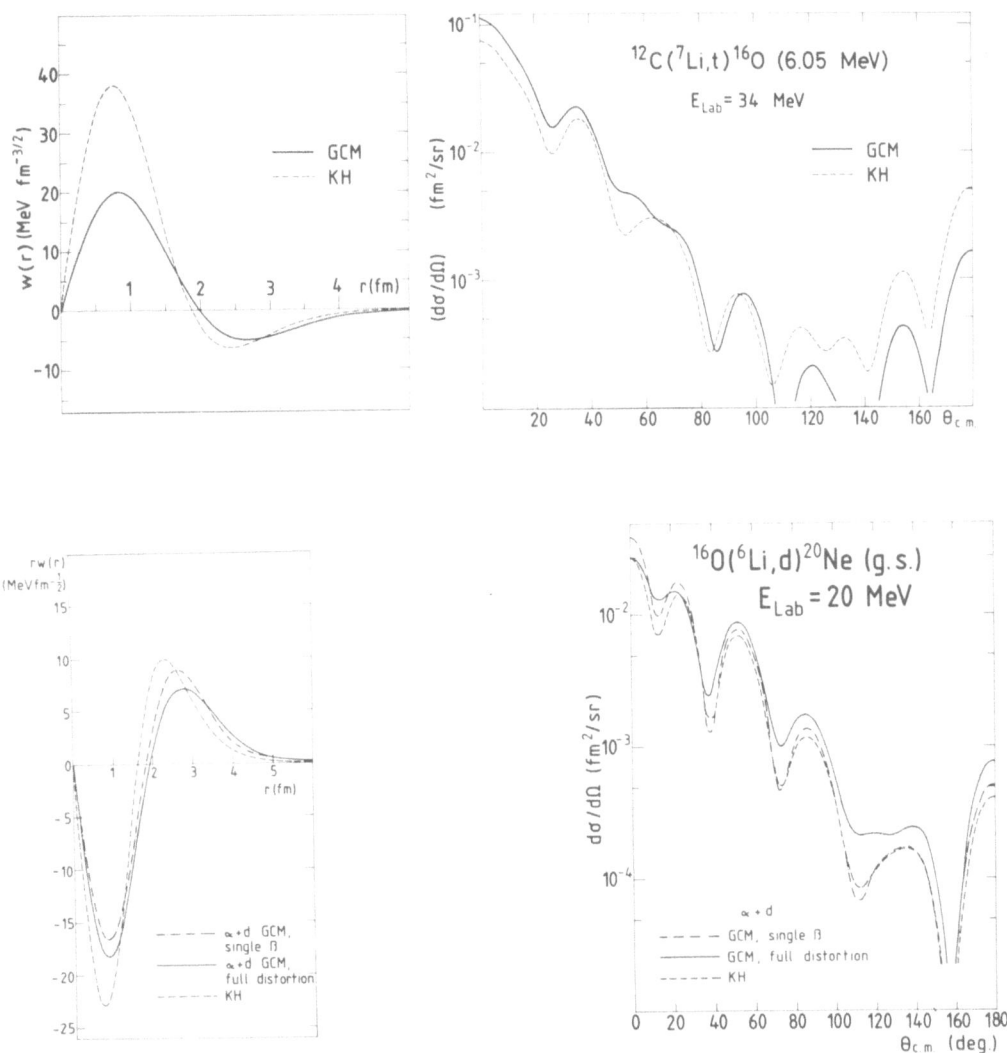

Figure 4. Microscopic [14,33] and phenomenological (KH: [38]) potential overlaps and the DWBA cross sections produced.

B=A+x. Our results however, show that their assumption of the α SF in [7]Li to be much smaller than unity is not well-founded. Furthermore, the analyses of Subbotin *et al.* [24] employing microscopic α+[16]O overlaps show a more coherent picture than the fully phenomenological analyses [40]. Finally, Shyam's analyses [41] show that the SF's extracted with our α+t potential overlap tend to be more consistent with the structure calculations than those extracted conventionally [42].

Beyond these examples, however, the microscopically well-founded form factors seem indispensable in forming a coherent total picture of cluster transfer reactions. Namely, as we shall see in sect. 6, we need such form factors to get reliable absolute SF's.

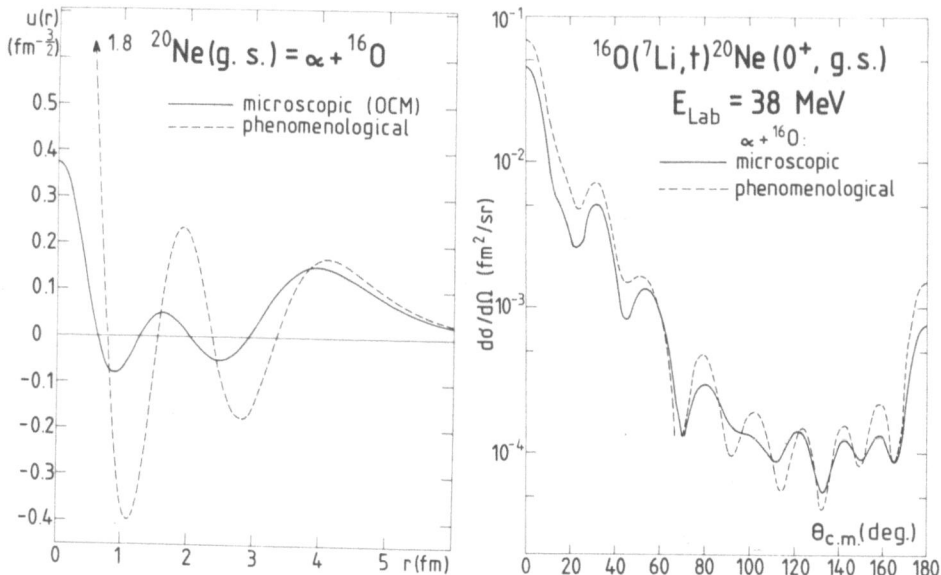

Figure 5. Microscopic [39] and phenomenological [40] $\alpha + ^{16}O$ overlaps and the DWBA cross sections produced.

## 6. LOCAL POTENTIALS AND THE DWBA

In sect. 5 I showed that some local potentials v that are used in the analyses in combination with (8a) and (7) fail to reproduce the microscopic u and w. Moreover, for the $\alpha+t$ system we optimized the parameters of v so as to get the best vu approximation to w, with very moderate success [14]. The fit was particularly bad when v was chosen to have a repulsive core to produce the single node of vu. Aoki and Horiuchi showed in general terms that such "shallow" potentials may be derived from the RGM interaction at the cost of a bad approximation [43]. Thus this family of potential overlaps, favoured as they are by fits to experiment [44], must be regarded as unphysical, and, to improve the agreement with the experimental cross section, other effects are to be sought.

On the other hand, Horiuchi has shown [4] the existence of a well-behaved potential $v'(r)$ to generate $\phi$ via (8b). Moreover, since the hermitean Hamiltonian whose eigenfunction is U depends explicitly on the energy [4], it is not likely that a well-behaved $v(r)$ that produces U in (8a) exists.

The foregoing considerations confirm the conjecture [2] that the conventional use of (8a) to produce U is as if U were replaced by $\phi$ in the DWBA form factor. The fact that this is not allowable [6,7] calls for an extensive use of microscopic form factors. In fact, for the reactions that involve $\alpha+t(^{3}He)$ [14], $\alpha+d$ [33] or $\alpha+^{16}O$ [6], there are realistic potential overlaps $W_a$ available. The targets, however, are too numerous and their structure too complicated, and hence the practitioners can hardly do without some compromise in constructing $U_B$.

In prescribing a recipe for such a compromise, it is reasonable to assume that $U_B(\vec{r}) \approx S'^{\frac{1}{2}} U(\vec{r})$, where $S'$ is a constant and $U$ is now the over-lap of a pure two-cluster problem, i.e. it is defined as $U_B$ in (5) with $\Phi_B$ replaced by $\Phi$ of (10a). Then $\phi$ of (12) may be determined by solving (8b), which yields $U$ through $U = N^{\frac{1}{2}}\phi$. Using this $U$ in the DWBA, $S'$ can be determined by comparison with the measured $\sigma(\exp)$: $S' = \sigma(\exp)/\sigma(DWBA)$. The SF is given by

$$S_B \equiv \langle U_B | U_B \rangle = S' \langle U | U \rangle. \tag{20}$$

The absolute SF's extracted via DWBA fits are generally much larger than those obtained with nuclear-structure calculations [45]. Fliessbach reduces this discrepancy by approving of the practice of the phenomeno-logical DWBA but changing its interpretation in nuclear-structure terms [5]. In his approach the nuclear-structure quantity to be compared with the DWBA SF is not $\langle U|U \rangle$ but $\langle \phi|\phi \rangle$, which is much larger than $\langle U|U \rangle$ in most cases. I shall now show that the correct DWBA, if applied consist-ently as described above, also tends to reconcile the two kinds of SF.

Fig. 6 shows u and the radial factor $\varphi$ of $\phi$ calculated in the pure cluster model for two states of $^{20}$Ne$=\alpha+^{16}$O as well as the DWBA cross sections of the $^{16}$O($^7$Li,t)$^{20}$Ne transitions to these states [6]. The cal-culations that use u correspond to the "correct" DWBA, whereas those using $\varphi$ represent the usual practice or, alternatively, Fliessbach's DWBA [5]. Since the two methods yield almost the same $\sigma(0°)$, they also produce the same value for $\sigma(\exp)/\sigma(DWBA)$. But while the usual practice would regard these numbers as the extracted values of the SF's, the cor-rect approach multiplies them by $\langle U|U \rangle$ of the pure cluster model, which range about 0.5 (table I). It is clear that the extracted SF is reduced just by $\langle U|U \rangle$ of the pure cluster model so long as $\sigma(0°)$ is the same in the two approaches. This, in turn, is likely to be so in general since $\sigma(0°)$ is primarily determined by the tail of u, which converges to $\varphi$. (In an h.o. model, however, this effect never shows up because u=const.$\times\varphi$ is assumed!) Thus the DWBA SF is reduced by the same factor whose recip-rocal enhances the theoretical SF in Fliessbach's approach, but, to substantiate this effect, we need not resort to dubious approximations to the DWBA. The reduction of the DWBA SF may be sizable: for heavier nuclei, where the discrepancy is also larger, the value of $\langle U|U \rangle$ is much smaller.

A further reduction may be gained by a similarly consistent treat-ment of the distorted waves. To see this, let us consider eq.(1), which gives the transfer transition amplitude between the channels described by the states

$$A_{Aa}\{\Phi_A \Phi_a \chi_{Aa}^+\} \text{ and } A_{Bb}\{\Phi_B \Phi_b \chi_{Bb}^-\}. \tag{21}$$

These are projections of two exact many-body states, and are usually viewed as solutions of two independent single-channel scattering prob-lems. Because of the antisymmetrization, the proper relative wave func-tions are not $\chi$, but rather

$$\eta_{Aa}^+ = N_{Aa}^{\frac{1}{2}} \chi_{Aa}^+ \text{ and } \eta_{Bb}^- = N_{Bb}^{\frac{1}{2}} \chi_{Bb}^-. \tag{22}$$

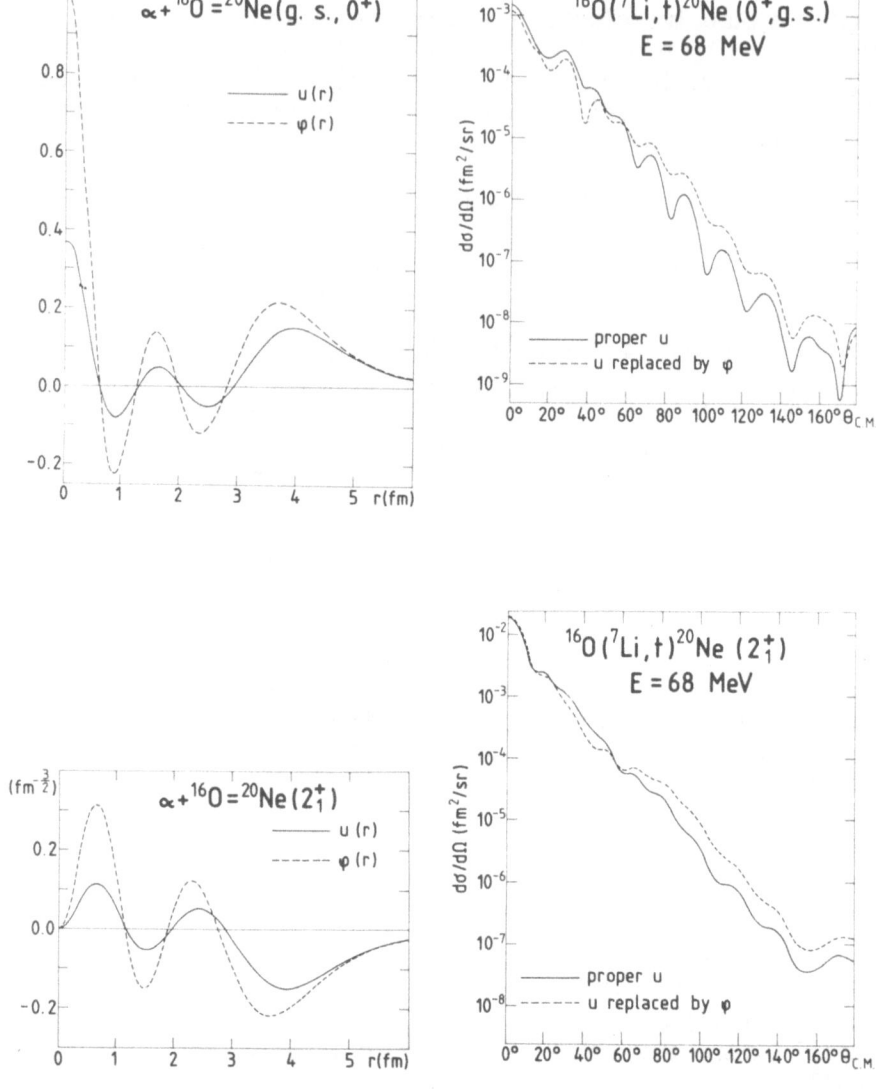

Figure 6. Overlaps $u$ and relative wave functions $\varphi$ for $^{20}\text{Ne}=\alpha+^{16}\text{O}$ and the results of their substitution for $u$ in the DWBA.

It is these functions that may be determined by solving effective s.p. problems, and $\chi$ can then be calculated by

$$\chi_{Aa}^{+} = N_{Aa}^{-\frac{1}{2}}\eta_{Aa}^{+} \quad \text{and} \quad \chi_{Bb}^{-} = N_{Bb}^{-\frac{1}{2}}\eta_{Bb}^{-}. \tag{23}$$

Just as $N^{\frac{1}{2}}$ reduces the functions, so does $N^{-\frac{1}{2}}$ blow them up. Thus the use of $\chi$ of (23) enhances the DWBA cross section, which, in turn, tends to reduce the DWBA SF. This effect is, however, likely to be smaller than that due to $\langle U|U\rangle$ since the scattering wave functions $\eta$ have less overlap with Pauli unfavoured states, and hence the effect of $N^{-\frac{1}{2}}$ on them must be less substantial.

## 7. CONCLUSION

We have seen that the realistic overlaps and potential overlaps differ appreciably from the phenomenological ones and their effect is seen mostly in the magnitude of the DWBA cross sections. The importance of these calculations is twofold. On the one hand, they should help us to obtain correct absolute spectroscopic factors. On the other hand, their use eliminates some free parameters from the analyses. Thus, when failing to reproduce the data, one can be sure that the blame is to be put on some other effects. The adaptation of the realistic potential overlaps and the recipe for the calculation of the overlaps in practitioners' DWBA analyses seem to be feasible.

## ACKNOWLEDGEMENTS

Most of the results I have reported on here have emerged from a collaboration of the Daresbury, Debrecen and Karlsruhe groups, with the participation of R. Beck, F. Dickmann, B. Gyarmati, M. A. Nagarajan, K. F. Pál, V. K. Sharma, Radhey Shyam and T. Vertse. I am particularly indebted to Dr. M. A. Nagarajan, who introduced me into this subject, and to Dr. K. F. Pál, who has been working with me throughout.

## REFERENCES

1. G. R. Satchler, *Direct nuclear reactions* (Clarendon, Oxford, 1983)
2. T. Fliessbach, *Z. Phys.* **A272** (1975) 39
3. R. G. Newton, *Scattering theory of waves and particles* (McGraw-Hill, New York, 1966) p. 487
4. H. Horiuchi, *Prog. Theor. Phys.* **71** (1984) 535
5. T. Fliessbach, *Z. Phys.* **A278** (1976) 353
6. K. F. Pál and R. G. Lovas, in *Proc. Int. Symp. on in-beam nuclear spectroscopy*, Debrecen, May, 1984, ed. Zs. Dombrádi and T. Fényes (Akadémiai Kiadó, Budapest, to appear in 1984)
7. R. G. Lovas, to be published
8. D. Kurath and I. S. Towner, *Nucl. Phys.* **A222** (1974) 1;
   B. Apagyi, G. Fái and J. Németh, *Nucl. Phys.* **A272** (1976) 303
9. I. V. Kurdyumov, V. G. Neudatchin and Yu. F. Smirnov, *Phys. Lett.* **31B** (1970) 426
10. J. Cook, *Nucl. Phys.* **A417** (1984) 477
11. P. Singh and A. K. Jain, *Phys. Rev.* **C25** (1982) 1705
12. B. Apagyi and T. Vertse, *Phys. Rev.* **C21** (1980) 779
13. A. Hasegawa and S. Nagata, *Prog. Theor. Phys.* **45** (1971) 1786;
    H. Horiuchi, *Prog. Theor. Phys.* **47** (1972) 1058
14. K. F. Pál, R. G. Lovas, M. A. Nagarajan, B. Gyarmati and T. Vertse, *Nucl. Phys.* **A402** (1983) 114
15. R. G. Lovas, K. F. Pál and M. A. Nagarajan, *Nucl. Phys.* **A402** (1983) 141
16. R. G. Lovas and K. F. Pál, *Nucl. Phys.* **A424** (1984) 143
17. H. Horiuchi, *Prog. Theor. Phys. Suppl.* **62** (1977) 90

18. R. Beck, F. Dickmann and A. T. Kruppa, contribution to this Conference

19. D. M. Brink, in *Proc. Int. School E. Fermi*, Course **XXXVI**, ed. C. Bloch (Academic, New York, 1966) p. 247

20. S. Saito, *Prog. Theor. Phys. Suppl.* **62** (1977) 11

21. H. Friedrich and L. F. Canto, *Nucl. Phys.* **A291** (1977) 249; H. Friedrich, *Phys. Reports* **74** (1981) 209

22. A. B. Volkov, *Nucl. Phys.* **74** (1965) 33

23. D. Wintgen, H. Friedrich and K. Langanke, *Nucl. Phys.* **A408** (1983) 239

24. V. B. Subbotin, V. M. Semjonov, K. A. Gridnev and E. F. Hefter, *Phys. Rev.* **C28** (1983) 1618

25. M. V. Mihailović and M. Poljšak, *Nucl. Phys.* **A311** (1978) 377; V. K. Sharma and M. A. Nagarajan, *Daresbury Laboratory report* DL/NUC/P124T (1980); R. Beck, R. Krivec and M. V. Mihailović, *Nucl. Phys.* **A363** (1981) 365

26. H. Kanada, T. Kaneko and Y. C. Tang, *Nucl. Phys.* **A380** (1982) 87

27. T. Kajino, T. Matsuse and A. Arima, *Nucl. Phys.* **A414** (1984) 185

28. B. Buck, C. B. Dover and J. P. Vary, *Phys. Rev.* **C11** (1975) 1803

29. D. M. Brink and E. Boeker, *Nucl. Phys.* **A91** (1967) 1

30. R. E. Brown and Y. C. Tang, *Phys. Rev.* **176** (1968) 1235

31. S. Saito, in *Proc. INS-IPCR Symp. on cluster structure of nuclei*, Tokyo, 1975 (IPCR Cyclotron Progress Report, Suppl. **4**) p. 141

32. H. Walliser and Y. C. Tang, *Phys. Lett.* **135B** (1984) 344

33. R. Beck, F. Dickmann and R. G. Lovas, to be published

34. R. Beck, F. Dickmann and R. G. Lovas, contribution to this Conference

35. G. Bertsch, J. Borysowicz, H. McManus and W. G. Love, *Nucl. Phys.* **A284** (1977) 399

36. T. Matsuse, M. Kamimura and Y. Fukushima, *Prog. Theor. Phys.* **53** (1975) 706

37. K.-I. Kubo, F. Nemoto and H. Bandō, *Nucl. Phys.* **A224** (1974) 573

38. K.-I. Kubo and M. Hirata, *Nucl. Phys.* **A187** (1972) 186

39. K. F. Pál and R. G. Lovas, to be published

40. N. Anantaraman, H. E. Gove, R. A. Lindgren, J. Tōke, J. P. Trentelman, J. P. Draayer, F. C. Jundt and G. Guillaume, *Nucl. Phys.* **A313** (1979) 445

41. R. Shyam, R. G. Lovas, K. F. Pál, V. K. Sharma and M. A. Nagarajan, *Daresbury Laboratory preprint* DL/NUC/P190T (1983)

42. F. D. Becchetti, E. R. Flynn, D. L. Hanson and J. W. Sunier, *Nucl. Phys.* **A305** (1978) 293

43. K. Aoki and H. Horiuchi, *Prog. Theor. Phys.* **69** (1983) 1154

44. F. D. Becchetti, D. Overway, J. Jänecke and W. W. Jacobs, *Nucl. Phys.* **A344** (1980) 336; D. R. Chakrabarty and M. A. Eswaran, *Phys. Rev.* **C25** (1982) 1933

45. F. D. Becchetti, in *Clustering aspects of nuclear structure and nuclear reactions*, ed. W. T. H. Van Oers *et al.* (AIP, New York, 1978) p. 308

STRUCTURE INFORMATION AND REACTION MECHANISM STUDIES IN THE ALPHA
TRANSFER

W. Oelert
Institut für Kernphysik, Kernforschungsanlage Jülich
Postfach 1913
D-5170 Jülich, W. Germany

ABSTRACT. Aspects of alpha transfer reaction studies investigated during the last few years are discussed. In the mass region of alpha decaying nuclei reduced widths extracted from alpha pickup reactions are
in very good agreement to those of alpha decay data. In the medium
mass range four nucleon transfer data are interpreted in the framework
of rather vivid models indicating a strong pair correlation. For lower
mass systems the observed experimental quantities are reasonably predicted by SU(3) and shell model calculations. The influence of the
interdependence of structure and reaction mechanism effects on amplitudes and shapes of angular distributions is discussed. The necessity
of using a coupled reaction channels mechanism and its success for the
description of natural and unnatural parity states is demonstrated for
the lower mass range of sd-shell nuclei.

1. INTRODUCTION

Since the early days of the observation of natural decay the alpha
particle has been often regarded as a constituent part of nuclear matter. Alpha transfer reaction studies have been intensively investigated with the aim of observing possible alpha-cluster effects. If in
fact clustered nucleons are preformed parts of nuclear matter a prefe-
rential candidate should be the alpha particle because of its symmetrical arrangement to two protons and two neutrons with large binding
energy.
    Jänecke and Becchetti[1] concluded that alpha cluster transfer
reactions are an important tool to establish alpha particle clustering
in the surface of medium heavy and heavy nuclei: "The data suggest
that in certain regions of the nuclear surface about 25 % of the charge (matter) consists of alpha particles". On the other side, already
in one of the earliest studies of few nucleon transfer reactions[2], it
was felt that "the spectroscopic information suggested by the analysis
245

J. S. Lilley and M. A. Nagarajan (eds.), Clustering Aspects of Nuclear Structure, 245–259.

of $(d, {}^6Li)$ and $(d, {}^7Li)$ reactions did not lend support to assumptions of predominant four-particle or alpha-clustering in the sd-shell".

For a long time the coexistence of cluster aspects and single-particle effects was regarded as a controversy. In his introductory talk at the Conference of Clustering Aspects of Nuclear Structure and Nuclear reactions in Winnipeg 1978 Arima[3] pointed out, however, that there is no conceptual difference between the two choices of "clustering being enhanced in nuclear matter" or of "no ground to consider clustering of nucleons in nuclei". It seems that this question is more one of semantics rather than of basic physics.

In the conference summary of the International Conference on Nuclear Physics Schiffer[4] was raising the challenge of simple understandable models. Especially for the discussions of four nucleon transfer data such kind of models were employed and formulated. These are known under the key-words of the quartet concepts[5], the cluster model[6], the SU(3) model[7], the pairing vibrational model[8], and the interacting boson model[9] with mutual overlap. The most general description is probably given by the shell model, which unfortunately has the drawback of less insight into the real understanding of the complex scheme.

Few nucleon transfer reactions provide the information of how much the few nucleon system in the projectile or ejectile resembles that which is transfered to or from the target mass. The degree of overlap between the nucleus A with the nucleus B = A ± (few nucleon system) is the obtainable information. The transfered system need not resemble the corresponding free nucleus. In that sense the phrase "alpha-transfer" used here should not necessarily be interpreted as a ${}^4He$ nucleus or alpha particle transfer.

Depending on the nuclear mass region the different models cited have been applied with different but increasing success during the last ten years. At high masses an impressive agreement of alpha-reduced widths extracted from ground state transition data in alpha transfer reactions and from alpha decay data has been observed. This is one of the rare cases of transfer reaction studies where experimental information obtained from different methods can be compared on an absolute scale. For alpha transfer data in the medium mass range the pairing-vibrational model has been proven to be rather useful, at least for the interpretation of $0^+$ final states. In the lower mass range the SU(3) and shell-model calculations have been widely used to compare experimental data with theoretical predictions.

## 2. REDUCED WIDTH FROM ALPHA DECAY AND ALPHA TRANSFER REACTIONS

Contrary to the situation of other nucleon transfer reactions the reaction theory of alpha particle transfer can be tested by comparing the experimental results to alpha decay information. The reduced widths from both experimentally studied alpha transfer and experimentally measured alpha decay provide independent information and therefore should give a measure of the quality of the extracted quantities.

The alpha transfer reaction probes the cluster wave function in the nuclear surface. The reduced width is defined in this case as:

$$\gamma_\alpha^2(s) = (\hbar^2 s/2\mu)\ S_\alpha\ \lfloor\Psi_G(s)\rfloor^2 \tag{1}$$

with: $\mu$ = reduced $\alpha$-particle mass,
$S_\alpha$ = the alpha spectroscopic factor,
$\Psi_G^\alpha$ = the alpha-nucleus wave function as the form factor in the DWBA calculations.

From the $\alpha$-decay life time the reduced width can be written as

$$\hbar\lambda_\alpha = 2\ \gamma_\alpha^2(s)\ P_L(Q_\alpha,s) \tag{2}$$

with $P_L(Q_\alpha,s)$ being the penetrability calculated from the same alpha-nucleus wave function as used in the DWBA.

This relates the reduced width to the cluster wave function as:

$$\gamma_\alpha^2(s) = \hbar\lambda_\alpha/2P_L(Q_\alpha,s) = (\hbar s/2kT^{\frac{1}{2}})\lfloor\Psi_G(S)\rfloor^2 \tag{3}$$

As a result, since $\Psi(r) = S_\alpha^{1/2}\ \Psi^{DWBA}(r)$ $\qquad$ (4)

and $d\sigma/d\Omega(exp) = S_\alpha\ d\sigma/d\Omega(DWBA) \propto S_\alpha\lfloor\Psi_G\rfloor^2$ $\qquad$ (5)

it follows that $d\sigma/d\Omega(exp)$ is approximately proportional to $\gamma_\alpha^2(s)$- The channel radius s, the value of which in principal is arbitrary, was chosen by Jänecke et al.[10,11] as s = 1.7 $A^{1/3}$ fm. This is supposed to be the region where the $\alpha$-transfer reaction occurs and hence the alpha-nucleus wave function and the reduced width are best determined. The spectroscopic factor $S_\alpha$ is known to be model dependent, the reduced width, however, calculated according to equation (1) is supposed to be rather model independent since the dependence of the $\alpha$-spectroscopic factor on the form factor is largely cancelled by the product of the spectroscopic factor with the form factor. The sensitivity to the optical model parameters, which are subject to considerable uncertainties, remains.

Using zero range (ZR) and finite range (FR) Distorted Wave Born Approximation (DWBA) calculations Jänecke et al.[10] analysed the $^{122}Te(d,^6Li)^{118}Sn$ and the $^{148}Sm(d,^6Li)^{144}Nd$ reactions employing different bound state potentials (A,B and C for the $\alpha$-target system and in addition K and W for the $\alpha$-d $\equiv$ $^6Li$ system in case of FR-DWBA calculations, see Table I.

Table I
Bound state wave functions
(from ref. 10)

| | V (MeV) | $r_R$ (fm) | $a_R$ (fm) |
|---|---|---|---|
| α in | A - 145 | 1.20 | 0.65 |
| Target: | B - 128 | 1.30 | 0.75 |
| | C - 115 | 1.40 | 0.65 |
| $^6Li$ = | K - 78 | 1.508 | 0.65 |
| d + α | with (N,L)=(1,0) | | |
| | W + 26 | 1.00 | 0.01 |
| | - 26 | 1.55 | 0.70 |
| | with (N,L)=(0,0) | | |

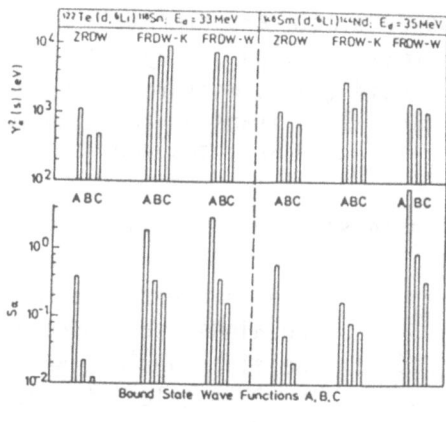

Fig. 1: Model dependent strengths of reduced width and spectroscopic factors.

Fig. 1 displays the extracted reduced width $\gamma^2(s)$ and spectroscopic factors $S_\alpha$ for both reactions. In each case (ZR-DW and FR-DW with bound state wave functions A, B, and C and with α-d wave functions K and W for the FR-DW calculations) the much stronger dependence of the spectroscopic factors rather than of the reduced width on the parameters of the bound state wave function is evident. Depending qualitatively on the bound state potentials the reduced widths vary by less than a factor of three, whereas the spectroscopic factors change typically by a factor of twenty.

Alpha transfer studies employing the $(^{16}O,^{12}C)$ stripping reaction in the lead region and the $(d,^6Li)$ pickup reaction on alpha decaying nuclei were performed by Davies et al.[12] and Jänecke et al.[10,11], respectively, in order to examine whether alpha decay widths and spectroscopic alpha transfer strength correlate to each other. A possible correspondence would give strong arguments that the reaction mechanism adopted would be reasonable and, hence, a direct one-step reaction mechanism would be rather realistic. The upper part of Fig. 2 shows that among nuclei with masses between 140 and 234 the reduced widths from α-decay data vary by two orders of magnitude, the half-life times for the target (residual) masses involved in the pickup (stripping) reaction as the alpha decaying nucleus change by 30 orders of magnitude (disregarding the stable $^{208}Pb$, the half-life time of which has been estimated by the reaction study[11,13] to be larger than $10^{127}$). The close agreement between decay and transfer reaction data (rather independent of the reaction type) is documented in the lower part of Fig. 2 showing that the ratio of the reduced widths extracted by both methods in close to unity for most cases and does not exceed three. The rather strong influence of optical potentials, details of the bound state potentials, and questions of the reaction mechanism might be the reason for this minor excess.

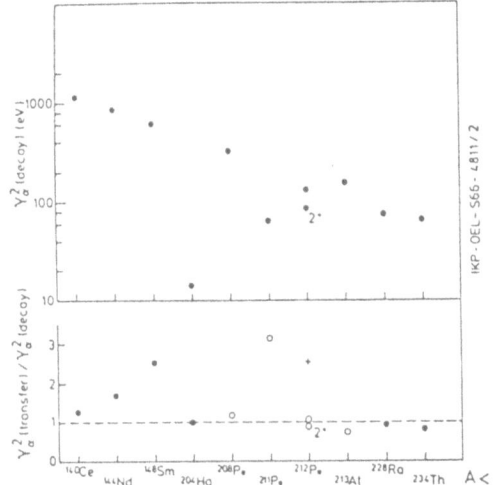

Fig. 2: Alpha decay reduced width and ratio of reduced width from transfer and decay data. Lower part of the figure: full points (d,$^6$Li)[11] data.

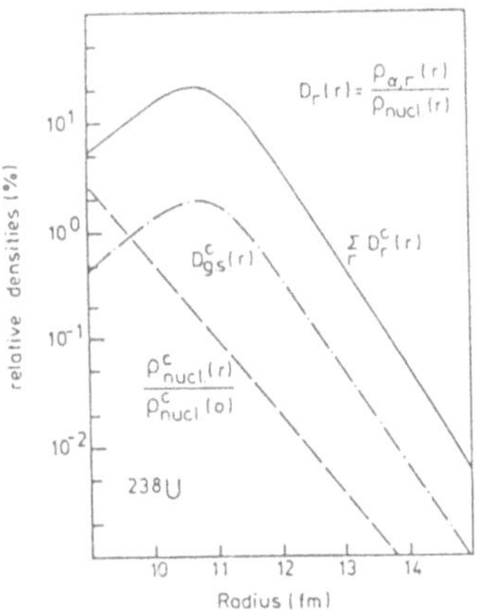

Fig. 3: Alpha-clustering probabilities D(r) in $^{238}$U$_{g.s.}$ extracted from (d,$^6$Li) studies[14] to $^{234}$Th ground state and sum of excited states.

Jänecke et al.[11,14] furthermore extracted probabilities for alpha cluster densities D, which are defined as the ratio of the nuclear matter (charge) density associated with alpha particles in a particular quantum state to the total nuclear matter (charge) density[14]. The probabilities of the charge densities versus the nuclear radius are displayed in Fig. 3 for the ground state transition (dashed dotted line) and the sum of the observed[14] transitions (solid line) to excited states in the final nucleus $^{234}$Th below 1.5 MeV excitation energy. The highest alpha cluster probability in the $^{238}$U ground state is reached at 10.65 fm (= about 1.7 x A$^{1/3}$ fm) with 1.8 % for the ground state transition and 20 % for the sum to the lowest excited states. The constantly decreasing curve (dashed line) shows that in the region of highest alpha-clustering probability the nuclear matter (or charge) density is only about 0.3 % of the central nuclear density. For other mass systems Jänecke et al.[11] found similar values for alpha-clustering probabilities in the ground state of the target nuclei ranging from 2 % to 14 % and from 7 % to 40 % for the ground state transition and for the sum including transition strengths to excited states, respectively. These results are shown in Fig. 4 and indicate that for in-

stance the nucleus $^{208}Pb = ^{204}Hg + \alpha$ has a low probability for alpha--
cluster density with $^{204}Hg$ in the ground state but a alpha-clustering
density probability of about 40 % of the nuclear matter density at
10.65 fm with $^{204}Hg$ being excited. On the other side the wave function
of the nucleus $^{212}Po$ has only a certain overlap with the system $^{208}Pb$
+ $\alpha$ with $^{208}Pb$ being in its ground state and not excited. Qualitatively this is plausible by the closed shell of $^{208}Pb$. However, a
quantitative description of alpha transfer spectroscopic properties by
simple shell model considerations among heavy nuclei is far of being
satisfactory. The discrepancy of a few orders of magnitude is reduced
to about one order of magnitude by introducing strong configuration
mixing or by renormalization due to total antisymmetrization. Since
these shell model applications became rather tortuous and hardly
feasible the challenge of creating more understandable, more simple
phenomenological models was taken.

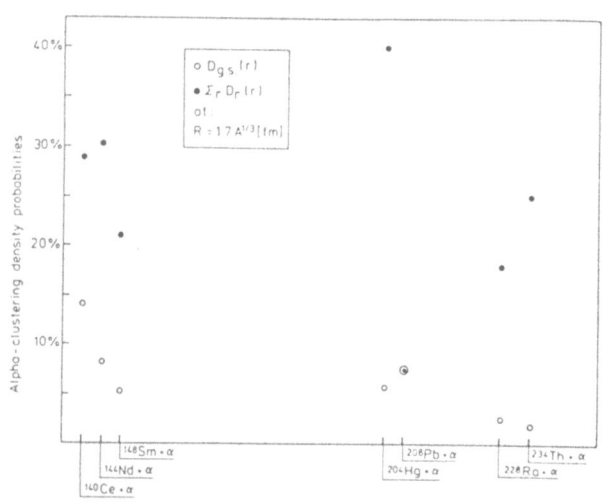

Fig. 4: Alpha-
clustering
probabilities D(r) to
final ground states
(open circles) and sum
of excited states
(full points), data
from Ref. 11).

## 3. TYPES OF ALPHA-TRANSFER REACTIONS

In the above section alpha-transfer data were compared to properties
of alpha-decaying nuclei. The types of transfer reactions studied
throughout the last two decades increased rapidly with the availibili-
ty of heavy ion acceleration. In review articles[15-18] much of the
early work in this field has been covered. The amount of data and information available is too tremendous and abundant to be reviewed
here.

     It remains a question of the particular intension and of feasi-
bility of performing the experiment and the interpretation, which kind
of alpha transfer reaction is selected at which optimum energy in

order to get an answer on the question asked. Q-value effects, momentum matching, reaction mechanism possibilities, structure features and strength of projectile/ejectile spectroscopic factor have to be taken into consideration beside the experimental convenience. Investigations on four nucleon correlations are certainly not only limited to transfer reactions as the pickup reactions: $(d,^6Li)$, $(^3He,^7Be)$, $(\alpha,^8Be)$, - $(^6Li,^{10}B)$, $(^{12}C,^{16}O)$, $(^{16}O,^{20}Ne)$ etc. or their inverse stripping reactions as $(^6Li,d)$, $(^7Li,t)$, $(^{13}C,^9Be)$, $(^{14}N,^{10}B)$, $(^{16}O,^{12}C)$, $(^{20}N,^{16}O)$ etc.. Knock out reactions like $(p,p\alpha)$, $(\alpha,2\alpha)$ or $(e,e\alpha)$ have been used as well. At the Triest Nuclear Physics Workshop[19] and at the Winnipeg conference[20] Becchetti and Jänecke pointed out that i) generally qualitatively reasonble agreement is observed at different investigations, taking the specific properties into account, ii) the decrease of cross section for the ground state transitions with increasing mass follows a $A^{-2}$ to $A^{-3}$ dependence which indicates a localization of the reaction at the nuclear surface and iii) the alpha transfer reaction proves to be rather selective provided the incident energy is sufficiently high. A recent investigation of the $^{230,232}Th(d,^6Li)$ reaction[21] demonstrates nicely the selectivity for

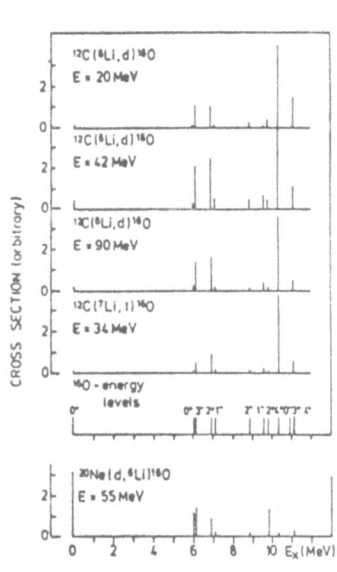

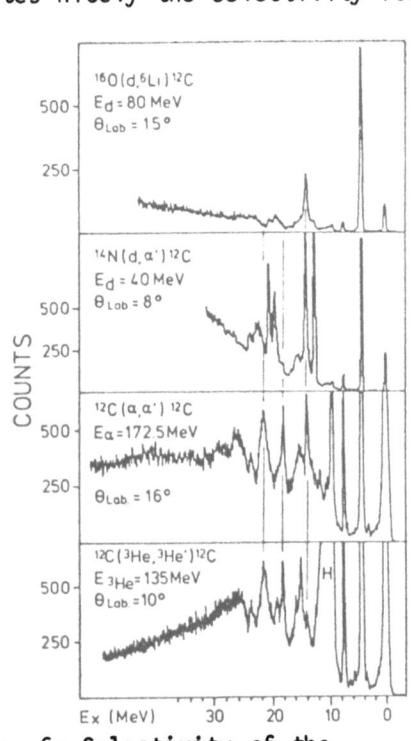

Fig. 5: Selectivity of the $^{16}O$ energy spectrum.

Fig. 6: Selectivity of the $^{12}C$ energy spectrum.

heavy mass systems. Fig. 5 (Ref. 20 and 22) shows for an example of light masses that the selectivity of the stripping reaction on $^{12}C$

leading to $^{16}O$ appears to be rather independent of the incident energy
and reaction. Certainly, the inverse pickup reaction to $^{16}O$ (lower
part of Fig. 5) does excite the individual states in $^{16}O$ according to
their structure with different strength, however, the reaction still
remains selective. The selectivity of the four nucleon transfer
reaction is further demonstrated in Fig. 6, which compares the (d,$^6Li$)
reaction to a two-nucleon transfer study[23] and inelastic scattering
investigations[24], all leading to energy spectra of $^{12}C$. These
inelastic scattering experiments have been performed to study giant
quadrupole resonance (GQR). Recent investigations of the decay of the
GQR[25] showed that: the observed direct decay contributions are the
manifestation of the specific structure of the GQR. As an example this
category of experiments Fig. 7 shows an experimental kinematic map of
the $^{20}Ne(\alpha,\alpha',\alpha_c)^{16}O$ reaction. The kinematic lines for the $\alpha$ decay to
the $^{16}O$ ground state ($\alpha_0$) and first not quite resolved group of low
excited states ($\alpha_1$) are clearly seen. Similar scatter plots can be

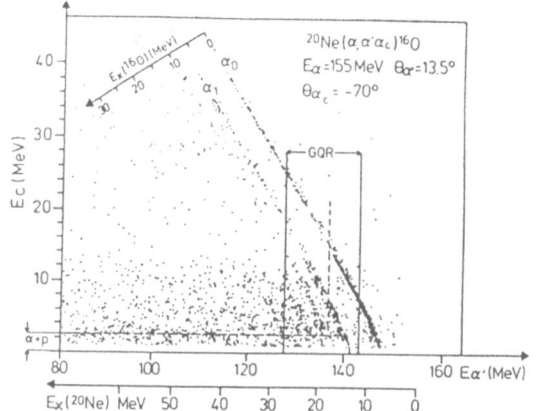

Fig. 7: Kinematic map of the reac-
        tion $^{20}Ne(\alpha,\alpha'\alpha_c)^{16}O$, Ref. 25.

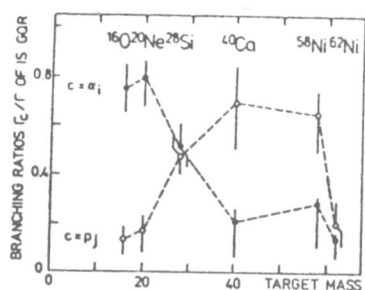

Fig. 8: Main components of
branching ratios for charged
particle decay, see Ref. 26
for more detail.

observed for other particles by which the GR decays. Depending on the
thresholds the GQR decays mainly by alpha particle, proton and neutron
emission. The distribution of the decay branches $\Gamma$ among these
different particles or to different energy levels in the final nucleus
depends strongly on the investigated mass system. For instance the
$\alpha_0$ branch is rather weak in $^{12}C$, $^{16}O$ and $^{28}Si$, but fairly strong
in $^{20}Ne$ and $^{22}Ne$, the $\alpha$:p ratio mass dependence is demonstrated by
the following values: 5 in $^{16}O$, 1 in $^{28}Si$, 0.25 in $^{40}Ca$ and 0.1 in
$^{58}Ni$, another presentation of these circumstances is given in Fig. 8,
see Ref. 26. Qualitatively some of these results have been explained
by Faessler et al.[27] in the framework of SU(3) calculations. Further,
experimental evidence has been reported that the deexcitation of the

GQR by particle emission shows features of direct decay contributions as for instance by the asymmetry about $90^0$ of the angular correlation functions, which could be explained by interference effects. Fig. 9 shows an example where the angular correlation function of the $^{20}Ne(\alpha,\alpha'\alpha_0)^{16}O$ process could be explained successfully by interference with quasi-free scattering amplitudes.

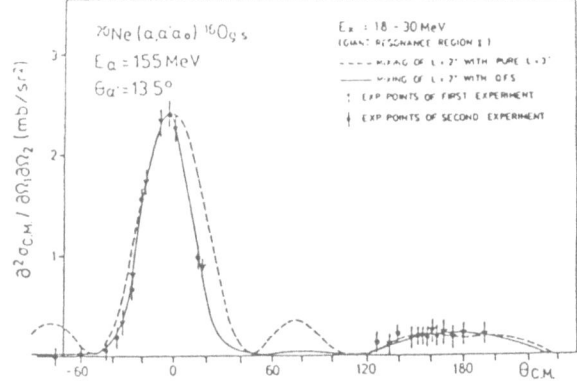

**Fig. 9:**
Angular correlation function of $\alpha_0$ decay from the GQR, see Ref. 25 for more detail.

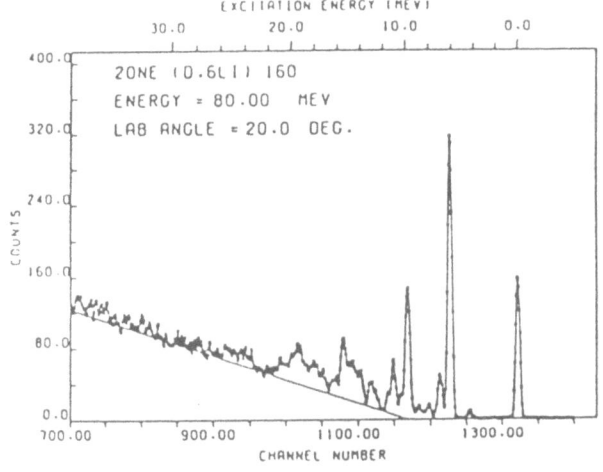

**Fig. 10:**
Energy spectrum of the (d,$^6$Li) reaction on $^{20}$Ne.

Above the close agreement between reaction and decay data has been pointed out. Therefore, one could imagine that in nuclear mass systems, where the GQR decays predominantly by alpha particle emission to low lying levels, GQR properties show up in alpha transfer studies. The $^{20}Ne(d,^6Li)^{16}O$ reaction shows rather compact excited level groups ranging from 13.5 MeV to 17 MeV and from 17 MeV to 22 MeV, see Fig.10. In a contribution to this conference Jänecke et al. show an energy spectrum of the same reaction with much better energy resolution, which again gives evidence for such kind of a structure. Whether this observed strength contains structure for which GQR configuration features are responsible remains to be investigated.

# 4. PHENOMENOLOGICAL MODELS

Stimulated by experimentally observed low lying excited $0^+$ states in medium and heavy nuclei and by the rather often found striking correspondence of spectroscopic strength between alpha particle and two-nucleon transfer (two-proton and/or two neutron) populating the same final nucleus the pairing vibrational model has been tried to be adopted for the interpretation of alpha transfer results (alpha pairing vibrations). The model has proven to be successful in the vicinity of nuclei with doubly closed shell structure. Details on the evaluation of spectroscopic strength by this model are given elsewhere[28]. R.A. Broglia et al.[29] predicted the ground state transition cross sections for the $(d,^6Li)$ reaction to six Cd, five Sn isotopes and the cross section to first excited $0^+$ states in three Sn isotopes, introducing the effect of admixtures of pairing modes due to phase transitions and the change of the Fermi level with increasing neutron number. Even though the quadrupole pairing vibrational model[30] has been moderately used for the interpretation of low lying $2^+$ states it seems not to be very promising for the considerations of four nucleon transfer correlations.

The interacting boson model (IBA) appears to be somewhat more successful in determining spectroscopic strengths for low lying excited states. In the mass range of 40 - 64 the general agreement between alpha transfer experimental results and IBA predictions[31] is very good for the transitions to the ground states and first excited $2_1^+$-states, for transitions to the first excited $4_1^+$ state it is not satisfactory in the center of this mass region. The data for the $2_2^+$ and $0_2^+$ states tend to be similar in character to the $4_1^+$ results. This is expected if those states are populated by two d-boson transfers differing only in their angular momentum coupling[31]. The IBA model takes the Pauli principle via the quantity $\Omega = j + \frac{1}{2}$ (boson capacity of the shells involved) into account. The early versions of the IBA did not distinguish between neutron and proton bosons, which might be a questionable approximation since it does not distinguish proton and neutron numbers of four nucleon excitations. Iachello and Jackson[32] have - within the framework of the IBA - constructed a suitable operator of low hindrance factor for alpha-decay. Alpha decay from the normal (s,d) boson configuration should lead to alpha-hindrance factors which are too large to permit for alpha decay. The introduced configurations account for low lying negative parity bands with small-alpha-hindrance factors. Direct comparison with the experiment, however, is not done. Theoretical models which predict reduced alpha-decay widths and hindrance factors are still under investigation[33,34] and not really mature.

## 5. ALPHA TRANSFER AT LOW MASS NUCLEI

Based on the assumption that the alpha transfer reactions behave like direct one-step processes DWBA calculations have been performed in such investigations in order to extract experimental spectroscopic factors by the ratio of experimental to DWBA cross sections. Though the input quantities of DWBA calculations are subject to considerable uncertainties, as e.g. discussed in contributions to this conference by Yamaya et al., generally rather fair agreement is observed between DWBA and exeprimental angular distribution shapes of investigations studied throughout the periodic table. There are, however, cases where significant differences appear which are of special interest when observed between angular distributions involving the same angular momentum transfer. Rather often the uncertainties of the input parameters entering into the reaction calculations have been blamed for such observations. Especially for the construction of bound state wave functions (geometrical parameters and quantum numbers enter extremely sensitive into the cross section, both in magnitude and in shape of angular distributions) several prescriptions have been suggested. These questions will not further be discussed here.

Experimental alpha spectroscopic factors given in the literature show a lot of scatter for ground state transitions as well as for transitions to excited final states. As shown by Fulbright[16] and Chung et al.[35] there is a systematic disagreement between experimental and theoretical spectroscopic factors for the ground state transitions of nuclei in the sd shell. One way to try to overcome this problem seems to be to instroduce[16] the heavy ion convention of the radius parameter of the real part of the $^6$Li optical model potential. Lezoch et al.[36] argue from systematic studies that the volume integral should be < 320 (MeV fm$^3$) for the most realistic $^6$Li optical potential. Fig. 11 shows at the example of data from Jülich[37] that the discrepancies are considerable removed in the upper part of the sd shell by using such potentials when comparing experimental and theoretical[35] spectroscopic factors. Still, rather significant deviations are observed for the lower mass region. Furthermore, for population of excited states involving the same $\Delta\ell$ between initial and final states significant variations of angular distribution shapes have been observed[38,39]. In addition, experimental results of different investigations are rather often inconsistent even on a relative scale and experimental and theoretical spectroscopic factors are only in limited agreement. This suggests that the adopted one-step DWBA formalism is not adequate. The strong collective character of the low lying states in the lower mass region of the sd shell give reasons for the important role of inelastic processes followed and/or preceded by an alpha transfer.

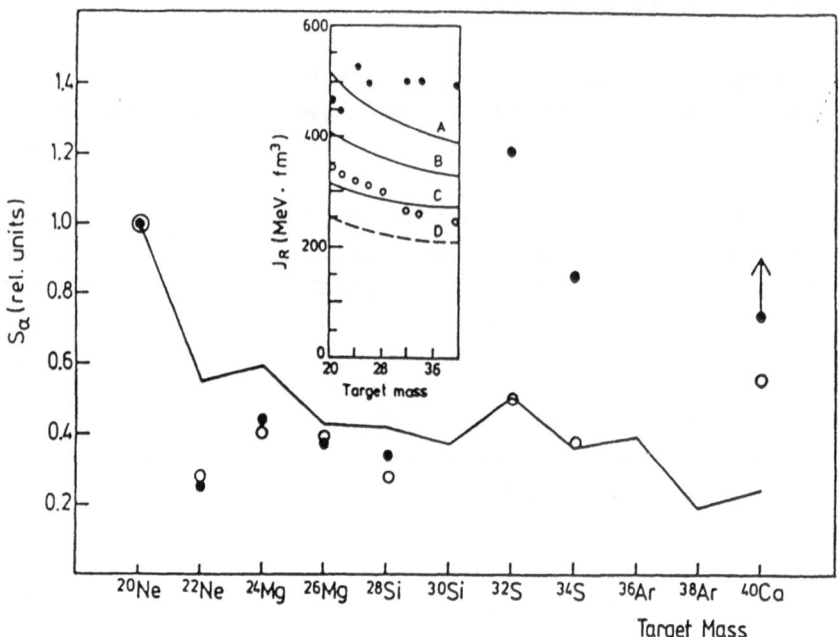

Fig. 11: Theoretical (solid line) and experimental spectroscopic
results. Open and full data points correspond to evaluations with
volume integrals of the [6]Li potential as given in the insert.

Theoretical alpha-spectroscopic amplitudes[6,40] as:

$$A_{NL} = \left(\frac{A}{A-4}\right)^{N+L/2} \cdot \sum \frac{\langle \Psi_A \| \Psi_\alpha(N,L) \| \Psi_{A-4}\rangle}{\sqrt{2J_A+1}\ \sqrt{2T_A+1}} \cdot \langle \Phi_\alpha(\xi)\phi_{NL}(R_\alpha)|\ \Psi_{SM}(\zeta_\alpha)\rangle \qquad (6)$$

were calculated between ground state[35] and excited target and residual nuclear states with the method described by Bennett[40] from complete sd shell-model wave functions generated by the Chung-Wildenthal
empirical particle-hole interactions[41] using the computer code of
Chung[41]. Details of the coupled channel calculations involving the
calculated spectroscopic factors with minor variations of their magnitude and adjustment of phases are given elsewhere[42]. The results are
presented in Figs. 12 and 13 showing angular distributions for
transitions leading to the lowest three members of the ground state
band in [20]Ne and [22]Ne, compared to the theoretical predictions in the
coupled reaction channel (CRC) framework. The difference in shape between both displayed experimental angular distributions to final 4[+]
states and the quality of description by the calculation is remarkable, demonstrating the unlike interplay between individual contributions in populating the final 4[+] states in these nuclei. Besides the
quality of the agreement between experimental and theoretical angular

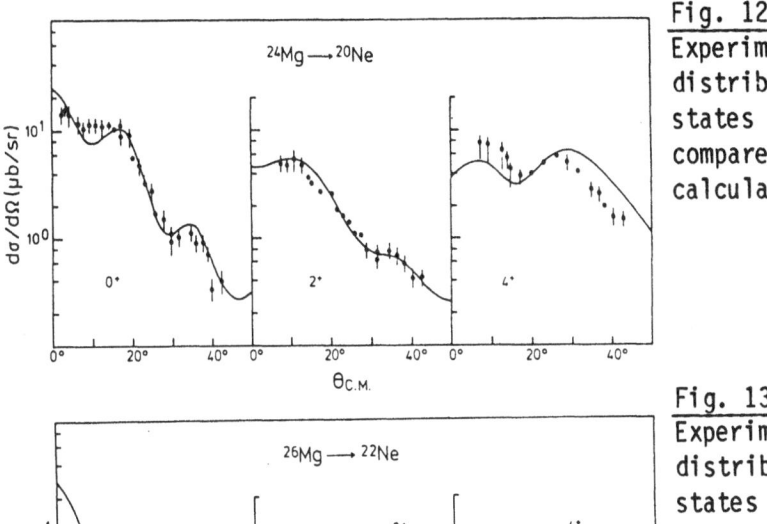

Fig. 12:
Experimental angular
distributions to
states in $^{20}$Ne,
compared to CRC
calculations.

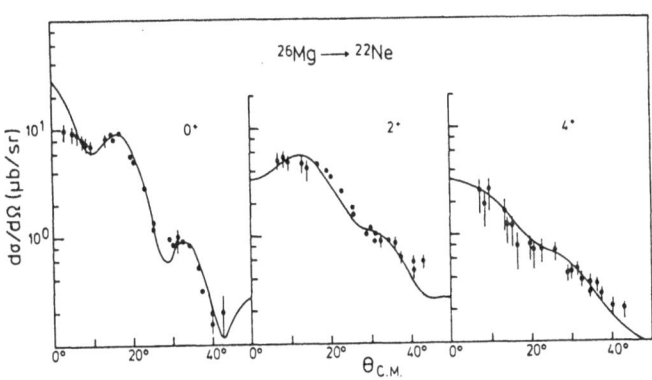

Fig. 13:
Experimental angular
distributions to
states in $^{22}$Ne,
compared to CRC
calculations.

distribution shapes the prediction of absolute cross section (using
only a zero range normalization constant) is satisfactory. Phases of
spectroscopic amplitudes are not meaningful as long as a single transition is regarded, only via the combination of the matrix elements
the relative phases become significant. It is not understood why the
CRC calculations seem to demand some phase changes relative to the
ones predicted by the shell model calculations, as discussed in
Ref. 42). Further indications of the CRC formalism to be an important
part of the reaction mechanism is demonstrated[43] in Fig. 14, comparing theoretical CRC results to experimental angular distributions
leading to the 3$^-$ and to unnatural parity state members of
the $K^\pi = 2^-$ band. Other reaction mechanisms as compound nucleus
contributions appear to be of no importance.

## 6. SUMMARY

Some topics of alpha transfer reaction investigations have been discussed without demanding for completeness. An impressive agreement
between reduced width extracted from reactions and decay has been observed. Different types of alpha-transfer reactions have been listed,
with the emphasis on the light ion stripping and pickup reactions and
features of GQR decay. Phenomenological models appear to give a reaso-

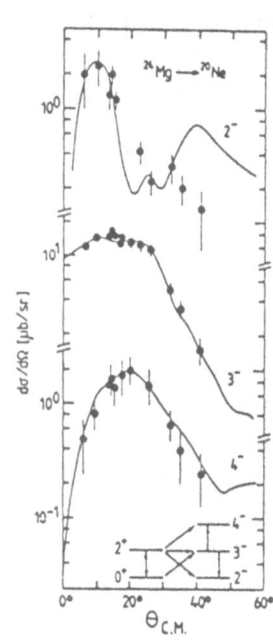

Fig. 14:
Experimental angular distributions to negative parity states in $^{20}$Ne, compared to CRC calculations.


nable account for spectroscopic quantities but unfortunately they are rather limited in application
to certain nuclear mass ranges or specific excited
states. In the lower part of the sd shell the
coupled reaction channel formalism does describe
the reaction mechanism and has to be used in the
analysis.

I would like to thank Mrs. A. Dohmen for her patience in typing the
manusript.


References

1) J. Jänecke and F.D. Becchetti, Proceedings of the Nuclear Physics
   Workshop, I.C.T.P., Trieste, Miramare, Italy, 5-30 October 1981,
   p. 253. Edited by C.H. Dasso, R.A. Broglia, A. Winther
2) L.J. Denes and W.W. Daehnick, Phys. Rev. 154, 928 (1966)
3) A. Arima, AIP Conference Proceedings No. 47, Clustering Aspects
   of Nuclear Structure and Nuclear Reactions, Winnipeg 1978, 1.
   Edited by W.T.H. van Oers, J.P. Svenne, J.S.C. McKee, W.R. Falk
4) J.P. Schiffer, Proceedings of the International Conference on
   Nuclear Physics, Munich, August 27 - September 1, 1973, Volume 2,
   p. 813. Edited by J. de Boer and H.J. Mang
5) A. Arima et al., Phys. Rev. Lett. 34, 24 (1971)
6) M. Ichimura et al., Nucl. Phys. A204, 225 (1973)
7) J.P. Elliott, Proc. Roy. Soc. A245, 128 and 1562 (1958);
   J.P. Elliott and M. Harvey, Proc. Roy. Soc. A272, 557 (1963);
   M. Harvey, Adv. in Nucl. Phys., 67 (1968)
8) A. Bohr, Nuclear Structure I.A.E.A. Vienna, 1968, p. 179;
   O. Nathan, Nuclear Structure I.A.E.A. Vienna, 1968, p. 191

9)  A. Arima and F. Iachello, Ann. Phys. 99, 253 (1976);
    A. Arima and F. Iachello, Phys. Rev. C16, 2085 (1977)
10) J. Jänecke et al., Nucl. Phys. A325, 337 (1979)
11) J. Jänecke et al., Nucl. Phys. A343, 161 (1980)
12) W.G. Davies et al., Nucl. Phys. A269, 477 (1976)
13) F.D. Becchetti et al., Phys. Rev. C19, 1775 (1979)
14) J. Jänecke et al., Phys. Rev. C23, 101 (1981)
15) K. Bethge, Ann. Ref. Nucl. Sci. 20, 255 (1970)
16) H.W. Fulbright, Ann. Rev. Nucl. Sci. 29, 161 (1979)
17) M.C. Mallet-Lemaire, AIP Conf. Proc. No. 47, see Ref. 3) p. 271
18) A.A. Ogloblin, Soviet J. Part. Nucl. 3, 467 (1973)
19) F.D. Becchetti and J. Jänecke, Proc. of the Nucl. Phys. Workshop,
    see Ref. 1) p. 235
20) F.D. Becchetti, AIP Conf. Proc. No. 47, see Ref. 3) p.308
21) A.M. van den Berg et al., Nucl. Phys. A422, 45 (1984)
22) J. Jänecke et al., contribution to this conference
23) A. van der Woude and R.J. de Meijer, Nucl. Phys. A258, 195 (1976)
24) P. Turek et al., Ann. Rep. 1976, KFA-IKP 10/77 (1977) 4
25) C. Sükösd et al., to be published in Nucl. Phys. and references
    therein
26) K.T. Knöpfle, Lecture Notes in Physics 109, 311 (1979)
27) A. Faessler et al., Nucl. Phys. A330, 333 (1979)
28) R.R. Betts, Proc. of the Nucl. Phys. Workshop, see Ref. 1), p.225
29) R.A. Broglia et al., Phys. Lett. 79B, 352 (1978)
30) R.A. Broglia et al., Nucl. Phys. A375, 217 (1982)
31) C.L. Bennett and H.W. Fulbright, Phys. Rev. C17, 2225 (1978)
32) F. Iachello and A.D. Jackson, Phys. Lett. 108B, 151 (1982)
33) T. Fliessbach and P. Manakos, Nucl. Phys. A324, 173 (1979);
    T. Fliessbach, Phys. Rev. C21, 919 (1980)
34) H. Sato, Proc. of the 1983 RCNP Symposium, Osaka 1983, p. 751.
    Edited by H. Ogata, T. Kammuri, and I. Katayama
35) W. Chung et al., Phys. Lett. 79B, 381 (1978)
36) P. Lezoch et al., Phys. Rev. C23, 2763 (1981)
37) W. Oelert et al., Phys. Rev. C28, 73 (1983) and references there-
    in
38) J.C. Vermeulen et al., Nucl. Phys. A262, 189 (1981)
39) W. Oelert et al., Phys. Rev. C22, 408 (1980)
40) C.L. Bennett, Nucl. Phys. A284, 301 (1977)
41) W. Chung, private communication
42) W. Oelert et al., to be published
43) G. Palla and W. Oelert, Ann. Rep., IKP-KFA, Jül-Spez 255, 7
    (1984),ISSN 0170-8937 and to be published

# THE STUDY OF CLUSTER STATES IN LIGHT NUCLEI
# VIA HEAVY-ION-ALPHA CORRELATION MEASUREMENTS

W. D. M.  Rae
Nuclear Physics Laboratory
University of Oxford
Keble Road, Oxford OX1 3RH
England

ABSTRACT.  Heavy-Ion-α correlation studies following direct heavy-ion reactions can yield information on both the spins of the states excited and the reaction mechanism.  Recent experimental correlation data are reviewed and the advantages of the different experimental arrangements compared.  It is shown that spin information is contained both in the correlation function itself and in the variation of the correlation function with the primary reaction angle.  The extraction of detailed information on the reaction mechanism is discussed.

## 1.  INTRODUCTION

Heavy-ion (HI) induced cluster transfer reactions have the potential of being a very powerful tool in nuclear spectroscopy.  What seems to be lacking is spin information and a better understanding of the reaction mechanism.  Particle-γ or particle-particle correlation studies can yield information both on the spin of the states excited and on the reaction mechanism itself.

The most common correlation measurements to have been made following direct heavy-ion induced reactions involve states which α-decay and the most common reactions to have been studied with this technique are α-transfer reactions.  In this talk I shall review these experiments.  I shall discuss the extraction of spin and reaction mechanism information.  In particular I shall describe a new technique for spin determination and emphasize the relationship between the semi-classical models and the DWBA.  Finally I shall show that coincidence techniques can be used to advantage to study a wide range of reactions including inelastic scattering.

## 2.  EXPERIMENTAL CONSIDERATIONS

For simplicity I shall discuss only α-decay, although all of the considerations will apply to any particle-particle correlation studies. I will not consider particle-γ studies.

*J. S. Lilley and M. A. Nagarajan (eds.), Clustering Aspects of Nuclear Structure, 261–275.*
© *1985 by D. Reidel Publishing Company.*

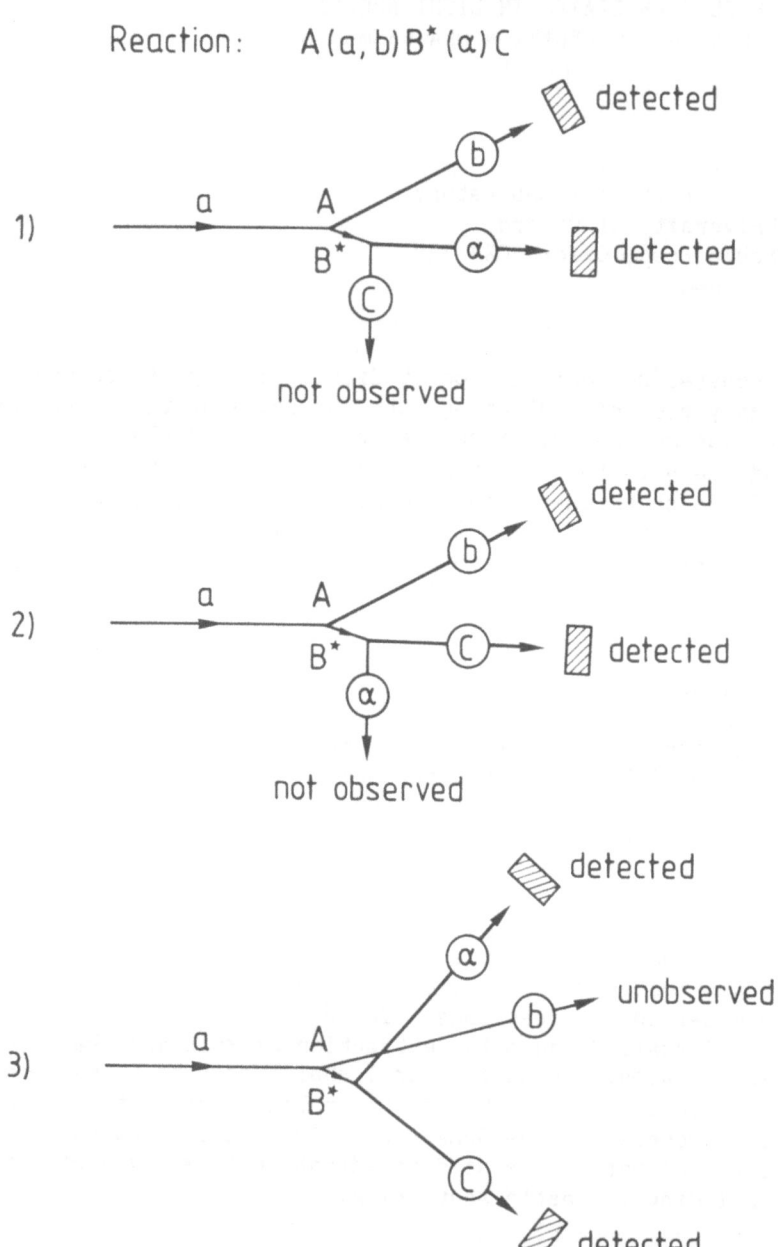

Experimental configurations

Reaction:    $A(a, b)B^*(\alpha)C$

1)

detected
detected
not observed

2)

detected
detected
not observed

3)

detected
unobserved
detected

Figure 1    Illustration of the three possible experimental
configurations.

Particle-particle correlation measurements involve a 3-body final state. If we denote the reaction by

$$A(a,b) \; B^* \; (\alpha) \; C$$

the three final state particles are b, C and the $\alpha$. Conservation of energy and momentum allow the reconstruction of all the kinematical variables from the energies and angles of any two of the three final particles. Thus there are three possible experimental configurations illustrated in fig.1:-

1) particles b and $\alpha$ detected
2)     "     b " C     "
3)     "     $\alpha$ " C     "

The most obvious approach is option 1). The measurement of b alone gives the inclusive cross-section, while the measurement of the $\alpha$ particles in coincidence is generally easy and yields the decay correlation directly. Since the $\alpha$ particles are generally emitted over $4\pi$ the efficiency is not enhanced by any focussing effects.

Option 2) is less common. The (heavier) recoils C are more difficult to detect; they generally have lower energies so there will be both angular scatter in the target and possibly significant energy loss and straggle. Since it is unlikely that any form of particle identification will be possible, ambiguities will exist if more than one decay channel is open. A significant advantage of this configuration is the efficiency gain, since the nuclei C are confined to a cone around the recoil direction - the direction of $B^*$.

Option 3) appears to have little to offer, as it suffers from both the problems associated with option 2), but also it is most likely that it would be impossible to identify the primary reaction uniquely, since neither decay product has enough energy to allow standard identification techniques to be used. However, by inverting the reaction option 3) can be made very attractive and has some important advantages. The reaction now reads

$$a \; (A,B^*) \; b$$

where $B^*$ decays into C and $\alpha$, both of which are now moving rapidly in the laboratory and so can be detected easily and identified using the same experimental configurations as are used in projectile breakup studies. In this configuration the excitation energy of $B^*$ must be calculated using geometry and good energy resolution requires good angular resolution for each fragment.

This technique has several advantages:-

1) A high efficiency from forward focussing of the decay products.
2) Easy identification of both products and hence the primary reaction and decay channel are unambiguously identified.
3) Access to very forward angles; $B^*$ may be produced at 0° but decay products can be detected away from 0°.
4) The possibility of very high resolution for the excitation spectrum of $B^*$ with the use of high-resolution PSD's.

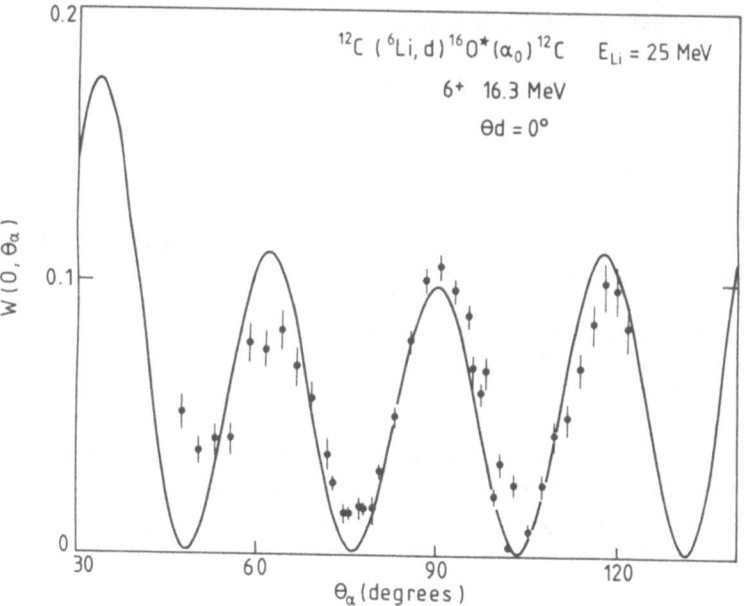

Figure 2  d-α angular correlation for the $^{16}O$ 16.3 MeV state of $^{12}C$, E $^6$Li-25 MeV, $θ_d=0°$.  From Cunsolo et al, ref.3).

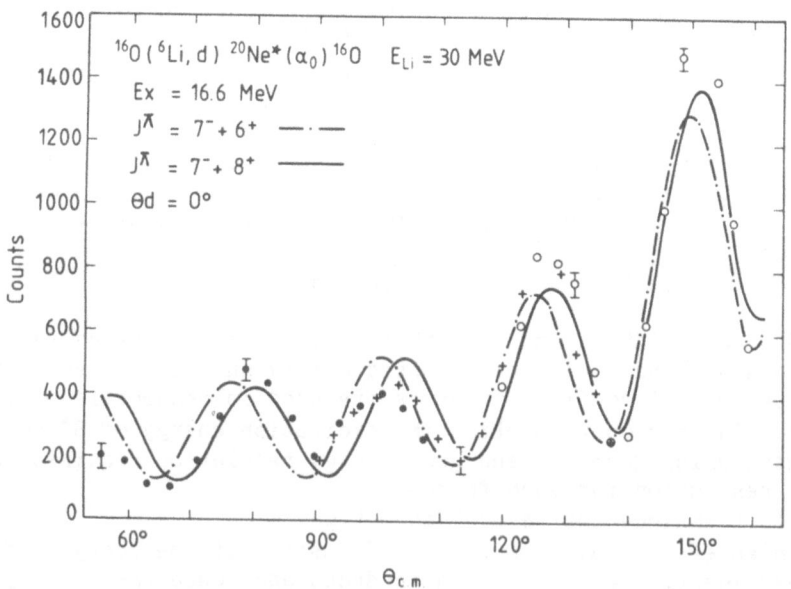

Figure 3  Angular correlation of α particles from the decay of the group at 16.6 MeV in $^{20}Ne$.  From Fou et al, ref.2).

A disadvantage of this approach is that the inclusive cross-section is not obtained simultaneously.  Also the data analysis is more involved and cannot generally be performed on-line.

Each of these experimental configurations has been used in particle-particle correlation studies and in the next section I review some of the experiments, their motivation and the results.


## 3.   REVIEW OF SOME EXPERIMENTAL RESULTS

Often the primary motivation of α correlation measurements is spin determination.  In such cases it is preferable to have the minimum model dependence.  For the special case of spin zero nuclei in the entrance and exit channels (except for the spin of the excited state of B*), the method of Litherland and Ferguson[1] is attractive.  This method requires B* and b to be travelling toward 0° or 180° in the laboratory.  This is experimentally a serious disadvantage but theoretically the results are easily interpreted.  The decay correlation for a single state of spin J should simply be

$$|P_J(\cos\psi)|^2$$

where $\psi$ is the angle of the relative velocity between the α and C with respect to the beam axis.  The detection of particles near 0° is difficult.  The solid angle available at exactly 0° is infinitesimally small:  the average value of $\theta^*$ for a detector at 0° will be some small but finite value $\theta_s^*$.  So long as other m-states can be neglected at $\theta_s^*$ the correlation will still be approximately $|P_J(\cos\psi)|^2$ but a model may be required to estimate the possible contribution from other m-states.

Measurements at 0° have been made by the Penn and Catania/Saclay groups[2,3] for ($^6$Li,d) reactions, where the spin of the $^6$Li is assumed to be equal to the spin of the deuteron which acts like a spectator – a model-dependent assumption.  Some experimental results from both groups are shown in figs.2 and 3.  Spins and parities can easily be extracted from these data, the Penn data also yielding some information on the mixing of overlapping resonances.

The Yale group used the ($^{12}$C,$^8$Be) α-transfer reaction[4] in a very beautiful experiment which takes advantage of the relatively much simpler task of detecting $^8$Be at 0° (cf the breakup technique discussed under option 3 above).  In this experiment the beam can be dumped on a small beam stop yet the α-particles from the decay of the $^8$Be can pass around this object.

Results from the Yale experiment are shown in fig.4.  Some shifting of a few of the correlations was observed, viz.  they required a correlation function of the form

$$|P_J(\cos[\psi+\psi_0])|^2$$

where $\psi_0$ is some small angle.  However in general the correlations were very well fitted by the square of a Legendre polynomial, and definite spin assignments could be made for states populated in both $^{16}$O and $^{20}$Ne.

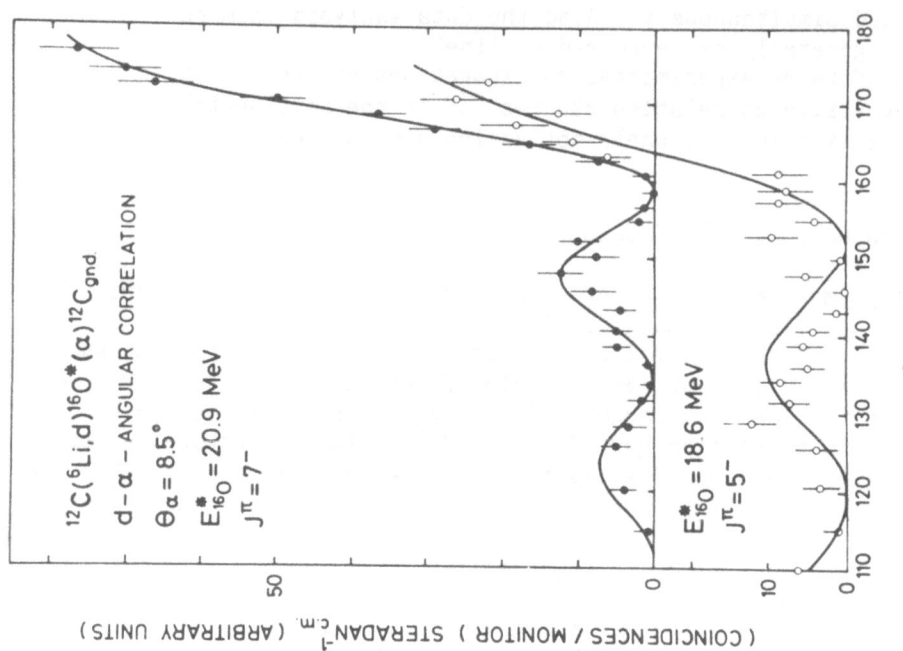

Figure 5 The d–α correlation functions for the reaction $^{12}C(^6Li,d)^{16}O^*(\alpha)^{12}C$ for the $^{16}O$ levels at 20.9 (7−) and 18.6 (5−). From Artemov et al, ref.9).

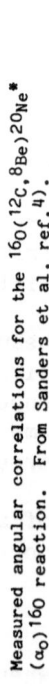

Figure 4 Measured angular correlations for the $^{16}O(^{12}C,^8Be)^{20}Ne^*$ $(\alpha_0)^{16}O$ reaction. From Sanders et al, ref.4).

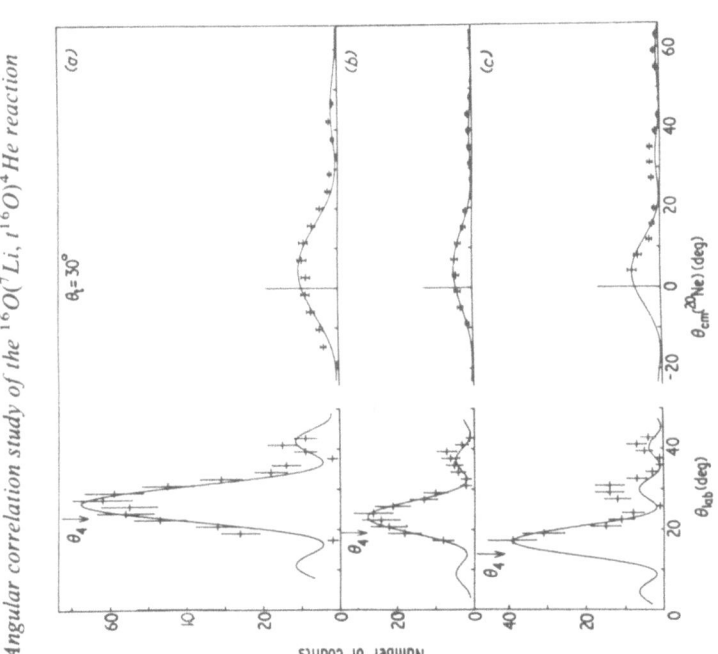

**Figure 6** (t, $^{16}$O) angular correlation for the reaction $^{16}$O($^{7}$Li, t $^{16}$O)$^{4}$He. From Panagiotou et al, ref.5).

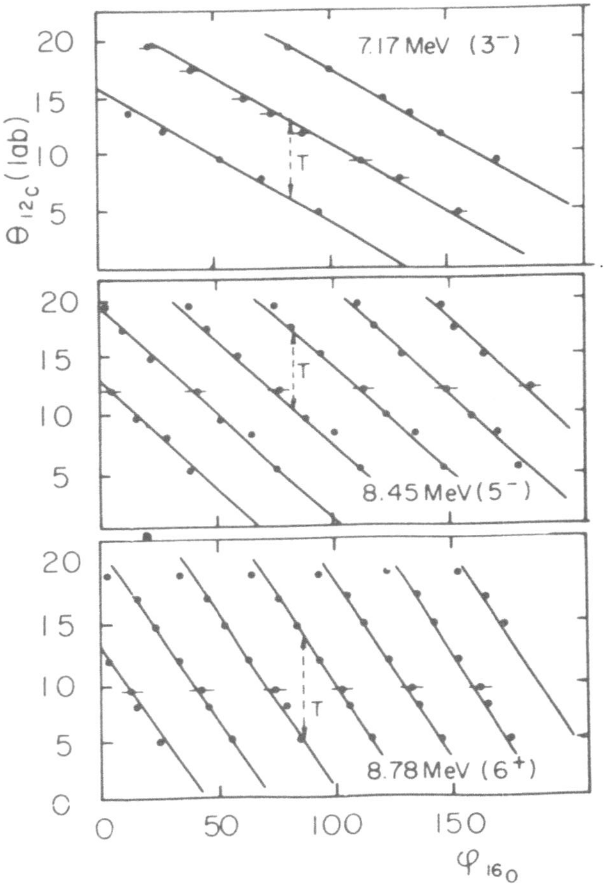

**Figure 7** The position of the minima in the correlation functions for different values of 12C lab angles. From Pougheon et al, ref.11).

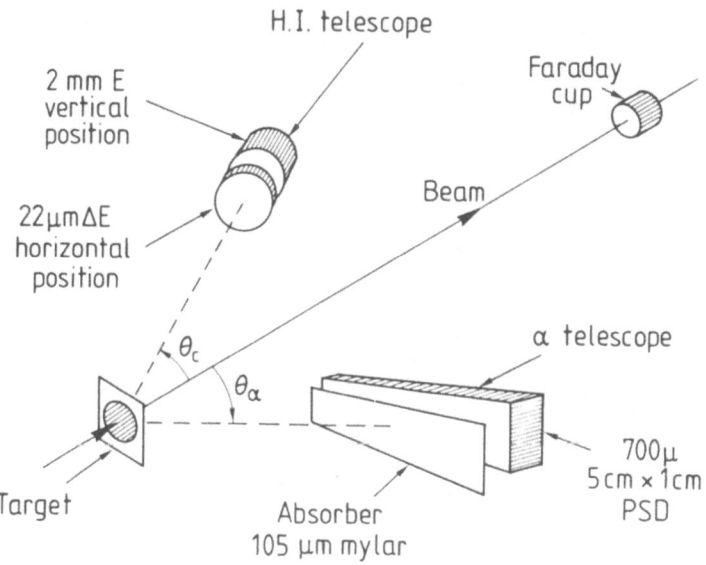

**Figure 8**  The breakup detector configuration used by the Oxford group ref.[15].

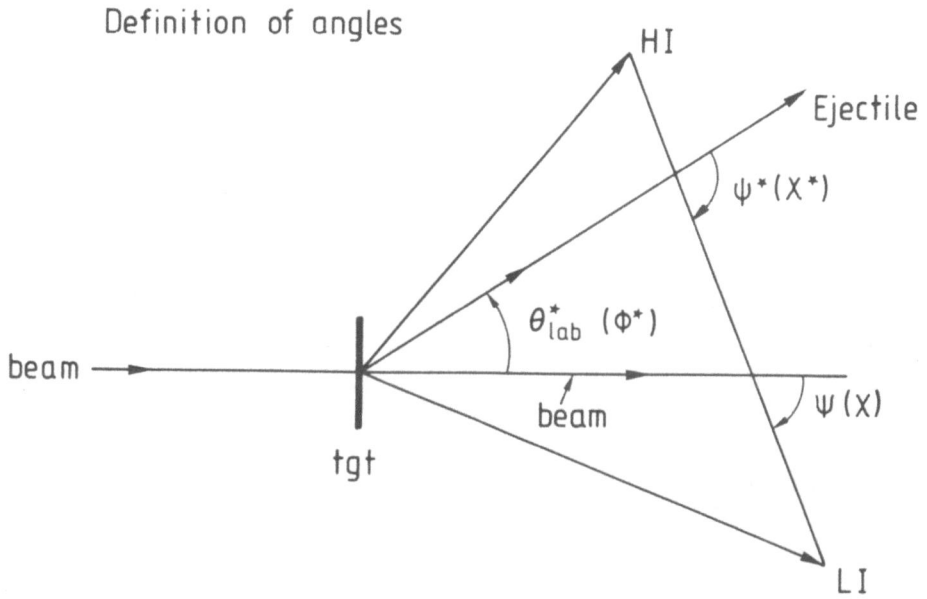

**Figure 9**  The co-ordinate system used by the Oxford group ref.[15].

It is more common to measure correlations for ejectile angles other than $\theta^* = 0°$. (The centre of mass angle of the primary reaction will always be denoted by $\theta^*$). In principle such measurements also contain information on the reaction mechanism. Several measurements have been reported for Li-induced $\alpha$-transfer reactions. At forward angles, where these measurements have been made the correlations still show the $|P_J(\cos\psi)|^2$ appearance but often significantly shifted with respect to the beam axis. This shift has been interpreted as a signature of the reaction mechanism[3,5]. The Catania-Saclay collaboration has shown that this shift is generally well predicted by DWBA calculations[6,7].

For most of the Li-induced reactions $(d,\alpha)$ or $(t,\alpha)$ coincidences have been measured. Such measurements have been made by many different groups[6-10]. An example from the work of Artemov et al[9] is shown in fig.5. Earlier the Oxford group had studied the reaction $^{16}O(^7Li,t^{16}O)\alpha$ by measuring $(t,^{16}O)$ coincidences[5] viz. option 2). This allowed them to measure the correlations very close to the recoil direction and make a detailed study of the shift of the correlations, fig.6.

Experimental configuration 2) was also employed by the Orsay group who performed one of the first systematic measurements of the evolution of the decay correlations for a wide range of primary reaction angles $\theta^*$ [11]. The reaction studied was $^{16}O(^{16}O,^{12}C)^{20}Ne$. The $^{12}C$ was detected in a doubly focussing magnetic spectrometer, and the recoiling $^{16}O$ from the decay of the $^{20}Ne$ was detected in coincidence in a 5cm PSD. The PSD was carefully positioned so that the whole decay correlation could be measured with one setting. This illustrates the advantage of detecting the heavier decay product. Decay correlations were obtained for several $^{12}C$ angles. If the positions of the minima in the correlations are plotted as a function of $\theta^*$ a rapid linear shift is observed (see fig.7). The spacing between these ridges in the $\theta^*$ direction is found to be independent of the spin of the state and characteristic of the spacing expected from a diffraction model. This implies that the slopes of these ridges (or the rate of shift) must depend on the spin of the state.

These detailed angular correlations allowed model-independent extraction of the m-state populations, which can be compared with various reaction models. The square of the amplitudes with $m=J$, $|p_{\pm J}^J|^2$, and the alignment, $|p_{\pm J}^J|^2 + |p_{-J}^J|^2$, were compared with DWBA calculations. The DWBA appeared to reproduce the alignment but not the polarization, which was extremely strong.

More recently the Oxford group[12] have studied the reaction $^{12}C(^{16}O,^{20}Ne^*)^8Be$ using the breakup detector configuration illustrated in fig.8, where the decay products of the $^{20}Ne$ were detected. Using PSD's this yields continuous double differential cross-sections in both the centre of mass angle $\theta^*$ and the decay angle $\psi$ (see fig.9). In some cases $\theta^* = 0°$ can be reached although the $\psi$ range is always limited. These data, illustrated in fig.10, again show the rapid shifting of the decay correlations with $\theta^*$ and the strong diffractive features alluded to above. The interpretation of these shifts and the extraction of reaction model information from the data is discussed in the next sections.

Similar results for the reaction $^{12}C(^{12}C,^{16}O^*)^8Be$ are reported in

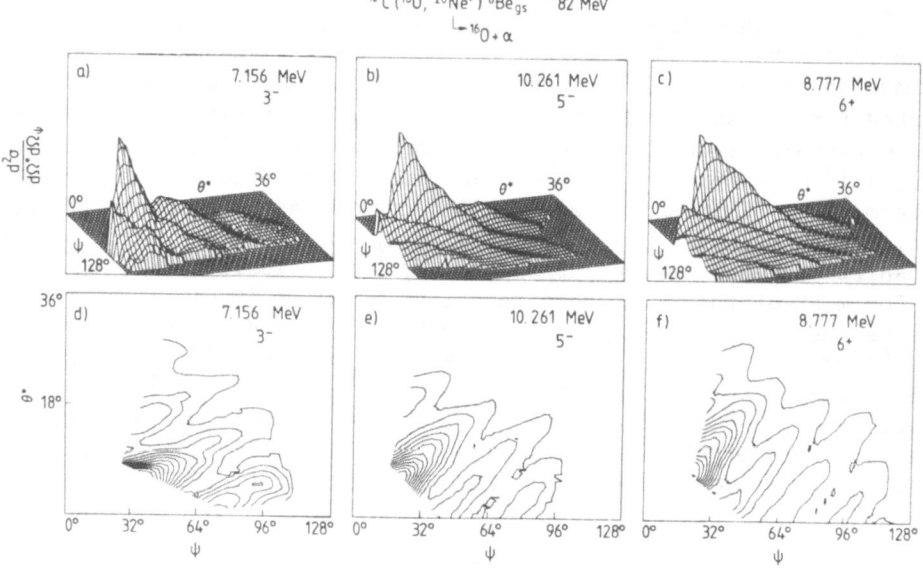

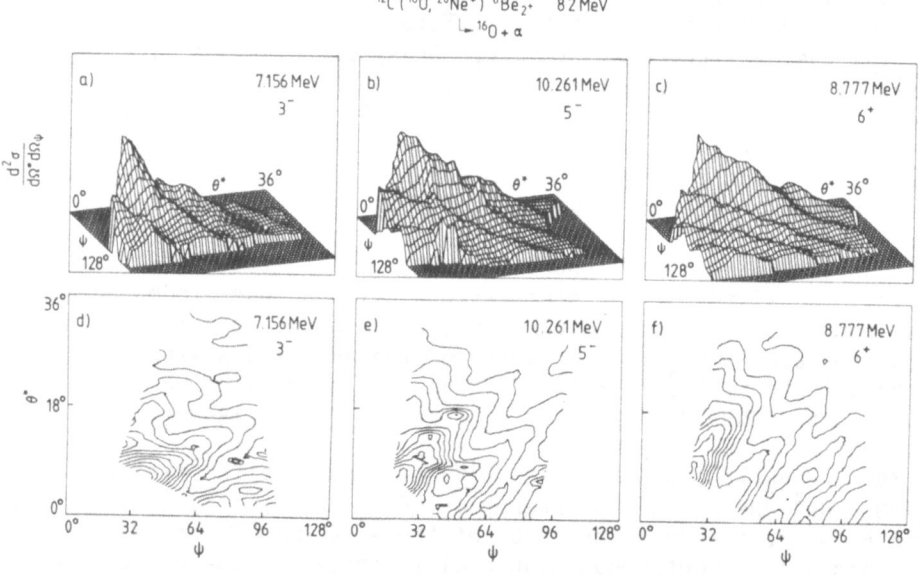

Figure 10 Double differential cross-sections for the reaction $^{12}C(^{16}O, ^{20}Ne^*)^8Be_{gs}$ at 82 MeV (top) and for the reaction $^{12}C(^{16}O, ^{20}Ne^*)^8Be_{2^+}$ at 82 MeV (bottom) leaving $^8Be$ in its 2+ state.

contributions to this conference by both the Oxford group (S.C.Allcock et al) and a group from Osaka University (S.Shimoura et al).


## 4.  THE DEPENDENCE OF THE CORRELATIONS ON THE REACTION ANGLE

It has been known for many years from the Li induced reactions that for $\theta^* \neq 0$, the so called "symmetry axis" of the correlation shifts away from the beam direction, which is clearly a symmetry axis for $\theta^* = 0°$. In a plane wave model (PWBA) a true symmetry axis does exist for all reaction angles $\theta^*$ and this axis coincides with the recoil direction - the direction of momentum transfer. This is also the prediction of a semi-classical spectator model where the transferred particle can carry no z-component of angular momentum along its direction of motion - again the recoil direction. However, for the Li-induced $\alpha$-transfer reactions, although the data show that an approximate symmetry axis may exist, this never coincides exactly with the recoil direction, and is generally shifted by several degrees[5]. These shifts are predicted by the distorted wave theories (DWBA) and thus have been attributed to distortion[6,7].

Da Silveira[13] proposed a semi-classical explanation of this shift for a situation in which there is strong absorption so that only a few partial waves contribute to the reaction, fig.11. In this model the change in direction of the spin of the excited nucleus can be calculated as a function of $\theta^*$ by conservation of total angular momentum. This explanation has been revived by the Oxford group who have found that it can explain the shift in their HI-$\alpha$ correlations following a heavy ion inelastic scattering reaction[14]. For negative Q-values it is easy to show that J will rotate with $\theta^*$ s.t. the shift in the correlation will be $\ell_g \theta^*/J$ where $\ell_g$ is the grazing partial wave. This is qualitatively consistent with the results of both the Oxford and Orsay groups. Comparison of Da Silveira's model with the PWBA result shows that there is a strong correspondence between the two. Indeed the model of Da Silveira can be obtained from the PWBA result by introducing a radius vector and hence converting the linear momenta to angular momenta.

The experimental results of the Oxford group[14,15] show that for a heavy ion reaction the slope of the ridges in their data - produced by the correlation pattern shifting with $\theta^*$ - is proportional to J, the spin of the state excited. The results from the Li-induced reactions are not consistent with either model, suggesting that the explanation of these data may lie somewhere between the two extreme models. This may be as a result of the weaker absorption in the (light-ion) exit channel in these reactions.


## 5.  REACTION MECHANISM INFORMATION

The results of the Orsay group[11] show that these correlation measurements are a very sensitive test of reaction models. However comparison of polarization and alignment with predictions of models is rather indirect and does not clearly indicate which features of the model may be inadequate.

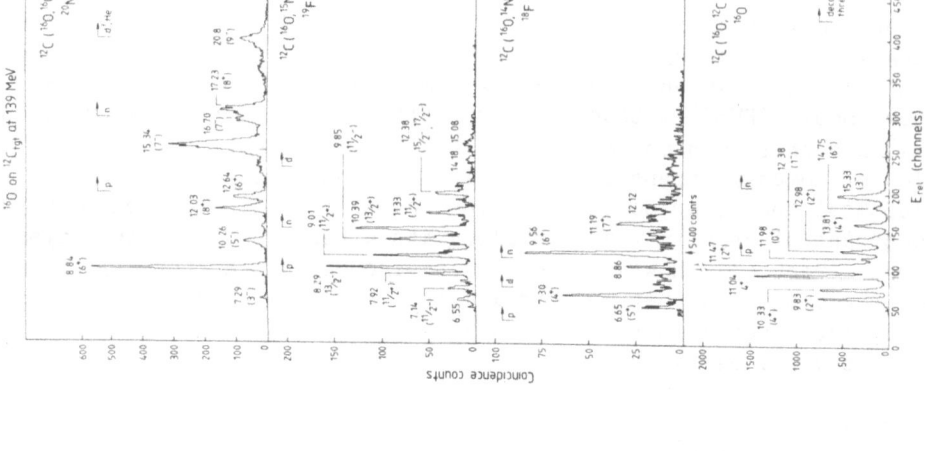

Figure 12 Reaction induced by $^{16}O$ on a $^{12}C$ target at 139 MeV in which the $\alpha$-decay of the ejectile has been measured, ref.23).

# Da Silveira's model

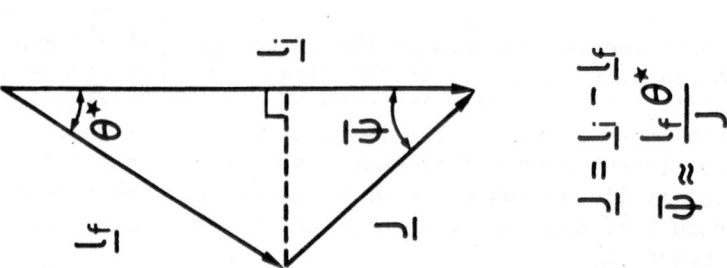

$$\underline{J} = \underline{l}_i - \underline{l}_f$$

$$\bar{\psi} \approx \frac{\underline{l}_f}{\underline{J}} \theta^\star$$

Figure 11 An illustration of Da Silveira's model which is derived from angular momentum conservation.

The Oxford group have illustrated that the angular correlations are sensitive to the details of the transition amplitudes in a very direct way[16]. Their approach is related to the work of the Brookhaven group[17,18] but follows more directly from the formalisms used in Strong Absorption Models and the Closed Formalisms, although these approximations need not be made[19,20,21].

The transition amplitude for a heavy ion reaction may be written quite generally (neglecting spin-orbit effects etc.) as[17]

$$\beta_{LM} = \frac{(4\pi)^{3/2}}{k_i k_f} \sum_{\ell_i \ell_f} i^{\ell_i - \ell_f - L} \langle \ell_f LM - M | \ell_i 0 \rangle \, I^L_{\ell_i \ell_f} \, Y^{\ell_f}_M (\theta^*, 0)$$

To obtain insight into the relation between the partial wave matrix elements $I^L_{\ell_i \ell_f}$ and the angular correlations it is only necessary to make a semi-classical approximation for the Clebsch-Gordan coefficient in terms of a rotation matrix[20], viz.

$$\langle \ell_f LM - M | \ell_i 0 \rangle \approx d^L_{-MK} (\tfrac{1}{2}\pi) \quad , \quad K = \ell_i - \ell_f.$$

This is a very good approximation for the grazing partial waves which are involved in heavy ion reactions. Introducing a change of variables to

$$K = \ell_i - \ell_f, \quad \bar{\ell} = (\ell_i + \ell_f)/2$$

we can write the transition amplitude as

$$\beta_{LM} = \frac{(4\pi)^{3/2}}{k_i k_f} \sum_{K, \bar{\ell}} i^{K - L} d^L_{-MK}(\tfrac{1}{2}\pi) \, I^L_{K\bar{\ell}} \, Y^{\ell_f}_M (\theta^*, 0)$$

To allow the physical interpretation of the transition amplitude in this form it is useful to consider angles $\theta^*$ s.t. $\theta^* > \sqrt{2} \, M/\ell_f$. Here the asymptotic expressions for the $Y^{\ell_f}_M$ can be used, and if finally a rotation is made through $\pi/2$ to a z-axis perpendicular to the reaction plane the transition amplitude becomes[19]

$$\beta_{LM=K} = \frac{(4\pi)^{\frac{1}{2}}}{k_i k_f} \frac{2}{(\sin\theta^*)^{\frac{1}{2}}} i^{K-L} \left\{ \sum_{\bar{\ell}} I^L_{K\bar{\ell}} \, e^{i[(\bar{\ell} - \frac{1}{2}K + \frac{1}{2})\theta^* - \pi/4]} \right.$$
$$\left. + (-)^K I^L_{-K\bar{\ell}} \, e^{-i[(\bar{\ell} - \frac{1}{2}K + \frac{1}{2})\theta^* - \pi/4]} \right\}$$

It can be seen that the amplitude has split into near- and far-side components $(e^{\pm i\ell\theta^*})$. But more importantly the m dependence of the transition amplitude (in this co-ordinate system) is directly related to the dependence of the partial wave amplitudes on $\ell_i - \ell_f$. This result depends only on the semi-classical interpretation of the Clebsch-Gordan coefficient. This provides a very direct physical interpretation of the partial wave matrix elements.

The diffraction observed in the experimental data is controlled by the near-side far-side interference, the overall characteristics being determined by the dependence of the partial wave matrix elements on the average partial wave $\bar{\ell} = (\ell_i + \ell_f)/2$, cf in the simpler case of elastic scattering. Further approximations[21] enable the K-dependence of the partial wave matrix elements to be associated with the semi-classical transfer amplitudes of Brink[22]. Thus the K-dependence should reveal all the features predicted by the Brink model viz. the Q-value and angular momentum matching.

Using this co-ordinate system with the z-axis perpendicular to the reaction plane the decay correlations are given by[11]

$$W(\theta^*,\psi) = \frac{2L+1}{4\pi} \left| \sum_K \beta_{JK}(\theta^*) \, (-)^K \, e^{-iK\psi} \, d^L_{K0} \, (\pi/2) \right|^2$$

Thus the $\psi$ dependence of the correlations directly reflects the $K=\ell_i-\ell_f$ dependence of partial wave amplitudes $I^L_{K\bar{\ell}}$.

The Oxford group have exploited this direct connection between the partial wave amplitudes and the decay correlations by parametrizing the amplitudes and using closed expressions to evaluate the sums over the average partial wave quantum no., $\bar{\ell} = (\ell_i + \ell_f)/2$. In principle it is possible to determine a great deal about the nature of the reaction amplitudes by fitting such correlation data. The techniques have become quite sophisticated for elastic scattering where only a one dimensional function, $S_\ell$, has to be determined. Here, the double differential cross-sections allow the partial wave t-matrix to be determined as a function of both its variables $\bar{\ell}$ and K or $\ell_i$ and $\ell_f$.

## 6.  SUMMARY

Particle-particle correlation measurements can provide detailed information on the spins of states excited in direct heavy ion reactions. These correlations also contain information on the reaction mechanism since they depend directly and sensitively on the t-matrix elements. To fully exploit the information available detailed correlations are required at many reaction angles $\theta^*$ and over a wide range of decay angles $\psi$. Measurements at one reaction angle would often be unable to distinguish between various reaction models (cf elastic scattering at one angle).

Most of the correlation data available so far is for Li-induced $\alpha$-transfer reactions. Some data is available for $^{12}C$- and $^{16}O$-induced $\alpha$-transfer reactions and some heavy-ion induced inelastic scattering. However data from the Berkeley group[23] show that HI-$\alpha$ coincidence data can be used to study a wide range of transfer reactions. Not only would the double differential cross-sections yield information on the spins and the reaction mechanism, but the study of a specific decay channel enhances the selectivity of these reactions when many decay channels are open, fig.12. This provides additional information on the structure of

the states which have been excited.  The high resolution attainable with
the breakup detector configuration should also be exploited.
    I'd hope that we shall see much more correlation data in the future
since there is great potential for enhancing our understanding of
heavy-ion induced cluster transfer reactions.

## REFERENCES

1)    A.E.Litherland and A.J.Ferguson, *Can.J.Phys.* $\underline{39}$ (1961) 788

2)    Cheng-Ming Fou et al, *Phys.Rev.* $\underline{C20}$ (1979) 1754

3)    A.Cunsolo et al, *Nuovo Cimento* $\underline{40A}$ (1977) 293

4)    S.J.Sanders et al, *Phys.Rev.* $\underline{C20}$ (1979) 1743

5)    A.D.Panagiotou et al, *J.Phys.* $\underline{A7}$ (1974) 1748

6)    A.Cunsolo et al, *Phys.Rev.* $\underline{C21}$ (1980) 2345

7)    A.Cunsolo et al, *Phys.Lett.* $\underline{112B}$ (1982) 121

8)    K.P.Artemov et al, *Phys.Lett.* $\underline{37B}$ (1971) 61

9)    K.P.Artemov et al, *Nucl.Phys.* $\underline{A320}$ (1979) 479

10)  M.C.Etchegoyen et al, *Nucl.Phys.* $\underline{A402}$ (1983) 87

11)  F.Pougheon et al, *Nucl.Phys.* $\underline{A325}$ (1979) 481

12)  S.C.Allcock et al, IOP Conference, Bradford (1984) Contributed
      Paper.

13)  E.F.Da Silveira, Proc.of the 14th Winter Meeting on Nuclear
      Physics, Bormio, 1976 (unpublished)

14)  R.K.Bhowmik, W.D.M.Rae and B.R.Fulton, *Phys.Lett.* $\underline{136B}$ (1984) 149

15)  W.D.M.Rae and R.K.Bhowmik, *Nucl.Phys.* $\underline{A420}$ (1984) 320

16)  A.E.Smith et al, *Nucl.Phys.A*, to be published

17)  S.Kahana et al, *Phys.Lett.* $\underline{50B}$ (1974) 199

18)  P.D.Bond, *Phys.Rev.* $\underline{C22}$ (1980) 1539

19)  F.J.W.Hahne, *Nucl.Phys.* $\underline{A104}$ (1967) 545

20)  W.E.Frahn, *Phys.Rev.* $\underline{C21}$ (1980) 1870

21)  H.Hasan and D.M.Brink, *J.Phys.* $\underline{G5}$ (1979) 771

22)  D.M.Brink, *Phys.Lett.* $\underline{40B}$ (1972) 37

23)  W.D.M.Rae et al, *Phys.Rev.C*, to be published.

# CLUSTER PRODUCTION IN MEDIUM-HIGH ENERGY REACTIONS USING HADRONS AND ELECTROMAGNETIC PROBES

CHAPTER PRODUCTION IN MEDIUM HIGH ENERGY REACTIONS
INDUCED BY PROTONS AND ELECTROMAGNETIC PROBES

# CLUSTERING ASPECTS OF MEDIUM AND HIGH ENERGY REACTIONS WITH LIGHT IONS

Philip G. Roos
Department of Physics and Astronomy
University of Maryland
College Park, Maryland 20742 U.S.A.

ABSTRACT. Measurements of clustering in nuclei using a variety of medium to high energy nuclear reactions are discussed. Particular emphasis is placed on quasifree knockout reactions with light ions. The discussion includes experimental tests of theoretical models, the sensitivity of the reaction to clustering for various probes and energies, polarization effects in proton-induced reactions, and the relationship to transfer reactions. In addition to light-ion projectiles the use of pions and electrons is considered.

## 1. INTRODUCTION

With relatively few new experiments and theoretical calculations oriented toward cluster structure since the last conference, there have been few, if any, significant advances in our quantitative knowledge of the importance of clustering in nuclei. In experimental physics the lack of advancement comes in part from the lack of new experimental equipment to improve both the quantity and quality of the experimental data. This may be alleviated in the near future, at least for three-body final state reactions, with the construction of dual-spectrometer systems for medium energy accelerators. From a theoretical standpoint most of the problems mentioned at the previous conference remain unsolved. Unfortunately this fact probably indicates a lack of interest in the subject by a large fraction of the theorists.

Because of the paucity of new results, part of this paper will use older results which were available at the previous conference. I will discuss only a few reactions (particularly quasifree knockout) which I believe have the highest probability of providing quantitative information. I will further limit the discussion to deuteron and α clustering, and will bring out some of the major problems with present day technology, both experimental and theoretical.

Some of the questions one would like to answer with our studies are the following: What is the probability of finding a cluster in a particular nucleus? What is the wave function for that cluster? Is the clustering significantly different from that predicted by nuclear

279

*J. S. Lilley and M. A. Nagarajan (eds.), Clustering Aspects of Nuclear Structure, 279–293.*
© 1985 by D. Reidel Publishing Company.

structure models, such as the shell model? Does clustering have a strong influence on hadron-nucleus reactions? To begin to answer these questions quantitatively we need to study reactions in which the initial and final nuclear states are simply connected by the one-step removal of a cluster. Such studies already lead to difficult reaction dynamics problems, without considering more complicated processes. The reaction measurements should also have the highest degree of exclusivity.

## 2.  ALPHA CLUSTERING

Many experiments have been devised to study $\alpha$-clustering in nuclei. In the early days of nuclear physics the high probability for ejection of $\alpha$ particles from the nucleus in inclusive hadron-nucleus reactions was viewed as evidence for the importance of clustering, but these data were unable to provide quantitative information. More recent studies of inclusive gamma-ray spectra from proton-nucleus[1] or pion-nucleus[2] reactions showed large yields for the removal of one, two, or more $\alpha$ particles.  In this case the effect is predominantly due to energetics, rather than to special clustering effects. Such inclusive studies do not contain sufficient detail to extract clustering information. One is therefore led to exclusive studies such as the quasifree knockout reactions. These I believe hold the most promise for obtaining quantitative information on $\alpha$-clustering in nuclei with medium energy reactions.

I will therefore review the present experimental and theoretical situation for quasifree $\alpha$-knockout. It is clear that these reactions are closely connected to alpha-transfer reactions at lower energies, so that I will discuss the similarities and differences between the two reactions.

Assuming an interaction with a preformed $\alpha$-cluster, the first order diagrams for $(p,p\alpha)$ and $(d,^6Li)$ suggest that both reactions measure identical cluster structure information. In first order the momentum of the $\alpha$-cluster before the collision is determined by the recoil momentum of the residual nucleus which is experimentally measured; thus the great hopes that such reactions will provide details of the cluster wave functions. Although distortion effects are crucial in the analysis of either reaction, we can use such simple models to compare the two types of reactions. In the case of the $(p,p\alpha)$ reaction low momentum clusters ($p_{recoil}=0$) can be observed independent of the bombarding energy. Thus one can choose a high bombarding energy to minimize the distortion and still study the cluster wave function starting at zero momentum. In the case of $(d,^6Li)$ reactions the momentum transfer is always rather large ($>250$ MeV/c at 50 MeV) and increases with bombarding energy. Thus, low momentum components of the cluster are sampled only via distortion effects, which makes the results quite sensitive to the distorting potentials. The knockout reaction because of its momentum matching capability and the use of the impulse approximation is more naturally a medium energy reaction.

Both types of reactions are analyzed using a distorted wave approximation. The knockout reaction studies invariably utilize the distorted wave impulse approximation (DWIA), which can be written[3] schematically as

$$\frac{d^3\sigma}{d\Omega_p d\Omega_\alpha dE_p} = KF \cdot \frac{d\sigma}{d\Omega}\bigg|_{p-\alpha} \cdot S_L^\alpha \cdot \sum_\Lambda |T_{L\Lambda}|^2, \qquad (1)$$

where KF is a known kinematic factor, $d\sigma/d\Omega$ is the half-off-shell p-$^4$He cross section, and $S_L^\alpha$ is the α-particle spectroscopic factor. T is the distorted momentum distribution

$$T_{L\Lambda} = (2L+1)^{-1/2} \int \chi_p^{(-)*}(\vec{r}) \chi_\alpha^{(-)*}(\vec{r}) \chi_p^{(+)}(\frac{B}{A}\vec{r}) \phi_{L\Lambda}(\vec{r}) d\vec{r}, \qquad (2)$$

where the χ's are distorted waves and $\phi_{L\Lambda}$ is the c.m. wave function of the α-cluster arising from an overlap integral of the four-particle nuclear wave function with a physical α particle.

Before presenting detailed (p,pα) results for cluster probabilities let me discuss some of the major difficulties with the analysis, ignoring any questions concerning the applicability of the distorted wave approximation.

(1) <u>The two-body cross section</u>. In general the half-shell cross section is replaced by a nearby on-shell cross section. Studies at 100 MeV using a p+$^4$He optical model potential[4] indicate that this approximation is rather good, assuming of course that the optical potential is a good description of the scattering. More serious perhaps is the factorization approximation of removing the p-$^4$He t-matrix from the full six-dimensional integral. There have been no attempts to study this approximation theoretically for cluster knockout — a very difficult computation. There have been some experimental tests, although they are limited to particularly favorable cases by count rate considerations. In Fig. 1 are shown the extracted half-shell cross sections (using Eq. (1)) versus the two-body p-$^4$He scattering angle for 100 MeV (p,pα) reactions on p-shell nuclei.[4] The data are for the zero recoil momentum point which is particularly favorable, since the yield tends to be concentrated at large radii where the particles more nearly have their asymptotic momenta. One observes that the agreement with p-$^4$He scattering is quite satisfactory, although close inspection of the data shows perhaps as much as a factor of two breakdown for $^{12}$C. Because of the higher binding energy for $^{12}$C a significant part of the yield may be coming from regions where the optical potentials are not small. In any case the accuracy of this approximation needs to be tested more fully, both theoretically and experimentally.

(2) <u>Other four-particle configurations</u>. The four-particle wave function in the nucleus contains many configurations in addition to those corresponding to the quantum numbers of an α-cluster. For example using a single particle oscillator basis for $^{40}$Ca(1d$_{3/2}$)$^4$ at most a few percent of the wave function corresponds to zero oscillator quanta for the internal motion. Are these other components important in knockout or transfer reactions? In the case of knockout one generally argues that such configurations correspond to excited states of

$^4$He so that (a) p+$^4$He inelastic scattering is required which should be significantly smaller than elastic scattering, and (b) the excited state cluster is more tightly bound in the nucleus and due to the strong absorption is concentrated in radial regions relatively unimportant in hadron-induced reactions. However, these are only hand-waving arguments.

One way to experimentally study such contributions is to look for transitions forbidden by direct α-particle knockout. Measurements of transitions to particular forbidden states will provide at least an upper limit on the importance of such configurations. It is an upper limit only since such transitions may also indicate a breakdown in the reaction mechanism. Again, presently available experimental facilities pretty much preclude such investigations.

A method which has been used is to compare quasifree elastic scattering with quasifree reactions. For example, assuming the (p,pα) and (p,d$^3$He) reactions both occur on an α-cluster and ignoring distortion effects, the ratio of the cross sections for the two reactions should be given by the ratio of the free two-body cross sections. For p-shell nuclei Cowley et al.[5] have found that the momentum distributions for the two reactions at 100 MeV are essentially the same. Furthermore, the extracted spectroscopic factors assuming the free two-body cross sections agree within experimental error. If other configurations were very important one might expect the results to differ significantly. Note that these data also provide confirmation of the factorization approximation.

Although not many open two-body reaction channels exist for light clusters, one can utilize elastic scattering as well by measuring the full angular distribution. For example, around 100 MeV the p+$^4$He elastic scattering rises at back angles because of triton exchange. It would be interesting to make measurements as in Fig. 1 over the full angular range to see if the exchange effects remain the same in quasifree scattering.

The understanding of the importance of such additional configurations requires new theoretical and experimental studies.

(3) <u>Cluster c.m. wave function</u>. Assuming knockout and transfer reactions to be dominated by the α-cluster component, one is still left with the treatment of the c.m. wave function of the cluster. Due to the distortion effects one does not obtain a direct measurement of the wave function as implied by the first order diagram. It is necessary to assume a radial wave function, carry out the DWIA calculation, and compare it to the experimental data. Standard practice in analysis of both knockout and transfer reactions is to use the wave function for an α-cluster bound in a Woods-Saxon potential whose depth is chosen to reproduce the α-particle separation energy. The quantum numbers of the cluster are based on the oscillator shell model, assuming all quanta of energy reside in the c.m. motion. The geometrical parameters of the well are chosen by the practitioner. The choices range from fixed geometrical parameters to attempts to include the finite size of the α particle either by modifying the radius parameter to $r(A^{1/3}+4^{1/3})$ or by carrying out a folding model calculation.[4,6,7] Unfortunately the DWIA calculations are very sensitive to the radius

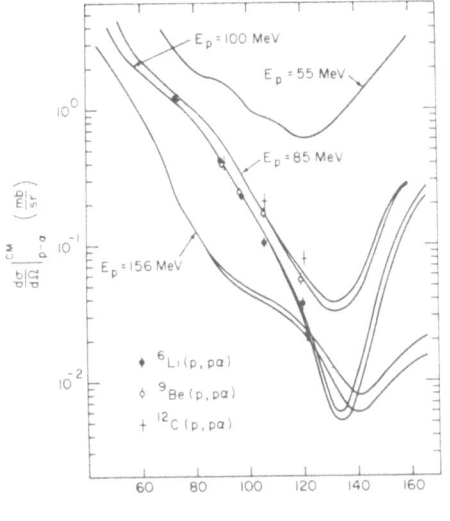

Fig. 1. Angular distribution for
(p,pα) at 100 MeV.[4]  The curves
represent free p+$^4$He scattering.

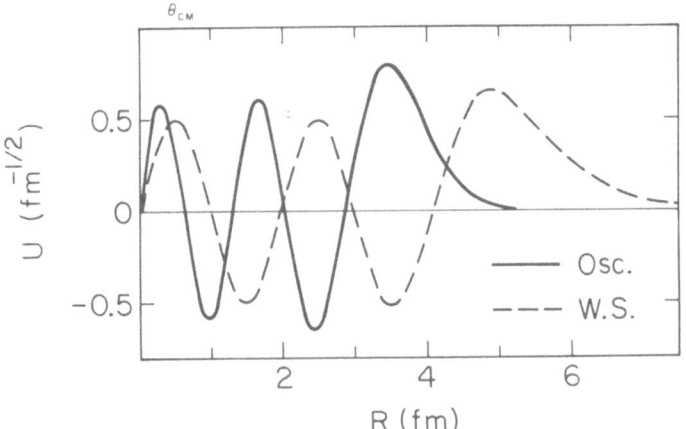

Fig. 2. Bound state wave functions for $^{40}$Ca for harmonic oscillator
         and Woods-Saxon potentials.

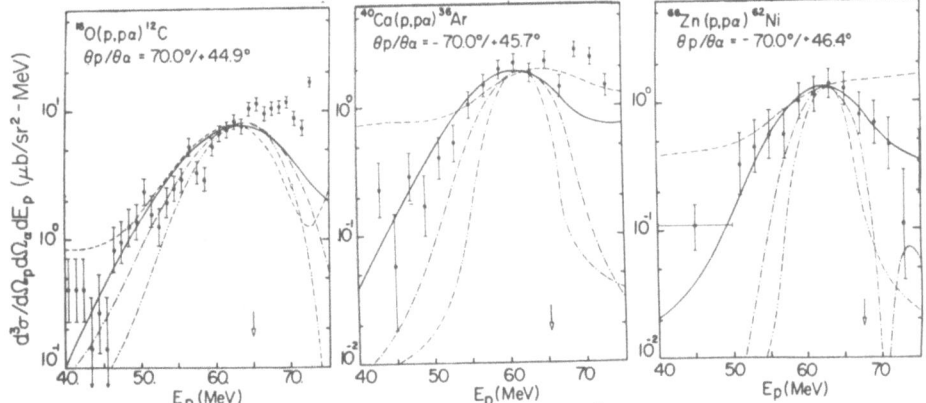

Fig. 3. (p,pα) energy sharing distributions.[7]  The curves are DWIA cal-
        culations with various bound state $r_0$ (---, .7; ——, 1.3;
        •-•-, 1.9; --•••, 2.5).

parameter. For example, for 100 MeV (p,pα) a modest 15% change in the radius parameter leads to almost a factor of three change in the cross section, and therefore a factor of three change in the extracted spectroscopic factor. I believe this to be the most severe limitation in present (p,pα) analyses, and is a problem which requires theoretical input. As an example of the uncertainties which arise suppose we consider a $(2s_{1/2})^4$ configuration in $^{40}$Ca. If we take an oscillator shell model representation, the transformation to relative and c.m. coordinates is straightforward using Moshinsky brackets.[8] Projecting out the zero oscillator quanta internal motion we have a 5S c.m. wave function. Plotted in Fig. 2 is a 5S oscillator cluster wave function using a single particle size parameter appropriate to $^{40}$Ca. Although this method is simple it leads to an asymptotic behavior which is unacceptable. The reactions are localized to the surface, and therefore it is essential to have the tail of the wave function well-described. Since we expect that at large radii the wave function should correspond to a binding energy tail, the use of a potential well to bind the α seems to be a reasonable alternative. For comparison shown in the figure is a typical Woods-Saxon wave function used in (p,pα) analyses. Clearly the two wave functions have quite different properties, and will therefore lead to different spectroscopic factors. Since the oscillator wave function does not have a proper asymptotic tail, one could "improve" the calculation using Woods-Saxon single particle wave functions and the transformation techniques developed for two-nucleon transfer. However one is left with the question of how to handle the approximately 28 MeV internal binding energy of the α particle.

The whole treatment of the bound state is a serious limitation in both knockout and transfer reactions for the extraction of quantitative spectroscopic information on α-clustering in nuclei. Various prescriptions employed by experimentalists can lead to factors of 5 to 10 differences in absolute spectroscopic factors for medium mass nuclei. Until much more theoretical guidance is provided such uncertainties will overide our ability to test for clustering phenomena beyond that predicted by shell model calculations.

Now that I have discussed some of the major problems which need attention, let me discuss some experimental data oriented toward the extraction of spectroscopic information. Studies of 100 MeV (p,pα) reactions on p-shell nuclei[4] show that DWIA calculations fit the data quite well, except for the high proton energy region where sequential α-decay processes are present. The extracted spectroscopic factors compare favorably with theoretical predictions. For $^9$Be and $^{12}$C the experimental values agree very well with shell model predictions,[9] suggesting that these nuclei contain no significant clustering beyond that contained in the shell model. The nuclei $^7$Li and 6Li both show evidence of enhanced cluster structure. The cluster structure in these cases is particularly favored because of the small α-particle binding energy.

For heavier nuclei in the (2s-1d) and (1f-2p) shells (p,pα) measurements have been carried out by Carey et al.[7] which overlap ($^6$Li,d) studies.[10] Shown in Fig. 3 are typical energy sharing distributions for three nuclei. Again DWIA calculations with $r_0$=1.3 do quite well in predicting the general shape. It is obvious that the statistics leave a lot to be desired, but such spectra require about 48 hours of accelerator time with conventional detector techniques. It is essentially impossible to obtain this kind of time at the few accelerators capable of carrying out this work. Thus it is crucial, if this type of work is to be continued, that experimental facilities be developed which can improve counting rates by roughly a factor of 100.

Using the DWIA calculations with $r_0$=1.3 the spectroscopic factors shown in Fig. 4 were extracted. Also plotted are the ($^6$Li,d) relative spectroscopic factors.[10] One observes that the general features for the two reactions agree very well and show maxima at closed shell nuclei. A reanalysis[11] of the ($^6$Li,d) data for heavier nuclei with different optical potentials produces the crosses, and shows the strong sensitivity of the reaction to the optical potentials. The (p,p$\alpha$) reaction is not as sensitive, and rather extreme variations in the optical potentials lead to only about 30% changes in the spectro-scopic factors. Another reanalysis[12] of these same ($^6$Li,d) data further depresses the spectroscopic factors for the heavier masses, but is inconsistent with the (p,p$\alpha$) data. Overall the agreement is encouraging and implies that both reactions at least measure a quantity related to the $\alpha$-cluster spectroscopic factor.

The right hand scale of Fig. 4 shows the absolute spectroscopic factors extracted from the (p,p$\alpha$) reaction, subject to the limitations of the choice of bound state wave function discussed previously. A comparison of the extracted spectroscopic factors to various theoretical predictions shows that for the lower part of the 2s-1d shell agreement is reasonably good. However, calculations limited to 2s-1d shell nucleons greatly underpredict the upper part of the major shell. These results show the importance of configuration mixing near $^{40}$Ca which contributes coherently to the reactions.

Finally we turn to (p,p$\alpha$) transitions to excited states. Carey[7] was able to measure transitions to a few well-separated excited states, but with very poor statistics, since the cross sections are about ten times smaller than the ground state transition. DWIA calculations are only moderately successful in reproducing the experimental data. Furthermore, the extracted spectroscopic factors are not in particularly good agreement either with theory or with transfer measurements. These results may indicate a problem with the DWIA treatment when attempting to predict cross sections an order of magnitude smaller than ground state transitions. Again, however, one needs significantly improved experimental data before reaching any specific conclusions.

Before discussing studies of the bound state, I should mention that there have also been studies of the ($\alpha$,2$\alpha$) reaction. These results have been discussed by Chant[13] at the previous conference, and there are several papers contributed to this meeting. These studies raise very interesting questions concerning either reaction dynamics or a high degree of clustering in the extreme low density region of the nucleus. Due to space limitations, I will not discuss them in this paper.

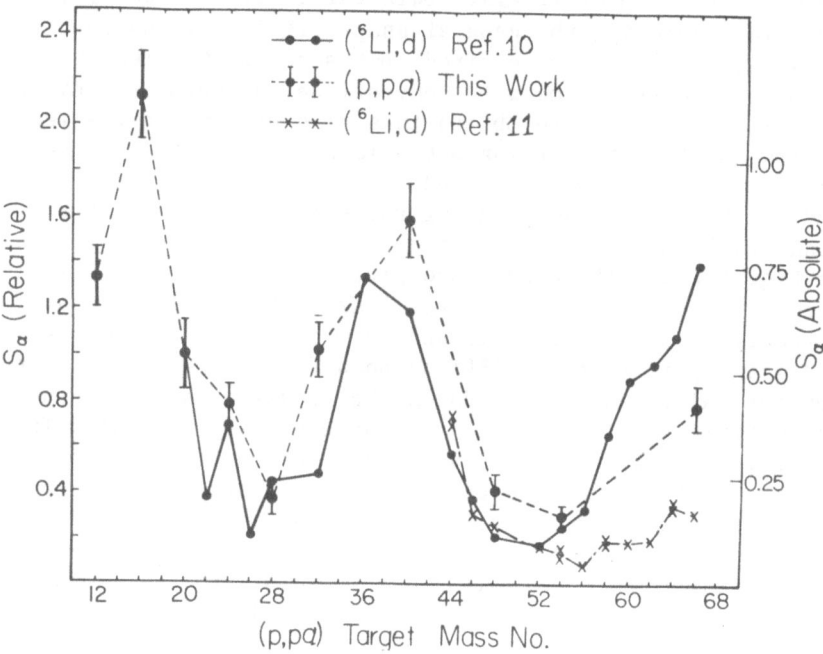

Fig. 4. Ground state α spectroscopic factors.[7]

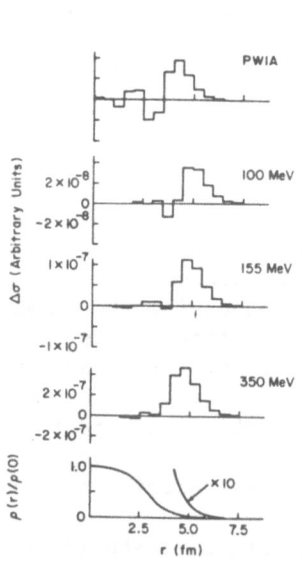

Fig. 5. Change in DWIA
$^{40}$Ca(p,pα) cross section vs.
cutoff radius for several
energies. Also shown is the
charge density.

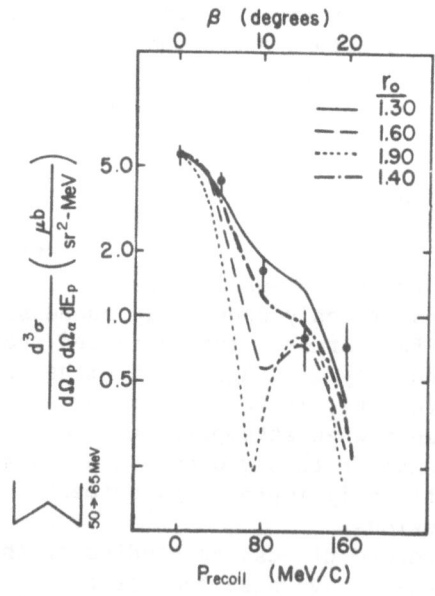

Fig. 6. Out-of-plane angular distribution for $^{40}$Ca(p,pα)$^{36}$Ar(g.s.)
at 100 MeV.[14] The curves are DWIA
calculations with different bound
state radii.

Experimentally there are ways of improving our knowledge of the bound cluster wave function. The shape of the energy sharing distribution is, in fact, somewhat sensitive to the radius parameter. In Fig. 3 are shown energy sharing distributions for 100 MeV (p,pα). The curves are DWIA calculations with different Woods-Saxon bound state radius parameters. We see that the width of the distribution provides a limit on this parameter (albeit loose because of the statistics) especially for the heavier nuclei. This sensitivity increases with bombarding energy as has been shown by Chant.[13]

Although such results place a limit on the model, they do not provide details of the wave function. To illustrate this in Fig. 5 are plotted the differences in the DWIA cross sections as a function of lower radial cutoff for several bombarding energies. These calculations illustrate the degree of surface localization for the (p,pα) reaction and show that one never measures the details of the wave function in the nuclear interior, only the asymptotic behavior of the clustering probability.

Another method which provides additional sensitivity to the bound state wave function is to measure the knockout cross section as a function of the out-of-plane angle for one of the particles. By measuring the cross sections for fixed energy of one particle, one creates a situation in which the p-$^4$He cross section and the distortions are nearly constant, while the recoil momentum varies. Such a measurement for $^{40}$Ca at 100 MeV[14] is shown in Fig. 6, along with DWIA calculations for different bound state radius parameters. This technique gives quite good sensitivity, but improved measurements are needed.

A quite interesting case, although not specifically α knockout, is the recent study of Warner et al.[15] of the $^6$Li(p,pd) reaction. Figure 7 shows the (p,pd) cross section as a function of out-of-plane angle for 120 MeV. The data show a clear minimum, as does the DWIA calculation, due to the node is the 2s cluster wave function. Due to distortion this minimum does not occur for the in-plane energy sharing distribution, even at 600 MeV.[16] I hope this is the first clear measurement of the location of the node in the $^6$Li wave function. However, further measurements at 200 MeV do not show a pronounced minimum. In spite of this the technique remains as a good method of controlling distortions effects in experimental studies.

## 2.1. New Types of Studies

Medium energy accelerators with intense polarized proton beams provide an opportunity for new studies of the spin observables in knockout reactions . Recently measurements of cross sections and analyzing powers for the $^9$Be(p,pα) reaction at 150 MeV have been carried out by Wang et al.[17] In Fig. 8 are presented the analyzing power data as a function of the two-body p-$^4$He scattering angle compared to free p-$^4$He scattering. The agreement is extremely good, although the magnitude at the peak is about 15% smaller. (This disagreement may be partially due to errors in the two-body data.[18]) These analyzing power data are again confirmation of the factorization approximation, but with the

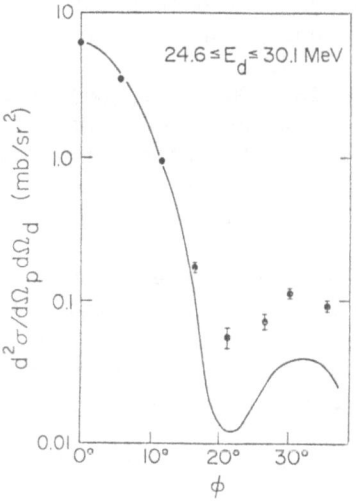

Fig. 7. Out-of-plane angular distribution for $^6$Li(p,pd)$^4$He at 120
MeV[15] and DWIA calculation.

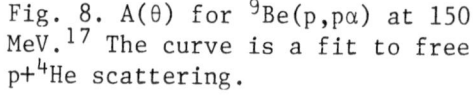

Fig. 8. A(θ) for $^9$Be(p,pα) at 150
MeV.[17] The curve is a fit to free
p+$^4$He scattering.

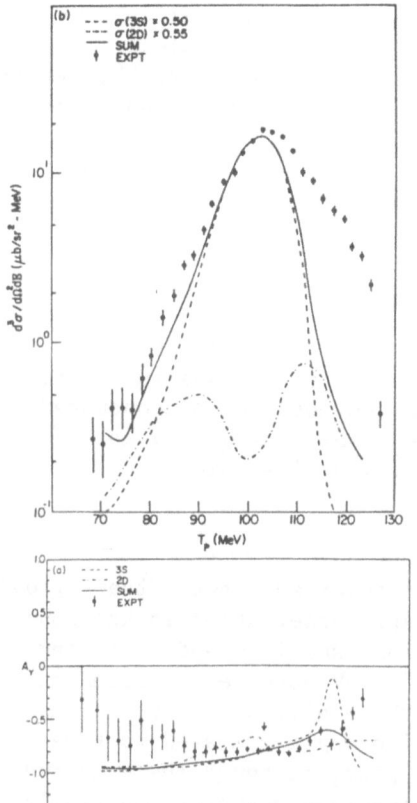

Fig. 9. Energy sharing distribution
for $^9$Be(p,pα) at 150 MeV.[17] The
curves are DWIA calculations.

same provisos expressed previously. Figure 9 shows the energy sharing distribution for one angle pair. Unfortunately, $^9$Be has contributions from both L=0 and L=2 and as one goes away from the quasifree peak, the L=2 contribution dominates. Overall the agreement is fairly good, although at low proton energies the measured analyzing power becomes quite small whereas the prediction remains approximately constant. This result may indicate the onset of more complicated multi-step processes in which the analyzing power is lost. More studies of this type, particularly for pure L transitions, will be very useful and should provide information on the accuracy of the DWIA direct knockout treatment.

Distortion effects play a major role in quasifree knockout reactions. One would like to know whether it is possible to reduce these signigicantly, either by increasing the bombarding energy of by using other projectiles. To look at this question I have carried out DWIA calculations of $^{40}$Ca(p,pα) at 800 MeV and $^{40}$Ca(e,eα) at 1 GeV. In both cases the outgoing α particle will dominate the distortion effects so that one would like to transfer significant energy to the α. On the other hand the two-body cross sections decrease with increased momentum transfer. I have therefore rather arbitrarily chosen the kinematics so that the outgoing α particles have approximately 120 MeV at the peak. Figure 10 shows the distorted momentum distributions for both reactions compared to the plane wave case. In both cases distortion is significant, broadening the distribution and removing the nodes. The reduction in magnitude at the peak is comparable. Clearly the benefits from the increased penetrability of the electron are not great, the strong absorption of the α particle limiting the reaction to the nuclear surface. The peak cross section for (p,pα) is approximately one $\mu$b/sr$^2$-MeV, whereas that for (e,eα) is approximately $10^{-11}$ $\mu$b/sr$^2$-MeV. Considering the benefits in terms of distortion it seems likely that the proton-induced knockout is the superior experiment.

## 3.  DEUTERON CLUSTERING

The discussion of deuteron clustering in nuclei is more concerned with the investigation of n-p correlations in nuclei. Particular types of reactions emphasize different types of correlations. To study these correlations we can use the same types of reactions discussed before. Assuming direct pickup of a pair of nucleons, (d,α) reactions populate states in which the n-p pair are removed from a relative T=0,S=1 (i.e., d) state. (p,$^3$He) reactions can proceed by the removal of either a T=0,S=1 pair or a T=1,S=0 pair. These reactions have been studied for many years and the DWBA combined with the shell model gives reasonable agreement for many states -- at least in terms of relative yields. In these cases enhanced correlations indicating pronounced clustering would manifest themselves primarily as an absolute magnitude effect, which is difficult to predict. As in the case of α-removal, transfer reactions at higher energies have large momentum transfer. Therefore in the plane wave limit the removed pair would be in a state with high c.m. momentum.

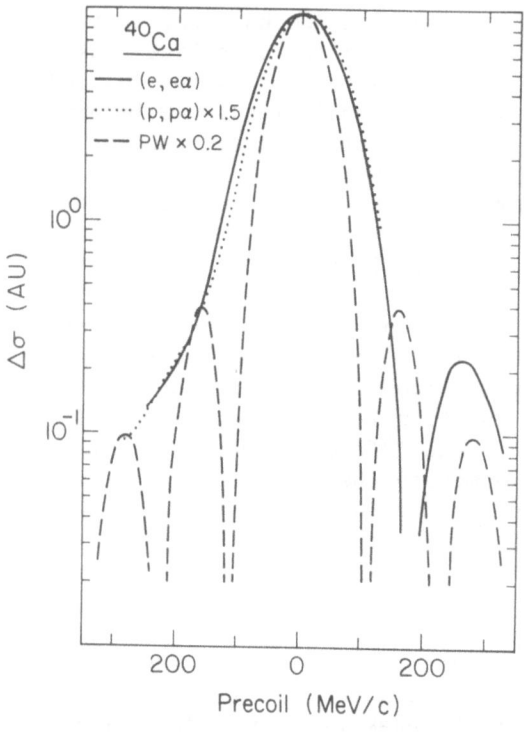

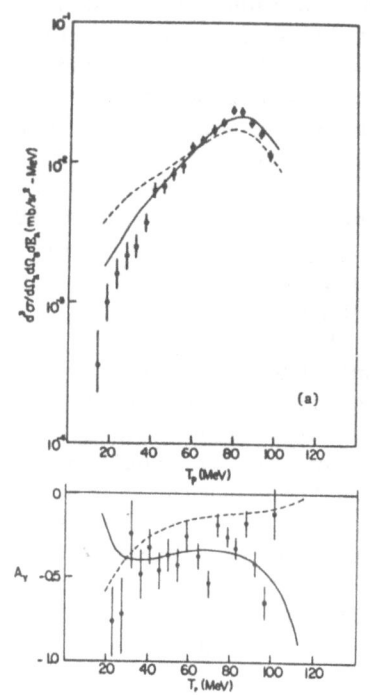

Fig. 10. Calculated distorted and undistorted momentum distributions for
1 GeV (e,eα) and .8 GeV (p,pα) vs.
recoil momentum.

Fig. 11. Energy sharing distribution for $^{12}$C(p,pd)$^{10}$B
(g.s.+.717) at 150 MeV.[20]
The curves are DWIA calculations.

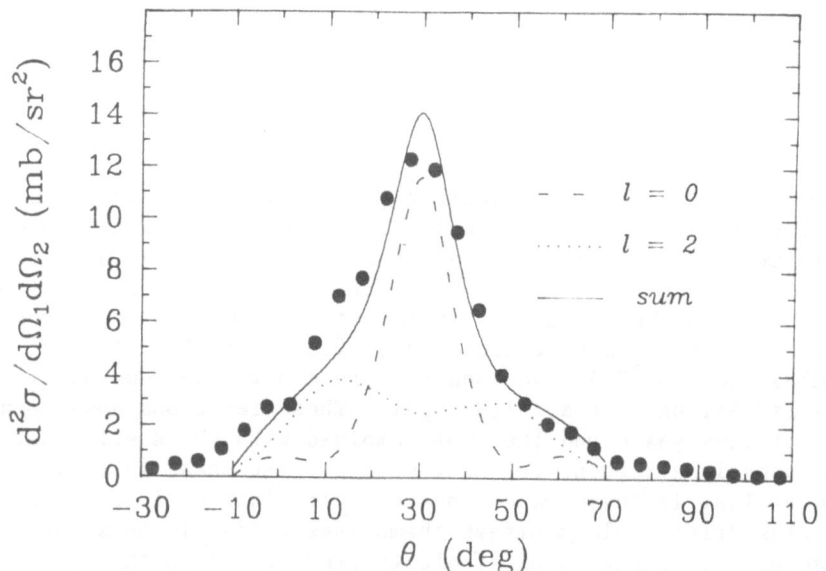

Fig. 12. Angular correlation for the $^{12}$C($\pi^+$,2p) reaction.[22] The curves
are DWIA quasideuteron calculations.[23]

As before, one can consider knockout reactions which can sample the low momentum components of the c.m. motion of the pair. In a (p,pd) reaction one again expects the d-component knockout to dominate, since the knockout of a singlet deuteron would correspond to inelastic scattering. However, one can in principle easily test for this contribution by studying $0^+$,T=0 to $0^+$,T=1 transitions in nuclei such as $^{12}C$ or $^{16}O$.

As in other cases the (p,pd) experimental data are very limited. Due to the large d-nucleus binding energy the cross sections are very small. As a result for nuclei other than $^6Li$, which most people believe has signigicant clustering, there are only a few experiments. Grossiord et al.[19] have studied (p,pd) on $^{12}C$ and $^{16}O$ at 75 MeV, a relatively low energy considering the large binding energy. Rees[20] has carried out limited measurements on the same nuclei with 150 MeV polarized protons. The lower energy measurements were able to separate the various final states, and in fact observed a small yield to the $0^+$,T=1 states which is consistent with expectations based on inelastic scattering. The higher energy experiment, although unable to resolve the $0^+$ states, is much less sensitive to distortion effects. The ratio of distorted wave to plane wave calculation is about 1/7 for 150 MeV, compared to 1/30 for 75 MeV.

Figure 11 shows an energy sharing distribution for 150 MeV, along with DWIA calculations. The calculations describe the cross section data reasonably well and give a spectroscopic factor within roughly a factor of two to three of expectations based on shell model calculations. Also shown are the measured analyzing powers. The DWIA calculations reproduce the magnitude of the analyzing power, although these data do not provide a stringent test.

Overall, one can only say that the agreement between theory and experiment for (p,pd) reactions is encouraging in terms of a direct knockout explanation. However, data of much higher quality are needed before any serious comparisons can be made.

Both the transfer and knockout reactions reactions tend to be sensitive to the longer range n-p correlations. As my last topic I want to discuss the $(\pi^+,2p)$ absorption reaction. This reaction is often mentioned as a method for studying shorter range correlations because of the high relative momentum of the two outgoing nucleons. A simple model for attempting to describe such reactions is a DWIA "quasideuteron" model in which the two body $\pi NN$ vertex is factored out of the integral. One can then use a theoretical model for the $\pi NN$ vertex.

In principle studies of such reactions can tell us about the n-p correlations in nuclei including the c.m. motion of the pair. However, again one must define the reaction dynamics to extract this information. Various $(\pi,NN)$ experiments have been carried out over the years, generally with insufficient resolution to separate the nuclear states. To examine the reaction dynamics let me consider a particular set of experiments. McKeown et al.[21] have measured inclusive proton spectra resulting from the capture of 160 MeV pions. By plotting contours of the invariant cross section on a rapidity plot, the authors conclude that for medium weight the capture process involves about four nucleons. This result immediately suggests clustering phenomena, but inclusive spectra contain little detail. A more recent experiment examining the angular correlation between two protons shows pronounced peaking at the $\pi+d\rightarrow2p$ kinematic angles.[22] This result is shown in Fig. 12. The curve is a theoretical calculation assuming capture purely on deuterons with the normalization given by shell model wave functions.[23] The agreement is remarkably good, and although a number of approximations were made since the data effectively had no missing mass resolution, the general agreement suggests that capture is dominated by n-p pairs. Experiments on $^3$He[24] as well as inclusive gamma ray measurements[25] show that capture on singlet pairs is very small.

The combined results of the experiments I mentioned suggests that the capture is dominated by d-type pairs and that additional nucleons are involved only through multiple scattering of the final state nucleons. This is in general agreement with DWIA calculations for specific states[26] which reduce PWIA calculations by a factor of five to ten, thereby implying large multiple scattering. I would point out, however, that the difference between capture on an alpha-cluster resulting in two final state protons and capture directly on two nucleons is not large and will require a high degree understanding of the reaction.

Once again the delineation of the reaction dynamics is a precursor to any understanding of the importance of correlations in nuclear structure. This will require much higher quality data than is presently available. In particular, it would be most useful to have a variety of experiments such as (p,$^3$He), (p,pd), and ($\pi^+$,2p) to attempt to obtain a consistent picture of the n-p correlations in nuclei.

## 4. CONCLUSIONS

I have attempted to review our present knowledge of clustering as observed in medium energy reactions. I believe that the message is quite clear. Little progress beyond our present understanding will occur until the quantity and quality of the experimental data (statistics, energy resolution, etc.) improves by at least an order of magnitude. I would hope that should this occur we experimentalist would be able to stimulate theorists into addressing some of the questions I raised earlier in my talk.

## ACKNOWLEDGEMENTS

I would like to acknowledge the invaluable contributions made to this research by Prof. N. S. Chant and to the preparation of this manuscript by Ms. Judy Myrick. This work was supported in part by the National Science Foundation.

REFERENCES

1. C. C. Chang, N. S. Wall, and Z. Fraenkel, Phys. Rev. Lett. 33, 1493 (1974), and references therein.
2. H. E. Jackson et al., Phys. Rev. Lett. 31, 1353 (1973), and references therein.
3. N. S. Chant and P. G. Roos, Phys. Rev. C 15, 57 (1977).
4. P. G. Roos et al., Phys. Rev. C 15, 69 (1977).
5. A. A. Cowley et al., Phys. Rev. C 15, 1650 (1977).
6. B. Buck, C. B. Dover, and J. P. Vary, Phys. Rev. C 11, 1803 (1975).
7. T. A. Carey et al., Phys. Rev. C 29, 1273 (1984); Phys. Rev. C 23, 576 (1981).
8. T. A. Brody and M. Moshinsky, Tables of Transformation Brackets, Mexico (1963).
9. K. Kurath, Phys. Rev. C 7, 1390 (1973).
10. N. Anantaraman et al., Phys. Rev. Lett. 35, 1131 (1975).
11. H. W. Fulbright et al., Nucl. Phys. A284, 329 (1977).
12. N. Anantaraman, H. W. Fulbright, and P. M. Stwertka, Phys. Rev. C 22, 501 (1980).
13. N. S. Chant, in Proc. of 3rd Intn'l Conf. on Clustering Aspects of Nucl. Structure and Nucl. Reactions, AIP Conf. Proc. No. 47, edited by W. T. H. Van Oers et al. (AIP, New York, 1978), p. 415.
14. A. Nadasen et al., Phys. Rev. C 23, 2353 (1981).
15. R. E. Warner et al., Nucl. Phys., to be published.
16. G. Landaud et al., Phys. Rev. C 18, 1776 (1978).
17. C. W. Wang, private communication.
18. A. M. Cormack et al., Phys. Rev. 115, 599 (1959).
19. J. Y. Grossiord et al., Phys. Rev. C 15, 843 (1977).
20. L. B. Rees, Ph.D. thesis, University of Maryland, 1983.
21. R. D. McKeown et al., Phys. Rev. Lett. 44, 1033 (1980).
22. A. Altman et al., Phys. Rev. Lett. 50, 1187 (1983).
23. B. G. Ritchie, N. S. Chant, and P. G. Roos, Phys. Rev. C, to be published.
23. D. Ashery et al., Phys. Rev. Lett. 47, 895 (1981).
24. C. E. Stronach et al., Phys. Rev. C 23, 2150 (1981).
25. P. G. Roos, L. Rees, and N. S. Chant, Phys. Rev. C 24, 2647 (1981).

# FROM NUCLEONS TO QUARKS IN NUCLEI

J.M. Laget
Service de Physique Nucléaire - Hautes Energies,
CEN Saclay,
91191 Gif-sur-Yvette Cedex, France.

ABSTRACT. Nuclear Physics has now evolved from the study of the many nucleon problem, to the study of the interplay of the degrees of freedom of such a complex system and the internal degrees of freedom of each of its hadronic constituents. Extensive studies of electronuclear reactions have allowed us to disentangle the basic mechanisms of the interaction between two baryons in a nucleus. The pion exchange mechanism, which dominates at large distance,has been singled out. The $N\Delta$ interaction, which enter the description of the intermediate range part, has been studied. Evidences for effects due to the quark structure of the nucleon have been found. They involve the short distance structure of the nucleus.

During a long time the description of nuclei in terms of nucleons only was a powerful and economical way to understand their structure, their symmetries and their interactions. This simple description of the ground state and of the low-lying excited states of nuclei has been beautifully checked by the systematic study of their electromagnetic properties. The measurement of elastic and inelastic form factors at high momentum [1], has made possible the accurate determination of the charge and the magnetization densities and has led to a good knowledge of the shape of the nuclei. The analysis of quasi free electron scattering (when the outgoing electron is detected in coincidence with the struck nucleon) has made possible the straightforward study of the shell structure of nuclei [2]. All these results have led to strong constraints on the self-consistent mean field description of nuclei.

However the use of probe of higher and higher energy, has forced us to go beyond this simple picture and to consider also the other degrees of freedom in nuclei. On the one hand, the increase of the momentum transferred to the nucleus allowed us to probe its spatial structure over distances comparable or smaller than the nucleon size,

J. S. Lilley and M. A. Nagarajan (eds.), Clustering Aspects of Nuclear Structure, 295–316.
© 1985 by D. Reidel Publishing Company.

and to study the details of the nucleon-nucleon interaction,or the nucleon deformation, in a nucleus. On the other hand, the increase of the available energy allowed us to create particles and to study their propagation in a nucleus. Among them the pion and Δ are of particular interest, since they are the basic ingredients of our modern under-standing of the nucleon-nucleon interaction (Fig. 1).

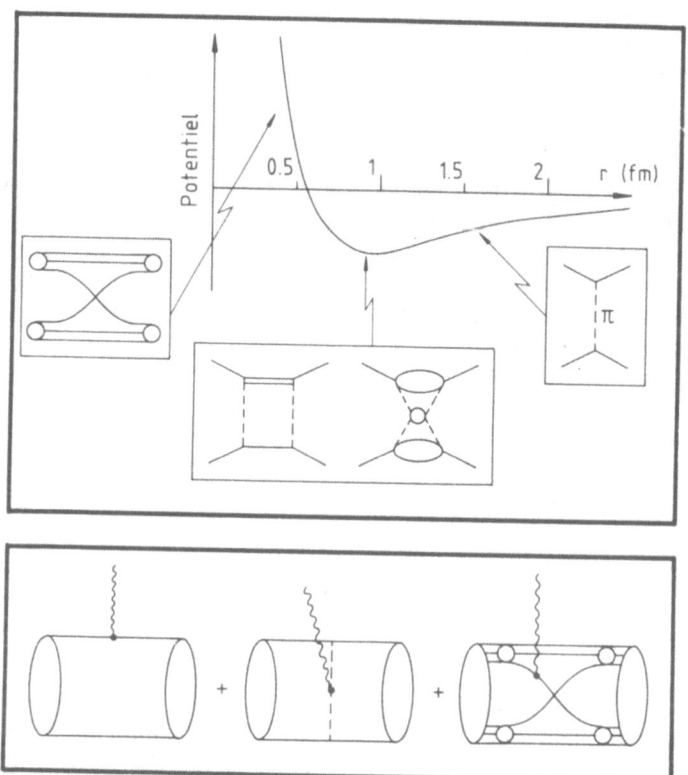

Figure 1. The Nucleon-Nucleon potential and the dominant driving terms. They must be iterated to obtain the full T-matrix. At long distance the pion exchange mechanism dominates. At intermediate distance a Δ can be created between the exchange of two pions. At short distance heavy mesons can be exchanged, but also the subnuclear degrees of freedom are expected to play a role : the Quark Interchange Mechanism is one pos-sible example. When the electromagnetic probe interacts with the nuc-leon current in nuclei, gauge invariance requires also its interaction with each charged particle which is exchanged in the driving terms.

The pion is the lighest particle which can be created on a nucleon and it is responsible for the long-range part of the nucleon force. The Δ is the first excited state of the nucleon and appears naturally in the description of the intermediate range part of the nucleon-nucleon interaction, when one of the nucleon (or both) is excited between the exchange of two correlated pions. Below the pion threshold virtual $\Delta$'s enter the description of the nucleon-nucleon interaction but close to this threshold the $\Delta$ can be created freely in a collision between two nucleons : the problems of the nucleon-nucleon interaction and the nucleon-$\Delta$ interaction should be solved at the same time in a coupled channel formalism. Since the $\Delta$ is unstable the only way to study the $N\Delta$ channel is to create the $\Delta$ in a nucleus and let it interact with the neighbouring nucleons. At small distance the exchange of vector mesons ($\rho$ and $\omega$) plays a role, but it is also here that the quark structure of the nucleon is expected to play a role. The relative importance of this two mechanisms, the double counting problem and the relevance of the description of the nucleus in term of quarks are still open questions.

Therefore the nucleus has now become a powerful laboratory to study the interactions between hadrons, and nuclear physics has smoothly evolved from the study of the many-nucleon problem to the study of a new domain where both the properties of the many-hadron system and the internal properties of each hadron are to be considered and are strongly interrelated.

Such complex hadronic systems can be made by smashing two nuclei against each other, or by strongly disturbing a nucleus with an high energy hadronic probe (proton, pion, etc...). But the simplest and the cleanest way is to look at a nucleus with an electromagnetic probe. Indeed most of the evidence of the role of the internal degrees of freedom of the nucleons in nuclei come from experiments performed with electrons and photons or, too a less extent, with weakly interacting probes.

On the one hand the coupling of a (real or virtual) photon with a nuclear system is well under control since it must satisfy the gauge invariance principle, and it is weak enough to be treated as a small perturbation. High energy electron beams are powerful microscopes which allow us to get detailed "Snap Shot" of a nucleus. For instance we shall see that the electromagnetic probe has been the best tool to disentangle the long range part of the nucleon-nucleon interaction, which is mediated by the exchange of one charged pion and the intermediate range part which is due to the exchange of two correlated pions of which the total charge is vanishing.

On the other hand, since it interacts weakly, the photon is not absorbed at the nuclear surface, like hadronic probes as the pion for instance, and sees the entire volume of the nucleus. It is therefore possible to create an unstable particle, like the $\Delta$, in the very center of the nucleus, and study its interaction in the final state. We shall see that this is the way to study the $N\Delta$ interaction which is a basic ingredient of the intermediate range NN interaction (Fig. 1).

All those mechanisms involve many nucleons (at least two), and their understanding rely upon a knowledge of the short range correlation between two or many nucleons in a nucleus. This is precisely the topic of this Conference and I will try to show, with a few examples, how the systematic study of electronuclear reactions, during the last ten or fifteen years, has helped us to understand the

various mechanisms of the interactions between two baryons in nuclei
and how their internal structure enters into the game.

    I will restrict myself to the few body systems. Their nuclear wave
functions are well known and well under control, and they provide us
with the best laboratory to study in detail the basic mechanisms of
photo-and electronuclear reactions. The case of heavier nuclei is very
similar [3-5]).

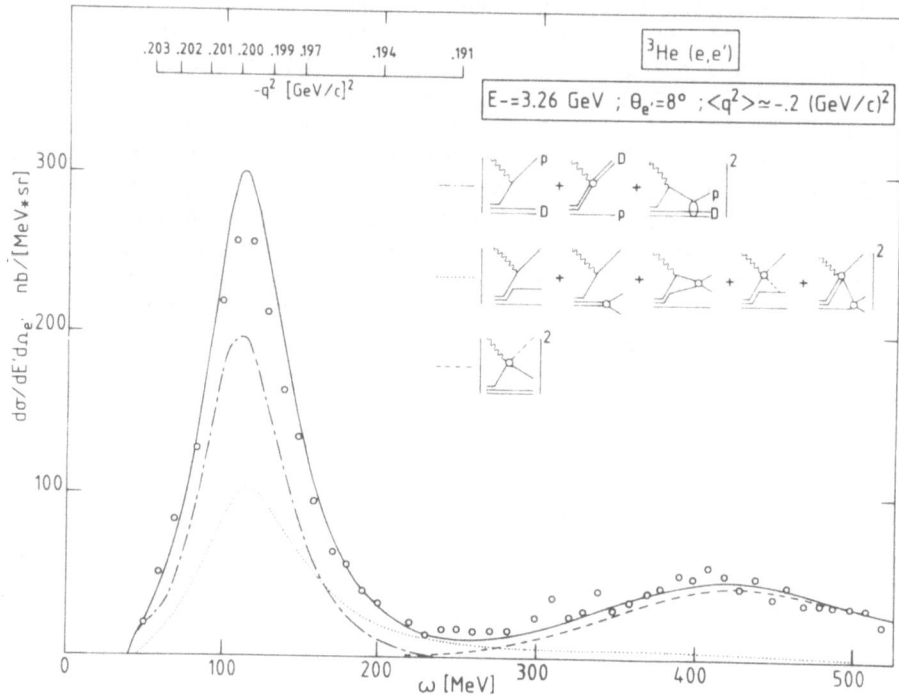

Figure 2. The spectrum of electrons inelastically scattered on $^3$He at
$\theta_{e'} = 8°$, when $E_- = 3.26$ GeV [ref. [6])]. The energy of the virtual
photon $\omega = E-E'$ and its squared mass $q^2$ are plotted on abcissa.

    The basic features of the absorption of a photon by a nucleus
appear clearly in Fig. 2, where the spectrum of the electrons inelastically scattered at $\theta_{e'} = 8°$ on $^3$He is shown. The experimental data
have been obtained at SLAC [6]). In spite of the high energy
$E_- = 3.26$ GeV, of the incoming electron beam, the momentum transfer
is small (the squared mass of the virtual photon varies little around
$q^2 = - .2$ (GeV/c)$^2$). The energy transfer is high enough to make it
possible to excite the $\Delta$ resonance, which is responsible for most of
the pions which are electroproduced on a quasi-free nucleon (the pions
created through the non resonant part of the electroproduction operator [3,4]) have also been taken into account). The range of
momentum and energy transfer is really that which is already allowed by
the present generation of high intensity electron machines (such as the
700 MeV Saclay linac for instance) : such spectra have already been

obtained for heavier nuclei and are systematically measured for the few-body systems [7]).

Besides pion electroproduction on quasi-free nucleons, the incoming electron may also scatter elastically on a quasi-free nucleon. The top of the peak, which appears for small value of the energy $\omega$ of the virtual photon, corresponds to the scattering of the electron on a nucleon at rest in the nucleus. Its width is due to the nucleon Fermi motion. The use of a good three-body wave function (the solution [8]) of the Fadeev equations in momentum space for the Reid potential [9]) makes it possible to compute separately the contributions of the two-body and the three-body break-up channels. They add up to give a fair agreement with the experimental data. While the shape of the quasi-elastic peak is directly related to the behaviour of the nucleon wave function, the shape of the quasi-free pion production peak, which appears in the high energy part of the spectrum in Fig. 2, is due to the internal degrees of freedom of the nucleons : one nucleon is changed into its first excited state, the $\Delta$, and the others nucleons are spectators.

Between the quasi-elastic scattering and the quasi-free pion electroproduction peaks, the excess of the cross section is well accounted for by the tail of three-body break-up channel, which is mainly due to the meson exchange mechanism : the pion (or the $\rho$) which is created at one nucleon is reabsorbed by an other, breaking up the residual nuclear system. This mechanism involves the short range correlation function between two nucleons, which is automatically contained in the three body realistic wave function [8]) which I use.

This spectrum is a good example of the interplay between the many body aspects (here the three-nucleon problem) and the internal degrees of freedom of the nucleon (the creation and the propagation of real or virtual pions, the excitation of the $\Delta$, etc...). But this is an integrated quantity which tells us only how the photon is absorbed by the nucleus. To go further we must perform more exclusive experiments, in order to single out each channel (pion photoproduction, photodisintegration, etc...) and study the various aspects of the nuclear dynamics. I have chosen an example in each of the three parts of this spectrum : the first one concerns the pion photoproduction at the top of the $\Delta$ resonance, the second concerns the creation of virtual pions and the mesons exchange currents and the third has to do with the behaviour of the low-energy side of the quasi elastic peak at very high momentum transfer.

Let me start with the $D(\gamma, p\pi^-)p$ reaction, which I have extensively discussed elsewhere [10]). Today I would like only to recall that in this experiment the kinematics is completely determined, and that each dominant term in the multiple scattering series (Fig. 3) has been singled out by looking for its singularities. This multiple scattering series converges quickly and is well under control, since the particles propagate near their mass shell, and there are no free parameters. The calculation relies heavily upon the free nucleon cross sections and the deuteron wave function, which are independently determined by other experiments. It reproduces a wide bulk of experimental data which have

been obtained at Saclay, and two deviations only remain today. They
appear in Fig. 4, where I show the Saclay data which have just been
reanalysed [11]).

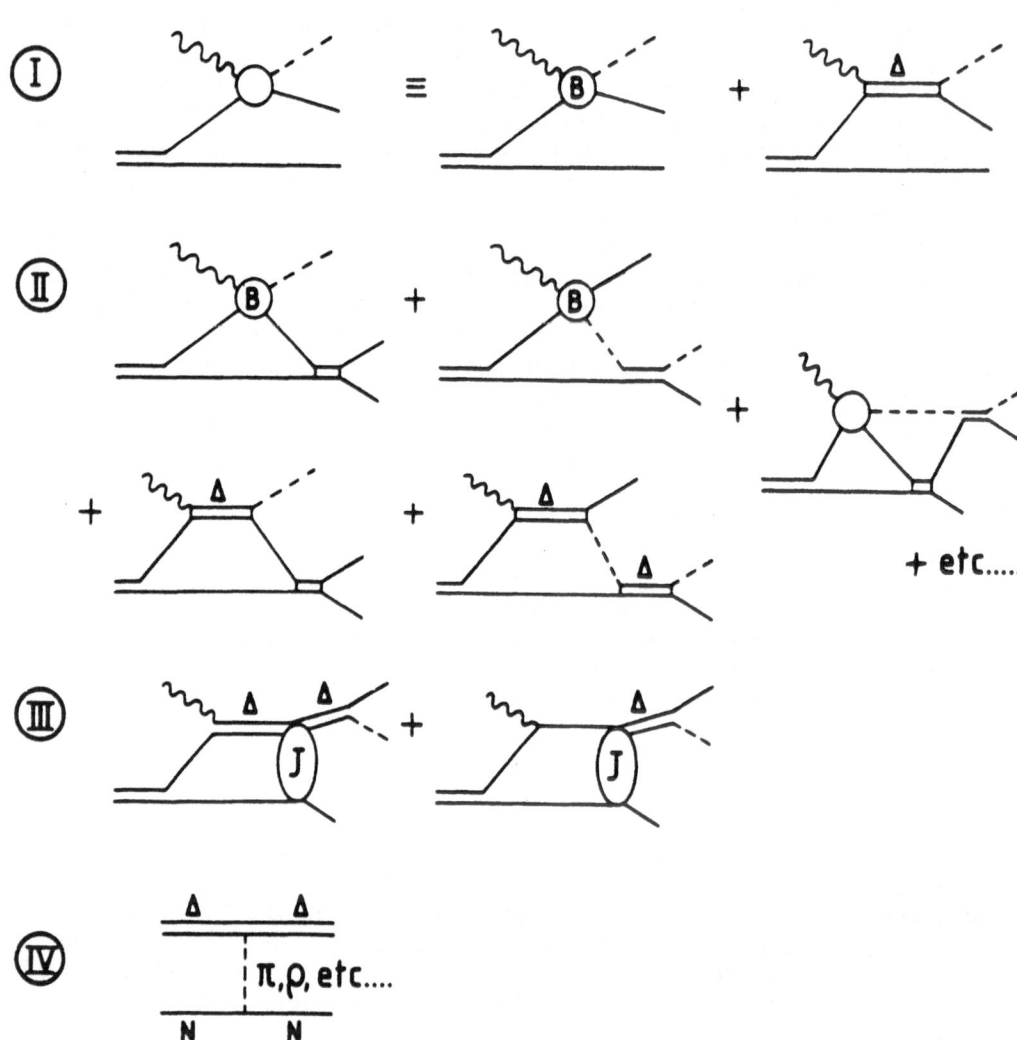

Fig. 3. The relevant diagrams in the analysis of the D($\gamma$,p$\pi^-$)p reaction. I : The quasi-free process where the elementary $\gamma$n$\rightarrow$p$\pi^-$ reaction
amplitude has been split into the non resonant Born terms and the $\Delta$
resonance production amplitude. II :  The dominant final state interaction diagrams which involve the rescattering of the $\Delta$ constituents.
III : The diagram which involve the part of the N$\Delta$ interaction which
does not reduce to the rescattering of one of the $\Delta$ constituents. IV :
A possible example of such a part of the N$\Delta$ interaction.

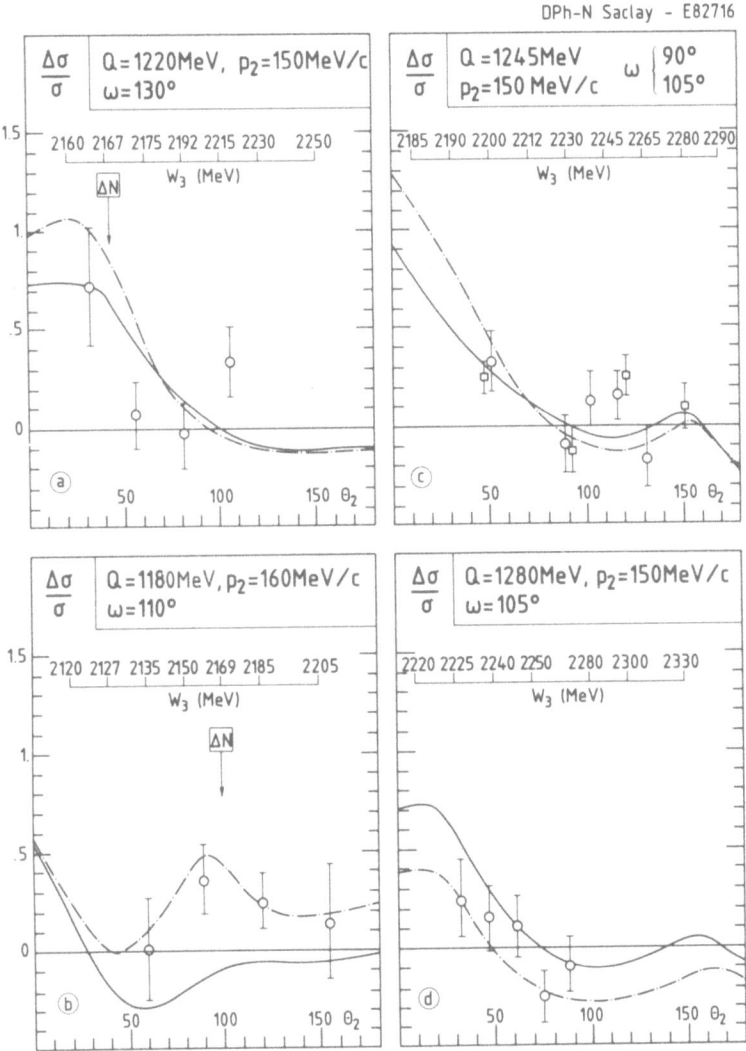

DPh-N Saclay - E82716

Fig. 4. The relative excess, with respect to the predictions of the quasi free mechanism, of the cross section of the reaction d(γ,pπ⁻)p as measured at Saclay [11]. The pion and one of the two protons were detected in coincidence, in such a way that the mass Q and the momentum of the undetected proton are kept constant. The abcissa is the angle $\theta_2$ of the undected proton. The full line curves include all the leading pion and nucleon rescattering graphs. The dashed line curves show the effects of the part of the NΔ interaction which does not reduce to the multiple scattering of the Δ constituents (see refs. [10] and [12]) for a detailed discussion).

The first one appears near the NΔ threshold (when the mass of the πNN system is $W_3 \simeq 2170$ MeV). The second one lies 80 MeV above ($W_3$ = 2250 MeV) but the signal is not very clear. They might be due to the part of the NΔ interaction (fig. 3) which does not reduce to a sequential two-body scattering of the constituents of the Δ, and which looks like the nucleon-nucleon interaction (and is due to the exchange of virtual mesons). A coupled treatment of the NN and NΔ channels, in the K-matrix approximation, leads to a good account for the structure which exists near the NΔ threshold, as well as the resonant behaviour of nucleon-nucleon $^1D_2$ phase shift [10,12]). But today the structure near $W_3$= 2250 MeV is still unexplained. This kind of experiment offers the best way to study the NΔ interaction by creating the Δ in a clean way with the electromagnetic probe but is at the limit of the capability of the present generation of low duty factor electron accelerators. Its development calls for a CW machine : a systematic study of this coincidence experiment will allow us to make a full partial-wave analysis of the final NΔ system.

Contrary to real photoproduction reactions, the analysis of virtual pion photoproduction reactions is not parameter free.

Although the method of calculation of the pion absorption graph and the pion rescattering graph is the same, the virtual pion which is reabsorbed is highly off its mass shell, and the free pion photoproduction operator, as well as the pion absorption operator, should be corrected. Two ways are usually followed to overcome this difficulty.

On the one hand, since it is far off-shell, the exchanged pion is sensitive to the finite size of the nucleon, and I have used at each pion-baryon vertex a monopole form factor

$$ F_\pi(q^2) = \frac{\Lambda_\pi^2 - m_\pi^2}{\Lambda_\pi^2 - q_\pi^2} $$

where $q_\pi^2$ is the squared mass of the virtual pion.

On the other hand, other virtual mesons can also be emitted and reabsorbed. Among them the ρ-meson exchange diagram plays an important role (in which case I use a dipole ρ-baryon form factor with a cut-off mass equal to two time the nucleon mass).

I have determined the values of the cut-off mass $\Lambda_\pi$ and of the ratio $G_\rho^2/G_\pi^2$ between the square of the ρ- and the pion-baryon coupling constants by fitting [13-14]) the 90° excitation function of the d(γ,p) reaction cross section (Fig. 5). It turns out that, in this reaction, the ρ-exchange mechanism is negligible below the pion production threshold, and only affects significantly the Δ-N → N-N transition in the resonance region. It is therefore possible to separately determine the cut-off mass $\Lambda_\pi$ = 1.2 GeV at low energy and the ratio $G_\rho^2/G_\pi^2$ = 1.6 in the Δ region. They lie in the range of the uncertainties of the currently accepted values.

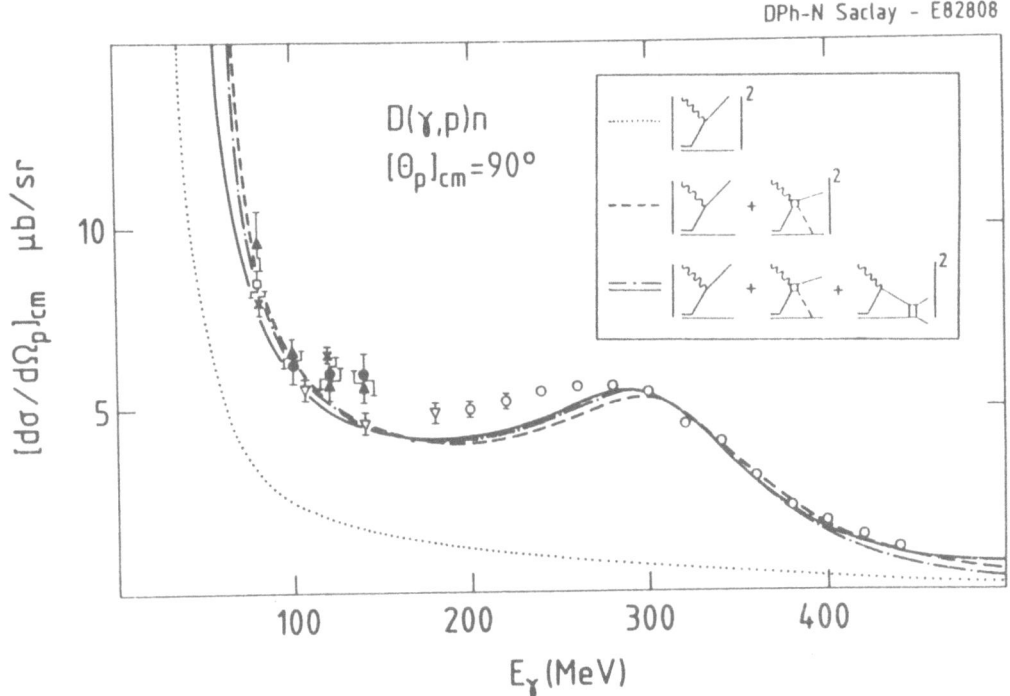

Fig. 5. The excitation function, at $\theta_p = 90°$, of the d($\gamma$,p)n reaction is plotted against the incoming photon energy $E_\gamma$. The high energy experimental points have been obtained at Bonn [16]). At low energy the references can be found in refs [13-14]). The dotted and dashed curves correspond to the plane-wave calculation without and with the exchange current contribution respectively. The dash-dotted and the full curves include also the neutron-proton final state interaction, in the S and S+P states respectively.

This model also provides us with a good representation of the angular distributions. As an example, I show in Fig. 6 the variations with the photon energy of the cross section at $\theta_p = 0°$, which has always been a puzzle, but which is very well reproduced by this model (more details are given in ref.[14]).

When this model is applied to $^3$He, the only change is the nuclear wave function [8]) and there are no free parameters. The total two-body photodisintegration cross section of $^3$He is plotted in Fig. 7, together with the total photodisintegration cross section of deuteron. In both case the contribution of the exchange current is essential to reproduce the data [13-17]).

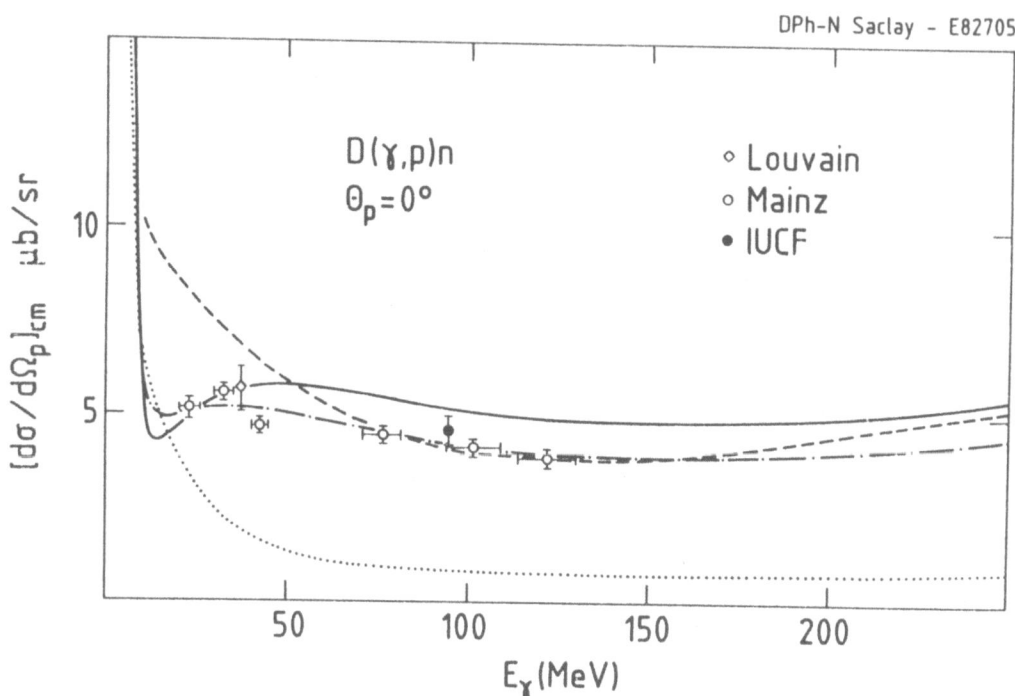

DPh-N Saclay - E82705

Fig. 6. The excitation function at $\theta_p = 0°$, of the d($\gamma$,p)n reaction is plotted against the incoming photon energy. The meaning of the curves is the same as in Fig. 5. The references for the experimental points can be found in ref. [14]).

    The two body electrodisintegration of $^3$He is also well reproduced by the same kind of calculation. As in the deuteron case [14]), the final state interactions in the emitted p-d pair as well as the meson exchange contribution help to accurately reproduce the data (fig. 8), which have been recently obtained at Saclay [18]).

    The same mechanism dominates also the electromagnetic form factors of the few-body systems. As an example Fig. 9 shows the magnetic form factor of $^3$He, which has been recently measured at Saclay [19]) up to $|q^2| = 32$ fm$^{-2}$ ($|q^2| \approx 1.2$ (GeV/c)$^2$).These data overlap nicely with the low momentum data which have been obtained at Bates [20]). It is clear that the meson exchange mechanisms are essential to obtain a fair agreement between the data and the most-up-to-date theory[21]). The ingredients of this calculation are the same as those which enter my calculation of the photo and electrodisintegration of the few-body systems. A realistic three-body wave function [20-22]) is used, a cut-off mass of the pion-baryon vertex is of the order $\Lambda_\pi = 1$ GeV, and the $\rho$-baryon coupling constants are comparable to those which I use.

    It is very nice and appealing that with the same set of parameters, this approach provides us with a consistent picture of a wide variety of experimental data. They represent the cleanest way to

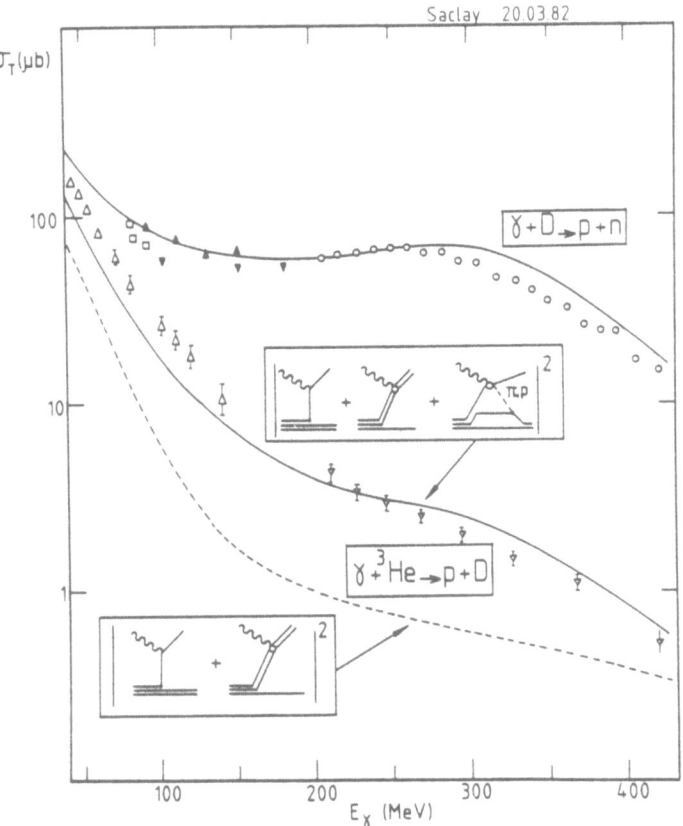

Figure 7. The integrated cross sections of the two-body photodisintegration of the deuteron and the $^3$He nucleus are plotted against the
energy of the incoming photon. The high energy data points have been
recently measured at Bonn [15,16]), and the low energy points come from
ref. [17]). The meaning of each curve is explained in the insets.


disentangle the long range mechanism of the nucleon nucleon force,
which is due to the exchange of charged pions, and the other mechanisms, which are due to the exchange of neutral object, such as the
two-pion exchange mechanism which dominates the intermediate range
part. In fact two strong constraints, the gauge invariance principle
and the Partially Conserved Axial Current (PCAC) hypothesis, fixe in an
unambiguous way the pion exchange matrix elements at low momentum
transfer : this picture is very reliable when the momentum transfer is
smaller than $|q^2|$ ~.3$(GeV/c)^2$ and the energy transfer does not exceed
500 MeV.
      However we may ask ourselves why such an approach works so well
for momentum transfer as high as 1 $(GeV/c)^2$. On the one hand, it is
basically non relativistic : only the first terms of an expansion in

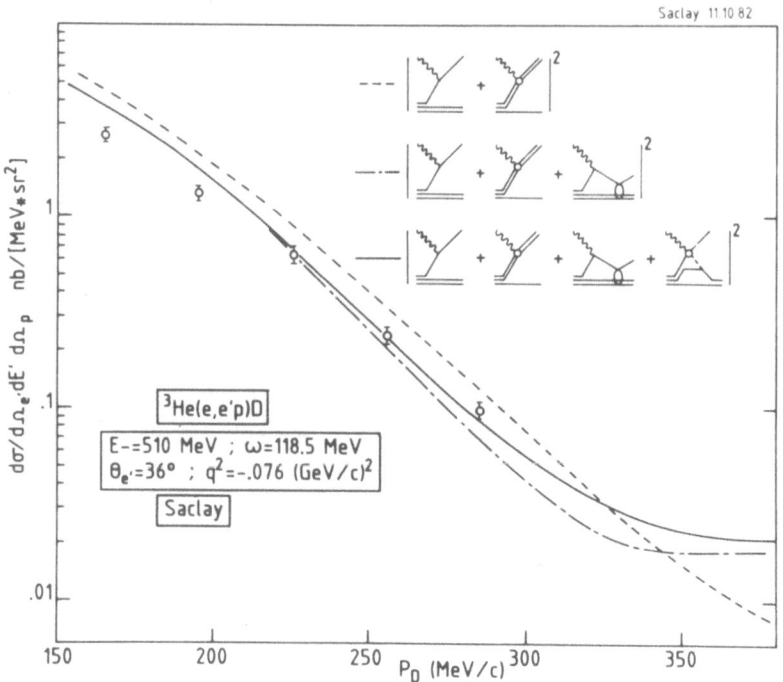

Fig.8. The variation of the $^3$He(e,e'p)D reaction cross section, recently measured at Saclay [18]), with the momentum of the final deuteron.

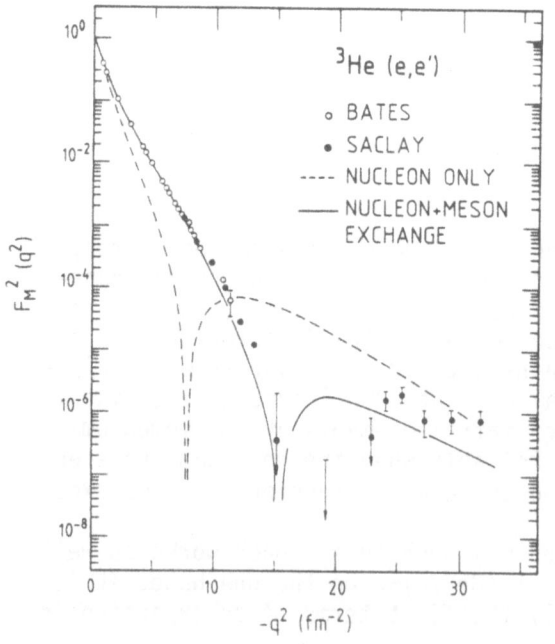

Fig. 9. The (squared) magnetic form factor of $^3$He. The low momentum data have been respectively obtained at Bates [20] and Saclay [19]).The full (dashed) curve takes (does not take) into account the meson exchange contributions [21]).

power of 1/m are retained in each elementary operator, and non-relativistic wave functions are used. At $1(GeV/c)^2$ a full relativistic
calculation is badly needed. On the other hand, for a momentum transfer
around 1 $(GeV/c)^2$, distances as small as .2fm are probed, and the quark
structure of nucleons should show up. In fact two parameters, the cut-
off mass $\Lambda_\pi$ and $\rho$-baryon coupling constant, have been adjusted to fit
the experiments. It is very likely that this procedure hides more fundamental mechanisms. For instance the use of a $\pi NN$ form factor is a way
to go beyond the description of a nucleus in term of point-like structure nucleons, and to take into account their finite size. The cut-off
mass of $\Lambda_\pi$ = 1.2 GeV, which is needed to reproduce the data, corresponds to a core radius of the nucleon of approximately .5 fm, very
close to the "little bag" radius [23]. If this were the reality, the
problem of quarks in nuclei would reduce to the understanding of the
various nucleon form factors. But if the bag radius is of the order of
1.fm, as in other current models [24], the cut-off mass is only
$\Lambda_\pi \sim$ 600 MeV, and the pion exchange mechanisms are strongly suppressed : room is left, even at low momentum transfer for more complex
mechanisms where the quarks of two distinct nucleons are mixed together
[25-26], and which give rise to six-quark clusters in the ground state
wave function.

An other criticism to this standard model has to do with the
$\rho$-exchange mechanism[27], which occurs at very short distance and might
simulate more subtle mechanisms which involve the quark degrees of
freedom of the nucleons.

It is always possible to play the game of adjusting the parameters
entering the meson exchange matrix elements and to add new mechanisms.
But, in order to have strong evidence of the quark degrees of freedom
in nuclei, we must first look for experiments which cannot be reproduced by varying the free parameters of this standard picture of nuclei
in terms of meson and nucleons. To my knowledge, the best evidence of
the breakdown of this approach occurs in the analysis of the low-energy
side of the quasi-elastic electron scattering in a nucleus at high
momentum transfer.

In Fig. 10, I have plotted on a logarithmic scale, the low energy
part of the quasi-elastic peak, which is seen in the $^3He(e,e')$ reaction [6], for two values of the squared mass of the virtual photon :
$q^2$ = -.2$(GeV/c)^2$ (as in fig. 2) and $q^2$ = -1.$(GeV/v)^2$. The arrows indicate
the kinematics of the elastic scattering of the electron on a nucleon
or a nucleon-pair at rest : respectively $-q^2/2m$ and $-q^2/4m$.

The pure quasi-elastic mechanism, where the virtual photon interacts with one nucleon and where the two others are spectators, is unable
to reproduce its low energy part. The final state interaction effects,
which correspond to the interaction of the virtual photon with a correlated nucleon pair, improve dramatically the agreement between the
model and data. The reason is simple : this active pair is almost at
rest in $^3He$ and the corresponding amplitude is sensitive to the low
momentum components of the wave function. Just near threshold the
strong low energy p-D scattering dominates the spectrum. However a
factor three still missing when $q^2 \sim$ 1 $(GeV/c)^2$.

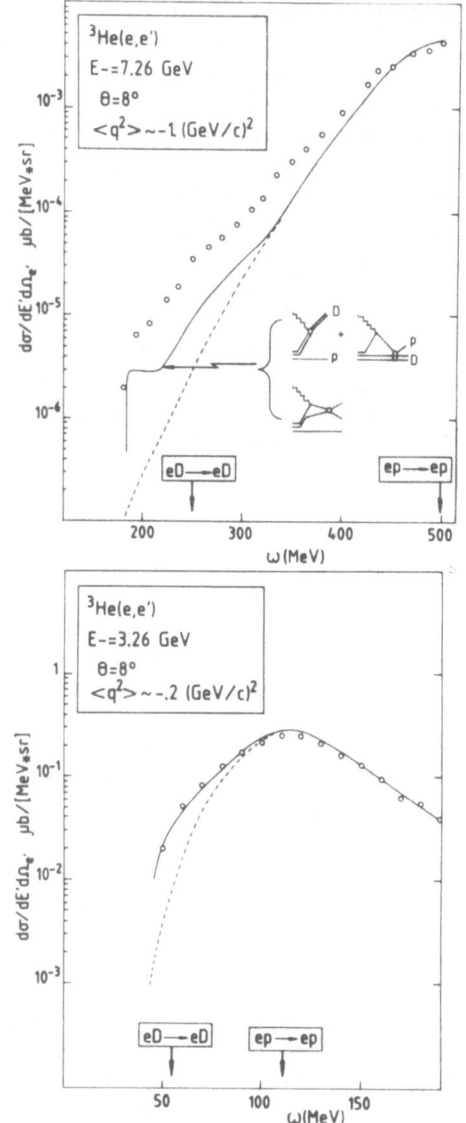

Fig. 10. The low energy part of the spectra of the electrons inelastic-ally scattered by $^3$He, when $q^2 = -.2$ $(GeV/c)^2$ and $q^2 = -1.(GeV/c)^2$, are plotted, on a logarithmic scale, against the energy of the virtual photon. The arrows correspond respectively to the elastic scattering of the elec-tron on a free proton or a free deuteron at rest. The broken curve takes only into account the quasi-elastic scattering of the electron on a nucleon. The full curve takes also into account the quasi-elastic scattering on a nucleon pair and the p-d rescattering for the two-body break-up channel, and the nucleon-nucleon rescattering in the active pair for the three-body channel.

The same disagreement occurs also in the D(e,e') reaction near $q^2 = 1$ $(GeV/c)^2$, and it is remarkable that the theoretical and the experimental ratio between the $^3$He(e,e') and D(e,e') cross sections are the same [28]. Two consequences immediately follow : in that energy and momentum range, the $^3$He(e,e') cross section is dominated by the two-nucleon mechanisms, and the source of the disagreement between the deuterium data or the $^3$He data and the model is the same.

It does not come from a lack of knowledge of the high momentum components in the wave function, since the same momentum range has also been probed in coincidence experiments which have recently been performed at Saclay [18,29]). We have seen in Fig. 8 that the same model leads to a good agreement, provided that the final state effects are also taken into account.

The main difference between the two experiments is the low value, $q2 = -.078$ (GeV/c)$^2$ (as compared to 1.(GeV/c)$^2$ ), of the mass of the virtual photon in the Saclay experiment. At the top of the quasi-free peak the nucleons are close to their mass shell and their on-shell electromagnetic form factor gives a good account of their internal structure. Far away from the quasi-free peak, they are highly off their mass shell, each of them being deformed and polarized by the proximity of the others. The free nucleon electromagnetic form factors are not a good description of their structure and a full description of the two-nucleon system in terms of its quark constituents must be used. However those effects occur only at short distance and only appear when the wave length of the virtual photon is small enough to resolve them : this is the case around $|q^2| = 1.$ (GeV/c)$^2$ but not below $|q^2| \sim .2$ (GeV/c)$^2$.

The first attempt to deal with these quark degrees of freedom in nuclei, is to treat them in a perturbative way, by taking advantage of the smallness of the interaction between two quarks at very short distance and therefore at high momentum. The basic assumption is that it is possible to factorize the soft and the hard scattering mechanisms which occur between the constituents of two interacting baryons. For instance, the evolution of a system of two nucleons is dominated by repeated soft interactions which insure the cohesion of each of them, but it happens that, during a very short time, a very hard collision between two quarks occurs in a small volume, and leads to the emission of fast particles . The asymptotic behaviour of the cross section is therefore driven by the hard scattering amplitude and is not modified by the soft scatterings which only fixe the overall normalization.

The dominant contribution to the hard scattering is given by the simplest Feynmann diagram where the transferred momentum is equally shared between each of the active quark. It is remarkable that this simple picture is strongly supported by the analysis of the scattering between elementary hadrons at high energy and high momentum transfer [30-32]).

As in the case of pion exchange mechanism, gauge invariance requires that the electromagnetic probe sees and interacts with the quarks which are moving from one nucleon to an other, during the hard scattering process. Therefore any evidence for the effects of the subnuclear degrees of freedom in electronuclear reactions should lead to strong constraints on the short range behaviour of the baryon-baryon interaction. For instance this game has been played in the analysis [33]) of the structure function $A(q^2)$ of the deuteron, which has been measured at SLAC [34]) up to $|q^2| \sim 6$ (GeV/c)$^2$.

The virtual photon interacts with one of the quark lines when two quarks are interchanged between the two nucleons (see Fig.1). Its momentum is shared between each of the two nucleons which remains a bound system of quarks in the initial and final states : the six quarks are only mixed during a small fraction of the time. Therefore the deuteron form factor is proportional to :

$$\sqrt{A(q^2)} = F_D(q^2) \sim \frac{F_N^2 \ (q^2/4)}{-q^2 + \frac{6}{5}\beta^2} \qquad (1)$$

where $F_N$ is the nucleon form factor and $\beta^2$ is related to the mean transverse momentum of the exchanged quarks. As can be seen in Fig.11,

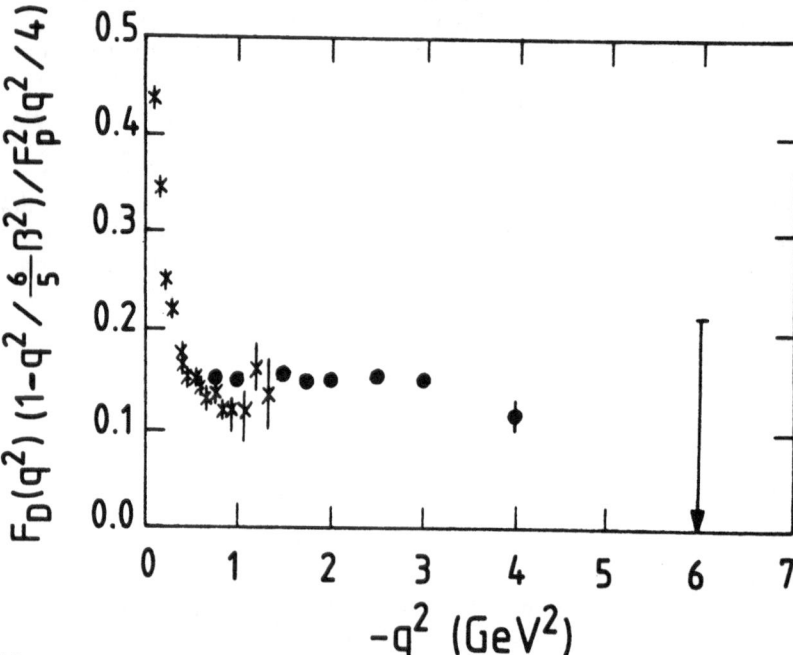

Fig. 11. The deuteron form factor $F(q^2) = \sqrt{A(q^2)}$ is divided by the right hand side of eq.(1). The high momentum data have been obtained at SLAC [34].

this model leads to a fair agreement with the experimental deuteron structure function, even at momentum as low as 1 $(GeV/c)^2$. This is surprising since this model has been tailored to analyse reactions at very high momentum transfer (perturbative approach). As in the case of pion exchange mechanism, it works far outside its range of validity.

Indeed it has been argued [35] that this perturbative treatment of the nuclear form factors is not valid at momentum transfer around few $(GeV/c)^2$. However it has only been shown that the nucleon form factors are dominated by the soft non-perturbative mechanisms. This does not mean at all that the quark interchange model [37] is ruled out, since these soft mechanisms have been factored out and parametrized by the nucleon form factors in eq. (1), leaving open the possibility that the photons sees the quarks at the very time which they are interchanged between the two nucleons. Moreover it should be stressed that this model predicts also the order of magnitude of the deuteron form factor : it is related to the short range behaviour of the deuteron wave function [33].

The second attempt to deal directly with the quark degrees of freedom is to assume that clusters made of more than three quarks are present in nuclei [26,36-37]. An example is again the analysis of the low energy side of the quasi elastic peak which appears in the $^3He(e,e')$ reaction around $q^2 = -1(GeV/c)^2$. The corresponding structure function $W_2(q^2,\omega)$ is plotted in Fig. 12 against the energy $\omega$ of the photon. It is related to the cross section which has been already plotted in Fig. 10 through :

$$\frac{d\sigma}{d\Omega\, dE'} = \sigma_M \left| W_2(q^2,\omega) + W_1(q^2,\omega^2)tg^2\, \theta/2 \right| \qquad (2)$$

where $\sigma_M$ is the Mott cross section which describes the scattering of an electron by a point-like object. The Bjorken variable $X = (-q^2/2\omega m)$ is also plotted on abcissa. At the top of the quasi free peak $X = 1$ since $\omega = -q^2/2m$. The six-quark clusters contribute mostly around $X = 2$, since $\omega = -q^2/4m$ when an electron scatters on a cluster at rest with a mass close to 2m. The nine-quark clusters contribute more likely in the vicinity of $X = 3$. This is exactly what I have found in my analysis (Fig. 10) of the same structure functions in terms of two or three nucleon final state interaction effects : this is kinematics.

However, the dynamics of the quark cluster model is different. The electron is assumed to interact directly with a quark of which the X distribution inside the cluster is folded with the probability of finding the cluster in $^3He$. While the X distribution is determined according to standard rules of perturbative QCD ref.[32], the probability of funding a six- or a nine-quark cluster is determined in ref. [37] from the geometrical overlap between bags (of which the radius R is adjusted to fit the experiment) and the correlation function of existing realistic Fadeev wave functions. A good fit is achieved when $R \sim .5$ fm, in which case the probability of finding three-, six- or nine quark clusters in $^3He$ is respectively .88, .11 and .01 (Fig. 12).

The analysis of the $X > 1$ part of the nuclear structure function, where the nucleon are very far from their mass shell since the scattering of an electron on a free nucleon at rest is kinematically forbidden here, has been extended [37-38] to its $X < 1$ part. Here the electron scatters on nucleons mainly at rest, and the nuclear structure function is expected to be basically equal to the sum of the nucleon structure functions. However, significant violations of this simple prediction have been recently found [39-40]. Between $X = .3$ and $X = .8$ the nuclear structure function is depleted by about 10 %, whereas it is enhanced below $X = .3$ by about 15 %. This behaviour can be explained in the framework of the quark cluster model, since the quarks confined in a bag share the total momentum : the more they are, the smaller the momentum which they carry. The nuclear structure function is therefore shifted towards the small X-value. Indeed Pirner and Vary [37-38] succeeded to reproduce the EMC [39] and SLAC [40] data whith the parameters, which they determined to fit the $^3He(e,e')$ data in the $X > 1$ region, slightly modified to take account of the change of the size and the density of the relevant nucleus. Other quark cluster models [41-42] have been proposed :

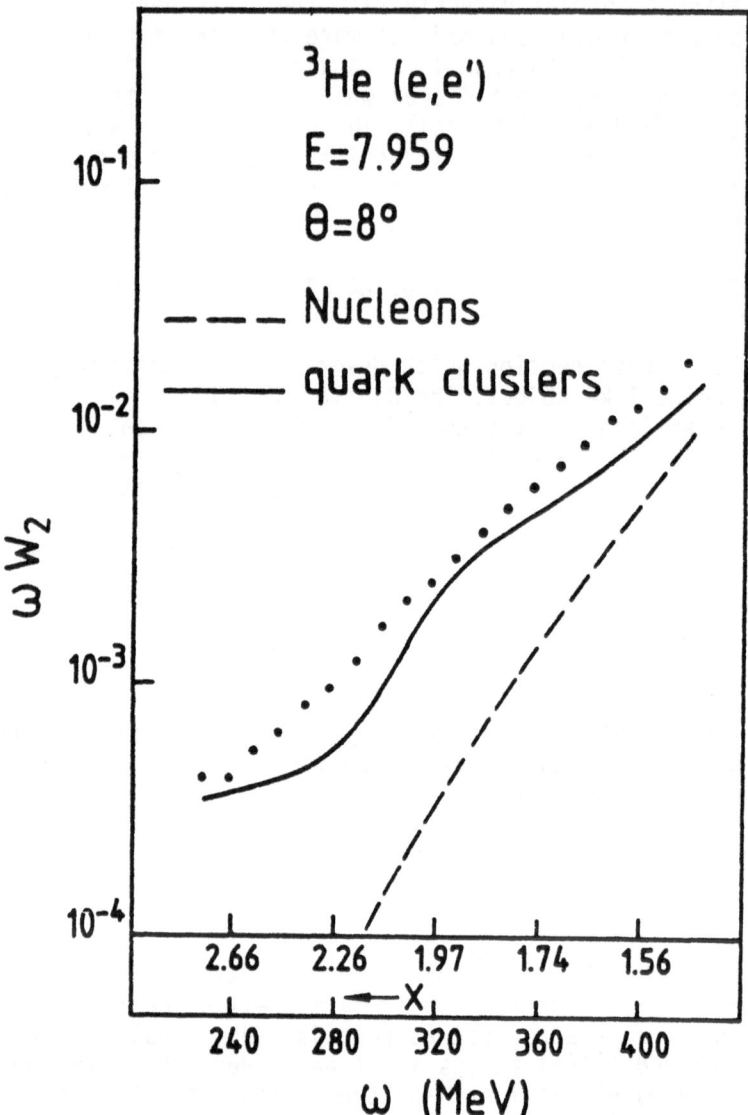

Fig. 12. The low energy side of the quasi-elastic peak of the $^3$He(e,e')
reaction is plotted against the photon energy $\omega$ and the scaling variable $X = -q^2/2m\omega$. The dashed curve corresponds to a conventional
calculation, where the photon interacts only with quasi-free nucleons
and where a modern three-body Fadeev wave function is used. It is very
similar to the dashed curves in Fig. 10. The full curve takes also into
account the interaction of the photon with the quarks confined in six-
and nine-quark clusters (adapted from ref. [37]).

they are based on the same basic assumptions although they depend upon slightly different parameters which are adjusted to fit the data.

All those model, assumes that the clusters are quasi free in the nucleus and neglect the final state interactions which can modify this simple picture. However it is known that they are responsible, through the so called higher twist effects, for the scaling violations in the behaviour of the nuclear structure functions.

The third attempt to deal directly with quark in nuclei is to start directly from the mechanisms which govern the evolution of the nucleon structure function. When its four momentum increases the photon sees more gluons and quark-antiquark pairs, which develop a chain as schematically depicted in Fig. 13. The net effect is to deplete the nucleon structure function at large X and enhance it at low X [43]

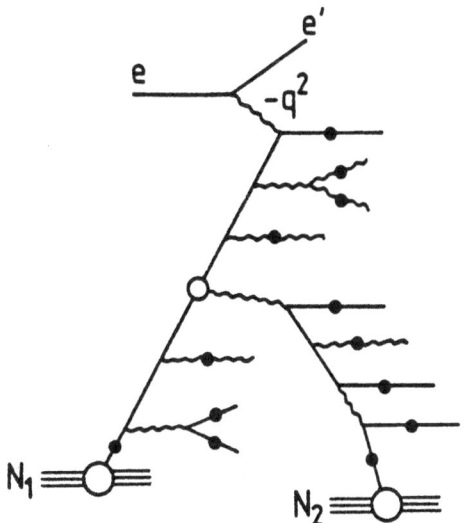

Fig. 13. The schematic description of the evolution of the nuclear structure function.

This is precisely the trend of the evolution of the nuclear response functions [39-40] when the mass number increases. It is therefore natural to assume that the vicinity of a second nucleon enhances the probability of creating more gluons or quark-antiquark pairs, since the chains which develop from each nucleon can match together. This idea has been put on a more quantitative bases by extending to the nucleus the methods which have already led to a good accounting for the evolution of the nucleon structure function when $q^2$ varies. Close et al. [44] start from the analysis of the moments of the structure function, in

the framework of the renormalization group, whereas Cohen-Tanhoudji
et al. [45]) start from the Altarelli-Parisi equations. Both groups are
able to reproduce the general trend of the data when the momentum $q^2$
and the mass number A vary. Although this approach relies also on phenomenology and some parameters are fitted to selected data, it provides
us with a picture which starts directly with our basic knowledge of
QCD.


These analyses of the nucleon response function, either within the
framework of the cluster model or of the scaling violation mechanisms,
imply that the mean free path of a quark is larger in the nucleus than
in the nucleon : since the nucleons are close together, they can overlap and the quarks can percolate. The effective confinement radius is
about 15 % larger in a nucleus than in a free nucleon [44]). The scaling
violations and the evolution of the nuclear structure function with the
mass number A have to do with the mechanisms which insure confinement
and which occur at long distance on the QCD scale, but at short distance on the nucleus scale. Therefore the study of the short range
properties of nuclear matter offers us with a powerful tool to study
these confining phenomena over a wide range of dynamical conditions,
which can not be achieved for a free nucleon. This is one of the most
important issues of nuclear physics to-day.
    However, these non perturbative mechanisms involve the modifica-
tion and the evolution of the sea quark distribution, and it has also
been proposed to describe the behaviour of the small X part of the
nuclear structure function in terms of the modification of the pion
field in the nucleus [46-48]) or of the exchange of Regge trajecto-
ries [42]). I will not enter into the details of these models since
A. Thomas is dealing with them during this Conference. But here also we
are faced with a remarkable duality of our description of the short
range mechanism in a nucleus. We have already noted that our description of the meson exchange currents allows for a good description of
electronuclear reactions at momentum transfer as high as $1.(GeV/c)^2$,
far beyond the domain where the corresponding models have been
taylored. The EMC data [39]) can also be reproduced if the muon is supposed to interact with a virtual pion of which the density is modified
by the spin isospin force in the nucleus [47]).
    The second important issue in nuclear physics is therefore to
understand this duality. What is the relationship between these two
descriptions of the nucleus (nucleons, mesons and $\Delta$, or quarks and
gluons) ? To what extent are they compatible ? How do they match together ? These are important questions which nuclear physicists must
now adress themselves.
    Obviously the study of electronuclear reactions at high momentum
transfer will allow us to probe the short range structure of the nucleus. It will be of great help to undertake the study of this two important issues in the same way as it has allowed us, as I have tried to
show it in this talk, to disentangle the dominant mechanisms of the
long and intermediate range parts of the interaction between two
baryons.

References

1) B. Frois, Lectures Notes in Physics 137 (1981) 55 ; Proc. Int. Nat.
   Conf. on nuclear physics, Florence, August 29-September 3 (1983), to
   be published.
2) J. Mougey, Nucl. Phys. A335 (1980) 35.
3) J.M. Laget, Nucl. Phys. A358 (1981) 275c.
4) J.M. Laget, Lectures Notes in Physics 137 (1981) 148.
5) J.M. Laget, Symp. on "Delta-Nucleus Dynamics", Argonne, May 2-4
   (1983), T.H.S. Lee ed., ANL-PHY-83-1, CONF-830588.
6) D. Day et al. Phys. Rev. Lett. 43 (1979) 1143.
7) P. Barreau et al., Nucl. Phys. A402 (1983) 515 and private
   communication.
8) C. Hadjuck et al. Nucl. Phys. A322 (1979) 329 ;
   R.A. Brandeburg et al., Phys. Rev. C12 (1975) 1368.
9) R.V. Reid, Ann. Phys. (N.Y.) 50 (1968) 411.
10) J.M. Laget, Phys. Report 69 (1981) 1.
11) P.E. Argan et al., private communication.
12) J.M. Laget, Nucl. Phys. A358 (1981) 329c.
13) J.M. Laget, Nucl. Phys. A312 (1978) 265.
14) J.M. Laget, Workshop on "Radiative processes in few nucleon
    systems" Vancouver, May 4-5, 1984, Can. J. Phys. (in press).
15) J. Arends et al., Nucl. Phys. A412 (1984) 509.
16) H.J. Gassen et al., Z. Phys. A303 (1981) 35.
17) D. Fallon et al., Phys. Rev. C5 (1972) 1926.
18) E. Jans et al., Phys. Rev. Lett. 49 (1982) 974.
19) J.M. Cavedon et al., Phys. Rev. Lett. 49 (1982) 986.
20) P.C. Dunn et al., Phys. Rev. C27 (1983) 71
21) R. Bornais, Thesis, University of Montréal;
    E. Hadjimichael et al., Phys. Rev. C27 (1983) 831.
22) J. Torre et al., Z. Phys. A300 (1981)319;
    C. Gignoux et al., Nucl. Phys. A203 (1973) 597.
23) G.E. Brown et al., Phys. today 36 (1983) 24 and references
    therein.
24) A.W. Thomas, Adv. Nucl. Phys. 13 (Plenum, N.J. 1983) 1 and
    references therein.
25) L.S. Kisslinger, Phys. Lett. 112B (1982) 307 ;
    G.A. Miller et al., Phys. Rev. C27 (1983) 1669.
26) H.J. Pirner et al., Phys. Rev. Lett. 46 (1981) 1376.
27) S.J. Brodsky, "New Horizons in Electromagnetic Physics",
    Charlotteville, April 21-24, 1982, J.V. Noble and R.R. Whitney eds.,
    p. 170.
28) J.M. Laget, "7ème Session d'études", Biennale de physique nucléai-
    re, Aussois, March 14-18, 1983, J.Meyer ed., LYCEN 8203 (IPN Lyon) ;
    J.M. Laget, "Fourth course of the Int. School of Intermediate
    Energy Physics", San Miniato, 1984, R. Bergère et al. eds., World
    Scientific.
29) M. Bernheim et al., Nucl. Phys. A365 (1981) 349.
30) S.J. Brodsky et al., Phys. Rev. D20 (1979) 2278 and references
    therein.
31) M. Chemtob, Nucl. Phys. A382 (1982) 317.

32) D. Sivers et al., Phys. Rep. 23 (1976) 1 ;
    S.D. Ellis et al., Rev. Mod. Phys. 49 (1977) 753 ;
    J. Jacob et al., Phys. Rep. 48 (1978) 285 ;
    D. Sivers, Annual Rev. Nucl. Part. Science 32 (1982) 149.
33) S.J. Brodsky et al., Phys. Rev. D14 (1976) 3003.
34) R.G. Arnold et al., Phys. Rev. Lett. 43 (1979) 1143.
35) N. Isgur et al., Phys. Rev. Lett. 52 (1984) 1080.
36) M. Chemtob, Nucl. Phys. A336 (1980) 299 ;
    and Nucl. Phys. A358 (1981) 57.
37) J.P. Vary, Nucl. Phys. A418 (1984) 195c.
38) H.J. Pirner, Comments Nucl. Part. Phys. 12 (1984) 199.
39) J.J. Aubert et al., Phys. Lett. B123 (1983) 275.
40) A. Bodek et al., Phys. Rev. Lett. 50 (1983) 1431 ;
    R.G. Arnold et al., Phys. Rev. Lett. 52 (1984) 727.
41) C.E. Carlson et al., Phys. Rev. Lett. 51 (1983) 261.
42) M. Chemtob et al., J. Phys. G, to be published.
43) F.E. Close, "An introduction to quarks and partons", (Academic
    Press,(1979).
44) F.E. Close et al., Phys. Lett. 129B (1983) 346 ;
    R.L. Jaffé et al., Phys. Lett. 134B (1984) 449.
45) G. Cohen-Tannoudji et al., Warsaw Symposium on Elementary Particle
    Physics, Kasimierz, June 1983.
46) C.H.Llewellyn Smith, Phys. Lett. B128 (1983) 107.
47) M. Ericson et al., Phys. Lett. B128 (1983) 112.
48) E.L. Berger et al. Phys. Rev. D29 (1984) 398.

QUARK STRUCTURE OF HADRONS, NUCLEI

# THE SHORT RANGE PART OF THE NUCLEON-NUCLEON INTERACTION AND THE QUARK MODEL

Amand Faessler
Institut für Theoretische Physik
Universität Tübingen
Auf der Morgenstelle 14
7400 Tübingen
West Germany

ABSTRACT. The resonating group method is used to calculate in the six quark model the $^3$S and $^1$S phase shifts of the nucleon-nucleon interaction. For large distances the model is supplemented by π, σ, ρ and ω-meson exchange. The role of the orbital $[42]_r$ symmetry for the short range repulsion is studied. It is shown that at short distances the orbital $[42]_r$ symmetry plays an important role which is even enlarged by the colour magnetic interaction. The $[42]_r$ symmetry enforces the short range repulsion by a node which it requests at short distances. The mechanism is complicated by the fact, that the orbital $[6]_r$ symmetry is admixed by about the same weight. We show that for meson exchanges which mediate the long range behaviour we can now use the $SU_3$ flavour ratios of the meson-nucleon coupling constants even for the ω-nucleon coupling. For the ω-meson one had to use in the OBEP's a ω-N coupling constant twice to three times as large as predicted by $SU_3$ flavour to describe the short range repulsion. We also comment on the different contributions to this conference about the quark-model and the NN interaction.

## 1. INTRODUCTION

The phase shift analyses[1] of the nucleon-nucleon (N-N) scattering yield phase shifts, which are characteristic for a strong short range repulsion. This short range repulsion has been described in the past by the exchange of the ω-meson between the nucleons[2]. But to obtain the observed repulsion one needed to increase the ω meson-nucleon coupling constant $g^2_{\omega NN}/4\pi$ to twice or even three times the value predicted by $SU_3$-flavour in relation to the ρ meson-nucleon coupling ($g^2_{\omega NN}/4\pi$ = 9·$g^2_{\rho NN}/4\pi$ = 4.5). All the other meson-nucleon coupling constants follow roughly the $SU_3$-flavour relations. The discrepancy for the ω meson-nucleon coupling reflects the fact that ω exchange is not the mechanism responsible primarily for the short range repulsion.

With the advent of the quark model and QCD one suggested that the intrinsic quark structure of the two interacting nucleons can explain this short range repulsion[3,4]. These early trials had serious short

319

*J. S. Lilley and M. A. Nagarajan (eds.), Clustering Aspects of Nuclear Structure, 319–332.*
© 1985 by D. Reidel Publishing Company.

comings: (i) They used[3,4,5] the Born-Oppenheimer approximation, which would be only justified, if the effective mass of the quarks is small compared to that of the nucleon. This is not the case in the constituent quark model. Even in the current quark model the energy eigenvalue of the quark is about one third of the nucleon mass. (ii) Another serious short coming of these calculations is the neglect of the orbital $[42]_r$ symmetry for the six quarks at distance zero between the two nucleons. The importance of the $[42]_r$ symmetry has first been pointed out by Neudatchin and coworkers[6,7]. The ansatz for the two nucleon wave function as three quarks in an oscillator potential at $\vec{r}/2$ and three quarks in another oscillator potential at $-\vec{r}/2$ suppresses the $[42]_r$ symmetry like[8]

$$F(r) = [1 - f^2 - f^4 + f^6]^{1/2}$$

$$f(r) = \exp(-r^2/4b^2)$$

$$b^2 = \frac{\hbar}{m\omega}$$

(1)

at small distances r. Harvey[5] included the $[42]_r$ symmetry even at small distances by allowing the excitation of two quarks in p states in the Born-Oppenheimer approximation. He obtained a large effect. The repulsive core[3,4] disappeared. This opened up again the question for the nature of the short range repulsion.

Oka and Yazaki[9] and Faessler and coworkers[8,10] showed that a non-adiabatic treatment with the help of the resonating group method including the $[42]_r$ symmetry yields hard core phase shifts for the $^1S$ and $^3S$ interaction between two nucleons. This short range repulsion is also strongly influenced by the colour magnetic interaction.

This talk is essentially devoted to understand the nature of the short range repulsion of two nucleons. In addition we are at the appropriate points mentioning and commenting the contributions submitted to this conference in the field of the NN interaction[11-14].

In detail we summarize shortly in chapter two the model and discuss a possible mechanism for the short range repulsion. In chapter 3 we give the results and in chapter 4 we summarize the main conclusions.

## 2. INTERACTION OF TWO NUCLEONS IN THE QUARK MODEL

At short distances between two nucleons we describe the nucleon-nucleon (NN) interaction by the exchange of gluons and quarks. The gluon exchange is determined by the quark gluon vertex.

$$L_{int}^{QCD} = i(g/2)\bar{q}_i \gamma^\mu \lambda_{ik}^{(a)} q_k G_\mu^{(a)}$$

(2)

Here $\lambda^{(a)}$ are the eight Gell-Mann colour $SU_3$ matrices. $q_i$ are the Dirac spinors of the quarks with colour i. Eq. (2) includes a sum over repeated indices $\mu = 0,1,2,3$, $a = 1,2,...8$ and i,k = red, blue, yellow. The one gluon exchange between two quarks in the non-relativistic reduction is given by[15]:

$$V^{OGEP}_{(i,j)} = \frac{\alpha_s}{4} \lambda^{(a)}_i \cdot \lambda^{(a)}_j [ \frac{1}{r_{ij}}$$

$$- \frac{\pi}{m_q} \delta(\vec{r}_{ij})(1 + \frac{2}{3}\vec{\sigma}_i \cdot \vec{\sigma}_j)] + \ldots \tag{3}$$

$m_q$ is the constituent quark mass and $\alpha_s = g^2/4\pi$ the strong fine structure constant. The dots indicate terms of tensor and two-body spin-
orbit nature of the quark-quark interaction. They don't play a role
since we restrict us to $^1S$ and $^3S$ interaction between nucleons. The
quark-quark interaction (3) is the leading term only for large momentum
transfer and therefore short distances between nucleons. But even there
one should not take (3) literally. It probably gives only the rough
dependence. Its quality is improved by fitting the parameters $m_q$ =
336 MeV and $\alpha_s$ = 1.3 to the nucleon and the $\Delta$ mass. Suzuki[11] suggests
that one should go in the non-relativistic expansion one term further
and include also a kinetic energy correction term proportional to $p^4_q$
and the colour Darwin term. He shows that the latter one lowers the
short range repulsion. We believe that refitting the parameters $\alpha_s$ and
$m_q$ with this terms yields similar results than without them.

The total six quark Hamiltonian

$$\hat{H}_6 = \sum_{i=1}^{6} [m_q + \frac{p^2_i}{2m_q}]$$

$$+ \sum_{i<j=1}^{6} [V^{OGEP}(i,j) - \lambda^{(a)}_i \cdot \lambda^{(a)}_j \, a \, r_{ij}] \tag{4}$$

must include also a colour confinement term. The parameter a = 41 MeV
fm$^{-1}$ is adjusted to the charge root mean square radius of the proton
including the pion cloud[16].

For large distances one can not exchange colour objects like
quarks and gluons. Thus we go back there to meson exchange. But we
have to guarantee asymptotic freedom. This is done by allowing the
coupling of the mesons to the quarks only near the surface of the
nucleon.

$$g_{qq\mu}(r) = C_\mu r^2 \tag{5}$$

The free parameters $C_\mu$ will be ajusted for each meson $\mu$ to the meson-
nucleon coupling constants[17] at zero momentum transfer $g^2_{\pi NN}(q=0)/4\pi$
= 14.1, $g^2_{\sigma NN}(q=0)/4\pi$ = 5.65 and $g^2_{\rho NN}/4\pi$ = 0.5. For the $\omega$-nucleon coupling we shall see that the flavour $SU_3$ value $g^2_{\omega NN} = 9 \cdot g^2_{\rho NN}$ yields a
good agreement for the NN phase shifts. This is opposed to the OBEP's[2,17]
where the $\omega$-nucleon coupling constant has to be blown up by a factor
two to three to describe the short range repulsion.

In a first step we include only the exchange of the two lightest
mesons ($m_\pi$ = 140 MeV; $m_\sigma$ = 520 MeV). The ansatz for the resonating
group wave function is:

$$\Psi_{6q} = A\{|\overline{NN}\rangle \, \chi_N(r)$$
$$+ |\overline{\Delta\Delta}\rangle \, \chi_\Delta(r) + |\overline{CC}\rangle \, \chi_C(r)\} \tag{6}$$

The Kohn-Hulthen variational principle

$$\langle \delta\psi_{6q}|\hat{H}_6 - E|\psi_{6q}\rangle = 0 \tag{7}$$

yields for $^1S$ and $^3S$ channels three coupled integral equations for the relative wave functions $\chi_N(r)$, $\chi_\Delta(r)$ and $\chi_C(r)$. If one describes the six quark wave function by three quarks in an oscillator at $-\vec{r}/2$ and by three quarks in an oscillator at $\vec{r}/2$, the antisymmetrized wave function has the form[8]:

$$A\{|NN,r\rangle\} = \frac{1}{3} D(r) |[6]_r \{33\}_{ST}[222]_C\rangle$$
$$+ \frac{2}{3} F(r) |[42]_r \{33\}_{ST}[222]_C\rangle - [42]_r \{51\}_{ST}[222]_C\rangle \tag{8}$$

with:

$$D(r) = [1 + 9f^2 + 9f^4 + f^6]^{1/2}$$

$$F(r) = [1 - f^2 - f^4 + f^6]^{1/2}$$

$$f = \exp(-r^2/4b^2)$$

Instead of (8) we are using a basis which allows that the variational principle chooses freely the amplitudes of the different orbital symmetries characterized by Young tableaux.

$$\begin{pmatrix} |NN\rangle \\ |\Delta\Delta\rangle \\ |CC\rangle \end{pmatrix} = A \begin{pmatrix} |(\{33\}_{ST}[222]_C) \, [\tilde{6}]\rangle \\ |(\{33\}_{ST}[222]_C) \, [\tilde{42}]\rangle \\ |(\{51\}_{ST}[222]_C) \, [\tilde{42}]\rangle \end{pmatrix}$$

$$\tag{9}$$

$$A = \begin{pmatrix} \frac{1}{3} & \frac{2}{3} & -\frac{2}{3} \\ -\sqrt{\frac{4}{45}} & -\sqrt{\frac{16}{45}} & -\sqrt{\frac{25}{45}} \\ \sqrt{\frac{4}{5}} & -\sqrt{\frac{1}{5}} & 0 \end{pmatrix}$$

The above basis states contain no spatial dependence. ST and color C are only coupled to the symmetry conjugate to the orbital one. The states $|\overline{NN}\rangle$, $|\overline{\Delta\Delta}\rangle$ and $|\overline{CC}\rangle$ contain also the internal spatial variables for each nucleon. A solution of the resonating group problem (7) yields for each symmetry its own radial dependence. The ansatz (9) includes quark excitations in higher orbital states. If one solves the relative wave function by a coupled system of Hill-Wheeler-Griffin equations.

$$\sum_{\beta=N,\Delta,C} \int dr' \, [<\alpha,r|\hat{H}_6 + V_\pi + V_0|\beta,r'>$$

$$- E<\alpha r|N|\beta,r'>] \cdot \chi_\beta(r') = 0,$$

(10)

it is useful to look to the wave function in the following form:

$$\tilde{\chi}_\alpha(r) = \sum_\beta \int <\alpha \quad r|N|\beta r'>^{1/2} \chi_\beta(r')dr'$$

(11)

According to (9) it can be decomposed into the symmetry basis. Fig. 1 shows such relative wave functions in the $^3S$ channel. One sees a node in the [42] orbital symmetries. The nodes are reflected in the norm-kernels in oscillator functions, which are diagonal in the symmetry basis.

$$<n,l=0|N_{[6]_r}|n,l=0> = 1 + 9(\tfrac{1}{3})^{2n}$$

(12)

$$<n,l=0|N_{[42]_r}|n,l=0> = 1 - (\tfrac{1}{3})^{2n}$$

One sees that the Os state is Pauli forbidden for $[42]_r$. This yields the node in the orbital [42] symmetric relative wave functions. If we would have only $[42]_r$ the node would guarantee a hard core phase shift.

The colour magnetic part of the quark-quark wave function is essential in enlarging the [42] orbital symmetry in the relative NN wave function.

At the end of this chapter we are hopefully convinced that the node of the $[42]_r$ symmetry is responsible for the hard core phase shift. But the $[42]_r$ symmetry has only his important position due to the colour magnetic part of the quark-quark force.

This result is also supported by table 1. It shows the decomposition of the relative wave function between two nucleons $\chi_{NN}(r)$ into the spatial symmetries [6] and [42] and their expansion into oscillator wave functions $|n \quad l=0$. The results with the color magnetic force (with CMI) shows the $[42]_0|1s>$ as the strongest admixture. While without CMI this role is played by $[6]_0|0s>$.

Table 1: Decomposition[20] of $\tilde{\chi}_{NN}(r)$ into the spatial symmetries [6] and [42] without and with the color magnetic interaction (CMI). The number given are the amplitudes squared $|<ns|\tilde{\chi}_{NN}[f]>|^2$ for the expansion into oscillator functions $|0s>$, $|1s>$, $|2s>$ and $|3s>$.

|           | [f]  | $|0_s>$ | $|1_s>$ | $|2_s>$ | $|3_s>$ |
|-----------|------|---------|---------|---------|---------|
| with      | [6]  | 0.450   | 0.045   | 0.005   | 0       |
| CMI       | [42] | 0       | 0.160   | 0.039   | 0.001   |
| without   | [6]  | 0.132   | 0.062   | 0.026   | 0.011   |
| CMI       | [42] | 0       | 0.219   | 0.188   | 0.088   |

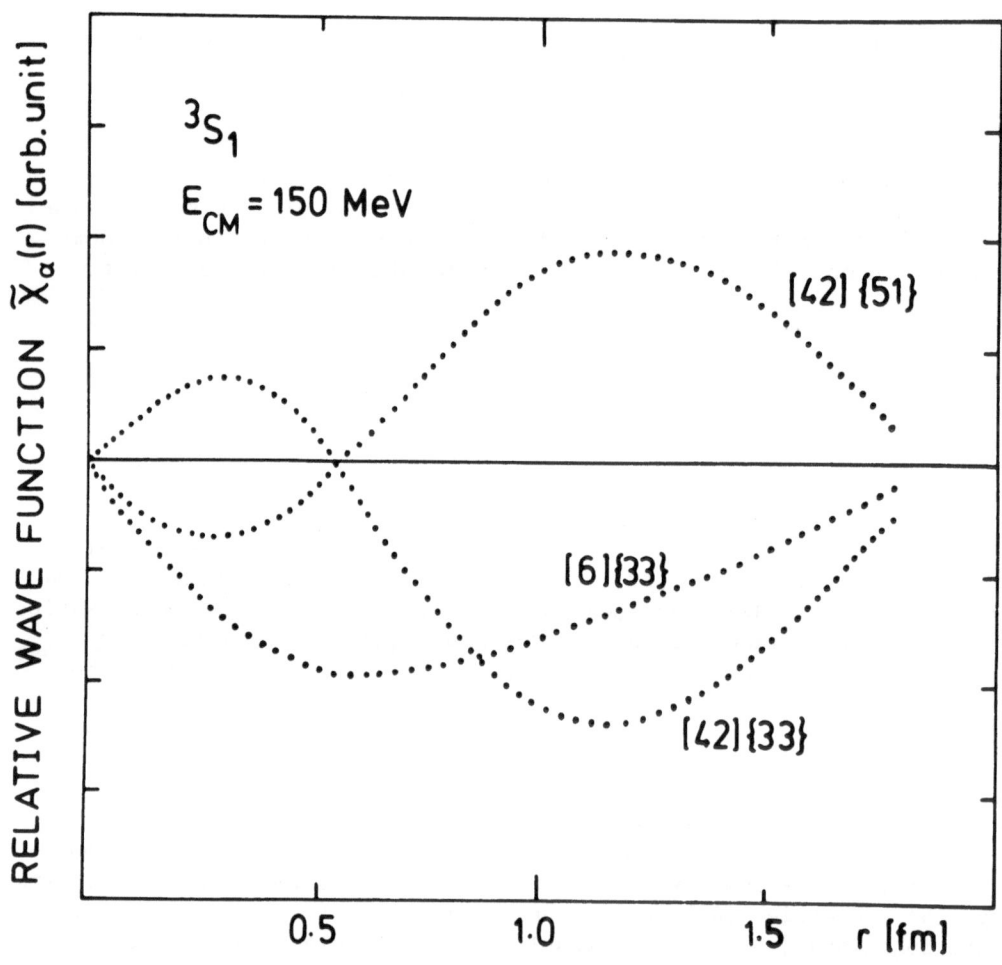

Fig. 1: Relative $^3$S wave functions in the symmetry basis at $E_{CM}$ = 150 MeV as a function of the distance r between the two nucleons.

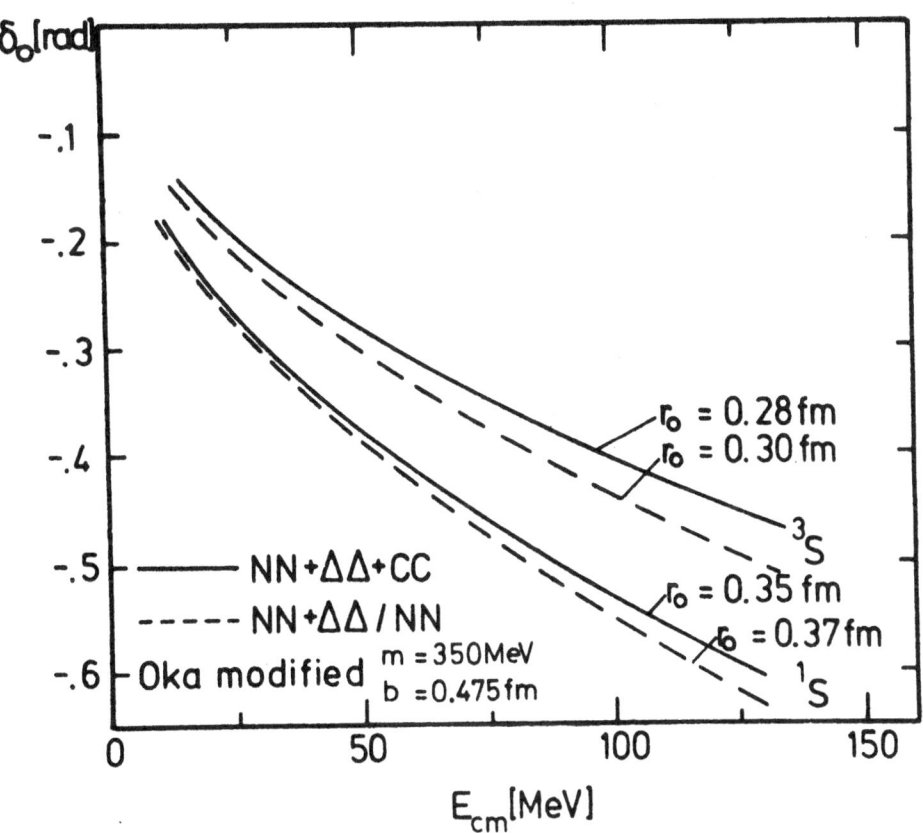

Fig. 2: $^1$S and $^3$S phase shifts calculated[18] without $\pi$ and $\sigma$ exchange in the three channel case (NN, $\Delta\Delta$, CC; solid line) and in the two channel case (NN, $\Delta\Delta$; dashed line). One obtains a typical hard core phase shift $\delta = -r_0 k = -r_0 [3 E_{CM} m_q]^{1/2}/\hbar c$ with the equivalent hard core radius $r_0$.

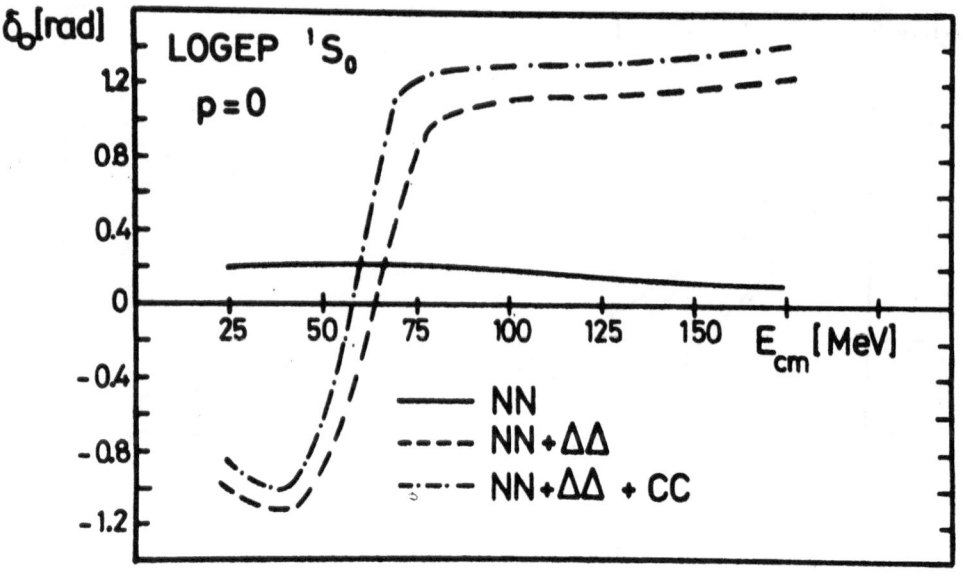

Fig. 3: $^1$S phase shifts calculated[18] without $\pi$ and $\sigma$ exchange
and with putting the colour magnetic interaction to
zero. Now one sees no hard core behaviour for the NN
phase shifts given here. This result is independent
if one takes only one (NN) or two (NN+$\Delta\Delta$) or three
(NN+$\Delta\Delta$+CC) channels.

## 3.  RESULTS

The solution of the coupled system of resonating group equations (10)
yields the relative wave-functions $\chi_N(r)$, $\chi_\Delta(r)$ and $\chi_C(r)$ for the channels with two asymptotic nucleons, two asymptotic $\Delta$'s and the six quark
hidden colour state, respectively. The asymptotic form of the relative
wave function in the NN-channel gives the phase-shift for the nucleon-
nucleon scattering. This calculation is performed including quark and
gluon exchange $\hat{H}_6$ and the exchange of $\pi$- and $\sigma$-mesons. The parameters
are adjusted as described in chapter 2. The results are given for the
$^1$S and $^3$S in figures 4 and 5, respectively.

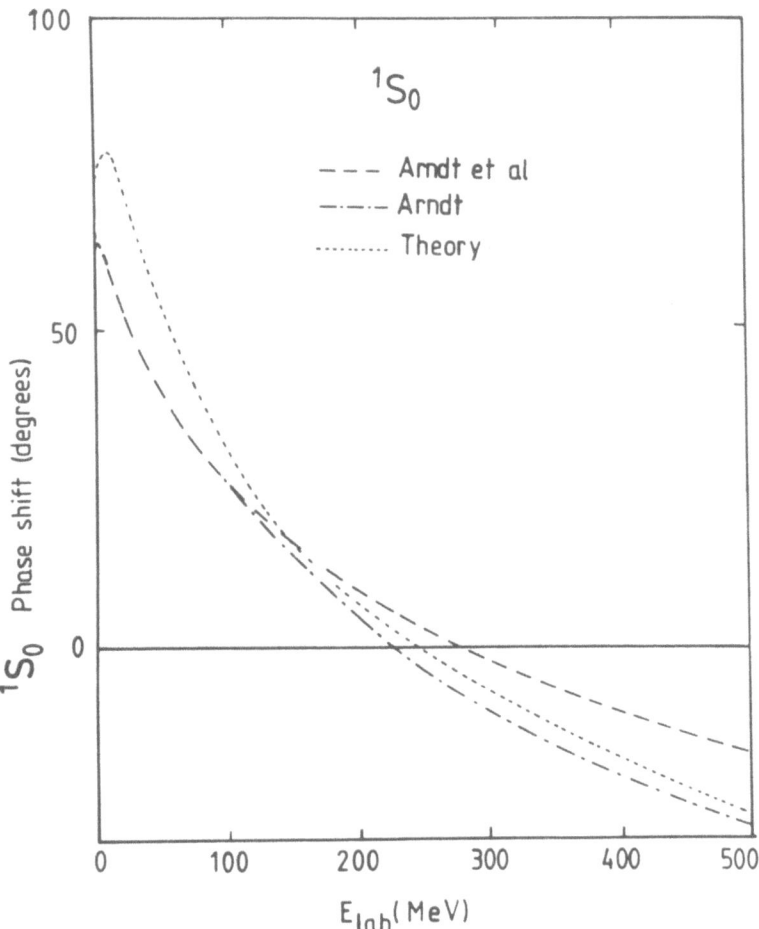

Fig. 4: $^1$S nucleon-nucleon phase shifts as a function of the laboratory energy. The dotted curve gives the results of the theory presented[16]. The dashed and the dashed-dotted line are two different sets of experimental phase shifts[19].

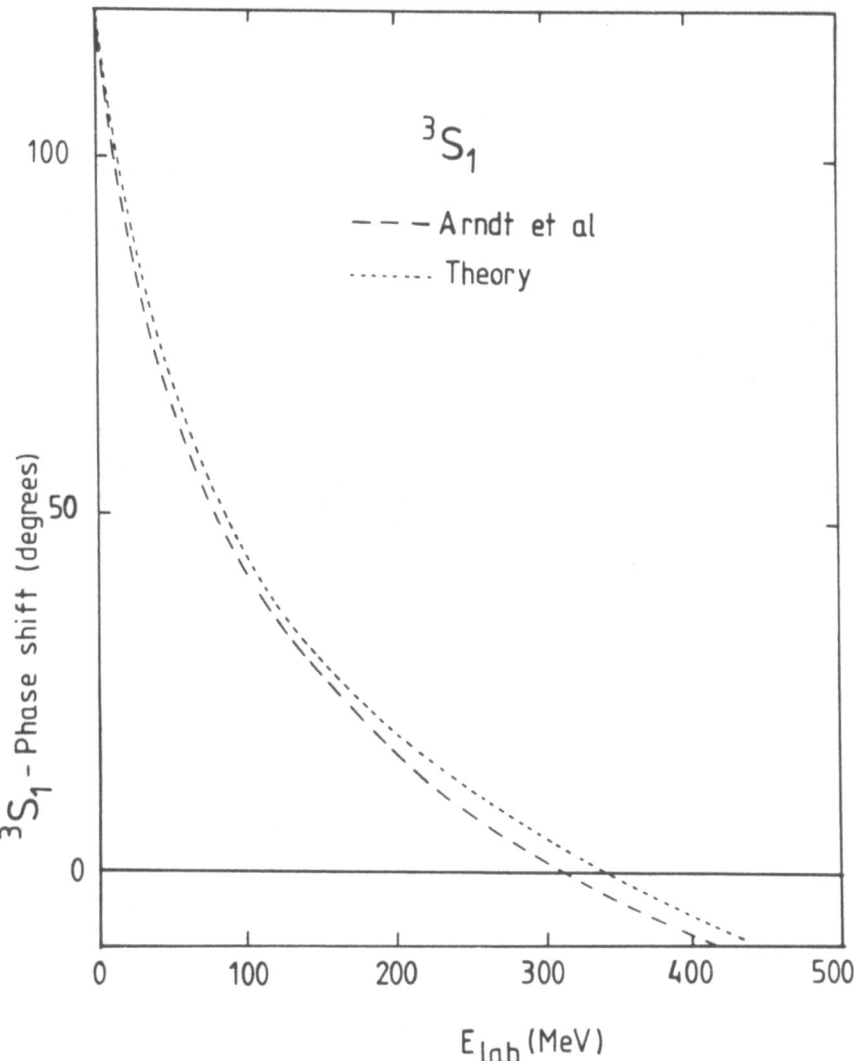

Fig. 5: [3]S nucleon-nucleon phase shifts as a function of the laboratory energy of the nucleon projectile. The dotted curve is the present calculation[16] and the dashed curve represents the experimental phase shifts[19].

In the results presented in figures 4 and 5 we included only the exchange of the two lightest mesons, the π- and σ-meson. It is interesting to see what happens if we include also in addition the $\rho$- and the $\omega$-mesons. This question is especially exciting since in the one meson exchange potentials the $\omega$-meson is solely responsible for the short range repulsion. Out of this reasons one adjusts in the OBEP's the $\omega$-nucleon coupling constant to a value which is by a factor two to three times larger than the flavour $SU_3$ value derived from the $\rho$-nucleon coupling. If we now have the correct nature of the short range repulsion, we should get a satisfactory fit to the phase shifts, if we use the flavour $SU_3$ value $g_{\omega NN}^2/4\pi = 4.5$. Figure 6 shows the $^3S$ phase shifts for different $\omega$-NN coupling constants in radians as a function of the nucleon bombarding energy. The factor $g_0$ is defined as the deviation from the flavour $SU_3$ value.

The figure shows that for a reasonable good nuclear radius one gets a good agreement for the phaseshifts for a value of the parameter $g_0 = 1$ in agreement with the flavour $SU_3$ symmetry. This results supports strongly our conviction that we found indeed the correct nature of the short range nucleon-nucleon repulsion.

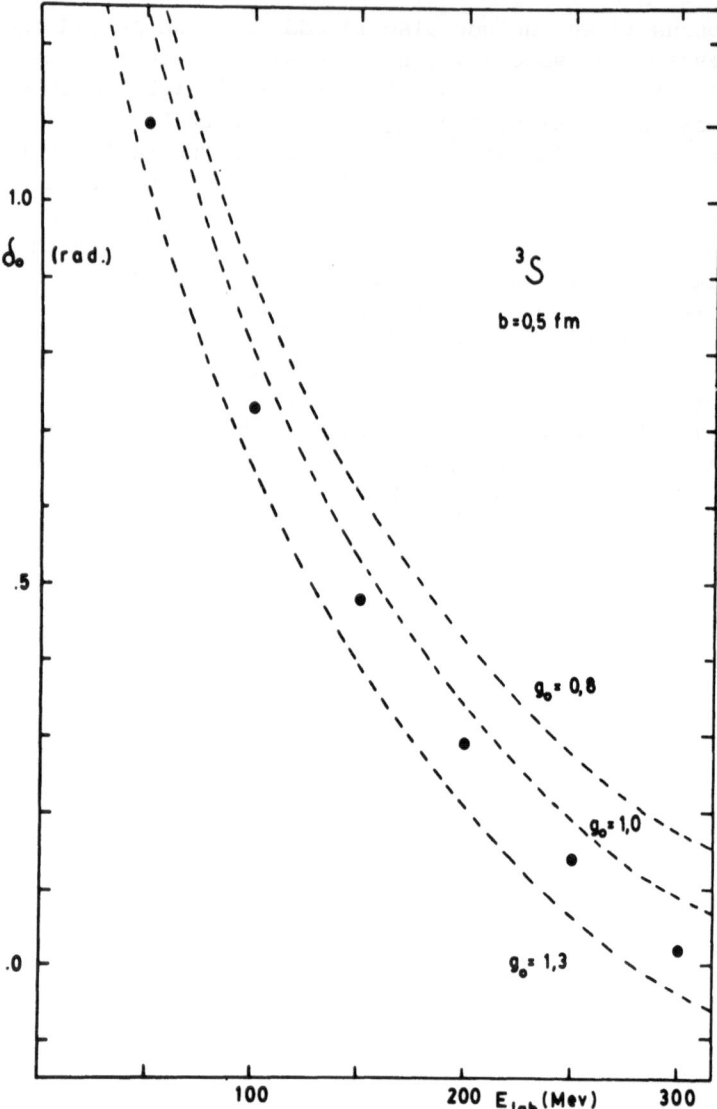

Fig. 6: $^3$S nucleon-nucleon phase shifts in radians as a
function of the laboratory energy for the oscillator
length b = 0.5 fm. The ω-nucleon coupling constant is
chosen as $g_{\omega NN}^2/4\pi$ = 4.5 $g_O$. The data are taken from
ref. 19.

## 4. CONCLUSIONS

The NN phase shifts are calculated using the quark model with a QCD inspired quark-quark force. The short range part of the NN force is given by quark and gluon exchange. The long range part is described by $\pi$ and $\sigma$-meson exchange. The data fitted in the model are five values connected with three quarks only: The nucleon mass, the $\Delta$ mass, the root mean square radius of the charge distribution of the proton including the pion cloud, the $\pi$-N and the $\sigma$-N coupling constant at zero momentum transfer. The $^1S$ and $^3S$ phase shifts are nicely reproduced. The short range repulsion is decisively influenced by the node in the $[42]_r$ relative wave function. Very important is the colour magnetic quark-quark force which enlarges the $[42]_r$ admixture.

In the OBEP's the short range repulsion is connected with the exchange of the $\omega$-meson. But to reproduce the short range repulsion one had to blow up the $\omega$-N coupling constant by a factor 2 to 3 compared to flavour $SU_3$. With quark and gluon exchange the best fit to the $\omega$-N coupling constant lies close to the $SU_3$ flavour value. This fact strongly supports the notion that we have found the real nature of the short range repulsion of the NN interaction.

I would like to thank Dr.'s Bräuer, Fernandez, Lübeck and Shimizu with whom I was working on the results reported above.

## REFERENCES

1.  R. Arndt et al., Phys.Rev. C 15 (1977) 1002 and R. Arndt, 'Nucleon-Nucleon Interaction', 1977 Vancouver conference, p.
2.  K. Holinde, Phys.Rep. 68 (1981) 191.
3.  D.A. Libermann, Phys.Rev.D 16 (1977) 1542.
4.  C.E. De Tar, Phys.Rev. D 17 (1978) 323.
5.  M. Harvey, Nucl.Phys. A 352 (1980) 301 and 326.
6.  V.G. Neudatchin, I.T. Obukhovsky, V.I. Kukulin, N.F. Golanova, Phys.Rev. C 11 (1975) 128.
7.  V.G. Neudatchin, Yu.F. Smirnov, R. Tamagaki, Progr.Theor.Phys. 58 (1977) 1072.
8.  A. Faessler, F. Fernandez, G. Lübeck, K. Shimizu, Nucl.Phys. A 402 (1983) 555.
9.  M. Oka, K. Yazaki, Progr.Theor.Phys. 66 (1981) 551 and 572.
10. A. Faessler, F. Fernandez, G. Lübeck, K. Shimizu, Phys.Lett. B 112
11. Y. Suzuki, 'Relativistic corrections to the short-range part of the NN interaction' and 'Relative importance of the quark-exchange kernels to the NN repulsive core"; contributions to this conference.
12. J. Burger, H.M. Hoffmann, 'Extension of the refined resonating group method to the six quark system'; contribution to this conference.
13. Y.E. Kim, M. Orlowski, 'Resonating group quark-cluster model of nuclei: two-nucleon systems'; contribution to this conference.

14. Wang Fan, He Yin, Zhuang Feiand, King Sing Nan, 'On the colour van der Waals force in the quark cluster model of the NN interaction"; contribution to this conference.

15. A. De Rujula, H. Georgi, S.L. Glashow, Phys.Rev. D 12 (1975) 147.

16. A. Faessler, F. Fernandez, Phys.Lett. B 124 (1983) 145.

17. K. Holinde, R. Machleidt, Nucl.Phys. A 256 (1976) 479.

18. E.G. Lübeck, Diplomarbeit, Universität Tübingen, February 1982.

19. R. Arndt et al., Phys.Rev. C 15 (1977) 1002; R. Arndt, 'NN interaction', 1977 Vancouver Conference, p. 117.

# IS QCD RELEVANT TO NUCLEAR PHYSICS

A.W. Thomas
Department of Physics, University of Adelaide
Adelaide South Australia 5001.

**ABSTRACT.**   We review recent work on baryon structure in a number of QCD-motivated models.   After establishing a prima facie case that the quark model should be relevant if we are to have a consistent description of the nucleus over a wide range of momentum transfer, we look for experimental confirmation.   The discussion includes the search for exotic states, for a six quark component of the deuteron, and an up to date report on the interpretation of the EMC effect.

## 1.   INTRODUCTION

Many of those working in the darkness of nuclear physics purport to have seen a great light.   This light is Quantum Chromodynamics, which for the first time gives us a firm theoretical foundation on which to build our microscopic understanding of strongly interacting systems.   Of course it will require a very powerful light indeed to penetrate the murk that lies between a truly microscopic theory of the strong inter- actions and the wealth of fascinating phenomena which constitute modern nuclear physics.   Even the most ardent advocates of QCD would accept that it will be a long time before sub-coulomb stripping (for example) can be adequately described at the quark level.   Indeed, few would argue that it would make sense to ever do so.   However, we believe very strongly that the quark model will provide a simpler and cleaner explanation of a variety of nuclear phenomena which already have a conventional explanation.   Even more exciting, we believe that new phenomena will be discovered in the next few years which can only be understood in a unified framework.

On the other hand it must be admitted that there are powerful voices in the nuclear community who, while accepting that QCD is the way to understand the properties of individual hadrons, would deny that it has any direct relevance to nuclear physics. Our purpose here is to review the state of play in this debate.   Clearly in fifteen pages it is not possible to do justice to any piece of work.   All we can hope is to touch the main issues in a fashion which although not unbiased at least acknowledges all points of view. Quite extensive references are given to enable the reader to follow up any question of particular interest.

*J. S. Lilley and M. A. Nagarajan (eds.), Clustering Aspects of Nuclear Structure, 333–352.*

The organisation of the review is as follows. After a brief confession that it is not QCD itself but phenomenological models supposed to model QCD which we are applying to nuclear systems, we review the most commonly used models in Sect. 2. In Sect. 3 we touch only briefly on how these models are being used to develop a quark level treatment of nucleon-nucleon (N-N) scattering. This question will be treated in a far more detailed fashion by Professor Faessler (1). Instead we concentrate on the phenomenological attempts to identify an effect which can only be associated with quark degrees of freedom – a search for the so-called "smoking gun" (2). We shall see that although there have been a couple of promising suggestions, there is a desperate need for new data.

The fourth section deals with real nuclei (A > 2). Particular attention is paid to the formal problems of constructing hybrid descriptions of nuclei involving both nucleons and quarks. In Sect. 5 we turn to deep-inelastic scattering (DIS) as a direct probe of the quark structure of nuclei. We examine the various explanations for the dramatic dependence of the structure function of a "nucleon" on the size of the nucleus containing it – the EMC effect (3). Even though there is no consensus on which of these explanations is correct, it is clear that this phenomenon is of central importance in any discussion of the relevance of QCD to nuclear physics.

Finally, it seems appropriate to call attention to some recent publications which provide useful background material (4-8).

## 2.   HADRON STRUCTURE

The essential motivation for talking of quarks in the context of nuclear physics arises from our deepening understanding of hadron structure. This does not mean that one can yet draw reliable conclusions about the structure of the nucleon from QCD. In spite of a tremendous amount of effort devoted to lattice calculations the state of the art still involves putting a nucleon inside a box about 1.5 fm on a side (on a $10^3$ grid), and neglecting the effect of $q\bar{q}$ creation and annihilation (9). The dependence on the boundary conditions (10) indicates that for baryons at least the box is not large enough. Furthermore, the neglect of $q\bar{q}$ creation and annihilation may be very poor if some of the ideas mentioned in Sect. 2.2 (or worse still, Sect. 2.3) are correct.

A lack of constraint from the underlying theory has led to a diversity of phenomenological quark models of hadron structure. However, whether these models involve constituent quarks (11,12,13) or current quarks in a bag (14) or confining potential (15), the majority of QCD motivated models involve three valence quarks which move freely over a volume comparable with the observed charge distribution of the proton ( $< r^2 >^{1/2} \sim 0.82$ fm ). Given that (at $\rho_0$) **on average** nearest neighbour nucleons are about 2 fm apart it is hard to believe that the structure of a nucleon will not be significantly altered for some sizeable fraction of the time it spends in a nuclear environment.

We shall return to this issue in Sects. 3, 4, and 5. For the present we catalogue the phenomenological models which are available and comment on possible experimental tests which might distinguish amongst them.

## 2.1    Constituent Quark Models

This approach assigns each of the non-strange quarks a mass of order $\frac{1}{3} m_N$, and assumes that they move non-relativistically. While having little in the way of a demonstrable connection to QCD (it arose almost a decade before QCD), it gives surprisingly good results for $\mu_p$ and $\mu_n$. By assigning a strange quark mass several hundred MeV higher than that of u and d, the elementary quark model mass formulas were given flesh and bones. However, the real power of this approach did not become clear until the mid-70's, when Isgur and Karl (16) made a serious numerical study of the ideas of De Rujula, Georgi and Glashow (11). As well as obtaining a remarkable overall description of baryon and meson spectroscopy, their calculations provided the first respectable explanation of why so many multiplets expected in the naive quark model have not yet been found (17).

Apart from the arguments of Ref. (11) the main theoretical support for this model as a consequence of QCD has come from Shuryak (13). Briefly he suggests that QCD sum rules indicate quite a different scale for chiral symmetry breaking (short range) as opposed to confinement. He identifies the constituent quarks as complicated clusters (radius of order 0.3 fm) formed around a current quark. Pions would move essentially freely between these clusters (18), in a way which is formally at least strikingly similar to the volume coupling version of the CBM (Sect. 2.2 and Refs. (19-22)). In the constituent quark model it was always necessary to assign a charge radius to the quark itself in order to reproduce the charge radius of the proton. Similarly the valon model (23) assigns each quark a structure function in order to confront DIS data. Overall perhaps the least satisfactory aspect of the constituent quark models is the large number of ad hoc parameters and adjustable assumptions.

From the point of view of phenomenological treatments of nuclear systems however, the non-relativistic quark model is ideal. With non-relativistic kinematics, harmonic oscillator wave functions and quark-quark interactions, all the familiar apparatus of nuclear many body theory can be used.

## 2.2    Bag Models

The pre-QCD origins of the bag model are well known (4, 24), but its popularity dates from the realization that both confinement and asymptotic freedom are expected consequences of QCD which the MIT bag (14, 25) incorporates. That the model necessarily violates chiral symmetry at the bag surface, and that this can be restored by coupling pions to the confined quarks at the bag surface has also been described at length elsewhere (4, 26, 27). One of these so-called hybrid chiral bag models (28), the Cloudy Bag Model (CBM) (29), has enjoyed enormous phenomenological success. (For comprehensive reviews of the CBM and the

very closely related work on relativistic potential models with chiral
symmetry see Refs. (4, 15, 22, 30)).

Much of the sharpest fighting in the early days of the debate about
quarks in nuclear physics came out of this work on chiral bag models.
The Stonybrook group (27,31) argued that the non-linear coupling of the
pion field to the nucleon bag would compress it to insignificance on the
scale of typical internucleon distances, (hence the little bag model).
On the other side, the West-coast argument (4, 22, 30) was that the pion
coupling should be treated in low order perturbation theory and should
complement, not destroy, the phenomenological successes of the MIT bag
model. Clearly in the absence of theoretical guidance from QCD itself,
the only way of choosing between alternative models such as these is to
test them against the widest variety of experimental phenomena. We
shall return to this in Sect. 2.4.

Non-topological solitons of the kind proposed by Friedberg and Lee
(32) and investigated numerically by Goldflam and Wilets (33) replace
the bag model with its boundary conditions by a scalar, confining
field. In many ways these models offer more attractive possibilities
for extending what are essentially bag model ideas to the nuclear many-
body problem (34). On the other hand the non-topological solitons are
quite different and we turn to those next.

## 2.3  Topological Solitons

The discovery of classical, solitary waves is over one hundred years
old, and the remarkable idea that baryons could be solitary waves
generated by interacting bose fields was first published by Skyrme about
twenty years ago (35). However, it is only in the past few years that
Witten and others (36) have claimed that these soliton solutions might
have some relationship to QCD. These claims are based on results for
QCD in the formal limit where the number of colours ($N_c$) becomes large
(36-38). In that limit QCD is supposed to look like a weakly coupled
field theory of mesons – see however Ref. (39) – which would allow
soliton like solutions.

Under the assumption that three is a good approximation to infinity,
the idea is to start with the Lagrangian density of the non-linear
σ-model  (f = 93 MeV is the pion decay constant)

$$L_2 = \frac{f^2}{4} \, \text{Tr} \, (\partial_\mu U \, \partial^\mu U^\dagger), \tag{1}$$

where

$$U = \exp (i \, \underset{\sim}{\tau} \, \underset{\sim}{\phi} \, (\mathbf{x})/f). \tag{2}$$

(here $\underset{\sim}{\phi}$ is the pion field) and look for soliton-like solutions. In
order to satisfy Derrick's Theorem (40), one must add a purely ad hoc
higher order term

$$L_4 = \frac{1}{2e^2} \, \text{Tr} \, [ \, (\partial_\mu U) \, U^\dagger, \, (\partial_\nu U) \, U^\dagger \, ]^2 \, , \tag{3}$$

where the dimensionless parameter e was introduced by Skyrme (35). The crucial observation now is that even though the Wess-Zumino term (41) vanishes for SU(2) x SU(2), the anomalous baryon number current which can be obtained from it does not vanish. Thus one has a topological charge which can be identified with baryon number (35), and the nucleon is then a soliton made of weakly interacting pions! It is remarkable that this extreme model has proven capable of giving a semi-quantitative description of the static properties of the nucleon.

The final proposal which we mention is a non-linear combination of the chiral bag and soliton ideas (38, 43-45). There the pions couple to the surface of a relatively small bag, and carry an appreciable fraction of the baryon number (all of it in the limit R → 0). This approach is really just a refinement of the little bag model described in Sec. 2.2.

## 2.4   Which Model Should One Believe ?

Although this is a question which is widely debated at the present time it is quite meaningless as it stands! QCD is a quantum field theory, and one cannot sensibly talk about the structure of a composite object without specifying the scale of momentum transfer at which you wish to examine it (46). Several years ago Jaffe and Ross (47) made a suggestion which goes to the heart of this question. They guessed that the MIT bag model, which describes the nucleon as simply three valence quarks with no sea, might represent the structure of the nucleon at some relatively low value of the momentum transfer, $\mu_0^2$. The comparison with the observed quark momentum distributions at higher $Q^2$ (typically beyond 5 GeV$^2$, where higher twist contributions to the data should be small), could then be made by evolving upwards using perturbative QCD.

In spite of the fact that the reference value of $\mu_0^2$ (0.75 ± 0.12 GeV$^2$) was too low to believe that the perturbative QCD calculation was totally reliable, good agreement with experiment was obtained for the moments of $F_3(x, Q^2)$ - the valence distribution. Thus not only does the bag model make rather sensible predictions for the static properties of the ground state baryons, but it also describes the twist two component of lepton-nucleon DIS at a momentum scale a little below 1 GeV. It is extremely difficult to see how the topological soliton could do so. Indeed, at face value the latter would seem to imply that the nucleon structure would be dominated by sea quarks, which is not the case at the momentum transfer accessible to present day experiments. This analysis strongly suggests that the bag model may provide a far more useful framework for interpreting the next generation of experiments at places like SLAC, NEAL (2), Saclay (48) and Bonn.

With reference to the hybrid chiral bag models, we note that as long as the bag radius is kept fairly large (R ≳ 0.8 fm), so that the non-perturbative effects of the pion field are small (49), we do not expect its inclusion to upset the analysis of Jaffe and Ross. Of course this pion cloud will contribute predominantly to the non-strange sea of the nucleon - a fact used in Ref. (5) to put a lower limit on the radius of the nucleon bag in the CBM of 0.87 ± 0.10 fm. It may therefore be that the reference scale, $\mu_0^2$, at which one could expect the twist-two contributions in the CBM to match data could be a little higher than Jaffe and Ross found. Because this would make the use of perturbative

QCD a little more believable this idea deserves, and is receiving further attention.

In conclusion, it seems that for some purposes (such as the calculation of static nucleon properties) the Skyrmion and bag model descriptions may be equivalent. Nevertheless, if one wants a unified picture of nuclear phenomena over range of momentum transfers and projectiles it is the bag model (or other explicit quark model) phenomenology which provides the best starting point.

## 3.    THE TWO-NUCLEON SYSTEM

There has already been a lot of effort put into the calculation of the short and intermediate range N-N force in various quark models. Within the non-relativistic quark model one can readily apply many of the techniques of conventional nuclear physics, such as the resonating group method.    Early calculations indicated a relatively weak short-range repulsion in s-wave – in comparison with the usual phenomenological potentials (51).    However, the strong mixing of a $(1s)^4$ $(1p)^2$ configuration with the spatially symmetric $(1s)^6$ introduces a node in the short distance wave function – simulating a hard core (52). Recent calculations along these lines have confirmed the early expectations and resulted in quite good fits to s-wave phase shifts (53).

The MIT bag model presents far more technical problems.    It was first applied to s-wave N-N scattering by De Tar (54), who found both strong short range repulsion and intermediate range attraction.    The latter was a spurious result of a colour coupling constant that was too big, and it has been claimed that the former was essentially spurious c.m. kinetic energy (55).    Recent work has produced semi-quantitative agreement with experiment after the addition of phenomenological σ – and π–meson exchange (56).

The P-matrix method of Jaffe and Low (57) is a less ambitious way to include the effect of an exotic bag state on two-body scattering.    For a summary of the many applications of this method to N-N elastic scattering we refer to the recent articles by Mulders and Simonov (58).    Because an overview of quark models of N-N elastic scattering is being given elsewhere at this meeting (59), we shall turn immediately to the phenomenological aspects of the problem.    That is, we wish to ask whether there exists any compelling, experimental evidence that a quark level description of the short-distance N-N system is necessary.

## 3.1    DIS

The deuteron is the cleanest nuclear system in which to look for a reason to explicitly consider quarks.    Almost eight years ago Kobushkin (60) proposed using the presence of quarks with a momentum fraction x ( $= Q^2/2m_N \nu$ ) greater than that allowed for a free nucleon (i.e. x > 1) as a measure of the 6q probability of the deuteron.    Unfortunately the data in this region is not yet as firm as one would like, and improving this should certainly have a high priority for the future. In addition, the calculation of fermi motion corrections is quite ambiguous for large x (61). For the present these problems preclude a firm conclusion about a 6-quark component in the deuteron.

A recent suggestion of Mulders and Thomas (62) overcomes the objection about fermi-motion corrections, but still demands new data. It is well known that the u and d distributions in the proton are not equal. In fact, by comparing electron (or muon) DIS from hydrogen and deuterium one finds

$$\tilde{d}(x)/\tilde{u}(x) \simeq 0.57 (1-x) \qquad (4)$$

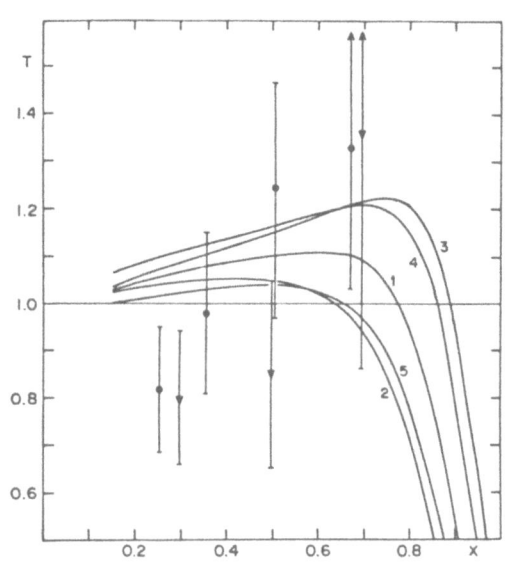

**Figure 1** The ratio of the true to effective d/u ratios calculated with a 5% 6-quark component in the deuteron. (62). Curves 1-5 correspond to reasonable guesses at the 6q valence distribution. The data is from Refs. (63) (dots) and (64) (triangles).

The tildas indicate that in our new way of looking at the deuteron, as being some of the time a six-quark state, the u and d distributions obtained under the old procedures may not be the true ones. The latter can be determined unambigously from $\nu$ and $\bar{\nu}$ experiments in hydrogen, but the low event rate means a lack of statistics for the moment. In Fig 1 we show the ratio, T, of the true to effective d/u ratios for a 5% six-quark component in the deuteron, in comparison with recent data (63, 64). The different theoretical curves correspond to a reasonable variation about the form of the six-quark valence distribution given by naive counting rules (65). Fermi motion has a negligible effect on T for x below 0.8 (62). While we can clearly draw no firm conclusion from Fig 1. at present, the high sensitivity of T (a 20-30% effect for a 5% admixture) suggests that accurate, new $\nu$ and $\bar{\nu}$ data on hydrogen would be extremely valuable.

## 3.2 Elastic Form Factors

At this stage it is almost canonical to recall the quark counting rules for a composite target, namely

$$F_N(Q^2) \sim (Q^2)^{-(N-1)} Q^2 \to \infty \tag{5}$$

and show that the data seems to yield N = 2 for the pion, N = 3 for the
nucleon and N = 6 for the deuteron (66). However, the data for
deuterium is less than convincing, with $Q^{-5.5}$ actually giving a better
fit to existing data (2). This strongly suggests that even 6 $GeV^2$ is a
long way from asymptotia. Further evidence to support this suspicion
comes from a recent study by Isgur and Llewellyn Smith (67), in which it
is claimed that the hard (perturbative QCD) contribution to $G_M(Q^2)$ for
the nucleon fails by almost
two orders of magnitude to
reproduce the data – see Fig
2. To summarise, it is
unlikely that the asymptotic
behaviour of elastic form
factors will provide
unambiguous evidence for a
multi-quark component in the
deuteron in the near future.
On the other hand, Carlson
and Gross have recently
proposed a rather
complicated spin measurement
for the deuteron which they
claim would constitute a
true "smoking gun" (2).
Briefly the ratio $P_x/P_{xz}$
uniquely determines the
ratio of the monopole ($G_c$)
and quadrupole ($G_Q$) form-
factors. A careful helicity
analysis in QCD suggests
that

$$R = \frac{4 M_d^2}{Q^2} \frac{G_c}{G_Q} \to \frac{2}{3}, Q^2 \to \infty. \tag{6}$$

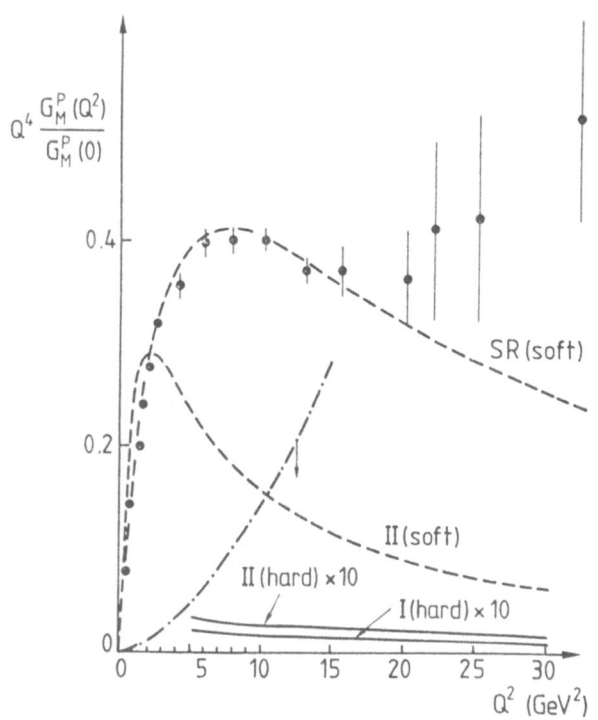

**Figure 2** Comparison of the
leading asymptotic contribution to
the proton magnetic form-factor
calculated with $\Lambda$ = 150 MeV
(labelled "hard"), with that based
on conventional quark model wave
functions ("soft") (67).

Most important they claim
that conventional meson
theory cannot reproduce this
result. Typically they find
$P_x/P_{xz} \in$ (-0.25, -0.50) at
$Q^2 \simeq 2 GeV^2$, in comparison
with + 0.25 from Eq. (6) –
see Fig. 3. If substantiated
by further theoretical
work[*] this measurement
should have a very high priority.

[*] One outstanding question is whether, in view of Ref (67) the
range of $Q^2$ considered is really asymptotic.

## 3.3   Exotic States

A common feature of all quark models is that in addition to the familiar q-q̄ mesons and qqq baryons, they predict a variety of so-called exotic states - such as 4q-q̄ baryons, 6q baryons etc.  Perhaps the greatest hobby-horse in intermediate energy physics has been the search for di-baryon resonances.  However, there is still not one jot of experimental evidence which can clearly not be explained in a more conventional way - e.g. as an N-Δ threshold effect (68).  The beauty of the original P-matrix work of Jaffe and Low was that it explained why the 6q states which couple to the N-N elastic channel (69) produce very little effect on N-N elastic scattering.

A far more convincing case for exotica could be made if one were to observe a state with quantum numbers impossible to make from a q q̄ or q³ configuration - for example a glueball (70), or an hermaphrodite (71).  It is even possible that such a state has recently been seen at LEAR (72), but we await more data.  There is also quite a bit of evidence for B=1, S = +1 Z* resonances in KN scattering, which could only be made as a 4q - q̄ state (73).  Finally, one once hoped to find exotic states (albeit with non-exotic quantum numbers) below the threshold for decay into two baryons - for example, the H-dibaryon (Λ Λ) (69).  Unfortunately, further work on c.m. and pionic corrections suggests that the H is most likely unbound (55, 74).  In any case there is no experimental evidence for it (75).

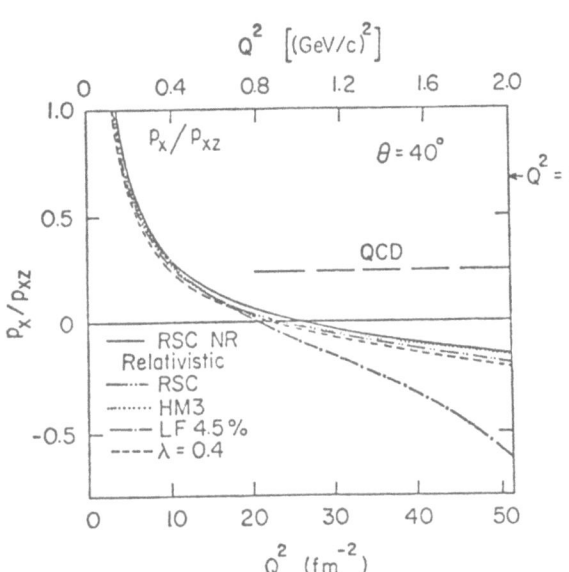

## 3.4   The Quark Model for Conventional Physics

If our thesis that a consistent microscopic picture of nuclear systems can only be built at the quark level is correct, then it is extremely important to develop the tools for

**Figure 3** Prediction of the ratio $P_x/P_{xz}$ for e-d elastic scattering in several meson theoretic models (2).

nuclear reactions in terms of the quark model.  The two-nucleon system is a simple, natural place to begin.  Already a number of calculations of this kind have been made.  These include deuteron electro-disintegration (76), parity violation in np → dγ (77) and N-N scattering (78), pion production (pp → π⁺d) (79) and even the asymptotic D/S ratio of the deuteron wave function (80).  For an excellent review of these matters we refer the interested reader to Ref. (81). Since these calculations, while very important, will not convince any unbiased observer of the necessity for a quark model description of nuclear physics, we shall not consider them further here.


## 4.    QUARK EFFECTS IN REAL NUCLEI


The classical game of nuclear physics was played according to a simple set of rules (82):

(a)    all nucleons are point-like with respect to strong interactions;

(b)    they interact through a two-body potential;

(c)    the interaction of a strong probe with the nucleus is to be calculated primarily in terms of nuclear structure information and the free projectile-nucleon potential.

One cannot help but be amazed at the successes produced by several decades of many-body theory within this frame- work. Nevertheless there are classic cases where the explicit inclusion of meson exchange currents is essential (7, 83, 84). More recently there has been a heated debate about the presence of virtual $\Delta$'s (and other baryon resonances) in the nucleus, particularly in connection with the discovery of Gamow-Teller resonances in (n,p) scattering (85).

A major problem with the conventional picture of baryon resonances and heavy meson exchange is that it is essentially non-convergent. The exchange of a $\rho$-meson already has a characteristic length scale of 0.2 fm, but there is nothing to suggest that it is more important than $\omega\pi$, $\phi\pi$, $\rho\rho$ or any more exotic mixture one can exchange. Imposing gauge invariance in the presence of meson-baryon form-factors which have no theoretical basis is a nightmare. Indeed, the moment one moves away from processes dominated by pion exchange, where PCAC provides strong constraints, the conventional approach is a mess (86). There is real hope that a reformulation of the problem at the quark level will be a great help in this regard.

### 4.1 How do we formulate nuclear physics at the quark level ?

An extreme answer to this question is being explored at Bonn. One simply puts all 3A quarks into one large bag. Absurd as it seems, some of the features of the usual nuclear shell model do emerge from this picture (87). While it deserves, and is receiving, further attention, we feel that this model is unlikely to be correct. Phenomenological models of nucleon structure all seem to yield a typical size for the quark confinement region of less than 1.1 fm, with a consensus somewhere between 0.7 and 1.0 fm. Even without dynamical effects which may tend to keep bags apart, it is hard to see how they could be permanently

merged even at nuclear matter density where the average separation of
nearest neighbours is a little over 1 fm.  In the nuclear surface (say
$\rho_0/2$) it is almost inconceivable.

With this in mind the logical approach is to develop a hybrid
description of the atomic nucleus.  Within a boundary condition model we
could talk about a nucleon when the separation distance between it and
all other nucleons is greater than some critical distance b.  Whenever
two nucleons come closer than that we would treat them as six quarks,
three nucleons as nine quarks, and so on.  Formally (88) we would define
the conventional uncorrelated shell model wave function to be

$$\Psi_N (1, 2.....A) = \mathbf{A} \left\{ \prod_{i<j}^{A+1} [ 1 - \theta (b-r_{ij}) ] \right.$$

$$\left. \prod_{i=1}^{A+1} \phi_{\alpha_i}(i) \right\}, \tag{7}$$

where $\phi_{\alpha_i}$ are the single-particle wave functions with quantum numbers
$\alpha_i$.  Equation (7) invites a systematic expansion in terms of the number
of overlapping nucleons.  To first order one would simply choose the
probability of the six-quark-(A-2)-nucleon component to give the correct
overall normalisation.

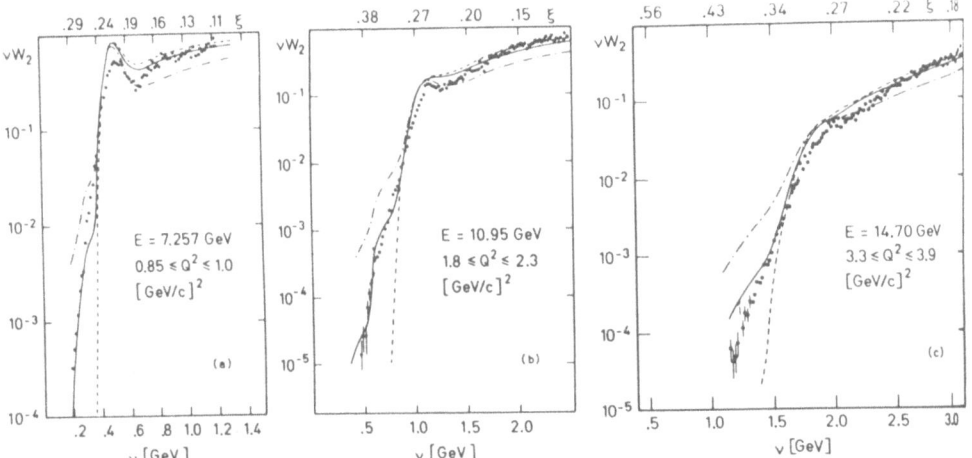

**Figure 4**      The large x region in $^3$He electrodisintegration.   The
calculations(89) correspond to normal impulse approximation (dash),
and the hybrid model with matching radius b = 0.9 fm (solid), and
1.8 fm (dot-dash).

No complete study of the convergence of the expansion (7) as a
function of b has been given for A > 3.  However, several studies of the

three-body system have been made. Almost three years ago, Pirner and Vary (89) suggested that quark admixtures were necessary to interpret data on the electro-disintegration of $^3$He (90).

They found that in the x-region forbidden for a free nucleon ($x \equiv Q^2/2m_N \nu > 1$), the usual impulse approximation failed by one or two orders of magnitude. However it was possible to obtain quite a good fit by introducing a boundary radius of order 0.9 fm within which the quark description was used – see Fig 4. (Incidentally, Pirner and Vary spoke of a bag radius $R = b/2 = 0.45$ fm for the best fit. However, we believe that such a simple interpretation is probably wrong. It seems likely to us that after two bags touch there would be a transition region, and only when there is a very significant overlap ($b \sim R$) would the system behave

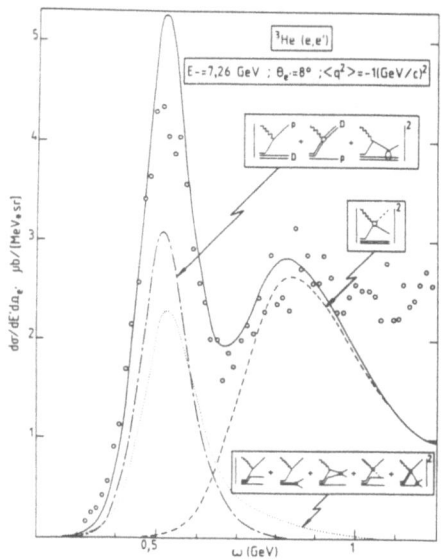

**Figure 5** Calculation by Laget (91) for the same data as Fig 4, but including a number of higher order corrections.

like a coherent six-quark state). A serious objection to this interpretation of the data is that the typical $Q^2$ was rather low. For large x the scaling region does not set in until quite high $Q^2$. Thus the interpretation of this inclusive measurement in terms of quarks must be viewed with suspicion. Indeed, a recent calculation by Laget (91), including a number of corrections associated with exchange currents and final state interactions (Fig 5) reproduced the same data rather well. It may well be that both of these calculations represent some part of the true physics of this reaction. However it is absolutely clear that this is not an example of a "smoking gun" which would show the necessity for the quark level description beyond doubt. Further experimental investigation to separate the transverse and longitudinal contributions would be most valuable, as the meson exchange contributions to

the latter are essentially zero (48). More recently the hybrid quark-nucleon picture has been used to estimate the $^3$H–$^3$He mass difference (88) and the corrections to their magnetic moments (92). As these are clearly not gold-plated examples of the need for the quark model we shall not describe them further here. It is worth mentioning however, that in A = 3 one can calculate exactly the probability of two- or three-nucleon overlap. Thus one has a good test case for the classical approximations which might be used in the A-body case. For example, let us define p:

$$p = \langle \phi \mid \theta (b-r_{12}) \mid \phi \rangle \tag{8}$$

to be the probability for a particular nucleon to overlap with a given second nucleon. Then we might classically expect that the probability for two and only two nucleons to overlap with a specified nucleon would be

$$P^v_{Q_2} = \binom{A}{2} p^2 (1-p)^{A-2}. \tag{9}$$

With b = 0.95 fm this gives 0.8% in $^3$He, which should be compared with the exact result of 1.6% (88). This seems quite poor, but has been shown to result from the anomalously strong centre-of-mass correlation in the A=3 system. A careful analysis of this effect led us to believe that, provided b is not too large, Eq. (9) should be a rather good approximation in heavier systems. In particular, we feel fairly confident in neglecting 9- and 12- quark configurations in heavier nuclei when b is less than 1 fm. This will be important when we consider the EMC effect in Sect. 5.

In concluding this section we mention the extensive work in the Soviet Union aimed at extracting the structure functions in heavy nuclei in the region x > 1 (93). As we mentioned in connection with similar data for deuterium, an unequivocal determination of the structure function in that region would be very valuable - albeit difficult to interpret. Unfortunately, for the present there is little understanding of the systematic errors involved in determining structure functions through (say) inclusive pion production at relatively low $Q^2$. Further experimental work may clarify this matter.

## 5. THE EMC EFFECT

The most direct evidence that the quark structure of the nucleon is fundamentally altered by being placed in a nucleus was provided last year by the European Muon Collaboration (94). For an elementary introduction to the quark parton model and the interpretation of the EMC effect we refer to ref (22).

### 5.1 The Experimental Data

For a number of years nuclear targets have been used in $\nu$ and $\mu$ experiments to determine structure functions, largely as a matter of convenience and event rate (22, 95). Only recently has the control of systematic errors been such that the implicit assumption that at $Q^2$ of order 200$(GeV/c)^2$ it would not matter whether a nucleon was inside or outside a nucleus been able to be tested. Early last year the EMC group (3) showed that the ratio of the structure function $F_2^\mu$ in Fe to that in D showed a significant deviation from unity (3). This claim was soon confirmed by a group which had worked at much lower $Q^2((1-10) (GeV/c)^2)$ as opposed to (19 - 270) $(GeV/c)^2$ at SLAC a decade earlier (96). Briefly the EMC data showed an enhancement of about 15% in the small-x region $(x \sim 0.1)$ and a 15% decrease at $x \sim 0.65$. This was immediately interpreted (97) as a substantial enhancement of the sea in Fe (by (20-60)%), and an even more dramatic depletion of the intermediate momentum part of the valence distribution (minus 15% versus plus 20% at $x = 0.65$ on the basis of standard estimates of fermi motion (61)). At very large $x$ $(x > 0.8)$ the ratio does finally rise above one (96) as expected because of the greater fermi motion in Fe (61).

Since these announcements a number of other experimental results using $\nu$- and e- beams have confused the situation. The BEBC group has reported on the ratio of the $\overline{\nu}$ cross sections of Ne and D - unfortunately with worse statistics (97). In the valence region these results are consistent with the softening of the valence distribution observed by EMC. However, for $x \in (0.05, 0.2)$ the BEBC analysis suggests no enhancement of the sea in Ne relative to D. Within the experimental errors one could perhaps tolerate an increase of the sea of about 50% of the lower limit of the systematic errors of EMC for Fe. Nevertheless this is worrying.

The CDHS group has also reported on the ratio of the anti-quark distributions for $\overline{\nu}$ reactions on Fe and H (98). At face value this experiment also contradicts the enhancement of the sea seen by EMC. However, because the reference is hydrogen rather than deuterium what is measured is actually $(\overline{u} + \overline{d} + 2\overline{s})_{Fe} / (2\overline{d} + 2\overline{s})_H$ , and an asymmetry of the $\overline{d}$ and $\overline{u}$ distributions, as suggested by the pion model (99, 100) is not ruled out. In this connection we note the recent work of Ericson and Thomas (101), who have shown that the slope of the rapidity distribution in proton-nucleus Drell-Yan is consistent with an enhancement of the nuclear sea by up to 30% (102).

Finally, we note that Bodek et al (103) have carried out a systematic study of the dependence of the EMC effect on atomic number. Unfortunately, because of limitations in beam time some sacrifices had to be made and a number of important questions were not answered. In particular the ratio of longitudinal to transverse cross sections, R, was not measured and therefore what is plotted are not ratios of $F_2$ but ratios of cross sections. Together with questions of shadowing at the relatively low values of $Q^2$ in this experiment this again confuses the issue of whether there is an enhancement of the nuclear sea. On the other hand the conclusions in the valence region (where R is known to be small and $Q^2$ is larger) are quite firm. The softening of the valence distribution goes like nuclear density (or approximately $\ell n$ A). Unfortunately while this rules out some extreme models, most predictions are linear in the effective density of the nucleus.

## 5.2   The Theoretical Interpretation

Perhaps the most inscrutable summary of the EMC data is that of Close, Roberts and Ross (104) who observed that for $0.2 < x < 0.8$ the distribution $F_2(x,Q^2)$ for Fe was equal to $F_2(x,\xi Q^2)$ for D, with $\xi \simeq 2$. Thus in everyday language it seems that there is a fairly dramatic "change of scale" when we go from a free nucleon to one imbedded in a nucleus. (A similar idea was proposed by Nachtman and Pirner (105), but they had a more specific model in mind). The beauty of this idea of a change of scale is that it **sounds** like QCD.  Unfortunately there was no suggestion of how to calculate the change of scale for a given nucleus.

More recently Jaffe et al (106) suggested a way around this limitation, by parameterizing the change of scale in terms of the probability of two clusters of quarks overlapping inside the nucleus. The suggestion that the EMC effect could be explained in terms of overlapping bags (exactly the scenario to which we were led in Sect. 2) had been suggested by Jaffe (107) and many others (108).  Clearly, any model which relates the effect to the probability of N-N pairs overlapping gives a result proportional to the effective nuclear density.  Because of the limitation to $x \geqslant 0.2$, there was no prediction about the enhancement (or otherwise) of the sea in the approach of Jaffe et al.

Amongst the alternative explanations of the EMC effect which have been proposed we mention:  a possible increase in the size of the nucleon itself (109);  extra stability of larger clusters of (say) 12-quarks (110);  the treatment of the nucleus as a whole as one bag (87, 111) - it was in this framework that Krzywicki actually predicted the enhancement of the nuclear sea;  and the possibility of an enhancement of the virtual pion field in the nucleus (99, 100, 112).  Apart from those models which predict an anomalously large effect in the α-particle, none of these models have been ruled out by existing data. Only the pion model relies on physics which can be tested without simply repeating the measurements of Refs (3, 98 and 103) with higher precision and more targets.  Since such a test was recently carried out (113) we shall end with a discussion of this model.

The pion cloud about the nucleon bag is the longest range piece of its structure.  It is the first aspect of nucleon structure which would be expected to alter in a nuclear environment.  Indeed it has long been expected that there could be a significant enhancement of the virtual pion field inside the nucleus in the region of momentum transfer near $2 \text{ fm}^{-1}$ (114, 115).  It is precisely this region which would be expected to contribute most strongly to the sea in DIS (50, 116).  These considerations lead unambiguously to the prediction of an enhancement of the non-strange component of the nuclear sea, which can be made to agree with the EMC data at small x (99, 100).  However, this result of the calculation is extremely sensitive to the Landau-Migdal parameters $g'_{NN}$, $g'_{\Delta\Delta}$ and particularly $g'_{\Delta N}$  (114, 115), which parameterise the short range piece of the spin-isospin force between two nucleons.  Thus to completely explore this model we cannot avoid the problem of overlapping bags!  Phenomenologically, we have been able to make an educated guess at what might result from a complete calculation by simply balancing momentum in the most naive way (99).  This leads to the results (117) shown in Fig 6, where we compare with the recent data of Arnold et al (103) for $^{12}C$ and $^{56}Fe$. (The g' parameter was adjusted to fit Fe, and the result for $^{12}C$ calculated using the linear dependence on effective nuclear density implied by this model).  Such a fit implies an enhancement of about a factor of two in the pion field at $q \sim 2 \text{ fm}^{-1}$,  $w \sim 80$ MeV, and it is this which is tested by Carey et al (113).  Following the suggestion of Alberico, Ericson and Molinari (118) they actually extract the ratio of the longitudinal to the transverse in $(\vec{p}, \vec{p}^1)$  on $^{208}Pb$ (and D) for $q \sim 1.7 \text{ fm}^{-1}$.  The experimental ratio is a factor of two under the theoretical expectations in the critical region w ~ 80 MeV, while it is as much as three or four times too small at w ~ 30 MeV. Contrary to the claims of ref (113) this is not a refutation of the pionic model for EMC. The discrepancy at small w ( $\gtrsim$ 30 MeV ) - which is irrelevant for EMC - reflects a problem in the fermi gas model of ref (118) which ignores nucleon binding energy. This is being corrected now. At large w the situation is confused by the presence of a sizeable isoscalar component in the transverse response. It may well be that after a more thorough analysis

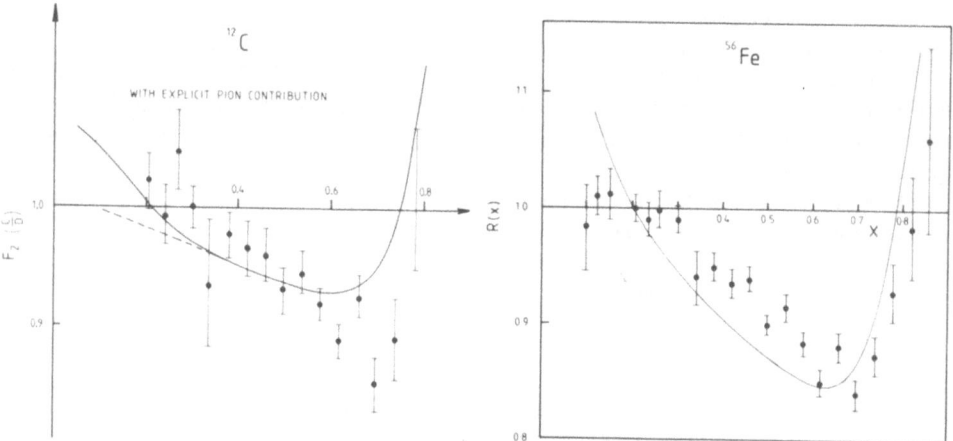

**Figure 6**    Comparison of the pionic model for EMC (117) with the recent data of Arnold et al (103).

this data actually confirms the enhancement of the virtual pion field in $^{208}$Pb.

In conclusion we note that some theoretical objections have been made to the pionic model (106), and these are answered elsewhere (117). In our view the fundamental question in this model is how to derive the intermediate and short-range spin-isospin force in the quark model, and hence make a consistent, microscopic description of the EMC effect.

## 6.    CONCLUSION

The next few years offer tremendous challenge to nuclear physicists. On purely theoretical grounds we expect quark degrees of freedom to be essential to a truly microscopic understanding of the atomic nucleus. Apart from the EMC effect, which is not yet understood, there is no experimental evidence to support this contention. There are, however, a number of crucial experiments which need to be carried out. If linked with a systematic theoretical effort to deal with the many open problems which we have mentioned, these experiments could go a long way to answering that challenge.

# REFERENCES

1. A. Faessler, invited talk at this Conference.
2. F. Gross, invited talk at the CEBAF (NEAL) Spectrometer Workshop, Williamsburg Oct 10-12, (1983).
3. J.J. Aubert et al, Phys. Lett. **123B** (1983) 275.
4. A.W. Thomas, "Chiral symmetry and the bag model: a new starting point for nuclear physics", in J.W. Negele and E. Vogt (Eds.) Adv. in Nucl. Phys. **13**, (1984) p. 1.
5. D.C. Fries and B. Zeitnitz (Eds.), Quarks and Nuclear Forces (Springer Verlag, Berlin, 1982).
6. Proc. of Indiana University Workshops, AIPCP **79** (1982), AIPCP (1983) and AIPCP **110** 1984).
7. Proceedings of the Erice schools on Nuclear and Particle physics, particularly 11, Sir Denys Wilkinson (Ed), (Pergamon, Oxford 1984).
8. B. Zeinitz (Ed), 'Few Body Problems in Physics' **Vols I and II** (Elsevier, Amsterdam, 1984).
9. See for example the review by J. Kuti in Ref (8).
10. H. Lipps et al, Phys. lett. **126B** (1983) 250.
11. A. De Rujula, H. Georgi and S. Glashow, Phys. Rev. **D12**, 147, (1975).
12. N. Isgur, Proceedings of Baryon 80 (Univ. of Toronto Press, 1980).
13. E.V. Shuryak, 'Non-Perturbative Phenomena in QCD...' CERN REPORT 83-01.
14. A Chodos et al, Phys. Rev. **D9**, (1974), 3471; ibid **D10** 2599, (1984).
15. W. Weise, in ref (7) p. 123.
16. N. Isgur and G. Karl, Phys. Rev. **D18**, (1978), 4187; ibid. **D19** (1979), 2653.
17. R. Koniuk and N. Isgur, Phys. Rev. **D21**, (1980), 1868.
18. Y. Nogami and N. Ohtsuka, Phys. Rev. **D26**, (1982), 261.
19. A.W. Thomas, J. Phys. **G7**, (1981), L283.
20. A. Szymacha and S. Tatur, Zeit Phys. **C7**, (1981), 311.
21. E.A. Veit et al, Phys. Lett **137B**, (1984), 415.
22. A.W. Thomas, in ref (7), p. 325.
23. R. Hwa, Phys. Rev. **D22** (1980), 759.
24. P.N. Bogolioubov, Ann. Inst. Henri Poincare **8**, (1967), 163.
25. C.E. DeTar and J.F. Donoghue, Ann. Rev. Nucl. and Part. Sci. **33** (1983), 235.
26. T. Inoue and T. Maskawa, Prog. Theor. Phys. **54** (1975), 1833; A. Chodos and C.B. Thorn, Phys. Rev. **D12** (1975), 2733.
27. G.E. Brown and M. Rho, Phys. Lett. **82B** (1979), 177; V. Vento et al, Nucl. Phys. **A345** (1980), 413.
28. R.L. Jaffe in 'Pointlike Structures Inside and Outside Hadrons' A. Zichichi (Ed.) (Plenum, New York, 1982).
29. S. Theberge, A.W. Thomas and G.A. Miller, Phys. Rev. **D22** (1980), 2838; **D23** (1981), 2106(E); Can. J. Phys. 60 (1982), 59.
30. G.A. Miller, Univ. of Washington preprint, 40048-02-N4 (1984), to appear in Int. Rev. of Nucl. Phys. **Vol 2** (World Scientific Publishing, Singapore, 1984).

31.    G.E. Brown, in Prog. in Particle and Nucl. Phys., Sir Denys Wilkinson (Ed.) (Pergamon, London 1982) p. 147.

32.    R. Friedberg and T.D. Lee, Phys. Rev. **D18** (1978), 2623.

33.    R. Goldflam and L. Wilets, Phys. Rev. **C25** (1982), 1951.

34.    L. Wilets. Univ. of Washington preprint 40048-29-N3 (1984), to appear in Proc. Adv. Course in Theor. Phys, Cape Town, (1984).

35.    T.H.R. Skyrme, Proc. Roy. Soc. **A260** (1961), 127.

36.    E. Witten, Nucl. Phys. **B160** (1979), 57; **B222** (1983), 433; A.P. Balachandran et al, Phys. Rev. Lett, **49** (1982), 1124; Phys. Rev. **D27** (1983), 1153.

37.    G.t'Hooft, Nucl. Phys. **B72** (1974), 461.

38.    M. Rho, in ref 7, p. 379.

39.    P. Di Vecchia and P. Rossi, CERN TH-3808 (1984).

40.    I am indebted to M. Lohe for a discussion of these matters.

41.    J. Wess and B. Zumino, Phys. Lett. **37B** (1971), 95.

42.    G.S. Adkins, C.R. Nappi and E. Witten, Nucl. Phys. **B228** (1983), 552.

43.    M. Rho, A.S. Goldhaber and G.E. Brown, Phys. Rev. Lett. **51** (1983), 747.

44.    J. Goldstone and R.L. Jaffe, Phys. Rev. Lett. **51** (1983), 1518.

45.    P.J. Mulders, MIT preprint CTP # 1138 (1984); S. Kahana, G. Ripka and V. Soni, Nucl. Phys. **A415** (1984), 351.

46.    F. Close, An Introduction to Quarks and Partons (Academic, New York, 1979). This matter is discussed at greater length in A.W. Thomas, invited paper at the 10th Int. Conf. on Particles and Nuclei (Heidelberg, 1984) – Univ. of Adelaide preprint ADP 312 (June 1984).

47.    R.L. Jaffe and G.G. Ross, Phys. Lett. **93B** (1980), 313.

48.    J.M. Laget, in these proceedings and A. Gerard and J.M. Laget, 'Rapport Prospectif sur la Physique Photo et Electronucleaire', Note CEA-N-2360 (1983).

49.    L.R. Dodd, A.W. Thomas and R.F. Alvarez-Estrada, Phys. Rev. **D24** (1981), 1961.

50.    A.W. Thomas, Phys. Lett. **126B** (1983), 97.

51.    D.A. Liberman, Phys. Rev. **D16** (1977), 1542.

52.    V.G. Neudatchin, Yu. F. Smirnov and R. Tamagaki, Prog. Theor. Phys. **58** (1977), 1072.

53.    M. Oka and K. Yazaki, Phys. Lett. **90B** (1980), 41; M. Harvey, Nucl. Phys. **A352** (1981), 326; A. Faessler et al. Phys. Lett. **112B** (1982), 201; M. Dey, J. Dey and P. Ghose, Phys. Lett. **119B** (1982), 198; Y.E. Kim and M. Orlewski, Purdue preprint PNTG-83-16 (1983); K. Maltman and N. Isgur, Phys. Rev. **D29** (1984), 952.

54.    C. de Tar, Phys. Rev. **D19** (1979), 1451.

55.    C.W. Wong, Phys. Rev. **D24** (1981), 1416.

56.    S. Furui and A. Faessler, Nucl. Phys. **A397** (1983), 413.

57.    R.L. Jaffe and F. Low, Phys. Rev. **D19** (1979), 2105; B.O. Kerbikov, ITEP preprint – 4, (1984).

58.    P. Mulders in Ref (8), p. 99; Yu. A. Simonov, in Ref (8), p. 109.

59.    A. Faessler, An invited contribution to this conference.

60.    A.P. Kobushkin, Kiev preprint ITP-76-I45E (1976).

61.   A. Bodek and J.L. Ritchie,  Phys. Rev.  **D23** (1981), 1070.
62.   P.J. Mulders and A.W. Thomas,  Phys. Rev. Lett.  **52** (1984), 1199.
63.   M.A. Parker et al,   BEBC TST Neutrino Collaboration, RL-83-059 (1983).
64.   G.R. Farrar, P. Schreiner and W.G. Scott,  Phys. Lett. **69B** (1977), 112.
65.   D. Sivers,  Ann. Rev. Nucl. Sci.  **32** (1982), 149.
66.   S. Brodsky and B. Chertok,  Phys. Rev.  **D14** (1976), 3003;  S. Brodsky in Ref (5).
67.   N. Isgur and C.H. Llewellyn Smith, Phys. Rev. Lett.  **52** (1984), 1080.
68.   I.R. Afnan in Ref (8), p. 257;  A.W. Thomas,  Nucl. Phys.  **A374** (1982), 40.,
69.   R.L. Jaffe,  Phys. Rev. Lett.  **38** (1977), 195;  (E) 615;  P.J. Mulders, A.T. Aerts and J.J. de Swart,  Phys. Rev.  **D21** (1980), 2653.
70.   R.L. Jaffe and K. Johnson,  Phys. Lett.  **60B** (1976), 201;  C.E. Carlson, R.H. Hansson and C. Peterson,  Phys. Rev.  **D27** (1983), 1556.
71.   T. Barnes, F. Close and F. de Viron,  Nucl. Phys.  **B224** (1983), 241.
72.   A. Angelopoulos et al,  CERN – EP/84-47.
73.   M.J. Corden et al,  Phys. Rev.  **D25** (1982), 720.
74.   P.J. Mulders and A.W. Thomas,  J. Phys **G9** (1983), 1159.
75.   G. d'Agostini et al,  Nucl. Phys.  **B209** (1982), 1.
76.   L.S. Kisslinger,  Phys. Lett.  **112B** (1982), 307.
77.   V.M. Dubovik and I.T. Obukhovsky,  Zeit. f. Phys. **10C** (1981), 123.
78.   L.S. Kisslinger and G.A. Miller,  Phys. Rev.  **C27** (1983), 1602.
79.   G.A. Miller and L.S. Kisslinger,  Phys. Rev.  **C27** (1983), 1669.
80.   T.E.O. Ericson, in Ref (8), p. 281;  P.A.M. Guichon and G.A. Miller, Phys. Lett.  **134B** (1984), 32.
81.   G.A. Miller, in Int. Review of Nucl. Phys. **Vol 2** (World Scientific Publishing, Singapore, 1984).
82.   A.W. Thomas,  Nucl. Phys.  **A354** (1981), 51.
83.   P.A.M. Guichon, M. Giffon and C. Samour,  Phys. Lett.  **74B** (1978), 15.
84.   M. Chemtob and M. Rho,  Nucl. Phys.  **A163** (1971), 1.
85.   C. Goodman,  in Ref (7), p. 474.
86.   I.S. Towner, in Ref (7), p. 91.
87.   Lecture Notes in Physics **197** (1984), 236.
88.   J.M. Greben and A.W. Thomas,  Alberta preprint THY-1-84, to appear in Phys. Rev. C (1984).
89.   H.J. Pirner and J.P. Vary,  Phys. Rev. Lett. **46** (1981), 1376.
90.   D. Day et al,  Phys. Rev. Lett.  **43** (1979), 1143.
91.   J.M. Laget,  Proc. Workshop on Electron Rings for Nucl. Phys. Res.  Eds. J.O. Adler and B. Schroder,  Lund 5-7 Oct. 1982.
92.   G. Karl, G.A. Miller and J. Rafelski,  to appear in Phys. Lett. (1984).
93.   A. Baldin,  private communication.
94.   J.J. Aubert et al,  Phys. Lett.  **123B** (1983), 275.
95.   F. Eisele,  Journ. de Physique.  Colloque **C3** C3-337, (1982).

96.   A. Bodek et al,  Phys. Rev. Lett.  **50** (1983), 1431.
97.   A.M. Cooper et al,  CERN preprint.  EP/84-37.
98.   H. Abramowitz et al,  CERN-EP/84-57
99.   C.H. Llewellyn Smith, CERN 83-02, p. 180 , (1983) and Phys. Lett.  **128B** (1983), 107.
100.  M. Ericson and A.W. Thomas,  Phys. Lett.  **128B** (1983), 112.
101.  M. Ericson and A.W. Thomas,  Contribution to this conference, Univ. of Adelaide preprint ADP-311 (June 1984).
102.  A.S. Ito et al,  Phys. Rev.  **D23** (1981), 604.
103.  R.G. Arnold et al,  Phys. Rev. Lett.  **52** (1984), 727.
104.  F.E. Close, R.G. Roberts and G.G. Ross,  Phys. Lett.  **129B**  346, (1983), 727.
105.  O. Nachtman and H.J. Pirner,  Z. Phys.  **C21** (1984), 277.
106.  R.L. Jaffe et al,  Phys. Lett  **134B** (1984), 449.
107.  R.L. Jaffe,  Phys. Rev. Lett.  **50** (1983), 228.
108.  C.E. Carlson and T.J. Havens,  Phys. Rev. Lett.  **51** (1983), 261; H.J. Pirner and J.P. Vary, Heidelberg preprint UNI-HD-83-02;  S. Date,  Prog. Th. Phys.  **70** (1983), 1682;  M. Chemtob and R. Peschanski, Saclay S. Ph.T/83/116;  A. Titov,  Dubna E2-83-460.
109   J.V. Noble,  Phys. Rev. Lett.  **46** (1981), 421;  E.M. Levin and M.G. Ryskin, Leningrad preprint 888 (1983); L.L. Frankfurt and M.I. Strikman, Leningrad 886 (1983);  M. Jandel and G. Peters, Stockholm preprint TRITA-TF4-83-28.
110.  L. Kondratyuk and M. Shmatikov,  ITEP-13 (1984);  H. Faissner and B.R. Kim, Aachen preprint (1983).
111.  A. Krzywicki,  Phys. Rev.  **D14** (1976), 152;  W. Furmanski and A. Krzywicki,  LPTHE Orsay 83/11.
112.  B. Friman, V.R. Pandharipande and R.B. Wiringa,  Phys. Rev. Lett  **51** (1983), 763;  E.L. Berger, F. Coester and R.B. Wiringa,  Phys. Rev.  **D29** (1984), 398;  G.B. West,  LASL Preprint (1983).
113.  T.A. Carey et al,  Los Alamos Preprint LA-UR-84-1207.
114.  M. Ericson,  in Ref (7), p. 277.
115.  E. Oset, H. Toki and W. Weise,  Phys. Rep. **83** (1982), 281.
116.  J.D. Sullivan,  Phys. Rev.  **D5** (1972), 1732.
117.  M. Ericson, C.H. Llewellyn Smith and A.W. Thomas,  (to be published).
118.  W.M. Alberico, M. Ericson and A. Molinari,  Nucl. Phys.  **A379** 1982), 429.

QUARK CONFIGURATIONS IN THE LIGHTEST NUCLEI.  CONNECTION WITH THE
NN-INTERACTION

V.G. Neudatchin, I.T. Obukhovsky and Yu. F. Smirnov
Institute of Nuclear Physics
Moscow State University
119899 Moscow
USSR

ABSTRACT.  This report presents our latest results and, on this basis,
our understanding of the future developments of this intricate field of
nuclear physics.

1.  THE RESONATING-GROUP CALCULATIONS OF THE NN-SYSTEM

The earlier calculations by the resonating-group method (RGM)[1,2] were
restricted to three channels NN + $\Delta\Delta$ + CC.  This is equivalent to
neglecting most of the spin-isospin (ST) symmetries in the 6q-system.
For example, from the expansion (S = 1, T = 0):

$$[42]_S \cdot [3^2]_T = [51]_{ST} + [3^2]_{ST} + [321]_{ST} + [41^2]_{ST} + [2^2 1^2]_{ST}$$
(1)

The only two Young schemes taken into account in the NN + $\Delta\Delta$ + CC
channels are $[51]_{ST}$ and $[3^2]_{ST}$, because these channels correspond to
the Young schemes:

$$[3]_{ST} \times [3]_{ST} = \begin{cases} [51]_{ST} + [3^2]_{ST} & \ell\text{-even,} \\ [6]_{ST} + [42]_{ST} & \ell\text{-odd} \end{cases}$$
(2)

It can be expected that in any variational calculation (the RGM as
well) the extension of the 6q-state basis will lead to additional
NN-attractive forces.  The matter is that the orbital symmetry $[42]_X$ is
compatible with all the Young schemes from the expansion (1).  At the
same time the colour-magnetic attraction of nucleons, as known from
ref.3, corresponds to the symmetry $[42]_X [42]_{CS}$.  However, the symmetry
$[42]_{CS}$ is nearly "orthogonal" to the system of channels NN + $\Delta\Delta$ + CC.
In particular, $\Delta\Delta$ is projected only onto the symmetry $[2^3]_{CS} + [21^4]_{CS}$.
As for the NN-system, its overlap with the symmetry $[42]_{CS}$ makes up a
small fraction of unity because the expansion of the triplet (S = 1)
states has the form

J. S. Lilley and M. A. Nagarajan (eds.), Clustering Aspects of Nuclear Structure, 353–369.
© 1985 by D. Reidel Publishing Company.

$$[2^3]_C \cdot [42]_S = [42]_{CS} + [321]_{CS} + [31^3]_{CS} + [2^3]_{CS} + [21^4]_{CS}$$

$$(3)$$

Therefore, the NN-system is distributed between all the states (3), except for $[21^4]_{CS}$ and the RGM calculations underestimate the contribution of the CM-attractive symmetry $[42]_{CS}$. There have been several studies using the complete basis[4-7]. Specifically, our work[7] is equivalent (in the sense of the symmetries taken into account) to the RGM-calculations including the channels NN + ΔΔ + N N, where $N^*_{1/2}$ or $N^*_{3/2}$ have the symmetry $[21]_{ST}$. All the Young schemes $[321]_{ST}$, $[41^2]_{ST}$ and $[2^21^2]_{ST}$ that are not taken into account in the NN + ΔΔ + CC approximation, enter the outer product $[21]_{ST} \times [21]_{ST}$ and hence we have used, in essence, the complete ST-basis or, more correctly, the equivalent complete CS-basis

$$\Psi^{S,T}_{3q-3q}(R) = C_0(R) \Big| [6]_X [2^3]_{CS} S,T \rangle + \sum_i C_i(R) \Big| [42]_X [f^{(i)}]_{CS} S,T \rangle$$

$$(4)$$

In complete analogy with Harvey's work[8] we used the harmonic oscillator wave functions of 3 q-clusters "centered" at the points ± 1/2R. The functions with the definite symmetry $[f]_X [f]_{CS}$ were constructed from the product $\Psi^{(+)}_N (R) \times \Psi^{(-)}_N (R)$ using the fractional parentage technique in the "quark shell model"[8,9]. The perturbative model Hamiltonian containing, alongside with the kinetic energy $H_{kin}$, the colour-electric ($H_C$) and colour-magnetic ($H_{CM}$) terms and the quadratic confinement potential $H_{conf}$ were taken from ref.10. At $R \to 0$ the lowest solution reduces to the mixing of the symmetries $[6]_X [2^3]_{CS}$ and $[42]_X [42]_{CS}$.

The results of the variational calculations for the triplet S-state are given in Table I.

We see that the CM-attraction of nucleons with the predominant configuration $[42]_X [42]_{CS}$ is possible only at a sufficiently "soft" $\hbar\omega$, otherwise, the configuration $[6]_X [2^3]_{CS}$ with the CM-repulsion will inevitably prevail.

Figs. 1 and 2 present the adiabatic potentials for the variants I-IV and the wave functions, respectively. The RGM wave functions[1,2] in the NN + ΔΔ + CC approximation are given in fig.3 for a comparison. The shape of the wave functions for variant I and II qualitatively agree with the RGM solution (fig.3) so that the node of the RGM functions is caused by the component $s^4 p^2 [42]_X$. However, the oscillation of the wave function is strongly suppressed due to the effective soft core (fig.1). On the whole, conspicuously enough, the situation is not radically changed as compared with the previous RGM calculations, despite the full allowance for the CM-attraction symmetries. In that, our conclusions are similar to the results of the work[5] that is also done with the complete basis, yet, without identifying the contribution of separate factors listed in Table 1 and figs.1-3.

Summing up some results we can say that the NN-interaction in the above-discussed triplet NN-channel can be interpreted either as the CM-attraction of nucleons with a nodal wave function in the NN-channel

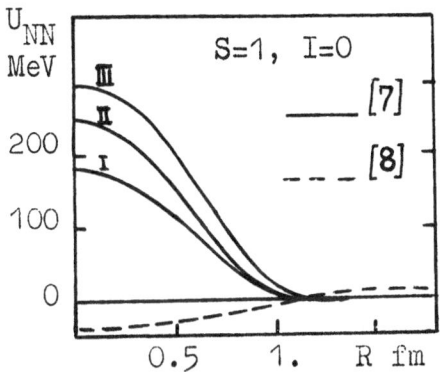

Figure 1. NN-potentials in the Harvey approach on the complete CS basis[7].

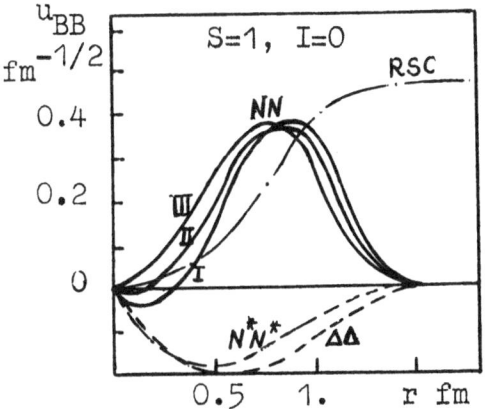

Figure 2. Wave functions of the I-III variants calculations. Projections to NN, ΔΔ and N*N* channels.

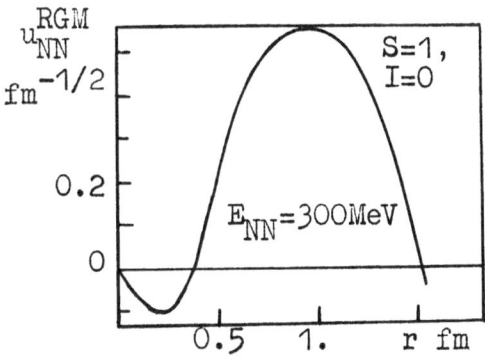

Figure 3. RGM wave function from ref.2.

Table 1

| Titles | Numerical constants | Variants | | | |
|--------|---------------------|--------|--------|--------|--------|
| | | I | II | III | IV |
| Parameters of the confinement potential | $K$ (MeV/fm$^2$) | 304.2 | 446 | 631 | 924 |
| | $C$ (MeV) | 358.3 | 442.3 | 533.5 | 653.8 |
| Oscillation frequency | $\hbar\omega$ (MeV) | 265 | 321 | 382 | 462 |
| Parameters of the cluster function | $C$ (fm) | 0.66 | 0.60 | 0.55 | 0.50 |
| Solution at the point R = 0 | $C_0(0)$ $C_I(0)$ | -0.247 0.969 | -0.502 0.865 | -0.822 0.570 | -0.934 0.332 |
| Mean value of | $\langle H_C\rangle$ (MeV) | 142 | 114 | 26 | -11 |
| individual terms of the Hamiltonian $H_c, H_{cm}, H_{kin}$ | $\langle H_{CM}\rangle$ (MeV) | -208 | -97 | 151 | 277 |
| | $\langle H_{kin}\rangle$ (MeV) | 249 | 240 | 124 | 52 |
| Energy above threshold | $E_{NN}$ (MeV) | 183 | 257 | 301 | 318 |

(i.e. in the spirit of the concept of a deep attractive NN-potential with "forbidden" bound-state) or as the potential with a soft repulsive core depending on the model and the parameters of the system. While the second interpretation has been widely studied in many papers[1,2,4,11] (the phenomenological description of the short-range distances + OBEP + $\pi\pi$-continuum) the first possibility is currently treated by our group only [3,7,12] (see also the papers discussed later). In this approach the NN-interaction should be constructed as a combination of the phenomenological short-range part ("superdeep attraction") + OBEP at medium and long distances. At present it is reasonable to follow both ways because the above-discussed perturbative QCD calculations are far from being perfect and their predictions should be regarded as tentative.

As for the NN-scattering phases, the corresponding calculations by the scheme "quark RGM at short distances + OBEP at long and medium distances"[1,2] are in qualitative agreement with the Paris potential's phenomenology[13]. Yet, the shortage of the noted quark approach is its purely numerical nature which is a typical drawback of the RGM. Yu. A. Simonov has noted in his P-matrix approach[14] that the repulsive core increasing with energy can be caused by the influence of the P-matrix pole connected with the primitive level $S^6[6]_x\ [2^3]_{CS}$ in the 6q-bag.

We find it useful to treat this problem on the basis of the general expression for the S-matrix containing the corresponding microscopic quantum numbers. Attacks at these notions are reasonable to start from the consideration of a simple illustrative example of the elastic scattering of nucleon clusters A + B such that to a given orbital momentum $\ell$ there could correspond to several Young schemes $[f]_X$ just as in our 6q-situation.

## 2. INTERFERENCE OF DIFFERENT YOUNG SCHEMES IN CLUSTER-CLUSTER AND NUCLEON-NUCLEON SCATTERING

The study of the cluster-cluster (A + B) scattering[15] revealed a strong dependence of the potentials between clusters upon the orbital symmetry $[f]_X$. For example, in the $^3$H-$^3$He scattering there are the Young schemes $[f]_X = [42]$ for the even orbital momenta $\ell$ and $[33]_X$ for the odd ones, and the potentials for these symmetries are rather different. The cases when several different Young schemes $[f]_X$ are possible at the same $\ell$ have not, however, been considered. Meanwhile, these cases really exist and we now shall discuss an interesting example d + d where in the even waves with the total spin S = 0 there are two orbital Young schemes $[f]_X = [4]$ and $[f]_X = [22]$ corresponding to two optical potentials with essentially different "strengths" $VR^2$.

The general expression for the A + B scattering amplitude is of the form

$$T_\ell = \sum_{S,T[v]} \langle t_A t_{3A}, t_B t_{3B} | T\, T_3 \rangle\ \langle s_A s_{3A}, s_B s_{3B} | S\, S_3 \rangle\ .$$

$$\langle [v_A] s_A t_A, [v_B] s_B t_B | [v] ST \rangle T_\ell^{[v]} \langle [v] ST | [v_A] s_A t_A, [v_B] s_B t_B \rangle\ .$$

$$\langle TT_3 | t'_A t_{3A}, t'_B t_{3B} \rangle\ \langle SS_3 | s'_A s_{3A}, s'_B s_{3B} \rangle \qquad (5)$$

and the contributions of different Young schemes $[v]$ interfere ($[v] = [\tilde{f}]_X$). Here we assumed that the NN-interaction is the SU(4) invariant as is usually done for the lightest nuclei. The isoscalar factor of the Clebsch-Gordan coefficient for the SU(4) group SU(4) $\langle [v_A] s_A t_A, [v_B] s_B t_B | [v] ST \rangle$ coincides with the corresponding spin-isospin fractional-parentage coefficient[16]. The main features of the d + d scattering process become evident from consideration of the squared amplitude (5) averaged over the orientations of the nucleon spins in the initial state and summed up on the polarizations in the final state (S = 0, $\ell$ - even):

$$|T_{\ell,e\ell}(S = 0)|^2 = \Big| \langle [\tilde{2}]10, [\tilde{2}]10 | [\tilde{4}]00 \rangle^2\ T_\ell^{[4]} +$$

$$+ \langle [\tilde{2}]10, [\tilde{2}]10 | [\widetilde{22}]00 \rangle^2\ T_\ell^{[22]} \Big|^2 =$$

$$= \Big| \tfrac{1}{2} T_\ell^{[4]} + \tfrac{1}{2} T_\ell^{[22]} \Big|^2 \qquad (6)$$

The values of the necessary coefficients are listed in Table 2. The spin summation in the differential cross section will be carried out in

Table 2

| $S_1 t_1$ | | $S_2 t_2$ | | [4] | [22] |
|---|---|---|---|---|---|
| 1 | 0 | 1 | 0 | $\sqrt{\frac{1}{2}}$ | $-\sqrt{\frac{1}{2}}$ |
| 0 | 1 | 0 | 1 | $\sqrt{\frac{1}{2}}$ | $\sqrt{\frac{1}{2}}$ |

complete analogy with the above procedure, except for the substitution of (6) by $T_\ell T_{\ell'}^*$, and so we can record

$$T_{\ell,el}(S = 0) = \frac{1}{2} T_\ell^{[4]} + \frac{1}{2} T_\ell^{[22]} \tag{7}$$

where each of the amplitudes corresponds to the potential scattering

$$T_\ell^{[f]} = \exp(2i\ \delta_\ell^{[f]}) - 1 \ , \ [f] \equiv [f]_X$$

The optical potentials $V_{[f]}$ obtainable by the RGM[17],[15] have essentially different "strengths" $VR^2$ for different [f]:
$V_{[4]}R_{[4]}^2 \approx 2.5\ V_{[22]}R_{[22]}^2$. The potential $V_{[4]}(r)$ is rather deep
$V_{[4]}(r = 0) \approx -45$ MeV, $R_{[4]} \approx 2$ fm. This potential corresponds to the ground-state of an $\alpha$-particle in the d + d channel, i.e. to the nucleon configuration $(OS)^4$, and is free of forbidden states ($\ell$-even). The potential $V_{[22]}$ corresponds to the configuration $(OS)^2(1p)^2[22]_X$, i.e. the lowest solution should be of the type $|2S\rangle$ and the ground state (the nodeless solution) is, thereby, forbidden (on the forbidden states see refs.18,15). Here the values of $\ell$ are also even. For the odd values of $\ell$ the symmetry appears to be $[f]_X = [31]$ with the potential $V_{[31]}(r)$ and the forbidden states are non-existent. The main result of the above discussion directly related to the symmetry of the six quark system NN is the fact that the A + B scattering is described by a set of amplitudes for different potentials, i.e. in a quite different, more complicated manner than the bound state AB corresponding to a single, energetically most favourable potential. This conclusion corresponds to the situation when the states with a definite total spin S of the system and the partial wave with a definite value of the orbital momentum $\ell$ have already been singled out by the usual polarization measurements. In order to describe both the bound d + d state and the scattering amplitude by the single potential $V_{[4]}(r)$ one should extract from the composite amplitude $T_\ell(S = 0)$ the part $T_\ell^{[4]}$ corresponding to $[f]_X = [4]$. To do this it is necessary to make a "generalized" polarization experiment on the measurement of the differential inelastic scattering cross section connected with the spin-isospin "flip-flop". Here it is appropriate to note that the problem at hand is a two-channel problem, as seen already from the formula (7) which can be recorded as

$$S_{\ell,el}(S = 0) = \frac{1}{2} \exp(2i\delta_\ell^{[4]}) + \frac{1}{2} \exp(2i\delta_\ell^{[22]}) \tag{8}$$

and this quantity differs from unity in absolute value

$$\bar{\eta}_\ell \equiv \left| S_{\ell,el}(S = 0) \right| = \cos(\delta_\ell^{[4]} - \delta_\ell^{[22]}) \tag{9}$$

Representing $S_{\ell,el}(S = 0)$ as

$$S_{\ell,el}(S = 0) = \bar{\eta}_\ell \exp(2i\bar{\delta}_\ell) \tag{10}$$

gives an interesting composition law for the phase shifts $\bar{\delta}_\ell$

$$\tan 2\bar{\delta}_\ell = \tan(\delta_\ell^{[4]} + \delta_\ell^{[22]}) \tag{11}$$

(here the upper bars mark the amplitude at the spin $S = 0$). The second channel, regarded by us as open, corresponds to two related amplitudes, one of which describes the transition of each deuteron to the singlet state, $d + d \rightarrow \bar{d} + \bar{d}$ (this is literally the spin-isospin "flip-flop")

$$S_{\ell,flip}(S = 0) = \langle 10,10 | 00 \rangle \left( \frac{1}{2} T_\ell^{[4]} - \frac{1}{2} T_\ell^{[22]} \right) \tag{12}$$

and the second charge-exchange amplitude corresponds to the final singlet pp and nn-states:

$$S_{\ell,ex}(S = 0) = \langle 11,1-1 | 00 \rangle \left( \frac{1}{2} T_\ell^{[4]} - \frac{1}{2} T_\ell^{[22]} \right) \tag{13}$$

Here, it is natural that

$$\left| S_{\ell,el}(S = 0) \right|^2 + \left| S_{\ell,flip}(S = 0) \right|^2 + 2\left| S_{\ell,ex}(S = 0) \right|^2 = 1 \tag{14}$$

The formulae (6-14) generalize the approach of A.M. Lane[19] to the case of optical scattering where the term $(\vec{T}.\vec{t})$ enabled one to couple the charge-exchange process and the elastic scattering, but without affecting the spin $S$. Operating on the whole, with the formulae (7-13) and the corresponding experimental data, we can separate, in principle, the amplitudes $T_\ell^{[4]}$ and $T_\ell^{[22]}$. If the energy $E$ is below the inelastic threshold, then at $S = 0$ the amplitudes $T_\ell^{[4]}$ and $T_\ell^{[22]}$ cannot be separated and, moreover, there will be a nonpotential amplitude due to the coupling with the closed channel $\bar{d} + \bar{d}$ that can have a resonance behaviour. All three factors together will determine the scattering phase $\bar{\delta}_\ell$ in a "simple" expression

$$T_\ell(S = 0) = e^{2i\bar{\delta}_\ell} - 1$$

True, there is an important facilitation typical of the system $d + d$ (and some other cluster systems $A + B$) that is unapplicable to the system NN. Namely, if the symmetry $SU(4)$ is valid, then, proceeding, for the case in question, to a different value of the total spin $S = 1$ (the corresponding scattering phase is $\delta_\ell'$) or $S = 2$ (the phase $\delta_\ell''$) we retain only one amplitude $T_\ell^{[22]}$ and will return to a single-channel situation because the symmetry $[f]_X = [4]$ is forbidden. In this case we have

$$\delta_\ell^{\,\prime} = \delta_\ell^{\,\prime\prime} = \delta_\ell^{[22]} \quad , \quad \ell\text{-even} ,$$

above and below the inelastic threshold. This circumstance, certainly, permits a detailed verification of the symmetry SU(4) of the Hamiltonian of the system. If required, it is easy to generalize the above aproach by introducing the SU(4) symmetry breakdown through the angle of mixing $\varepsilon$ between the schemes [v] and [v'], just as is done, for example, in the case of tensor forces. However, this point does not seem important for the lightest nuclei. Proceeding to the system NN we realize at once, in the light of this experience, that the composite structure of nucleon provides good reasons to separate the description of bound states of the lightest nuclei $^2$H, $^3$H-$^3$He, $^4$He from the description of NN scattering. Indeed, as will be discussed later, the energetically most favourable symmetries for the bound states are, seemingly, the attractive colour-magnetic (CM) quark symmetries where they are, hence, realized in a "pure" form. Meanwhile, in the NN-scattering there occurs the interference of many possible symmetries (the real sensitivity to the NN-structure in the "quark region" $r_{NN} < 1$ fm is exhibited by the scattering phases at sufficiently high energies $E_{lab} \gtrsim 500\text{-}800$ MeV). For example, as was already noted, the allowed symmetries for the triplet NN scattering in the even waves are $[6]_X[2^3]_{CS}$, $[42]_X[2^3]_{CS}$, $[42]_X[31^3]_{CS}$ with the CM-repulsion, the "neutral" symmetry $[42]_X[321]_{CS}$ and the single symmetry $[42]_X[42]_{CS}$ with the intensive colour-magnetic attraction of nucleons (see ref.3). The symmetry $[42]_X[21^4]_{CS}$ is not realized in the channel N + N.

   For the singlet NN-scattering in the even waves there also occurs the interference of many symmetries, but the symmetry with the colour-magnetic attraction has the form $[42]_X[33]_{CS}$. And, finally, the allowed symmetries for the singlet odd waves are $[3^2]_X[411]_{CS}$ and $[3^2]_X[3^2]_{CS}$ with the CM-attraction and $[3^2]_X[221^2]_{CS}$ and $[51]_X[221^2]_{CS}$ with the CM-repulsion.

   How good are the "generalized polarization measurements" for identifying individual components of the S-matrix? The corresponding general expression for the scattering amplitude has the form

$$\left\langle \binom{S}{S_3}_1 \binom{t}{t_3}_1, \binom{S}{S_3}_2 \binom{t}{t_3}_2 \middle| T_\ell \middle| \binom{S'}{S_3'}_1 \binom{t'}{t_3'}_1, \binom{S'}{S_3'}_2 \binom{t'}{t_3'}_2 \right\rangle =$$

$$= \sum CGC \left\langle \begin{matrix} [f_1]_C & [f_2]_C \\ C_1=0 & C_2=0 \end{matrix} \middle| \begin{matrix} [f]_C \\ C=0 \end{matrix} \right\rangle^2 \left\langle \begin{matrix} [f_1]_{CS} & [f_2]_{CS} \\ [f_1]_C\, S_1 & [f_2]_C\, S_2 \end{matrix} \middle| \begin{matrix} [f]_{CS} \\ [f]_C\, S \end{matrix} \right\rangle \times$$

$$\times \left\langle \begin{matrix} [f_1]_{CST} & [f_2]_{CST} \\ [f_1]_{CS}\, t_1 & [f_2]_{CS}\, t_2 \end{matrix} \middle| \begin{matrix} [f]_{CST} \\ [f]_{CS}\, T \end{matrix} \right\rangle T_\ell \left\langle \begin{matrix} [f]_{CST} & [f]_{CS} \\ [f']_{CST} & [f']_{CS} \end{matrix} \right\rangle \times$$

$$\times \left\langle \begin{matrix} [f']_{CST} \\ [f']_{CS}\, T \end{matrix} \middle| \begin{matrix} [f_1'']_{CST} & [f_2'']_{CST} \\ [f_1'']_{CS}\, t_1' & [f_2'']_{CS}\, t_2' \end{matrix} \right\rangle \left\langle \begin{matrix} [f']_{CS} \\ [f']_C\, S \end{matrix} \middle| \begin{matrix} [f_1''] & [f_2'']_{CS} \\ [f_1'']_C\, S_1' & [f_2'']_C\, S_2' \end{matrix} \right\rangle$$

$$\tag{15}$$

where "CGC" is an obvious product of the spin and isospin Clebsch-Gordan coefficients, a double bar stands in the isoscalar factors of different groups and $[f_{CST}] = [\tilde{f}_x]$. The group reduction chain chosen corresponds to the colour-magnetic interaction of quarks

$$\sum_{i<j} (\lambda_i \cdot \lambda_j)(s_i \cdot s_j).$$

As the first example, we consider the inelastic channel NN → ΔΔ. This process is compatible only with $[2^3]_{CS}$ because in the channel ΔΔ $[1^3]_{CS} \times [1^3]_{CS} = [1^6]_{CS} + [21^4]_{CS} + [2^21^2]_{CS} + [2^3]_{CS}$. The algebra in the formula (15) enables one to connect the differential NN → ΔΔ cross section (in the ideal case this requires the measurements of fourfold coincidences) with the cross section of the corresponding elastic scattering NN → NN. The energies needed here are $E_{lab} \gtrsim 1$ GeV. Of course, this experiment is extremely difficult, yet, very interesting (see below on the quasielastic knock-on of the Δ-isobar). The second example is furnished by the inelastic process NN → N* $([21]_x)$N*$([21]_x)$, which, in even waves, singles out the orbital Young scheme $[f]_x = [42]$ and permits the reconstruction of the corresponding part of the elastic amplitude by means of the same expression (15). However, the well-known level N*$([21]_x)$ with L = 1, $J^\pi = 3/2^-$ corresponds to the mass 1520 MeV and, therefore, the inelasticity in question can be studied only at $E_{lab} \gtrsim 2$ GeV. On the whole, the possibilities of identifying the amplitudes for individual symmetries are, of course, not very promising at present. It will also be noted that one more difficulty in the interval 1 GeV $\lesssim E_{lab} \lesssim 3$ Gev is the large cross sections of the inelastic processes that are not caused by the quark effects (the "bremsstrahlung" of soft π-mesons etc.). Therefore, for the lowest orbital momenta $|S_\ell| = \eta_\ell \approx 0.3 - 0.6$, and an as-yet-disregarded problem to be treated is how the strong absorption affects the general behaviour of the real part of the potential scattering phases (the validity of Levinson's theorem, the region of a Born-like behaviour of the phases if it is present, etc.). It is a very severe problem, yet, not hopeless. The interesting attempt by N. Hoshizaki to extend the phase analysis of NN-scattering up to $E_{lab} \approx 3.1$ GeV seems to be rather promising[20].

It is appropriate to conclude this section with a sceptical observation. It was noted above that at $E_{lab} \gtrsim 500-800$ MeV when the scattering amplitudes reflect the structure of the "quark region", the concept of a NN-potential becomes meaningless because the scattering is determined by the interference of many amplitudes corresponding to different quark symmetries and different potentials. Outwardly, this can be expressed in that the "resultant potentials reconstructed in a pure formal way by the RGM become essentially E-, L-, J- and T-dependent (see the E-dependent repulsive core of the Paris potential). Of course, it is not excluded that only a few selected symmetries will actually dominate (for example, owing to algebraic factors) and the potential-description of NN-scattering at relatively high energies will retain its meaning, yet, it is hardly probable.

3.   QUARK SYMMETRIES FOR THE BOUND STATES OF THE LIGHTEST NUCLEI

Now we proceed to the bound states of the lightest nuclei. As we have
already noted, a very interesting possibility can occur here, i.e. the
realization, in a pure form, of a single energetically most favourable
quark symmetry or quark configuration. From the same point of view,
the dibaryon and three-baryon resonances are, of course, both interesting as "separators" of the symmetries.

Let us discuss, on this plane the results of our paper[21] where the
evolution of the CM-attraction of nucleons in the nuclear series $^2$H,
$^3$H-$^3$He and $^4$He was considered in the MIT bag model. The energetically
most favourable are here the excited quark configurations
$s^4p^2[42]_X[42]_{CS}$, $s^5p^4[52^2]_X[621]_{CS}$ and $s^6p^6[63^3]_X[82^2]_{CS}$, respectively.
In the $^2$H system the situation is reminiscent of the degeneration with
the configuration $s^6[6]_X[2^3]_{CS}$ and for the $^3$H-$^3$He and especially $^4$H
nucleus the listed configurations of the CM-attraction are shifted to
lower energies. An interesting circumstance is the possibility of
representing the wave function for the noted configurations in such a
way that to each pair of 3q + 3q clusters there will correspond the
configuration $s^4p^2[42]_X[42]_{CS}$ (see ref.22). In other words, in the
nucleon sector (3N or 4N) the wave function of the $^3$H-$^3$H and $^4$He nuclei
has a node in the relative co-ordinate of each pair of nucleons and
this oscillation is not suppressed, as opposed to the case of
NN-scattering, because there are no competing symmetries.

Thus, in the $^3$H-$^3$He and $^4$He nuclei it is possible to describe the
nuclear structure wholly in the 3N or 4N sector with the correct inclusion of the central quark region efects by means of the two-particle
potentials with forbidden states (FSP). The parameters of the FSP in
the quark region $r_{NN} \lesssim 1$ fm should certainly, be defined phenomenologically in as much as the discussion of the 3Aq-configuration is based on
a rather imperfect perturbative scheme. Yet, qualitatively, the
appearance of this very type of interaction in the quark consideration
is of great importance. This conclusion will be used in the next
section and now we shall dwell upon the further evolution of quark
configurations with increasing A. Of importance in this case is the
competition between the kinetic energy $E_{kin} = \langle T \rangle$ which (without competition) would determine the usual ("regular") structure of the Fermi
sphere characteristic of an ideal Fermi-gas (the most symmetrical
orbital Young schemes $[f]_X$), and the colour-magnetic qq-interaction
$\langle V_{CM} \rangle$ which singles out the most symmetical Young schemes $[f]_{CS}$. The
colour-electrical interaction is not, seemingly, of particular concern
here because only one symmetry $[f]_C = [A^3]$ is possible for every
nucleus A. The qualitative investigation shows that the states with
the most symmetrical among all possible, scheme $[f]_{CS} = [f^M]_{CS}$ and the
most symmetrical orbital scheme $[f]_X = [f^M]_X$ compatible with $[f^M]_{CS}$ are
distinguished by a substantially lower energy. The corresponding gain
in energy, as compared with the "regular configuration", related to one
NN-pair, is for the same model as ref.21 nearly 150-170 MeV (the same
as for the $^4$He nucleus) starting from A = 15-16 whereas between the $^4$He
and $^{16}$O nuclei there is a "gap" in this quantity (fig.4). For example,
for the $^{16}$O nucleus we have $[f^M]_{CS} = [32,8^2]$, $[f^M]_X = [6^42^{12}]$ which

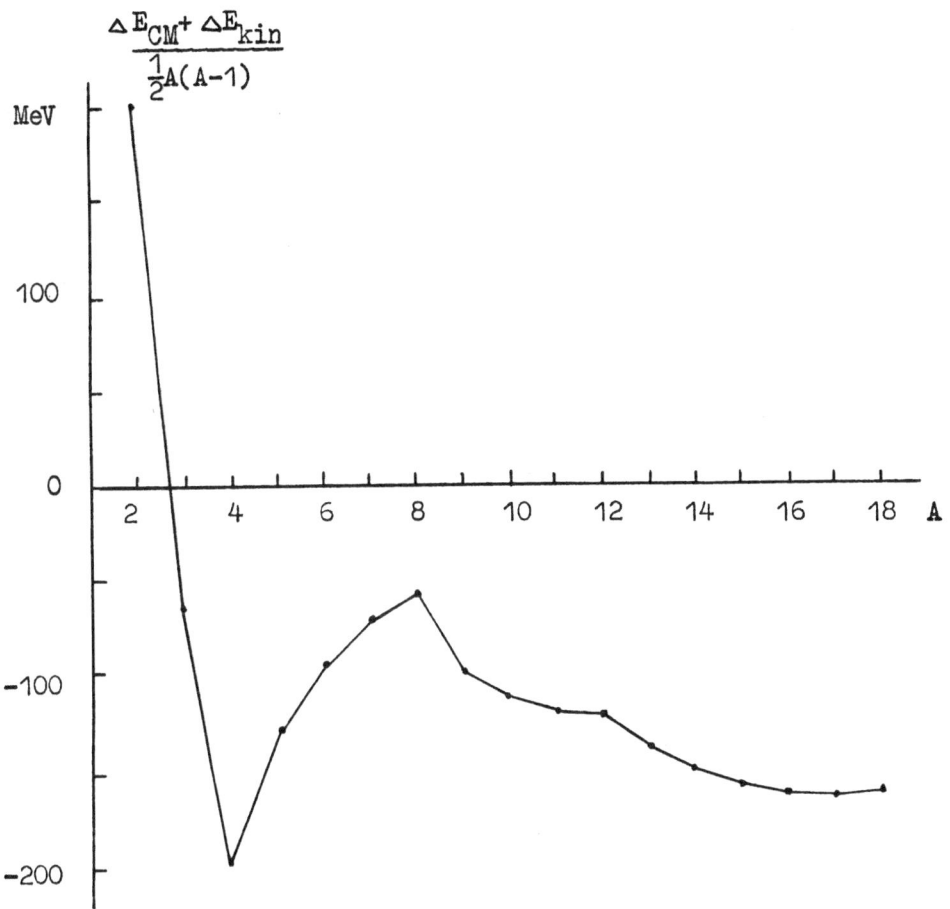

Figure 4. The lowest energy levels of 3Aq systems referred to a single NN pair. The point E = 0 corresponds to the AN threshold.

corresponds to the configuration $(OS)^6(1p)^{18}(2s - 2d)^{12}(3f - 3p)^{12}$. All these characteristics radically differ from the "regular" symmetry $[f^R]_{CS} = [8^6]$, $[f^R]_X = [12^4]$ and the regular configuration $(OS)^{12}(1p)^{36}$. This risky and purely qualitative extrapolation, which, for example, does not take into account the dependence of the matrix elements upon quantum numbers $n\ell$ of quarks, raises, nevertheless, a very interesting question of the "multihole" structures of quark matter (discussed first by C.W. Wong[23] with the radical simplification $[f^M]_X = [1^{3A}]$). This peculiarity leads, as we see, to a substantial increase of the Fermi momentum $P_F$ and to a more hard momentum distribution of quarks. The thermodynamical properties of quark matter will also alter. It will be emphasized, however, that for the scheme in question there exists a serious danger, the nonperturbative effects of the QCD whose action upon the multiquark system is only in the initial stage of investigation[24]; we are faced here with many vague points and, nevertheless, the baryon excitation spectrum permits us to realize how important various symmetry factors are.


## 4.    CALCULATIONS ON THE BASIS OF THE POTENTIALS WITH FORBIDDEN STATES

We have already explained that owing to the composite structure of a nucleon the data on NN-scattering at high energies cannot be used in some simple way to determine the short-range part of the NN-potential corresponding to the energetically  most favourable 6q-symmetry of the bound states of nuclei.  Therefore, in view of our suggestion to try out, along with other possibilities, the potentials with forbidden states (FSP), the phenomenology of the FSP in the quark region $r_{NN} \lesssim 1$ fm must be defined and verified on the basis of indirect data on the bound states of the lightest nuclei.  At the distances $r \gtrsim 1$ fm (or, may be, 0.8 fm) the potential must, certainly, change over to the OBEP and, so, the FSP must describe the phases of NN-scattering up to 500 MeV (at higher energies the descriptions of the NN-phases by the potentials with the repulsive core and by the FSPs will be rather different, reflecting differences in the interpretation of the quark region $r \lesssim r_c$).

The three-parameter FSP for the NN-scattering was first constructed in our paper[23] using, as an example, the singlet even waves.  The description up to energies $E_{lab}$ = 500 MeV was the same as in the case of Reid's potential with the soft core[25].  Then followed the construction[26] of the simplest two-parameter Gaussian central potentials for the singlet and triplet S-waves.  The triplet potential has the form

$$V^{31}(r) = - V_0^{31} \exp(- \beta_{31} r^2) \tag{16}$$

$$V_0^{31} = 944.0 \text{ MeV}, \quad \beta_{31} = 1.20 \text{ fm}^{-2} \tag{17}$$

and describes likewise the Reid's potential, the deuteron binding energy, the scattering length, the effective radius, the vertex constant of the virtual deuteron decay and the $^3S_1$-phase up to energies $E_{lab}$ = 450 MeV.  The singlet potential with the parameters

$$V_0^{13} = 1128.6 \text{ MeV}, \quad \beta_{13} = 1.55 \text{ fm}^{-2} \tag{18}$$

describes all analogous characteristics, except the effective radius (2.18 fm instead of 2.83 fm in experiment), though there seems to be an uncertainty in the experimental values of the singlet radius $\rho(0,0)$ the values of $\rho(0,0)$ = 2.2-2.5 fm are also discussed, see, for example, ref.27.  But, of course, these remarkably efficient potentials have no asymptotics of the OBEP.

Recently our group has considered[28] the tensor plus central interaction of nucleons with the asymptotics extending to the OBEP, which leads to a good description of all even waves by the FSP.  True, this interesting result has a special point: the node is present not only in the S-wave but also in the D-wave which is incompatible with two excitation quanta in the 6q-system that are typical of the CM-attraction configuration $S^4 P^2 [42]_X [42]_{CS}$. This point is to be elucidated in future.

It will be emphasized once again that the forbidden state in the noted potentials is due not to the Pauli-exclusion principle but to the "principle of dynamical exclusion" - the attraction of nucleons arises only in the face of the symmetry requiring the excited quark configurations, i.e. a node of the relative motion wave function at $r_{NN} \approx r_c$.

The situation with the energetically favourable quark configurations arising from the CM-attraction in the lightest nuclei has a good analogue in nuclear physics. It is a 3 $\alpha$-system that is furnished with the results of the calculations on the basis of the $\alpha\alpha$-potentials with the repulsive core[29] and also the attractive $\alpha\alpha$-potentials with forbidden states[30,22]. The latter potentials[15] give quite a different, as compared to the former[31], description of the $\alpha\alpha$-scattering phases at high energies and overestimated by 15 MeV binding energy of the 3 $\alpha$-system.

In a similar way in collaboration with the colleagues from the Kalinin University, we have made a tentative attempt[32] to test the FSP in the calculation of the binding energy and size of the $^4$He nucleus. The calculation was carried out with the effective central interaction (16-18). To include correctly the additional conditions of orthogonality to the forbidden state

$$\langle \Psi \mid \phi_0(ij) \rangle \tag{19}$$

is represented as

$$\Psi = \prod_{i>j} \beta_{ij} \, \tilde{\Psi} = \hat{\beta} \, \tilde{\Psi} \tag{20}$$

where the two-particle correlators $\chi_{ST}(r) = r\beta_{ST}(r)$ satisfy the equation

$$\chi_{ST}''(r) - [V^{ST}(r) - E_{ST}^{cut}] \, \chi_{ST}(r) = 0 \ , \ r \in [0, a_{ST}]$$

$$\chi_{ST}(o) = 0 \ , \ \frac{d}{dr} \ell n \, \chi_{ST}(r) \Big|_{r=a_{ST}} = a_{ST}^{-1} \tag{21}$$

and the solution chosen is with the node at $r \approx r^c$ which provides a rather good fulfilment of the condition (19) at the level $10^{-4}$. The D- and P-states of the NN-system are switched off. The "smooth" function $\tilde{\Psi}$ is determined from the equation

$$[t + V_{cut}] \, \tilde{\Psi} = - \tilde{E}_0 \tilde{\Psi}$$

$$V_{cut} = \sum_{i>j} [v_{cut}^{10}(r_{ij}) \, P_S^+(ij)P_T^-(ij) + v_{cut}^{01}(r_{ij}) \, P_S^-(ij)P_T^+(ij)]$$

$$v_{cut}^{ST} = \begin{cases} E_{ST}^{cut} \ , \ r \in [0, a_{ST}] \\ v^{ST}(r), \ r > a_{ST} \end{cases} \tag{22}$$

$P_S^+(P_S^-)$ is the projector onto the triplet (singlet) state.

The expansion of $\Psi_0$ and $\tilde{E}_0$ in the hyperspherical basis[33] rapidly converges. Recording

$$\hat{H}\,\hat{\theta}\,\Psi = \hat{\theta}(\hat{T} + \hat{V}_{cut})\,\Psi + \Delta\tilde{H}\,\Psi$$

where $\Delta H$ contains the commutators of the $\beta$-operators, we represent

$$E_0 = \tilde{E}_0 + \Delta E_0$$

$$\Delta E_0 \approx \langle\hat{\theta}\,\Psi \mid \Delta\hat{H} \mid \Psi\rangle$$

and obtain that the system is minimized at $a_{01} = 1.13$ fm, $a_{10} = 1.33$ fm. In the end, $\Delta E_0 \approx 3$ MeV with an error less than 1 MeV, $E_0 \approx 42$ MeV and the radius of nuclear matter R = 1.19 fm.

This tentative result inspires us with optimism because it is known that the inclusion of the tensor interactions in the calculations with "realistic" forces decreases the binding energy by 8-9 MeV and, correspondingly, increases the nuclear radius. And, of course, it is interesting to verify to what extent the difficulties encountered in the description of the lightest and light nuclei and, also, nuclear matter by means of the "realistic" NN-potentials can be surmounted with the help of the FSP (the underbinding of the $^4$He-nucleus by $\approx$ 4-5 MeV if the nuclear size is correctly described, etc.). But the corresponding accurate calculations for the $^3$H-$^3$He and $^4$He nuclei on the basis of the FSP which will, undoubtedly, be carried out in the near future, involve a number of new problems to be solved even if the calculation results are good. First, the assumption of the two-particle NN-interaction of the type FSP is only the "minimum" assumption compatible with the structure of a "central quark drop" in the $^3$H-$^3$He, $^4$He nuclei. The triple, fourfold ... short-range forces between nucleons are also quite possible, as suggested by E. Schmid[34]. Of particular interest, on this plane, is the experimentally revealed "collective co-ordination" of all three, and respectively, four nucleons in the $^3$He-$^3$H, $^4$He nuclei which manifest itself in the decreased density of the nuclear centre[35]. In principle, this can be a manifestation of the two-particle FSPs, in as much as all three (four) nucleons are "allowed" to be at "one point" (i.e. in the quark drop), but the probability of this event is small. Therefore, the density is low here and the wave function has a great positive gradient at $r \approx r_c$, (see fig.1 and 2, ref.12). To verify this moment it is certainly necessary to take accurate account of the above discussed correlators $\beta(ij)$ to all orders which is not easy. But the decreased density in the centre of nucleus may reflect the triple and fourfold forces between nucleons having the "quark" origin.

Second, it should be proved that the quark configuration $s^4p^2[42]_x[42]_{CS}(S = 1, T = 0)$ or $s^4p^2[42]_x[33]_{CS}(S = 0, T = 1)$ and not $s^6[6]_x[2^3]_{CS}$ is realized in the nucleon-nucleon pairs at short distances. There are three ways about it. The first is to use, for example, the quasielastic knock-out $^2$H(e,e'$\Delta$)$\Delta$. As noted above, the symmetry $[42]_{CS}$ or $[33]_{CS}$ is incompatible with the $\Delta\Delta$ channel and, so if the NN-pairs in the lightest nuclei correspond to the quark configurations $s^4p^2[42]_x[42]_{CS}(S = 1, T = 0)$ or $s^4p^2[42]_x[33]_{CS}(S = 0, T = 1)$ in the noted experiment one will observe only the "soft" recoil momenta to $\Delta$-spectator $q < r_c^{-1}$ corresponding to the formation of $\Delta\Delta$-admixture in the meson "cloud"[36] and not in the quark "drop". This exclusive experiment on coincidence of five particles (e',N,N,$\pi$,$\pi$) is rather difficult and, in addition, the separation of the recoil momenta into "soft" and "hard" type is not defined in a good quite quantitative manner.

The second way is to study the elastic electron form-factors in the region of "medium" values of momentum transfer q, $5$ fm$^{-2} \lesssim - q^2 \lesssim 50$ fm$^{-2}$. Since the calculations[37] show that the quark exchange effects in these form factors are, seemingly, weak it is expected that the influence of the node of a wave function upon the form factor will be noticeable, especially in the experiments with polarized electrons, when the S-state contribution is identifiable.

The third way is connected with analysis of the momentum distribution of nucleons in the deuteron $^2$H and helion $^3$He obtained by different means[38]. This distribution is presently known up to the momentum k = 0.8 GeV/c and is characterised by an explicit excess, by 2-3 times, as compared with that provided the Paris potential, at k > 0.3 GeV/c. It is clear that the use of the wave function possessing a node in the NN-channel with its large gradient at the nodal point will serve to improve this point.

A joint use of all three ways is, seemingly, the only means of clarifying the situation in question.


ACKNOWLEDGEMENT

The authors express their gratitude to V. Iskra, A.I. Mazur and Yu.I. Kharitonov for assistance with the preparation of the materials for the report.

# REFERENCES

1. Oka M., Yazaki K., Prog. Theor. Phys. **66** (1981) 572; Nucl. Phys. **A402** 477.
2. Faessler A., Fernandez F., Lubeck G., Shimizu K., Nucl. Phys. **A402** (1983) 555.
3. Obukhovsky I.T., Neudatchin V.G., Smirnov Yu.F., Tchuvil'sky Yu.M., Phys Lett. **B88** (1979) 231; Neudatchin V.G., Smirnov Yu.F., Tamagaki R., Prog. Theor. Phys. **58** (1977) 1072.
4. Cvetic M., Golli B., Mankoc N., Rosina M., Nucl. Phys. **A395** (1983) 349.
5. Storm M.H., Watt A., Nucl. Phys. **A408** (1983) 397.
6. Ohta S., Oka M., Arima A., Yazaki K., Phys Lett. **B119** (1982) 35.
7. Obukhovsky I.T., Kusainov A.M., XXIV Conf. on Nuclear Spectroscopy and Nuclear Structure, Alma-Ata, 1984, Extended Abstracts.
8. Harvey M., Nucl. Phys. **A352** (1981) 301, 326.
9. Obukhivsky I.T., Z. Phys. **A308** (1982) 253.
10. Bhaduri R.K., Cohler L.E., Nogami Y., Phys. Rev. Lett. **44** (1980) 1369.
11. Hecht K.T., Suzuki Y., Phys. Rev. **C28** (1983) 1458.
12. Neudatchin V.G., Obukhovsky I.T., Kukulin V.I., Golovanova N.F., Phys. Rev. **CII** (1975) 128.
13. Lacambe M., Loiseau B., Richard J.M., Vinh Mau R., Pires P., de Tourreil R., Phys. Rev. **C21** (1980) 861; Cote J., Pires P., de Tourreil R., Lacombe M., Loiseau B., Vinh Mau R., Phys. Rev. Lett. **44** (1980) 1031.
14. Simonov Yu.A., Phys. Lett. **107B** (1981) 1; Preprint ITEP-93, Moscow 1983; Yad. Fiz. **38** (1983) 1542.
15. Kukulin V.I., Neudatchin V.G., Smirnov Yu.F., Nucl. Phys. **A245** (1975) 429; ECh AYa **10** (1979) 1236.
16. Neudatchin V.G., Smirnov Yu.F., Nucleon Clusters in Light Nuclei. Nauka, Moscow 1969 (in Russian).
17. Barrett R.F., Baldock R.A., Robson B.A., Nucl. Phys. **A381** (1982) 138; Horiuchi R., Prog. Theor. Phys. **69** (1983) 516.
18. Saito S., Prog. Theor. Phys. **41** (1969) 705.
19. Lane A.M., Nucl. Phys. **35** (1962) 676.
20. Higuchi Y., Hoshizaki N., Nakao H., Suzuki K. Talk at 1983 INS Int. Sym. High Energy Photonuclear Reactions and Related Topics, Tokyo, Sept. 1983.
21. Neudatchin V.G., Obukhovsky I.T., Smirnov Yu.F., Tkalya E.V., Z. Phys. **A313** (1983) 357.
22. Smirnov Yu.F., Obukhovsy I.T., Chuvil'sky Yu.M., Neudatchin V.G., Nucl. Phys. **A235** (1974) 289.
23. Wong Ch.Wa., Phys. Lett. **100b**(1982) 383.
24. Vento V., Rho M., Nucl. Phys. **A412** (1984) 413.
25. Reid R.V., Ann. Phys. (N.Y.) **50** (1968) 411.
26. Dubovitchenko S.B., Jusupov M.A., Izv. Akad. Nauk Kazakh. SSR, Ser. Fiz.-Mât., **N6** (1982) 34.
27. Kolesnikov N.N., Tarasov V.I., Yad. Fiz. **35** (1982) 609.

28.  Kukulin V.I., Pomerantsev V.N., Krasnopol'sky V.M., Sazonov P.B.,
     Phys Lett. **135B** (1984) 20.
29.  Aguilera-Navarro V.C., Portilho O., Ann. Phys. **107** (1977) 126;
     Nakaichi S., Akaichi Y., Tanaka H., Prog. Theor. Phys. **57**
     (1977) 1086.
30.  Horiuchi H., Prog. Theor. Phys. **51** (1974) 1266; G. Bertsch and
     B. Poon, Preprint, Michigan St. University, 1981.
31.  Ali S., Bodmer A.R., Nucl. Phys. **80** (1966) 99.
32.  Bursak A.V., Gorbatov A.M., Kukulin V.I., Neudatchin V.G.,
     Rudyak B.V.  All-Union Conference on the Theory of Few-Particle
     Systems with Strong Interactions.  Leningrad 1983.  Extended
     abstracts, p.103.  Leningrad St. University Publ. House, 1983.
33.  Efros V.D., Yad. Fiz. **15** (1972) 226; Gorbatov A.M., Yad. Fiz.
     **18** (1973) 1017.
34.  Schmid E., Review Talk presented to X Int. Conf. on Few-body
     Problems in Physics, Karlsruhe, Aug.21-27, 1983.
35.  McCarthy J.S., Sick I., Whitney P.R., Phys.Rev. **C15** (1977)
     1395.
36.  Weber H.J., Arenhövel H., Phys. Repts. **366** (1978) 277.
37.  Obukhovsy I.T., Tkalya E.V., Yad. Fiz. **35** (1982) 288.
38.  Bosted P., Arnold R.G., Rock S., Szalata Z., Phys. Rev. Lett.
     **49** (1982) 1380.
     Ableev V.G., Strunov L.N., Kobushin A.P. et al.  Proc. II
     All-Union Symp. "Nucleon-nucleon and hadron-nuclear interactions
     at intermediate energies".  Leningrad Institute of Nuclear Physics,
     April 1984.

ASTROPHYSICS AND ANOMALONS

# CLUSTERING ASPECTS IN ASTROPHYSICS AND RELATIVISTIC HEAVY-ION REACTIONS

Y. C. Tang
School of Physics
University of Minnesota
Minneapolis, Minnesota 55455
U.S.A.

ABSTRACT. Selected topics in the fields of relativistic heavy-ion reactions and nuclear astrophysics will be considered. These topics are: anomalon production and asymmetric fission in the former field and radiative-capture reactions in the latter field. Experimental situations will be reviewed, and clustering aspects, necessary for the explanation of the observed phenomena, will be discussed.

## 1. INTRODUCTION

In this presentation, I shall discuss some selected topics in the fields of relativistic heavy-ion reactions and nuclear astrophysics. As is quite evident from the nature of these two fields, the relationship between them is certainly not intimate, but there does seem to exist one common feature. This feature is that, for an explanation of the phenomena to be discussed, clustering aspects of nuclear structure will need to be carefully considered.

In the field of relativistic heavy-ion reactions, the topics selected for discussion are:

(i) Possible existence of anomalous components in projectile fragments [1]. Experimentally, some evidence seems to exist that, when relativistic heavy ions interact with target nuclei in nuclear emulsion, resulting projectile fragments may contain entities which have anomalously short mean free paths (mfp) [2]. These entities are now referred to as anomalons.

(ii) Asymmetric fission of nuclei with $N < 132$. When relativistic Au nuclei ($Z = 79$, $N = 118$) were sent into nuclear emulsion, it was observed [3,4] that there occurred a number of binary-fission events, out of which an appreciable fraction appears to be asymmetric-fission events. This is rather surprising, since it was previously believed [5,6] that symmetric fission predominates if the fissioning nucleus has fewer than 132 neutrons.

For astrophysics, my discussion will be confined to the nuclear aspects of the solar-neutrino problem [7]. As is well known, this problem is concerned with the discrepancy between calculated and measured solar-neutrino capture rates for $^{37}Cl$

*J. S. Lilley and M. A. Nagarajan (eds.), Clustering Aspects of Nuclear Structure, 373–396.*

detector. By studying the accuracy of the experimental results and the reliability of the theoretical calculations, I shall show that this problem does not likely have a nuclear-physics origin and its resolution should be sought elsewhere.

Anomalons and asymmetric fission will be discussed in sections 2 and 3, respectively. The discussion here is essentially a follow-up of my review talk [8] given at the Second Workshop on Anomalons. Nuclear aspects of the solar-neutrino problem will be considered in sect. 4. Finally, in sect. 5, concluding remarks will be made.

## 2. ANOMALONS

### 2.1. Experimental situation

The experimental situation is not very clear at this moment. Both positive and negative evidence for the existence of anomalons seems to be present. Here I shall review some of the experimental results and attempt to offer some clarification.

Experiments to detect anomalons have been performed either in nuclear emulsions or in various types of plastic media. The results are summarized as follows:
(a)  Experiments in nuclear emulsions.

Local-mfp results (i.e., variation of mfp with distance D from the point of creation) for Z = 2 projectile fragments (PF) are depicted in fig. 1. In this figure, the right side shows the results of El-Nadi et al. [9] for selected $N_h$ =0 or white-star events ($N_h$ is the number of heavy target-fragmentation tracks) initiated by 3.7A-GeV $^{12}$C projectiles, while the left side shows the results of Ismail et al. [10] and Karant et al. [11] for events initiated by 2A-GeV $^{40}$Ar and $^{56}$Fe projectiles. The quantity $\tilde{R}$ denotes the local mfp normalized to the mfp-value of about 20 cm for primary He-projectiles in nuclear emulsion.

The interesting features shown in fig. 1 are: (i) the value of $\tilde{R}$ obtained by El-Nadi et al. in white-star interactions varies slowly with D and is consistently smaller than 1, the value expected when there are no anomalous components in the He projectile fragments, and (ii) from the combined results of Ismail et al. and Karant et al., one notes that $\tilde{R}$ is close to 1 for all D-values considered. At first sight, these results appear to be contradictory. However, it should be pointed out that the projectile species and beam energies are different in these experiments. In addition, one should perhaps also keep in mind that the result of El-Nadi et al. has rather low statistical accuracy, since the number of secondary interactions observed was only 223.

To achieve some understanding of the variation of $\tilde{R}$ with D, we assume that the PF contain two components, a normal component with mfp $\lambda_1$ = 20 cm and an anomalous component with mfp $\lambda_2$. With this assumption, $\tilde{R}$ is given by

$$\tilde{R}(D) = \frac{1}{\lambda_1} \sum_{i=1}^{2} f_i \exp(-D/\lambda_i) / \sum_{i=1}^{2} (f_i/\lambda_i)\exp(-D/\lambda_i) , \qquad (1)$$

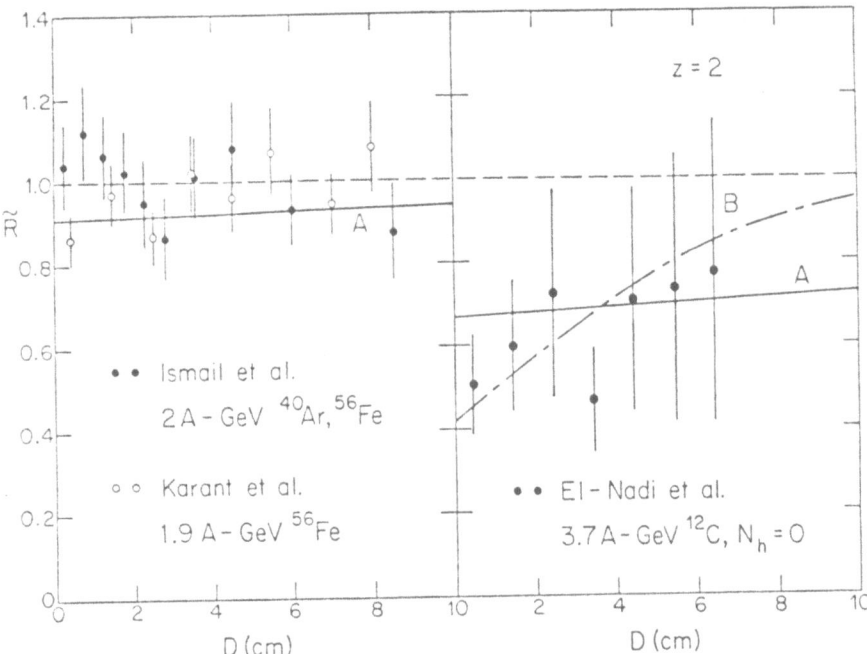

Fig. 1: Normalized local mfp for Z=2 projectile fragments in emulsion. (Adapted from refs. 9-11).

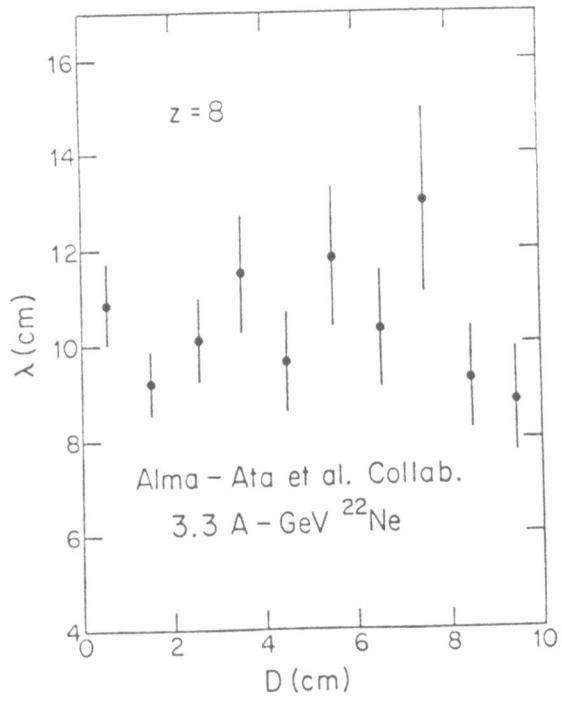

Fig. 2: Local mfp for Z=8 projectile fragments in emulsion. (Adapted from ref. 12).

with $f_i$ denoting the initial fraction of the $i^{th}$ component. In the
right side of fig. 1, curve A is obtained by choosing $\lambda_2$ = 10 cm and
$f_2$ = 50%, while curve B is obtained by choosing $\lambda_2$ = 2.5 cm and
$f_2$ = 20%. As is seen, both curves fit the data reasonably well. This
shows that, because of the crudeness of the experimental result of
El-Nadi et al., it is in fact not possible to learn the detailed nature
of the anomalous component. A similar analysis can also be performed
to study the combined data of Ismail et al. and Karant et al.     In
this case, curve A shown in the left side of fig. 1 is obtained
with $\lambda_2$ = 10 cm and $f_2$ = 10%. Here one notes that, although the
experimental result is consistent with the non-existence of anomalons,
the possibility of the presence of a small fraction (less than 10%) of
anomalous component with $\lambda_2$ around 10 cm cannot be ruled out.

For $3 \leqslant Z \leqslant 9$, local-mfp effects have been examined by the Alma-
Ata et al. Collaboration [12], using 3.3A-GeV $^{22}$Ne projectiles.
The results, illustrated for Z = 8 in fig. 2, do not exhibit any
significant local-mfp effect, thus casting doubt on the existence
of anomalons in this Z-range. A similar conclusion for Z = 3 and 4,
based on smaller statistical samples, has also been reported by Ismail
et al. [10].

The initial evidence for the existence of anomalons was provided
by the emulsion studies of Jain and Das [13] and Friedlander et al.
[14] with 2A-GeV beams of $^{16}$O, $^{40}$Ar, and $^{56}$Fe, and the emulsion
study of Barber et al. [15] with cosmic rays. In these studies, the
statistical significance was increased by pooling the mfp data for
individual charges. The pooling was achieved by observing that, for
primary nuclei, the empirical mfp values can be approximately
represented by the power-law relation

$$\lambda_z = \Lambda_{beam} \, Z^{-b} \quad . \tag{2}$$

Utilizing the information contained in this relation, one can then
experimentally determine the values of a charge-independent mfp
parameter $\Lambda$ and its variation with respect to D [16]. The results
for $\Lambda/\Lambda_{beam}$ obtained by these three groups are shown in fig. 3. The
salient feature is that $\Lambda/\Lambda_{beam}$ is significantly smaller than 1
within the first few centimeters from the points of creation. Together
with the information gleaned from the higher-statistics data of the
Alma-Ata et al. Collaboration mentioned above, one can conclude that
there may exist higher-Z anomalous components within the projectile
fragments.

Based on the results from an emulsion experiment utilizing 1.5A-
GeV $^{84}$Kr projectiles, Jain et al. [17] recently reported that they
have obtained some evidence for the existence of anomalons with $Z \geqslant 15$.
In this experiment, however, they have also observed 13 secondary
interactions of PF which occur within about 50 μm of the production
points. These are strange events which have not been clearly
explained. My viewpoint is that, until the nature of these events
becomes better understood, the physical significance of this experiment
should perhaps not be over-emphasized.
(b) Experiments with plastic media

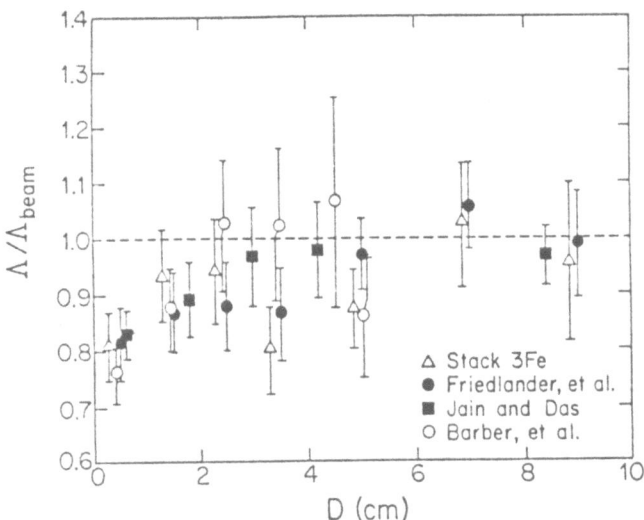

Fig.3: Normalized local mfp in emulsion as a function of distance from the point of creation. (Adapted from refs. 13–15).

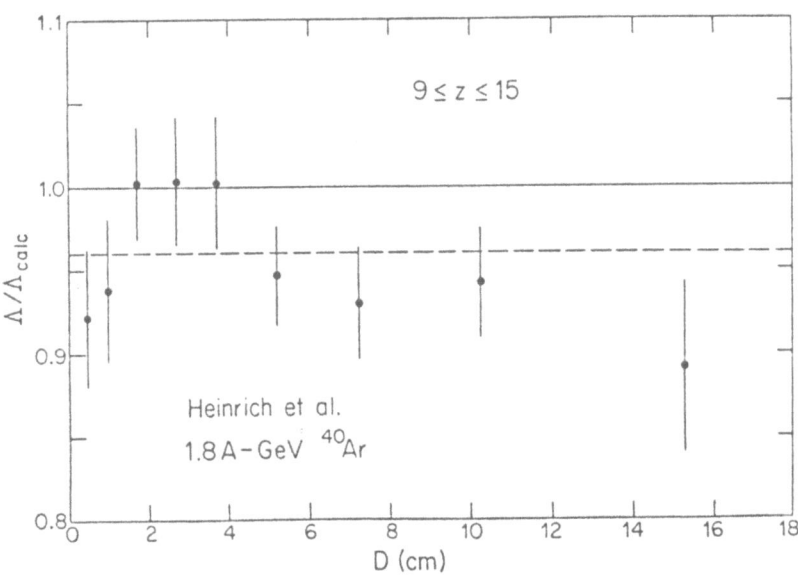

Fig. 4: Normalized local mfp in CR-39 plastic. (Adapted from ref. 18).

Experiments using plastic media have been performed by many groups. In comparison with emulsion studies, the main advantage of these experiments lies in the fact that the data acquisition rate is much higher. But there are disadvantages as well. As has been pointed out by Jain et al. [17], experiments with plastic detectors do not yield any useful information about target fragmentation. In addition, these detectors can detect only those events with $\Delta Z \geqslant 1$ and are limited to the detection of PF with $Z > 7$. Also, because of the sheet thickness (about 600 μm), the vertex positions of interactions cannot be accurately determined.

Using CR-39 plastic track detectors and 1.8A-GeV $^{40}Ar$ projectiles, Heinrich et al. [18] observed 6444 fragment interactions with $9 \leqslant Z \leqslant 15$. To increase statistical significance, they have again pooled the data from different fragment charges by normalizing the mfp for each charge to the calculated value for the interaction mfp. The results for $\Lambda/\Lambda_{calc}$ are shown as a function of D in fig. 4. To determine whether there exist anomalon effects or not, one must compare the measured results with the value expected in case of no anomaly. Because of uncertainty resulting from the use of calculated mfp, it is not immediately clear whether the data should be considered with respect to the solid line corresponding to $\Lambda/\Lambda_{calc} = 1$, or to the dashed line corresponding to $\Lambda/\Lambda_{calc} = 0.96$. By utilizing some experimental information on charge-changing interaction mfp for beam particles in CR-39 and by observing that the measured values of $\Lambda/\Lambda_{calc}$ are consistently smaller than 1 for large values of D, Heinrich et al. decided that it may be more reasonable to adopt the dashed line as a standard for comparison. In this way, they came to the conclusion that, under their experimental condition, no statistically significant anomalous local-mfp effect was observed.

The conclusion reached by Heinrich et al. contradicts that reached by Tincknell et al [19]. With same types of detectors and incident projectiles, these latter authors found anomalon effects in the Z-range between 11 and 17. It should be noted, however, that this conclusion was obtained with a much smaller sample consisting of only 612 fragment interactions. In view of this, one should perhaps not be overly concerned with this apparent contradiction between these CR-39 results.

Other experiments utilizing plastic media have also been reported. With 1.8A-GeV $^{40}Ar$ and $^{56}Fe$ beams and Lucite plastic strips, Symons et al. [20] electronically detected Cherenkov radiation from the fragments and obtained information concerning fragment charges with $Z \geqslant 11$. In fig. 5, their results for $\Lambda/\Lambda_{>}$ in the Ar case are illustrated, with $\Lambda_{>}$ denoting the average value of the mfp parameter for D > 3 cm. As is seen from this figure, the value of $\Lambda/\Lambda_{>}$ fluctuates around 1, and there seems to be no anomalous local-mfp effect. Based on such findings, Symons et al. then concluded that they have found no evidence for the existence of a short mfp component in Fe and Ar secondaries with $Z = 13 - 24$ and $Z = 11 - 16$, respectively.

A similar high-statistics electronic experiment has also been performed by Stevenson et al [21]. Here the conclusion is again that

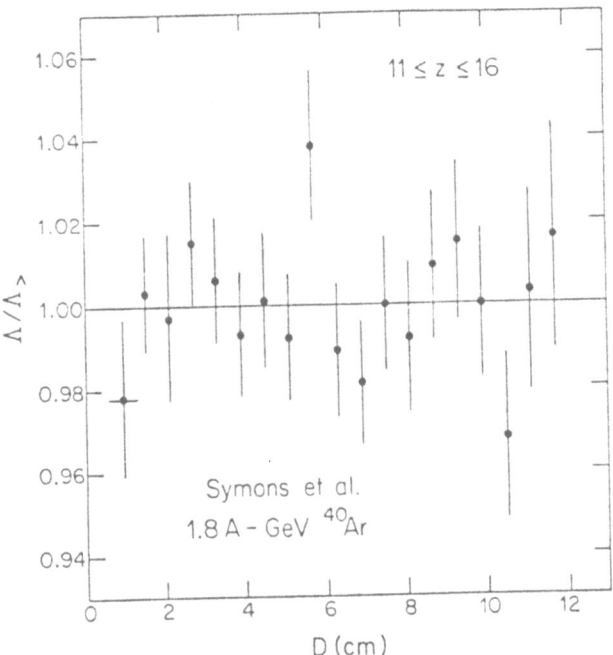

Fig. 5: Normalized local mfp in Lucite plastic. (Adapted from ref. 20).

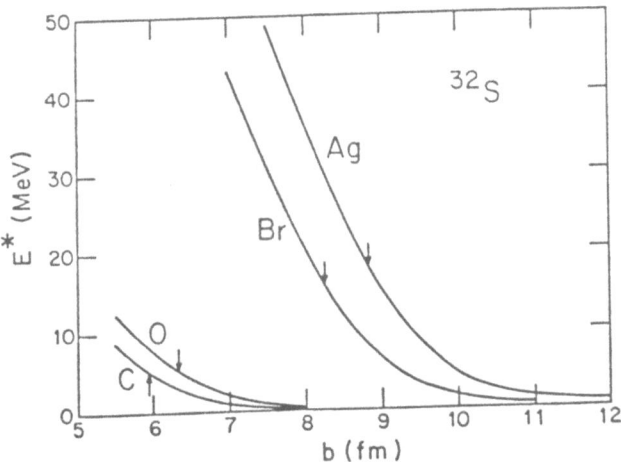

Fig. 6: Excitation energy of $^{32}$S as a function of the impact parameter.

there appears to be no anomalon effect for fragment charges with
Z = 13 - 17.
(c)   Summary
        Emulsion studies seem to indicate that anomalon effects may be
present in selected Z = 2 fragments produced by specific projectiles
and in fragments with Z-values larger than at least about 10.  However,
because of low statistical accuracy, the existing evidence is by no
means definitive.  My opinion is that, for a more convincing
demonstration for the existence of anomalons, refined experiments
should be further performed to supplement these studies.
        On the other hand, results from experiments with plastic media
suggest that, for fragment charges with Z larger than about 8,
anomalous mfp effect, if it exists, must be rather weak.  In
conjunction with the emulsion results mentioned above, this indicates
that PF with anomalously short mfp may be more likely to be produced by
interactions in nuclear emulsion than by interactions in plastic.  In
the following, it will be shown [22] that this suggestion is consistent
with the impulse mechanism which was proposed for anomalon production
several years ago [8,23].  The main point will be that the impulse
mechanism produces anomalons more copiously when projectile nuclei
interact with heavy target nuclei such as Ag or Br in emulsion, than
with light target nuclei such as H, C, or O in plastic.

## 2.2.   Theoretical considerations

(a)   General properties of anomalons
        Anomalons, if they exist, should exhibit the following general
properties:
(i)  Large geometrical size.  At relativistic energies, reaction cross
sections depend mainly on the geometrical sizes of the nuclei involved
[24].  Thus, anomalons can have anomalously short mfp only if they turn
out to be significantly larger in geometrical size than the
corresponding normal nuclei.
(ii) Long lifetime.  Local-mfp studies indicate that anomalons survive
for an appreciable distance from the points of creation.  This shows
that they must have a proper lifetime longer than $10^{-11}$ sec.
(iii)Appreciable production probability.  This follows again from
local-mfp studies.  If the production probability is very small,
anomalon effects would have escaped detection in existing emulsion
experiments which have generally rather small sample sizes.
        In searching for candidates which might qualify as anomalons, it
is my viewpoint that one should take a conservative approach and first
stay within the proper realm of conventional nuclear physics.  Only
when such an endeavor fails that one should venture out and look for
explanations in terms of more exotic phenomena such as hadroids, quark-
gluon effects, and so on [1,25].
        It is well known in cluster physics [26] that there exist in
light, medium and heavy nuclear systems quasi-molecular structures
(QMS) which are composed of two or more relatively well-separated
nucleon clusters [8].  In the nucleus $^{16}O$, for example, simple
arguments [27] indicate that highly deformed 4 $\alpha$  linear-chain
structures may occur with rather high excitation energies.

Experimentally [28], it has indeed been determined that, in the
excitation region between about 17 and 20 MeV, resonance states do
exist which form a rotational band with a very large moment of inertia,
similar to that expected for a $4\alpha$   linear-chain intrinsic structure.
    When $^{16}O$ is in one of these $4\alpha$   linear-chain states, it will have
an extended spatial configuration, but will not be anomalon-like.  The
lifetimes of these states, corresponding to the observed widths, are
quite long on the nuclear time scale, but still too short to meet the
criterion stated above.  On the other hand, it is interesting to
mention that, up to a certain limit determined by the stability with
respect to the spontaneous fission into two shorter linear chains [29],
the linear-chain structure is expected to become more stable as the
number of $\alpha$  clusters becomes larger.  This will especially be true, if
the $\alpha$  clusters are further interposed by a few neutrons.  In addition,
as has been pointed out previously [27,30], other linear-chain
structures may also exist, which consist of not only  $\alpha$ clusters but
also larger clusters such as $^{12}C$ cluster, $^{16}O$ cluster, and so on.
    The above discussion is merely intended to demonstrate the large
variety of nucleon configurations which may occur in nuclear systems.
In the following, I shall concentrate on QMS in medium systems which
are perhaps more relevant to the anomalon problem.  Also, I shall
discuss briefly the important role played by neutron-rich isotopes in
explaining the observed phenomena regarding Z = 2 projectile fragments.
(b)  QMS in medium-weight systems
    In medium-weight systems, the properties of QMS consisting of more
than two clusters are little known at this moment; therefore, the
discussion here will be confined to QMS with binary clusters.   By
utilizing symmetry arguments [31,32], one can show that, if the
compound nucleus of an A + B system has more than six protons or six
neutrons beyond the filled 1s and 1p shells, the potential- or
deformation-energy curve will generally have two minima (see fig. 4 of
ref. 8).  The states in the first well, having shell-model-like
character and possessing weak clustering correlations, are referred to
as class-I states, while those in the second well, having quasi-
molecular structure and containing strong clustering features, are
referred to as class-II states.
    It is my belief that anomalons may be QMS of rather low angular
momentum which lie deep in the second potential well [33].  This belief
is reached in view of the following observations: (i) the QMS is highly
deformed and, therefore, has an extended geometrical configuration, and
(ii) its lifetime can be very long, since the spatial overlap between
class-I and class-II configurations is expected to be very small.
    To support the above assertions, I shall use the   $^{32}S$ system as an
illustration.  For this particular system, the best calculation has
been performed by Schultheis and Schultheis [34], utilizing the
microscopic  $\alpha$-cluster model.  The result showed that a QMS band does
appear, with the bandhead at an excitation energy of 7.5 MeV and a
rotational constant of only 77 keV.  The intrinsic state was found to
have a strong degree of $^{16}O + ^{16}O$ clustering, with the density
distribution showing a distinct necked-in configuration and the proton
intrinsic quadrupole moment having a very large value of 208 fm$^2$ .  The
overlap between the intrinsic states for the QMS band and the ground-

state band turned out to be very small, being only equal to about
$4 \times 10^{-9}$ .

Next, I shall briefly discuss the mechanism for the excitation of
projectile fragments into quasi-molecular states [8,23]. In a
peripheral collision between projectile and target nuclei, a few
nucleons or nucleon clusters are abraded and fragmented off, and the
remaining PF feels a short-range attractive nuclear force and a longer-
range repulsive Coulomb force. Due to the high relative speed of the
target and projectile, the nucleus does not have sufficient time to
respond significantly during the encounter. Thus, the effect on the PF
of these forces is an impulsive one. The net result of the nuclear and
Coulomb components of this impulse is the excitation of the PF into a
state of transverse motion relative to its mass center. This
transverse motion could carry the fragment into the deformation region
associated with quasi-molecular states.

Based on the above mechanism, excitation energies $E^*$ of various PF
have been computed. In fig. 6, $E^*$ of $^{32}S$ at 1A-GeV as a function of
the impact parameter b is illustrated [22]. To study the relative
effectiveness of the various target nuclei in exciting the $^{32}S$ quasi-
molecular states, we consider the $E^*$ value at an impact parameter $b_c$
for grazing interaction, defined as

$$b_c = R_p + R_t - 2a ,\qquad (3)$$

where $R_p$ and $R_t$ represent the equivalent uniform-density radii for
the projectile and target nuclei, respectively, and a is the
diffuseness parameter for the matter-density distribution. As is seen
from this figure, the values of $E^*$ for grazing collisions (indicated
by arrows) are equal to 5.1, 5.4, 15.7, and 17.9 MeV when the target
nuclei are C, O, Br, and Ag, respectively. The important point to note
here is that, at the grazing distances, the excitation energies caused
by the interactions with the heavy elements Br and Ag are about three
times larger than those caused by the interactions with the light
elements C and O.

Since QMS in $^{32}S$ are expected to occur in the excitation-energy
region above 7 MeV, it is clear that the range of impact-parameter
values for effective excitation to quasi-molecular states is much
larger when the target nucleus is a heavy element than when the target
nucleus is a light element. This suggests that the production of
anomalons is mainly effected by peripheral interactions with the heavy
nuclei in the emulsion. Interactions with light nuclei may have some
significance, but the significance is likely minor. Thus, it is
perhaps not surprising that anomalon effects seem to be stronger in
experiments employing nuclear emulsions than in experiments utilizing
plastic media.

(c) Neutron-rich isotopes

Even for a fixed value of Z, projectile fragments do not have a
homogeneous composition. In addition to nuclei within the valley of
stability, these may also appear in PF long-lived neutron-rich isotopes
such as $^9Li$, $^{16}C$, and so on. Because of either low abundances or the
fact that isotopes do not generally differ greatly in geometrical size,
the effect on mfp arising from the presence of such isotopes is

frequently not expected to be of major significance. However, there
are special situations where a careful account of the isotope effect
should be made. One such situation is the Z = 2 case [35], where the
presence of the isotope $^6$He may cause the appearance of an anomalous
behavior, similar to that discussed in subsect. 2.1(a).

The reason for the importance of the isotope effect in the He case
is quite simple. The isotope $^6$He is a lightly bound, particle-stable
system and the isobaric analogue of the 3.56 MeV, T = 1 excited state
of $^6$Li. Its mfp in emulsion should be substantially shorter than that
of $^6$Li (T = 0 ground state), which is about 14.5 cm. In fact, a
simple calculation [36], utilizing an $\alpha$ + 2n cluster model, yielded a
value of 12.6 cm. This latter value is appreciably smaller than the
mfp value of about 20 cm for either $^3$He or $^4$He in emulsion, and is
similar to the value adopted in plotting curve A in fig. 1. Thus,
with the assumption of a large admixture (around 50%) of $^6$He in
the PF, one can in fact explain the seemingly peculiar data obtained
by El-Nadi et al [9].

The important question is therefore: can $^6$He be abundantly
produced in white-star interactions of relativistic $^{12}$C projectiles in
emulsion? My opinion is that this is not improbable. According to the
impulse production mechanism described above, it is not unlikely that
the combined effects of nuclear and Coulomb impulses during a white-
star interaction may cause the $^{12}$C to be excited to a $^6$He + $^6$Be
binary-cluster configuration, which subsequently decays to yield one
$^6$He, one $^4$He, and two protons. Thus, it is conceivable that the
production probability of $^6$He in such interactions may be quite large.
A crucial test of this explanation would be a detailed measurement of
the relative numbers of $^4$He and $^6$He nuclei produced in nuclear
emulsion by a $^{12}$C beam with a kinetic energy around 3.7A-GeV.

(d)  Discussion

On the theoretical side of the anomalon problem, many questions
remain to be answered. The production probability of anomalon has not
been quantitatively studied and the lifetimes of quasi-molecular states
have not been accurately calculated. Quite obviously, these are
complicated many-nucleon problems, the solution of which would require
careful considerations of nuclear structure and reaction dynamics. My
personal viewpoint is that, until the experimental picture becomes
clarified, one should refrain from going into quantitative details. At
the present moment, it is perhaps appropriate to deal only
qualitatively with this subject by offering explanations which are at
least plausible and self-consistent.

## 3.  ASYMMETRIC FISSION

### 3.1. Experimental situation

Recent experiment by Waddington and Freier [3,4] showed that, when
relativistic  Au nuclei with an initial kinetic energy of about 1A-GeV
interact in nuclear emulsion, 7% of the interactions results in the
occurrence of binary fission events, with each event accompanied by the
simultaneous appearance of either no other charged-particle tracks or a

very small number of proton and/or alpha-particle tracks. Out of these fission events, a substantial fraction (about 30%) appears to be asymmetric-fission events. This is an astonishing discovery, since the previous belief [5,6] has been that, for N < 132, strong asymmetric deformations are not to be expected and symmetric fission is supposed to be the dominant fission mode.

Experimental evidence for the presence of significant asymmetric-fission contribution is depicted in fig. 7. In this figure, each circle represents an event characterized by the charge numbers of the heavy and light fragments (events initiated by Au nuclei with kinetic energies greater and smaller than 150 MeV/nucleon are represented by solid and open circles, respectively). The two solid lines are drawn to denote charge losses by the Au nuclei of 0 and 5 units. Here one finds that there does appear a group of asymmetric-fission events, with one fragment having a proton number around 26 and another fragment having a proton number around 50 (the uncertainty in charge identification is about 2 or 3 units).

A more interesting demonstration is given in fig. 8, where the charge-yield distribution in the Au-case is qualitatively compared with the symmetric mass-yield distribution (properly mass-shifted and normalized) obtained by Plasil et al. [37] for the fission of $^{209}$Bi by 36-MeV protons. From this figure, one can reasonably conclude that for fission in the Au-region, both symmetric and asymmetric modes seem to be of similar importance.

In contrast, fission in emulsion initiated by $^{238}$U projectiles with kinetic energies up to about 1A-GeV seems to show a more expected behavior. Recently, Jain et al. [38] reported that, in this case, both the charge- and mass-yield distributions turn out to be asymmetric, with characteristics quite similar to those found in conventional low-energy fission of uranium induced by the absorption of a photon, a neutron, or a charged light or heavy ion [5]. This indicates that the fissioning nuclei must have rather low excitation energies and originate from peripheral interactions of $^{238}$U in the nuclear emulsion.

There are some characteristic differences between fission in emulsion initated by $^{238}$U and that initiated by $^{197}$Au. In the $^{238}$U case [38,39], a large fraction (about 50%) of all interactions resulted in binary fission and about 40% of these fission events are of type $F_0$ with no accompanying light charged-particle or target-fragmentation tracks. This is rather different from the findings in the $^{197}$Au case, where it was reported [4] that only 7% of all interactions resulted in binary fission and the fraction of $F_0$ type is small (about 10%). As will be discussed below, these characteristic differences can in fact be readily explained in terms of the impulse mechanism (impulse fission) described in subsect. 2.2(b) and by the fact that the fission-barrier heights in the uranium and gold regions are quite different.

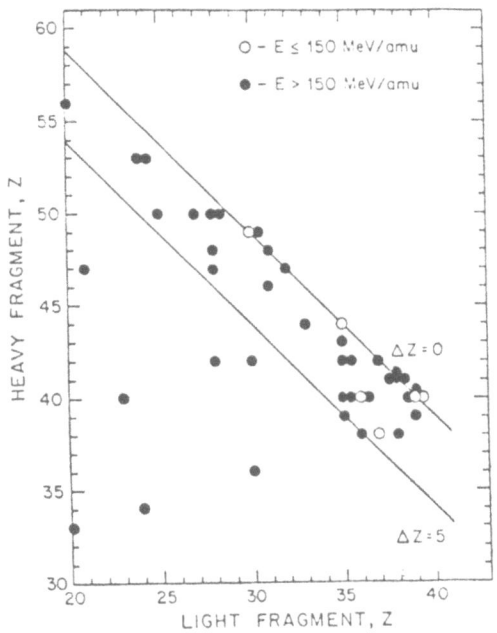

Fig. 7: Symmetric and asymmetric fission events observed in the Au experiment. (From ref. 4).

Fig. 8: Qualitative comparison of fragment-yield distributions for fission in the Au and Bi regions. (From ref. 4).

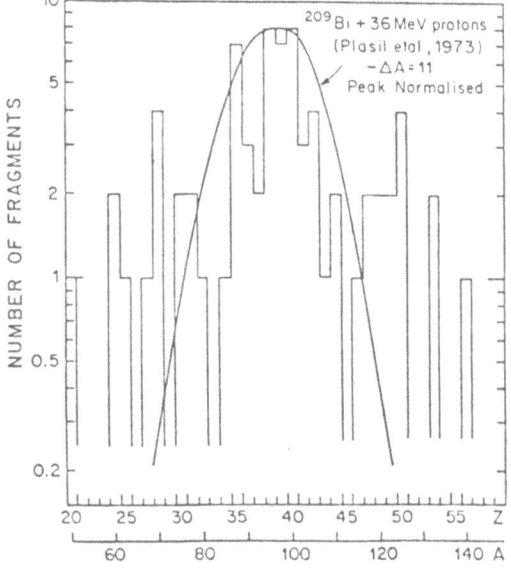

It should be mentioned that there was, in fact, experimental evidence for asymmetric fission even in lighter elements with Z ≈ 58 [40,41]. The 190-MeV and 600-MeV proton-induced fission of Ce showed that the mass-yield distribution is asymmetric with mass peaks at about 50 and 80 mass units. Thus, it would be of interest to further investigate the fission characteristics in this region of the periodic table by studying the interactions initiated by a relativistic beam with Z around 58 (e.g., La beam with Z = 57) in nuclear emulsion.

## 3.2. Theoretical considerations

General properties of fission can be reasonably explained in terms of the two-mode hypothesis proposed by Gönnenwein et al [6]. In this hypothesis, one fission mode (cluster mode) is determined by energetically-favored cluster correlations, while the other fission mode (liquid-drop mode) is due to liquid-drop effects only. For all nuclei lighter than Fm (Z = 100), the cluster and liquid-drop modes correspond, respectively, to asymmetric and symmetric fission. Proceeding from this viewpoint, these authors have further given convincing arguments that, for fissioning nuclei with relatively low excitation, the cluster mode should be important if favored closed-shell structures, corresponding to the cluster configuration at the second minimum of the deformation-energy curve, can be built into the heavy and light clusters. In the case of uranium fission, for example, the observed asymmetric mass distribution can be readily understood by noting that strong asymmetric deformation can be present because of the possibility of having closed-shell structures with Z = 50 and N = 82 in the heavy cluster and the closed-shell structure with N = 50 in the light cluster.

The two-mode theory can be used to achieve a qualitative understanding of the properties of fission in the Au-region. For this purpose, one must look for binary-cluster state which is responsible for the asymmetric part of this fission process. In constructing this state, it is important to keep in mind energetically-favored substructures and symmetry-energy consideration [26]. The general guidelines are: (i) closed-shell cluster configurations with a few outside nucleons play a dominant role in determining the characteristics of asymmetric fission, and (ii) the neutron-to-proton ratio in each cluster must be nearly the same as that in the original compound nucleus, i.e.,

$$N_h/Z_h \approx N_\ell/Z_\ell \approx N/Z \quad , \tag{4}$$

where (N,Z), (N_h, Z_h), (N_ℓ, Z_ℓ) represent the neutron and proton numbers of the compound nucleus, the heavy cluster, and the light cluster, respectively. By assuming in addition that, near the scission configuration, the surplus nucleons form the neck joining the clusters and that there is nearly equal scission probability anywhere along the neck, one can then obtain a reasonable account of the essential features of the primary fragment mass and charge distributions.

The neutron and proton configurations of the heavy and light clusters will now be discussed. First, it is noted that, for nuclei around Au, the neutron-to-proton ratio is about 1.5. Then, to comply as closely as possible with the guidelines mentioned above, we choose $(N_h, Z_h) = (82,50)$ for the heavy cluster and $(N_\ell, Z_\ell) = (28,20)$ for the light cluster. With this choice, the neutron-to-proton ratios are $N_h/Z_h = 1.64$ and $N_\ell / Z_\ell = 1.40$. This will leave a neck that is proton-rich, except in the unlikely event that, during the peripheral collision between the Au projectile and an emulsion target nucleus, most of the abraded nucleons were protons. To produce these doubly-closed-shell clusters and a proton-rich neck requires some redistribution of the original nuclear charge. The important point to note here is that this charge redistribution could be facilitated as a consequence of the sharp nuclear and Coulomb impulses received by the compound nucleus. In conventional experiments where low-energy particles initiate the formation of fissioning compound nuclei, such a mechanism is not readily present. As a result, the probability of exciting class-II configurations in these experiments is not appreciable and asymmetric fission yield in the Au region becomes quite small.

After scission takes place across the neck, the resulting light fragment will have a proton number between 20 and about 28, in agreement with the experimental finding of Waddington and Freier [3,4]. This indicates that the new features discussed here, namely, impulse excitation of class-II configurations and charge redistribution by Coulomb effects, are important and should be incorporated into any complete theory of nuclear fission.

In the case of asymmetric fission for nuclei with Z around 58, one can make a similar analysis. Here the doubly-closed-shell configurations, which are expected to be involved in the fission process, are likely those with $(N_h, Z_h) = (50,28)$ and $(N_\ell, Z_\ell) = (28,20)$.

Dynamic effects arising from the impulse mechanism will now be considered. In fig. 9, we show the excitation energy $E^*$ of either $^{197}$Au or $^{238}$U when it makes a peripheral interaction with Ag or C target nucleus. For convenience in interpreting experimental results, the impact parameter is implicitly specified in terms of the number $\tilde{N}$ of projectile nucleons abraded. This latter quantity is then estimated by using the soft-spheres model of Karol [24]. For the calculation of $E^*$, the combined effect of nuclear and Coulomb impulses is taken into account.

Experimental results obtained with relativistic $^{238}$U and $^{197}$Au beams can then be explained as follows:
(i) For $^{238}$U, the fission barrier height is about 6 MeV. According to fig. 9, both Ag and C target nuclei can excite $^{238}$U above this barrier, with the abrasion of no more than one and two nucleons, respectively. Since $F_0$ events can be associated with the emission of one or two neutrons, we see that the occurrence of $F_0$-type $^{238}$U fission is not unlikely. On the other hand, the fission barrier height for $^{197}$Au is about 20 MeV for symmetric fission and somewhat less for asymmetric fission. From fig. 9, one finds that $^{197}$Au can receive this

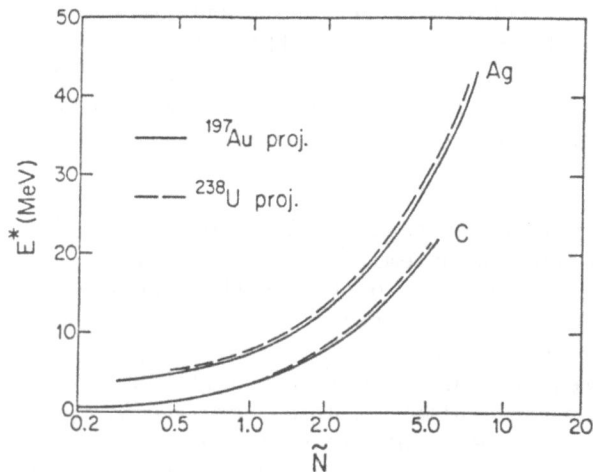

Fig. 9: Excitation energy of Au or U as a function of the number of abraded projectile nucleons.

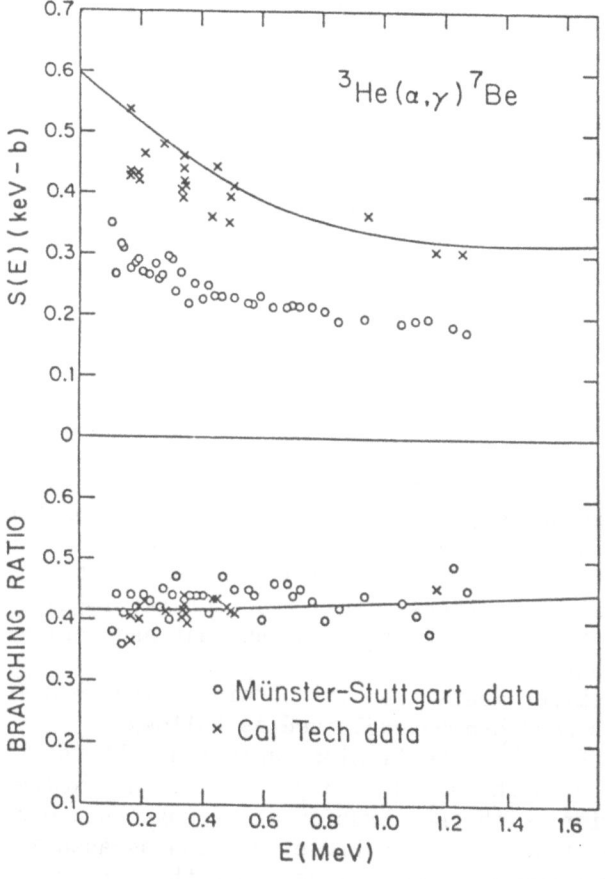

Fig. 10: Cross-section factor and branching ratio as a function of the c.m. energy. (Adapted from ref. 49).

much excitation energy only in association with the abrasion of at least 3 and 5 nucleons for Ag and C targets, respectively. Thus, we expect $^{197}$Au to have a lower fission probability than $^{238}$U, and a smaller percentage of $^{197}$Au fission should be of type F₀.
(ii) For binary-fission events, the average number of projectile nucleons abraded is rather small. Even in the Au case, this number is only around 4 or 5 (see fig. 7). From fig. 9, one notes then that $E^*$ is generally smaller than about 30 MeV. For such low excitation energies, it is experimentally known [5] that asymmetric-fission mode should significantly contribute.

In summary, it is my viewpoint that the mechanism of <u>impulse fission</u> at relativistic energies discussed here can qualitatively explain the observed phenomena. As in the anomalon case, quantitative investigations are still lacking at the present moment. However, my opinion is that, in view of the complicated nature of the fission problem, such effort should perhaps be postponed until further and more revealing experimental information becomes available.

## 4. NUCLEAR ASPECTS OF THE SOLAR-NEUTRINO PROBLEM

### 4.1. Experimental situation

The solar-neutrino problem is concerned with the fact that there is considerable discrepancy between calculated and measured solar neutrino capture rates for $^{37}$Cl detector. Based on a standard solar model, this problem has recently been carefully re-examined by Bahcall et al. [7] with regard to all of its relevant aspects. My purpose here is much more modest; the discussion will be confined to some key nuclear reactions which are closely connected with the production of neutrinos important in the $^{37}$Cl experiment.

The $^{37}$Cl detector is sensitive to electron neutrinos with energies higher than 0.81 MeV. These are the neutrinos produced by the reactions pep, $^{7}$Be electron capture, and $^{8}$B, $^{13}$N, $^{15}$O, and $^{17}$F beta decays. By using flux considerations from standard solar models and by noting that the detection is based on the endoergic reaction $^{37}$Cl($\nu$,e⁻)$^{37}$Ar, it has been shown [7] that the $^{7}$Be electron-capture and $^{8}$B beta-decay neutrinos provide the large majority of events observed in the $^{37}$Cl experiment. Since these nuclei $^{7}$Be and $^{8}$B are the end products of the reactions $^{3}$He($\alpha,\gamma$) $^{7}$Be and $^{7}$Be(p,$\gamma$) $^{8}$B, I shall concentrate on discussing these particular radiative-capture reactions here.
(a)   $^{3}$He($\alpha,\gamma$) $^{7}$Be radiative capture reaction.

Because of the Coulomb penetration factor, the total capture cross section $\sigma$ at a c.m. energy E decreases rapidly at very low energies. Therefore, it is convenient to define a slowly energy-dependent cross-section factor $S_{34}(E)$ as

$$S_{34}(E) = E\ \sigma(E)\ \exp\ (2\pi\eta\ )\ ,\tag{5}$$

where $\eta$ is the Sommerfeld parameter. Since astrophysical reactions proceed at energies around 1 to 20 keV, the basic quantity of interest is the zero-energy factor $S_{34}(0)$, which can be obtained by performing an extrapolation procedure [42] utilizing measured results in the low-energy region.

Within the last few years, the importance of this reaction has prompted many groups to re-measure the total capture cross sections at low energies. The measurements were of two types. The first type involves the observation of prompt $\gamma$-rays (direct measurement), while the second type involves the measurement of the activity of $^7$Be by observing the 478-keV $\gamma$-ray following the electron capture of $^7$Be to the first excited state of $^7$Li (activation measurement).

Results of direct measurements for $S_{34}(E)$ and the branching ratio for capture to the ground and first excited states are shown in fig. 10. In this figure, the experimental data are those obtained by Kräwinkel et al. [43] (labelled as Münster-Stuttgart or MS data; open circles) and by Osborne et al. [44] (labelled as Cal Tech or CT data; crosses). As is seen, there is a large discrepancy between these two sets of results. By extrapolating to zero energy, the $S_{34}(0)$ values for the MS and CT data are equal to 0.31 ± 0.03 and 0.52 ± 0.03 keV-b, respectively. In this respect, it should be mentioned that, for these two direct measurements, there is no significant disagreement about the branching ratio and the energy dependence of the measured cross section; it is only the absolute normalization that seems to be in discrepancy.

The absolute normalization of the MS experiment has, in fact, been checked later by Volk et al. [45] with an activation measurement. The result obtained showed that the value of $S_{34}(0)$ reported by Kräwinkel et al. should be revised upward by a factor of 1.88. In other words, the $S_{34}(0)$ value resulting from this measurement is equal to 0.58 ± 0.03 keV-b.

There exist also two other activation measurements, performed by Robertson et al. [46] and by Osborne et al. [44]. The results for $S_{34}(0)$ from these measurements are equal to 0.63 ± 0.04 and 0.54 ± 0.04 keV-b, respectively. These values are in reasonable agreement with the value obtained by Volk et al.

(b) $^7$Be$(p,\gamma)$ $^8$B radiative capture reaction

The total cross section for the reaction $^7$Be$(p,\gamma)$ $^8$B has recently been measured by Filippone et al. [47] in the c.m. energy range from 117 to 1230 keV. In this measurement, the cross section was determined by detecting delayed $\alpha$ particles following the beta decay of $^8$B. Using the measured values and performing an extrapolation procedure based on the direct-capture calculation of Tombrello [48], these authors concluded that the zero-energy cross-section factor $S_{17}(0)$ is equal to 0.0216 ± 0.0025 keV-b.

Also, there existed, prior to 1978, many older measurements of the $^7$Be$(p,\gamma)$ $^8$B cross sections at low energies. The extrapolated $S_{17}(0)$ values range from 0.016 ± 0.006 to 0.045 ± 0.011 keV-b (see Ref. 47). A weighted average of these values and the value mentioned in the preceding paragraph was quoted by Filippone et al. [47] to be $\bar{S}_{17}(0) = 0.0238 ± 0.0023$ keV-b.

In the solar-model study of Bahcall et al. [7], the value used for $S_{17}(0)$ was 0.029 ± 0.010 keV-b. This value was arrived at by averaging some of the older results. By taking into account also the newer value of Filippone et al., it seems to me that a lower value

$$S_{17}(0) = 0.024 \pm 0.007 \text{ keV-b,} \qquad (6)$$

with effective 3 σ-uncertainty, should perhaps be adopted.

## 4.2. Theoretical considerations

Microscopic studies [49,50] of the $^3$He$(\alpha,\gamma)^7$Be capture reaction exist at the present moment. In these studies, resonating-group wave functions are employed and cluster correlations are properly taken into consideration. For the reaction $^7$Be$(p,\gamma)$ $^8$B, on the other hand, only macroscopic direct-capture calculations [48,51-53] have been performed. The reason for this is simple. Because of the strong clustering property of $^7$Be, the p + $^7$Be system can be adequately described only within the framework of a three-cluster formulation [54]. As is well known, reliable calculations of this type are not easy to carry out due to complexities in computation.

In the following, I shall briefly describe the results obtained from the calculations for these reactions:
(a) $^3$He$(\alpha,\gamma)$ $^7$Be radiative capture reaction.

The microscopic calculation of Walliser et al. [49] will be discussed here. In this calculation, the wave functions from a coupled-channel calculation of Kanada et al. [55], involving cluster configurations α + $^3$He and α + $^3$He$^*$, are employed. The spatial structure of the α cluster is assumed to have a simple one-Gaussian form, with the width parameter adjusted to yield the correct rms matter radius. For the $^3$He and $^3$He$^*$ clusters, a more flexible form consisting of a sum of two Gaussian functions is adopted, with the parameters determined by minimizing the expectation value of the three-nucleon Hamiltonian and by demanding that the excited $^3$He$^*$ function be orthogonal to the ground-state $^3$He function.

The seven-nucleon coupled-channel calculation of Kanada et al. [55] yields not only the correct $^3$He + α separation energies in the ground and first excited states of $^7$Be, but also a very good fit to the low-energy $^3$He + α scattering cross sections. This leads me to believe that the subsequent capture calculation, utilizing the wave functions obtained in this investigation, is likely reliable, and uncertainties resulting from the simplifying assumptions are probably not major.

The resultant values for $S_{34}(E)$ and the branching ratio are depicted by the solid curves in fig. 10, where it is seen that there is a reasonable agreement with the CT data. The zero-energy cross-section factor $S_{34}(0)$ turns out to be 0.60 keV-b, which may be slightly too large.

The calculation may be somewhat improved. For the α cluster, it would be desirable to use a more flexible spatial wave function, with the parameters determined by a variational procedure. In addition, we have noticed that the effect of specific distortion, determined in the calculation of Kanada et al., is not as large as that determined in a more recent seven-nucleon calculation [56] employing a very extensive model space. This indicates that the spectroscopic factors for $\alpha$-clustering in the ground and first excited states of $^7$Be [57], obtained recently with the wave functions of Kanada et al., are somewhat overestimated. An improved calculation, utilizing a more flexible $\alpha$-cluster wave function and taking specific distortion effects into better account, may result in a slightly reduced value for $S_{34}(0)$, estimated to be around 0.56 keV-b.

Kajino and Arima [50] have performed a similar microscopic study and obtained a value for $S_{34}(0)$ equal to 0.51 keV-b. Apart from different choices of nucleon-nucleon potential and cluster wave functions, there is also one difference which is worth mentioning. The $^3$He + $\alpha$ reduced mass used in their calculation has a value corresponding to $\hbar^2/2\mu$ = 12.095 MeV-fm$^2$. This latter value is smaller than the correct value of 12.155 MeV-fm$^2$ used in the calculation of Walliser et al [49]. By revising the value of $\hbar^2/2\mu$ in their calculation, it is almost certain that the calculated value of $S_{34}(0)$ will slightly increase, thus leading to a better agreement between the results of the two microscopic calculations.

Based on various calculated and experimental results, my suggestion is that, in a solar-neutrino calculation, one should use

$$S_{34}(0) = 0.55 \pm 0.08 \text{ keV-b} , \tag{7}$$

where the uncertainty is given at the effective 3$\sigma$-level. This suggested value is slightly larger than the value adopted by Bahcall et al. [7] in their extensive study of the solar neutrino problem.

(b) $^7$Be(p,$\gamma$) $^8$B radiative capture reaction

For this capture reaction, only macroscopic calculations exist, the latest of which being that performed by Barker [53]. In this type of calculation, one assumes that the reaction proceeds by direct capture only, and that the initial and final states of the system can each be described by a simple single-particle model of a proton moving in an optical potential which represents its interaction with $^7$Be in its ground state.

As was shown by Filippone et al. [47], such a model calculation can yield satisfactorily the energy-dependent behavior of the off-resonance data, including the upturn in $S_{17}$ near zero energy. On the other hand, there is one major shortcoming in that the absolute normalization of the S-factor cannot be determined. For this latter purpose, a microscopic treatment of the structure of the compound system has to be undertaken.

## 4.3. Summary

With the values of $S_{17}(0)$ and $S_{34}(0)$ suggested in eqs. (6) and (7), the predicted solar-neutrino capture rate for $^{37}$Cl detector needs to be re-calculated. This is, however, a simple procedure, since the necessary information for re-calculation has been provided by Bahcall et al. [7].  With the only modification from the input parameters of Bahcall et al. being in these two cross-section-factor values, the new value for the capture rate is 6.9 ± 2.5 SNU (effective 3σ errors), which should be compared with the latest measured value of 1.9 ± 0.3 SNU (1σ error) reported by Davis et al. [58].

As far as nuclear-reaction input parameters are concerned, the largest uncertainties are associated with uncertainties in the cross sections for p + p and p + $^{7}$Be reactions.  Presently, however, these uncertainties are already small enough that more refined experiments may not help greatly in reducing the discrepancy.  It is possible that, for a resolution of the solar-neutrino problem, one may eventually have to look for modifications in standard solar models such as turbulence mixing and/or change in the behavior of neutrinos over astrophysical distances such as neutrino oscillation.

## 5.  CONCLUDING REMARKS

In this talk, I have considered several selected topics in the fields of relativistic heavy-ion reactions and nuclear astrophysics. What I have essentially done was to review the experimental situation and discuss theoretical calculations or interpretations.

For relativistic heavy-ion reactions, the topics selected are anomalons and asymmetric fission in lighter elements.  Here my main point is that, for a satisfactory explanation of the experimental data which are rather sparse, the existence of quasi-molecular structures in nuclear systems may have to be kept in mind.

In the field of astrophysics, I have concentrated on the nuclear aspects of the solar-neutrino problem.  Two key radiative-capture reactions are carefully discussed and analyzed.  From this analysis, it appears that the present uncertainties associated with nuclear-reaction input parameters are already small enough that further experiments may not be of much help in resolving the discrepancy between predicted and measured solar-neutrino capture rates for $^{37}$Cl detector.  For an eventual resolution of this problem, it is possible that one may have to examine more complicated problems such as modifications in standard solar models and/or the change in neutrino behavior during its transit from the sun.

In conclusion, my opinion is that, for all three subjects discussed here, puzzling aspects remain, and much theoretical and experimental work has yet to be done in order to achieve a clear picture and a general understanding.

## ACKNOWLEDGMENT

I would like to take this opportunity to express my gratitude to B. F. Bayman, P. J. Ellis, P. S. Freier, S. Fricke, and C. J. Waddington for valuable discussion and constructive criticisms.

In particular, I wish to thank C. J. Waddington for the permission in quoting his results prior to publication.

This work was supported in part by the U. S. Department of Energy under Contract No. DOE/DE-AC02-79 ER 10364.

References:

1.  Proceedings of the Sixth High Energy Heavy Ion Study and Second Workshop on Anomalons, Lawrence Berkeley Laboratory, University of California, 1983 (LBL Report 16281).
2.  See, e.g., I. Otterlund, Ref. 1, p. 607.
3.  P. S. Freier and C. J. Waddington, Ref. 1, p. 301.
4.  C. J. Waddington and P. S. Freier, "The Interactions of Gold Nuclei in Nuclear Emulsions", preprint, June 1984.
5.  R. Vandenbosch and J. R. Huizenga, <u>Nuclear Fission</u> (Academic, New York, 1973).
6.  F. Gönnenwein, H. Schultheis, R. Schultheis, and K. Wildermuth, Z. Phys. <u>A278</u>, 15 (1976).
7.  J. N. Bahcall, W. F. Huebner, S. H. Lubow, P. D. Parker, and R. K. Ulrich, Rev. Mod. Phys. <u>54</u>, 767 (1982).
8.  Y. C. Tang, Ref. 1, p. 149.
9.  M. El-Nadi, O. E. Badawy, A. M. Moussa, E. I. Khalil, and A. El-Hamalawy, Phys. Rev. Lett. <u>52</u>, 1971 (1984). In this paper, it is reported that non-white-star events exhibit no anomalous mfp behavior.
10. A. Z. M. Ismail, M. S. El-Nagdy, K. L. Gomber, M. M. Aggarwal, and P. L. Jain, Phys. Rev. Lett. <u>52</u>, 1280 (1984).
11. Y. J. Karant, H. H. Heckman, and E. M. Friedlander, Ref. 1, p. 23.
12. Alma-Ata <u>et al</u>. Collaboration, in Proceedings of the Eighteenth International Cosmic-Ray Conference, Bangalore, India, 1983.
13. P. L. Jain and G. Das, Phys. Rev. Lett. <u>48</u>, 305 (1982).
14. E. M. Friedlander, R. W. Gimpel, H. H. Heckman, Y. J. Karant, B. Judek, and E. Ganssauge, Phys. Rev. Lett. <u>45</u>, 1084 (1980). This experiment was repeated and the result was reported in Ref. 11.
15. H. B. Barber, P. S. Freier, and C. J. Waddington, Phys. Rev. Lett. <u>48</u>, 856 (1982).
16. For a more detailed description of this pooling procedure, see, e.g., ref. 13.
17. P. L. Jain, M. M. Aggarwal, and K. L. Gomber, Phys. Rev. Lett. <u>52</u>, 2213 (1984).
18. W. Heinrich, H. Drechsel, W. Trakowski, J. Beer, C. Brechtmann, J. Dreute, and S. Sonntag, Phys. Rev. Lett. <u>52</u>, 1401 (1984).
19. M. L. Tincknell, P. B. Price, and S. Perlmutter, Phys. Rev. Lett. <u>51</u>, 1948 (1983).
20. T. J. M. Symons, M. Baumgartner, J. P. Dufour, J. Girard, D. E. Greiner, P. J. Lindstrom, D. L. Olson, and H. J. Crawford, Phys. Rev. Lett. <u>52</u>, 982 (1984).
21. J. D. Stevenson, J. A. Musser, and S. W. Barwick, Phys. Rev. Lett. <u>52</u>, 515 (1984).

22. B. F. Bayman, P. J. Ellis, S. Fricke, and Y. C. Tang, "Anomalon Production by Impulsive Excitation in Relativistic Heavy-Ion Collisions", to be published.
23. B. F. Bayman, P. J. Ellis, and Y. C. Tang, Phys. Rev. Lett. 49, 532 (1982).
24. P. J. Karol, Phys. Rev. C11, 1203 (1975).
25. W. C. McHarris and J. O. Rasmussen, Scientific American 250, 58 (1984).
26. K. Wildermuth and Y. C. Tang, A Unified Theory of the Nucleus (Vieweg, Braunschweig, 1977).
27. H. Horiuchi, K. Ikeda, and Y. Suzuki, Prog. Theor. Phys. (Suppl.) 52, 89 (1972).
28. P. Chevallier, F. Scheibling, G. Goldring, I. Plesser, and M. W. Sachs, Phys. Rev. 160, 827 (1967).
29. I. N. Mishustin, "Anomalous Fragments and Quasi-One-Dimensional Nuclear Systems", preprint, 1984.
30. Y. Fujiwara, H. Horiuchi, K. Ikeda, M. Kamimura, K. Katō, Y. Suzuki, and E. Uegaki, Prog. Theor. Phys. (Suppl.) 68, 29 (1980).
31. M. Harvey, in Proceedings of the Second International Conference on Clustering Phenomena in Nuclei, College Park, Maryland, 1975.
32. P. Kramer, in Proceedings of the International Symposium on Clustering Phenomena in Nuclei, Tübingen, 1981.
33. In the following discussion, the term QMS will be used to refer only to class-II states lying deep in the second potential well. Class-II states which occur close to the top of the Coulomb plus centrifugal barrier have short lifetimes and are irrelevant for our present purposes.
34. H. Schultheis and R. Schultheis, Phys. Rev. C25, 2126 (1982).
35. Another interesting situation may be the $Z = 1$ case, since the mfp in emulsion of d or t is significantly shorter than that of the proton.
36. B. F. Bayman, S. Fricke, and Y. C. Tang, "Interaction of Relativistic Helium Projectile Fragments in Nuclear Emulsions", to be published.
37. F. Plasil, R. L. Ferguson, F. Pleasonton, and H. W. Schmitt, Phys. Rev. C7, 1186 (1973).
38. P. L. Jain, M. M. Aggarwal, M. S. El-Nagdy, and A. Z. M. Ismail, Phys. Rev. Lett. 52, 1763 (1984).
39. E. M. Friedlander, H. H. Heckman, and Y. J. Karant, Phys. Rev. C27, 2436 (1983).
40. F. D. Becchetti, J. Jänecke, P. Lister, K. Kwiatowski, H. Karwowski, and S. Zhou, Phys. Rev. C28, 276 (1983).
41. H. Å. Gustafsson, G. Hyltén, B. Schrøder, and E. Hagebø, Phys. Rev. C24, 769 (1981).
42. R. D. Williams and S. E. Koonin, Phys. Rev. C23, 2773 (1981).
43. H. Kräwinkel, H. W. Becker, L. Buchmann, J. Görres, K. U. Kettner, W. E. Kieser, R. Santo, P. Schmalbrock, H. P. Trautvetter, A. Vlieks, C. Rolfs, J. W. Hammer, R. E. Azuma, and W. S. Rodney, Z. Phys. A304, 307 (1982).

44. J. L. Osborne, C. A. Barnes, R. W. Kavanagh, R. W. Kremer,
    G. J. Mathews, J. L. Zyskind, P. D. Parker, and A. J. Howard,
    Nucl. Phys. A419, 115 (1984).
45. H. Volk, H. Kräwinkel, R. Santo, and L. Wallek, Z. Phys. A310, 91
    (1983).
46. R. G. H. Robertson, P. Dyer, T. J. Bowles, R. E. Brown, N. Jarmie,
    C. J. Maggiore, and S. M. Austin, Phys. Rev. C27, 11 (1983).
47. B. W. Filippone, A. J. Elwyn, C. N. Davids, and D. D. Koetke,
    Phys. Rev. Lett. 50, 412 (1983).
48. T. A. Tombrello, Nucl. Phys. 71, 459 (1965).
49.  H. Walliser, H. Kanada, and Y. C. Tang, Nucl. Phys. A419, 133
    (1984).
50. T. Kajino and A. Arima, Phys. Rev. Lett. 52, 739 (1984).
51. A. Aurdal, Nucl. Phys. A146, 385 (1970).
52. R. G. H. Robertson, Phys. Rev. C7, 543 (1973).
53. F. C. Barker, Aust. J. Phys. 33, 177 (1980) and Phys. Rev. C28,
    1407 (1983).
54. Y. Fujiwara and Y. C. Tang, University of Minnesota Report
    UM-RGM2.
55. H. Kanada, T. Kaneko, and Y. C. Tang, Nucl. Phys. A380, 87 (1982).
56. Y. Fujiwara and Y. C. Tang, "Multi-Configuration Resonating-Group
    Study of the Seven-Nucleon System with Realistic Cluster Wave
    Functions", to be published.
57. H. Walliser and Y. C. Tang, Phys. Lett. 135B, 344 (1984).
58. R. Davis, Jr., B. T. Cleveland, J. C. Evans, Jr., and
    J. K. Rowley, Bull. Am. Phys. Soc. 29, 731 (1984).

# CONCLUDING REMARKS

# CONCLUDING REMARKS

Akito Arima
Department of Physics
Faculty of Tokyo
Bunkyo-ku, Tokyo
Japan

ABSTRACT.
I do not think that I need give any summary talk, since Allan Bromley
gave such a superb introductory talk which covered everything discussed
in this conference. His talk was almost a summary talk.

The United Kingdom is certainly the best place for us to discuss
the clustering aspects of nuclei, because there have been many important
activities. Needless to refer to the great Rutherford in Manchester,
or to Massey who applied in the early days the Resonating Group Method
for studying scattering of light ions, or Brink who rediscovered the
Margenau model and gave it reality, and Philips who found the clustering
nature of light nuclei.

It is my great pleasure to summarize this conference at such an
appropriate place as Chester near Daresbury which is one of the most
active centres of nuclear physics. I would like to divide the subjects
of this Conference into the following six groups;

1. POTENTIAL BETWEEN NUCLEI
2. CHANNEL COUPLING
3. MOLECULAR RESONANCES
4. ALPHA TRANSFER REACTIONS
5. BREAK-UP PROCESSES
6. MESON EXCHANGE, QUARKS AND ANOMALONS

## 1. POTENTIAL BETWEEN NUCLEI

Horiuchi showed that double folding potentials with strong attraction
are favoured by converting highly non-local potentials obtained from
the Resonating Group Method into their equivalent local forms. He
pointed out that the diabatic process is more probable than the adiabatic one which produces a shallow potential in nuclear collisions.
His prescription is given as follows: One calculates first the direct
parts $V_D$ of the equivalent potentials by double folding and then adds
to $V_D$ the exchange contributions $V_E$ which are essentially determined by
the knock-on exchange, one-particle exchange and two-particle exchange
processed.

*J. S. Lilley and M. A. Nagarajan (eds.), Clustering Aspects of Nuclear Structure, 399–406.*
© *1985 by D. Reidel Publishing Company.*

In the Resonating Group Method, one has three relative wave functions $\chi(\vec{r}_{AB})$, $\phi(\vec{r}_{AB})$ and $U(\vec{r}_{AB})$. The $\chi$ appears in the wave function

$$\psi = A\,\Phi_A(\xi_A)\,\Phi_B(\xi_B)\chi(\vec{A}_{AB}),$$

where A is the antisymmetrizer, and $\Phi_x(\xi_x)$ is the internal wave function of the nucleus x with its internal coordinates $\xi_x$. The other two wave functions are

$$\phi = \sqrt{N}\,\chi$$
$$U = N\,\chi\,,$$

where N is the norm kernel

$$N(\vec{r},\,\vec{r}') = \langle\,\Phi_A(\xi_A)\,\Phi_B(\xi_B)\delta(\vec{r}-\vec{r}_{AB})|A\,\Phi_A(\xi_A)\,\Phi_B(\xi_B)\delta(\vec{r}'-\vec{r}_{AB})\rangle\,.$$

Among these wave functions $\phi$ satisfies the proper orthonormality

$$\langle\phi_E|\phi_{E'}\rangle = \delta(E-E'),$$

and is well approximated by solving the following Schrödinger equation with a local potential $V^{eff}(r)$, for which one may use a double folding potential:

$$(T+V^{eff})\phi = E\phi\,.$$

Lovas, however, argued that the function U rather than $\phi$ should be used as a microscopic form factor for cluster transfer reactions.

It is very interesting to investigate the imaginary part of the nucleus-nucleus interaction. We have heard a nice talk, given by Pollarolo, who showed two important origins of the imaginary potential. These are phonon excitations in both target and incident nuclei, and particle exchange channel couplings between the two nuclei. In a private conversation, he told me that his calculation predicts the value of the diffuseness parameter a to be $0.5 \sim 0.6$ fm, and $W(r)/V(r) \cong 0.3 \sim 0.2$ for the $^{16}O + {}^{40}Ca$, $^{16}O + {}^{88}Sr$ and $^{16}O + {}^{208}Pb$ elastic scattering. Here W(V) is the imaginary (real) part of his calculated optical potential. This ratio $0.3 \sim 0.2$ is very interesting for me, because, as will be discussed later, this value is very close to the number needed by Kamimura in his channel coupling calculation.

Before leaving this subject, I would like to mention that Buck has succeeded in explaining many properties of light nuclei using a simple cluster model which is essentially the same as the Orthogonality Condition Model of Saito. Here we require that 2N+L must be larger than that of the Pauli forbidden states, i.e. it has to satisfy the Wildermuth condition. There is, however, an ambiguity in the value of 2N+L for core excited states such as the first $0^+$ excited band in $^{16}O$. Usually we assume that 2N+L $\geq$ 8, believing that this is a 4p-4h band. There is, however, no clear reason to reject 2N+L $\geq$ 6.

## 2. CHANNEL COUPLING CALCULATIONS

The elastic scattering of $^6$Li, $^7$Li and $^9$Be from $^{40}$Ca, $^{208}$Pb etc has been analyzed by using double folding potentials with two renormalization factors $N_R(N_I)$ for their real (imaginary) part. The results showed that $N_R \sim 0.5$. Thompson and Nagarajan found that channel couplings are important for bringing $N_R$ back to unity. Kamimura extended similar calculations by keeping $N_R$ as 1 but using $N_I$ as a parameter. Taking $^7$Li as an example, he made out several different calculations. He took into account
i) only the ground state, ii) the ground and the first excited state, ii) these two states plus the second $7/2^-$ and third $5/2^-$ states as resonant states, and iv) these four states plus non-resonant breakup states. He obtained beautiful agreement with data. He, however, still needs $N_I$ which is about 0.3. This imaginary part must be due to weakly coupled channels such as those discussed by Pollarolo. Then I asked a question: Why do these states not affect $N_R$ ? A possible answer is as follows: The second order correction due to the channel couplings is written as

$$\sum_n \frac{|\langle n|v|0\rangle|^2}{E_n - E + i\varepsilon} = \Delta V_R + i\Delta V_I$$

where $|n\rangle$ is the n-th channel with energy $E_n$. Contributions from states with $E > E_n$ tend to cancel those with $E < E_n$, while $\Delta V_I$ is finite and can be large, because it is proportional to $\sum_n \delta(E-E_n)|\langle n|v|0\rangle|^2$.

It would be interesting to know what happens in Kamimura's calculation when both $N_R$ and $N_I$ are taken as adjustable parameters. I would like to know especially which parts of the channel coupling can be accounted for by the DWBA by adjusting the optical potential parameters. He assumes the prior form while the Groningen group, as reported by de Meijer, takes the post form. It would be interesting to know how much Kamimura's conclusions depend on the assumption of the form.
The importance of the channel coupling mechanism can be seen very clearly in vector and tensor analyzing powers as discussed by Johnson. The vector analyzing power $iT_{11}$ for the $^{58}$Ni($^7$Li, $^7$Li)$^{58}$Ni reaction cannot be well explained without channel couplings between the ground state, the $1/2^-$, $7/2^-$ and the $5/2^-$ state of $^7$Li. The Birmingham - Surrey experiment using tensor analyzing power data for the $^{40}$Ca(d, $\alpha$) $^{38}$K reaction showed clear evidence of a D-state admixture in the alpha particle wave function. I feel that the vector and tensor analyzing powers are sensitive to very new aspects of reaction mechanisms and structure.

Mihailovich reported that he obtained a large energy gain by admixing a few different cluster configurations for $^7$Be; $|^3H + ^4He> + |^6Li + n>$ + etc. My question is : Can we take into account this coupling by renormalizing the effective interaction, because the difference among these configurations comes from $2\hbar\omega$ and higher excited shell model states ?  In the $0\hbar\omega$ shell model space, these configurations are identical. Another question is: Where do we really need these couplings ?  One answer to this question is obvious: we need them for the $^3H + ^4He \rightarrow ^6Li + n$ reaction cross section.  Do we have any other indications ?

## 3.   MOLECULAR RESONANCES

As soon as Bromley and his collaborators discovered some interesting resonances in $^{12}C + ^{12}C$, which are now called molecular resonances, Nogami in Tokyo suggested that they could be explained as the result of coupling with the first $2^+$ excited state in $^{12}C$.  Imanishi took up this idea and could explain the resonances.  Later Kondo, Matsuse and Abe extended this idea and introduced the Band crossing Model, which has succeeded in explaining many characteristic properties of molecular resonances.  Mosel at this Conference reported his interesting study of these resonances.  In his picture he concluded that almost the same result was produced by the following four models a) Austern - Blair, b) DWBA, c) Strong Channel Coupling and d) Weak Channel Coupling.  The common feature of these models is that the structure is due to the aligned coupling channel.  He then asked: Can we learn anything from this comparison ?  He used Feshbach's Projection Method and pointed out that the DWBA does not necessarily mean weak coupling.  I think that he is right, but the potential obtained would be very complicated for strong coupling.  There is a comment on his study:  He took into account the coupling of two channels; $0^+ \otimes 0^+ \otimes L$ and $2^+ \otimes 2^+ \otimes (L-4)$, where $0^+(2^+)$ stands for the ground (first excited) state and L is the angular momentum between the two $^{12}C$ nuclei.  He does not seem to consider the effect of $[2^+ \otimes 2^+]^{(2)} \otimes (L-2)$, which may distinguish the four models from one another.

When McVoy and I plotted the energies of McVoy's Regge poles of the $^{12}C + ^{12}C$ elastic scattering as a function of I(I+1), where I is the spin of a pole, I was surprised that they were on a straight line. We then compared this line with the V. M. I. prediction provided by S. Goldhaber.  We then realized that they were different from each other.  In those days, in 1972, there were no data except for very low energies.  (A. Arima, McVoy and G. S. Goldhaber, Phys. Lett. 40B(1972) 15).  Now we know that many resonances have been observed in the $^{12}C + ^{12}C$ scattering.

Betts told us that resonances occur more generally than only for $^{12}C + ^{12}C$ and $^{16}O + ^{16}O$.  He confirmed that narrow structures in

$^{28}$Si + $^{28}$Si are not statistical fluctuations. Using $^{29}$Si + $^{28}$Si and $^{30}$Si + $^{28}$Si scattering, he showed that the addition of one or two neutrons suppresses the narrow resonances. This observation tells us that this addition makes the imaginary part strong. Thus in the $^{28}$Si + $^{28}$Si system, the absorption should be weak, and then the potential resonances remain narrow.

The resonances occur at Ex $\sim$ 70 MeV with $J^{\pi} \sim$ 40. The widths of those resonances are surprisingly narrow $\Gamma \sim$ 100 Kev. I believe that optical resonances, which have broader widths than observed ones, are spread over many special channels.

The origin and nature of the molecular resonances are still very challenging open problems. Adding to the band-crossing and alignment, the super-deformation predicted by Bengtson et al and Greiner et al, as reported by Bromley,    **might be** a possibility to explain the molecular resonances.

It will be very useful to measure gamma transitions between two resonances to understand the nature of the molecular resonances. Daresbury is one of the best places to do such measurements, and especially "TESSA" can play a very decisive role.

Bromley pointed out a very intriguing phase change in the molecular resonances. The moment of inertia below the Coulomb barrier is larger than that above the barrier. The reduced width into the entrance channel is large below the barrier, while it is small above it. This may indicate the presence of a complex structure such as superdeformation above the Coulomb barrier.


4.   ALPHA TRANSFER REACTIONS

Taking the following experiment

$$A(a, b)B^*$$
$$\hookrightarrow c + \alpha \qquad ,$$

Rae explained how to determine the spins and parities of the intermediate states B*. Applying this technique to the reaction $^{14}$C($^{6}$Li, d)$^{18}$O* $\rightarrow \alpha_0 + ^{14}$C, Cunsolo succeeded in assigning $8^{+}$ to a state at 17.6 MeV in $^{18}$O. Comparing with Buck's calculation, he concluded that this state is a member of the 2h-4p band. Katori reported that peaks found in the $^{12,13}$C($^{16}$O, $\alpha$) reactions were not due to the massive transfer reaction mechanism.

Oelert showed the importance of two-step processes in alpha transfer reactions such as $^{26}$Mg(d, $^{6}$Li)$^{22}$Ne. Both the angular distribution and absolute cross section are affected by two-step processes. Thus one should be careful when extracting spectroscopic information from observed cross sections.

Roos discussed the $(p, \alpha)$ reaction in medium and high energy regions. He concluded that shell model calculations by Cohen and Kurath are alright. It should, however, be pointed out that the calculated cross section depends very sensitively on the assumed bound state wave functions.

Hodgson reported the result of his calculation concerning the Knock-on process of the $(p, \alpha)$ reaction. According to his calculation enhancement factors of $10^2 \sim 10^3$ are needed to explain the observed data. The calculated pick-up process also needs enhancement factors of similar magnitude. He assumed that preformation factors were unity. If one takes into account the smallness of the theoretical values for the preformation factors, one has to have extremely large enhancement factors. I am suspicious of the reaction mechanisms.

Spectroscopic factors extracted from $\alpha$ transfer reaction data seem to be larger than those calculated by a factor as large as $10^3 \sim 10^4$. Oelert, however, mentioned that $\gamma_\alpha^2$ obtained from alpha decays correlate very well with those from alpha transfer reactions. Thus he convinced us that those two have a common feature.

The alpha decay and alpha transfer reactions are still interesting problems. Before worrying about the structural aspect, one must understand the reaction mechanisms.

Iachello found an interesting correlation between the excitation energy of the lowest $1^-$ state and alpha decay widths in the Actinide nuclei. If the $1^-$ is low, the alpha decay width is wide. He then introduced the p-boson together with s-and d-bosons, which are responsible for describing the quadrupole collective degrees of freedom. The p-boson produces negative parity levels which are alpha molecular type states. Because of the p-boson, he predicts that both E1 transitions and alpha decay transitions are enhanced.

The Actinide region is very exciting to study from the nuclear spectroscopic point of view.

## 5.   BREAK-UP PROCESS

Shotter and Meijer reported on the break-up process which makes up a very large part of the total cross sections. In order to obtain spectroscopic information on giant resonances for example, one must understand the break-up process as well and as simply as possible. The Groningen group can tell us the relative importance of individual final states such as p+t, d+d and n+$^3$He in the break-up of an incident alpha particle. The group analyzed its data using the Impulse approximation and absorption. I hope that the analyses can be made more quantitative in order to be useful for spectroscopic studies.

## 6.   MESON EXCHANGE CURRENTS, QUARKS AND ANOMALONS

According to Laget, the role of meson exchange currents is clearly seen in the magnetic form factor of the $^3$He(e, e)$^3$He. I agree with him. The best nuclei to study the role of meson exchange currents are the deuteron and $^3$He, because their wave functions are calculated accurately in the non-relativistic approximation. The role of the six quark configuration in the EMC effect, for example, is very controversial. Pirner and Vary reported a 12% admixture of this state while Kim's number is 2.2 ∿ 5.5%. Magda Ericson and her collaborators on the other hand claimed that they can explain the EMC effect by increasing pions in nuclei.

Faessler gave a nice review on the nucleon-nucleon interaction based on the quark point of view including the results of his group, as well as that of Yazaki and Oka and others. My personal feeling is that once one understands the origin of the repulsive core between two nucleons, one does not need to evoke quarks because the nucleons avoid each other very well and keep their identities. A similar report was given by Hofmann who takes into account tensor and LS interactions between quarks. There are however, two difficulties in their approach, though I have no idea how to solve them; namely "How good is the non-relativistic approximation ?", and "How to avoid the Van der Waals force ?".

Thomas reviewed the present state of the art in the quark picture of the nucleon. This was very educational and I enjoyed his talk very much.

Tang and Kim argued about the Anomalon. Tang said that the anomalon is a molecular resonance, while Kim said that it is a multi-quark resonance. I am still waiting for more convincing data concerning the anomalon.

Finally, Tang discussed very briefly the radiative capture reaction $^3$He + $^4$He → $^7$Be + γ. All theoretical calculations now seem to give a value of 0.5 ∿ 0.6 KeV-b for the S-factor of this reaction;

| | | |
|---|---|---|
| Tang et al. | 0.56 KeV-b | |
| Kajino and Arima | 0.51 | Phys.Rev.Lett.$\underline{52}$(1984)739. |
| Buck | 0.47 | , |

while the observed values are

| | |
|---|---|
| Cal. Tec. | 0.52 ± 0.03 |
| Münster | 0.31 ± 0.03 |

Here the definition of the S-factor is

$$S = E\sigma(E) \exp 2\pi\eta ,$$

where η is the Sommerfeld parameter and E is the incident energy. These theoretical results seem to suggest that the S-factor should be closer to the Cal. Tec. value. Cluster calculations will be very useful for astrophysical applications of nuclear physics. This concerns, for example, the reactions $^{12}$C + α → $^{16}$O + γ and

$$^{16}O + \alpha \rightarrow {}^{20}Ne + \gamma \, .$$

CONCLUSION

My personal assessment of our present understanding is as follows:

Potential between nuclei

        Real part                    good

        Imaginary part          fair

Cluster states in light nuclei (very) good

Alpha spectroscopic factor      very bad

Alpha transfer (relative)       good

               (absolute)       very bad

Alpha decay in heavy nuclei     very bad

Molecular resonances

        Existence           clearly seen

        Nature              fair

The coexistence of the cluster aspect and the single particle aspect in nuclei is very unique. Thus we must understand these features and at the same time apply our knowledge to astrophysics for example.

As the last speaker, I would like to thank the organizers Dr. J. Lilley, Dr. Nagarajan, Mrs. S. Lowndes and Mrs. C. A. Lloyd for their efforts to organize this Conference. I also thank Ann Ireland, Gaynor Brown, Connie Clark and Noel Clarke for their help in making this Conference lively.

ABE, Y., Service de Physique Theorique, Centre D'Etudes Nucléaires de Saclay, B.P. No.2, F-91191 Gif-sur-Yvette Cedex, France

ÅBERG, S., Dept. Mathematical Physics, LTH, PO Box 725, S-220 07 Lund 7, Sweden

ALLCOCK, S.C., Nuclear Physics Laboratory, University of Oxford, Keble Road, Oxford OX1 3RH

ARIMA, A., Dept. of Physics, University of Tokyo, 7-3-1 Hongo , Bunkyo-Ku, Tokyo 113, Japan

ASHER, J., Nuclear Physics Division, B477, AERE Harwell, Didcot, Oxon OX11 0RA

ASSENBAUM, H.J., Inst. Theoretische Physik, Universität Münster, Domagkstrasse 71, D-4400 Münster, Germany

AUGER, F., C.E.N. Saclay, DPH N-BE, F-91191 Gif-sur-Yvette Cedex, France

BADER, R., Inst. Theoretische Physik, Universität Tübingen Auf der Morgenstelle 14, D-7400 Tübingen, German Fed. Rep.

BALDOCK, R.A., Dept. Theoretical Physics, University of Oxford, 1 Keble Road, Oxford OX1 3NP

BARKER, F.C., Dept. Theoretical Physics, Australian National University, G.P.O. Box 4, Canberra A.C.T. 2601, Australia

BARNETT, A.R., Dept. Physics, University of Manchester, Manchester M13 9PL

BARRETT, B.R., Max-Planck-Institut f. Kernphysik, Postfach 10 39 80, D-6900 Heidelberg 1, German Fed. Rep.

BARRETT, R.C., Dept. Physics, University of Surrey, Guildford, Surrey GU2 5XH

BAYE, D., Université Libre de Bruxelles, Physique Théorique et Mathématique CP 229, Campus de la Plaine, B-1050 Brussels Belgium

BECK, C., Centre de Recherches Nucléaires, Groupe PNIN, B.P. 20, F-67037 Strasbourg Cedex, France

BECK, R., KFZ Karlsruhe, Institut für Kernphysik III, Postfach 3640, D-7500 Karlsruhe, German Fed. Rep.

BETTS, R.R., Chemistry Div., Argonne National Laboratory, 9700 South Cass Ave., Argonne, Illinois 60439, USA.

BILWES, B., CEN Laboratoire des Basses Energies, B.P. 20 CRO, 23 Rue du Loess, F-67037 Strasbourg-Cedex, France

BLANN, M., Lawrence Livermore Laboratory, E-Division/L-405, P.O. Box 808, Livermore, California 94550, USA

BLYTH, C., Dept. Physics, University of Birmingham, P.O. Box 363, Birmingham B15 2TT

BOND, P., Kernfysisch Versneller Instituut, Zernikelaan 25, N-9747 AA Groningen, The Netherlands

BONETTI, R., Ist. di Fisica Generale Applicata, Universita di Milano, Via Celoria 16, I-20133 Milan, Italy

BRINK, D.M., Dept. Theoretical Physics, University of Oxford, 1 Keble Road, Oxford OX1 3NP

BROMLEY, D.A., A.W. Wright Nuclear Structure Laboratory, Yale University, P.O. Box 6666, 272 Whitney Avenue, New Haven, Connecticut 06511, USA

BROWN, J., Indiana University Cyclotron Facility, Milo B. Sampson Lane, Bloomington, Indiana 47405, USA

BUCK, B., Dept. Theoretical Physics, University of Oxford, 1 Keble Road, Oxford OX1 3NP

CHANT, N.S., Dept. Physics and Astronomy, University of Maryland, College Park, Maryland 20742, USA

CSEH, J., Inst. Nuclear Research of the Hungarian Academy of Sciences, BEM TER 18/C, P.O. Box 51, H-4001 Debrecen, Hungary

CUNSOLO, A., Ist. Dipartimentale di Fisica, Universita di Catania, INFN Sezione di Catania, Corso Italia 57, I-95129 Catania, Italy

DALMAS, J., CEN Bordeaux-Gradignan, Université de Bordeaux I, Le Haut-Vigneau, F-33170 Gradignan, France

DAVINSON, T., University of Edinburgh, Kings Buildings,
Mayfield Road, Edinburgh EH9 3JZ

DE MEIJER, R.J., Kernfysisch Versneller Instituut,
Zernikelaan 25, Rijksuniversiteit Groningen, NL-9747 AA
Groningen, The Netherlands

DESCOUVEMONT, P., Université Libre de Bruxelles, Physique
Théorique et Mathématique CP 229, Campus de la Plaine, B-1050
Brussels, Belgium

DICKMANN, F., KFZ Karlsruhe, Inst. für Kernphysik III,
Postfach 3640, D-7500 Karlsruhe, German Fed. Rep.

DURAND, M., Inst. Sciences Nucléaires, Université de
Grenoble, 53 Avenue des Martyrs, F-38026 Grenoble Cedex,
France

EASTHAM, D.A., SERC Daresbury Laboratory, Daresbury,
Warrington WA4 4AD

ENGLAND, J.B.A., Dept. Physics, University of Birmingham,
P.O. Box 363, Birmingham B15 2TT

FAESSLER, A., Inst. Theoretische Physik, Universität
Tübingen, Auf der Morgenstelle 14, D-7400 Tübingen 1, German
Fed. Rep.

FATEMIAN, M., Dept. Theoretical Physics, University of
Oxford, 1 Keble Road, Oxford OX1 3NP

DA HSUAN FENG, National Science Foundation, Washington, D.C.
20550, USA

FERRERO, J.L., Inst. Fisica Corpuscular, Facultad de Fisicas,
Avienda Dr. Moliner 50, Burjasot, Valencia, Spain

FIELD, G., Dept. Nuclear Physics, University of Oxford,
1 Keble Road, Oxford OX1 3NP

FOTI, A., Ist. di Fisica, Universita di Catania, Corso Italia
57, I-95129 Catania, Italy

FRIEDRICH, H., Technische Universität München, James Frank
Strasse, D-8046 Garching BEI München, German Fed. Rep.

FULTON, B.R., SERC Daresbury Laboratory, Daresbury,
Warrington WA4 4AD, U.K.

GARRETT, J.D., Niels Bohr Institute, Blegdamsvej 17, DK 2100
Copenhagen, Denmark

GIRAUD, B.G., Theoretical Physics Div., CEA-CEN Saclay,
B.P. No.2, F-91190 Gif-sur-Yvette, France

GOMEZ-CAMACHO, J., SERC Daresbury Laboratory, Daresbury,
Warrington WA4 4AD, U.K.

GREEN, L.L., SERC Daresbury Laboratory, Daresbury,
Warrington WA4 4AD, U.K.

HERING, W.T., Sektion Physik, Universitat München, Am
Coulombwall 1, D-8046 Garching, Fed. Rep. Germany

HODGSON, P.E., Dept. Nuclear Physics, University of Oxford,
1 Keble Road, Oxford OX1 3NP

HOFMANN, H.M., Inst. Theoretische Physik, Universität
Erlangen-Nürnberg, Glückstrasse 6, D-8520 Erlangen, Germany

HOPKINS, P.J.B., Dept. Theoretical Physics, University of
Oxford, 1 Keble Road, Oxford OX1 3NP

HORIUCHI, H., Dept. Physics, Kyoto University, Kyoto 606,
Japan

IACHELLO, F., A.W. Wright Nuclear Structure Laboratory, Yale
University, P.O. Box 6666, New Haven, Connecticut 06511, USA

IKEDA, K., Dept. Physics, Faculty of Science, Niigata
University, Ikarashi 2-8050, Niigata 950-21, Japan

INOUE, T., Dept. Physics, College of Humanities and Sciences,
Nihon University, 3-25-40 Sakurajosui, Setagaya-Ku, Tokyo,
Japan

INSOLIA, A., Ist. di Fisica, Universita di Catania, Corso
Italia 57, I-95129 Catania, Italy

IOANNIDES, A.A., Dept. Physics, The Open University, Milton
Keynes MK7 6AA

JAIN, B.K., Nuclear Physics Div., Bhabha Atomic Research
Centre, Trombay, Bombay 400 085, India

JOHNSON, R.C., Dept. Physics, University of Surrey,
Guildford, Surrey GU2 5XH

KAMIMURA, M., Dept. Physics, Kyushu University, Fukuoka 812,
Japan

KARBAN, O., Dept. Physics, University of Birmingham, P.O. Box
363, Birmingham B15 2TT

KATO, K., Dept. Physics,  Hokkaido University, Kita 10 Nishi 9, Sapporo 060, Japan

KATORI, K., Dept. Nuclear Studies, Osaka University, Toyonaka, Osaka 560, Japan

KIM, Y.E., Dept. Physics, Purdue University, West Lafayette, Indiana 47907, USA

KONDO, Y., Dept. Theoretical Physics, Australian National University, Research School of Physical Sciences, GPO Box 4, Canberra ACT 2601, Australia

KRAUS, L., CRN Basses Energies, B.P. 20 CRO, 23 Rue du Loess, F-67037 Strasbourg Cedex, France

LAGET, J.M., DPH.N/HE, CEN Saclay, B.P. No.2, F-91190 Gif-sur-Yvette Cedex, France

QING-RUN LI, TRIUMF, 4004 Westbrook Mall, Vancouver, British Columbia, Canada V6T 2A3

LILLEY, J.S., SERC Daresbury Laboratory, Daresbury, Warrington WA4 4AD, U.K.

LINCK, I., CEN Basses Energies, B.P. 20 CRO, 23 Rue du Loess, F-67037 Strasbourg Cedex, France

LISLE, J.C., Dept. Physics, Schuster Laboratory, University of Manchester, Manchester M13 9PL

LISTER, P.M., Dept. Physics, Randall Laboratory, University of Michigan, Ann Arbor, Michigan 48109, USA

LIU, Q.K.K., Bereich Kernphysik, Hahn-Meitner-Institut für Kernforschung, Glienickerstrasse 100, D-1000 Berlin 39, German Fed. Rep.

LOVAS, R.G., Institute of Nuclear Research, Hungarian Academy of Sciences, BEM TER 18/C, P.O. Box 51, H-4001 Debrecen, Hungary

MACKINTOSH, R.S., Dept. Physics, The Open University, Milton Keynes MK7 6AA

MARSH, S.M., Dept. Nuclear Physics, University of Oxford, 1 Keble Road, Oxford OX1 3NP

McKEE, J.S.C., Dept. Physics, University of Manitoba, Winnipeg, Canada R3T 2N2

MERCHANT, A.C., SERC Daresbury Laboratory, Daresbury,
Warrington WA4 4AD, U.K.

MIHAILOVIC, M.V., Jozef Stefan Inst., University of
Ljubljana, P.O. 199/IV, 6100 Ljubljana, Yugoslavia

MILLER, D.W., Dept. Physics, Indiana University, Bloomington,
Indiana 47405, USA

MORRISON, G.C., Dept. Physics, University of Birmingham, P.O.
Box 363, Birmingham B15 2TT

MOSEL, U., Inst. Theoretische Physik, Universität Giessen,
Heinrich Buff-Ring 16, 63 Giessen, German Fed. Rep.

MOTOBAYASHI, T., Dept. Physics, St. Pauls (Rikkyo)
University, 3 Nishi-Ikebukuro, Toshima, Tokyo, Japan

NAGARAJAN, M.A., SERC Daresbury Laboratory, Daresbury,
Warrington WA4 4AD

NISHIOKA, H., Dept. Physics, University of Surrey, Guildford,
Surrey GU2 5XH

O'DONNELL, J., Dept. Physics, University of Bradford,
Bradford BD7 1DP

OELERT, W., Inst. Kernphysik, KFA Jülich IKP, Postfach 1913,
D-5170 Jülich, German Fed. Rep.

OHKUBO, S., Dept. Applied Sciences, Institute for Physics,
Kochi Women's University, Kochi 780, Japan

PADE, D., Inst. Experimentalphysik, Ruhr-Universität Bochum,
Universitätsstrasse 150, Gebäude NB/3 O.G., Postfach 102418,
4630 Bochum, German Fed. Rep.

PARK, J.Y., Dept. Physics, North Carolina State University,
Box 8202, Raleigh, North Carolina 27695, USA

PARKER, D.J., Nuclear Physics Div., B477, AERE Harwell,
Didcot, Oxon OX11 0RA

PHILLIPS, G.C., T.W. Bonner Nuclear Labs., Rice University,
Houston, Texas 77251, USA

PHILLIPS, W.R., Dept. Physics, Schuster Laboratory,
University of Manchester, Manchester M13 9PL

PISENT, G., Dept. Physics, Universita di Padova, Via Marzolo
8, I-35100 Padova, Italy

POLLAROLO, G., Ist. Fisica Teorica, Universita di Torino, Corso Massimo D'Azeglio 46, I-10125 Torino, Italy

PRICE, H.G., SERC Daresbury Laboratory, Daresbury, Warrington WA4 4AD, U.K.

PROVOOST, D., Inst. Kernphysik, KFA Jülich, Postfach 1913, D-5170 Jülich, Germany

RAE, W.D.M., Dept. Theoretical Physics, University of Oxford, 1 Keble Road, Oxford OX1 3NP

RAI, G., Dept. Physics, University of Birmingham, P.O. Box 363, Birmingham B15 2TT

SRINIVASA RAO, K., MATSCIENCE, Inst. Mathematical Sciences, (T.T.T.I. Tharamani S.O.), Madras - 600 113, India

REBEL, H., KFZ Karlsruhe, Institut für Kernphysik III, Postfach 3640, D-7500 Karlsruhe, German Fed. Rep.

ROBSON, D., Dept. Physics, Florida State University, Tallahasse, Florida 32306, USA

ROHWER, T., Physikalisches Inst. Tübingen, Universität Tübingen, Auf der Morgenstelle 14, D-7400 Tübingen, Germany

ROOS, P.G., Dept. Physics and Astronomy, University of Maryland, College Park, Maryland 20742, USA

ROWLEY, N., SERC Daresbury Laboratory, Daresbury, Warrington WA4 4AD, U.K.

SARMA, N., Nuclear Physics Div., Bhabha Atomic Research Centre, Trombay, Bombay 400 085, India

SHIMOURA, S., Dept. Physics, Faculty of Science, Kyoto University, Kyoto 606, Japan

SHOTTER, A.C., Dept. Physics, University of Edinburgh, James Clerk Maxwell Building, Mayfield Road, Edinburgh EH9 3JZ

SKOURAS, L., NRC Demokritos, Agia Paraskevi, Attika, Athens, Greece

SMITH, A.E., Dept. Nuclear Physics, University of Oxford, 1 Keble Road, Oxford OX1 3NP

SPENCER, J.T., Dept. Physics, Schuster Laboratory, University of Manchester, Manchester M13 9PL

SRIVASTAVA, D.K., KFZ Karlsruhe, Inst. für Kernphysik III, Postfach 3640, D-7500 Karlsruhe, German Fed. Rep.

STAUDT, G., Physikalisches Inst. Universität Tübingen, Auf der Morgenstelle 14, D-7400 Tübingen, German Fed. Rep.

STRATTON, R.A., Dept. Theoretical Physics, University of Oxford, 1 Keble Road, Oxford OX1 3NP

TANAKA, H., Dept. Physics, Faculty of Science, Hokkaido University, Kita 10 Nishi 9, Sapporo 060, Japan

TANG, Y.C., School of Physics, University of Minnesota, Minneapolis, Minnesota 55455, USA

TANIFUJI, M., Dept. Physics and Research Center of Ion Beam Technology, Hosei University, 17-1.2-Chome, Fujimi, Chiyoda-Ku, Tokyo 102, Japan

THOMAS, A.W., Dept. Physics, University of Adelaide, P.O. Box 498 GPO, Adelaide SA 5001, Australia

THOMPSON, I.J., SERC Daresbury Laboratory, Daresbury, Warrington WA4 4AD, U.K.

THORSTEINSEN, T.F., Dept. Physics, University of Bergen, Allegt. 55, N-5000 Bergen, Norway

TRAUTMANN, W., Brookhaven National Laboratory, Physics 901A, Upton, New York 11973, USA

TRIPATHI, R.K., Inst. Physics, Sachivalaya Marga, Bhubaneswar-751005, India

TWIN, P.J., SERC Daresbury Laboratory, Daresbury, Warrington WA4 4AD, U.K.

VANZANI, V., Universita di Padova, Ist. di Fisica "Galileo Galilei", Via Marzolo 8, I-35131 Padova, Italy

VITTURI, A., Universita di Padova, Ist. di Fisica "Galileo Galilei", Via Marzolo 8, I-35131 Padova, Italy

WADA, T., Dept. Physics, Faculty of Science, Kyoto University, Kitashirakawa, Kyoto 606, Japan

WALKER, P.M., SERC Daresbury Laboratory, Daresbury, Warrington WA4 4AD, U.K.

WANG, C., Dept. Physics and Astronomy, University of Maryland, College Park, Maryland 20742, USA

WATSON, D.L., Postgrad. School of Studies in Physics, University of Bradford, Bradford BD7 1DP

WEIGUNY, A., Inst. Theoretische Physik I, Universität Münster, Domagkstrasse 71, D-4400 Münster, German Fed. Rep.

WHITEHEAD, R.R., Dept. Natural Philosophy, University of Glasgow, Glasgow G12 8QQ, Scotland

WILDERMUTH, K., Inst. Theoretische Physik, Universität Tübingen, Auf der Morgenstelle 14, D-7400 Tübingen, German. Fed. Rep.

WINDHAM, G., Dept. Physics, University of Surrey, Guildford, Surrey GU2 5XH

WRIGHT, I.F., Dept. Physics, Schuster Laboratory, University of Manchester, Manchester M13 9PL

416